KV-422-399

Analytical Applications of Bioluminescence and Chemiluminescence

Academic Press Rapid Manuscript Reproduction

Based on the Proceedings
of the Third International Symposium
on Analytical Applications
of Bioluminescence and Chemiluminescence

Analytical Applications of Bioluminescence and Chemiluminescence

Edited by

L. J. Kricka

Department of Clinical Chemistry
University of Birmingham
Edgbaston
Birmingham
England

P. E. Stanley

Consultant Scientist
Cambridge
England

G. H. G. Thorpe
and T. P. Whitehead

Department of Clinical Chemistry
Wolfson Research Laboratories
Queen Elizabeth Medical Centre
Edgbaston
Birmingham
England

1984

ACADEMIC PRESS, INC.

(Harcourt Brace Jovanovich, Publishers)

London Orlando San Diego New York
Toronto Montreal Sydney Tokyo

ACADEMIC PRESS, INC.
Orlando, Florida 32887

United Kingdom Edition published by
ACADEMIC PRESS, INC. (LONDON) LTD.
24/28 Oval Road, London NW1 7DX

LIBRARY OF CONGRESS CATALOG CARD NUMBER: 84-45603

ISBN 0-12-426290-2

PRINTED IN THE UNITED STATES OF AMERICA

84 85 86 87 9 8 7 6 5 4 3 2 1

Senior Contributors

Numbers in parentheses indicate the pages on which the author's contributions begin.

R. C. Allen (279), Department of Pathology and Clinical Investigation, Brooke Army Medical Center, Fort Sam Houston, Texas, USA

S. Ansehn (63), Department of Clinical Bacteriology, University Hospital, 581 85 Linkoping, Sweden

T. O. Baldwin (101), Department of Biochemistry and Biophysics and The Texas Agricultural Experiment Station, Texas A&M University, College Station, Texas 77843, USA

M. Barak (373, 377), Department of Microbiology, Technion Faculty of Medicine and Rambam Medical Center Technion - Israel Institute of Technology, Haifa, Israel

G. Barnard (159), Department of Obstetrics and Gynaecology, King's College School of Medicine, Denmark Hill, London, SE5 8RX, UK

J. Barth (381), Department of Internal Medicine, Kiel University, 2300 Kiel, FRG

B. Beckers (67), Department of Medical Microbiology, Medical Faculty, University of Aachen, Aachen, FRG

F. Berthold (457), Laboratorium Prof Dr Berthold, D-7547 Wildbad, FRG

H. Berthold (265), Institut fur Virologie, Universitat Freiburg, Hermann Herder Str 11, FRG

G. Bertoni (385), Institute of Virology, University of Zurich, Switzerland

B. E. Bordalen (577), Central Institute for Industrial Research, PO Box 350, Blindern, 0314 Oslo 3, Norway

W. Bogl (573), Institute for Radiation Hygiene of the Federal Health Office, Neuherberg, FRG

S. E. Brolin (479, 515), Department of Medical Cell Biology, Biomedical Center, University of Uppsala, Uppsala, Sweden

K.-H. Buscher (327), Department of Medical Microbiology and Immunology, Ruhr-Universitat, Bochum, FRG

A. L. Caldini (171), Endocrinology Unit, University of Florence, Italy

P. T. Conroy (369), Department of Cardiothoracic Anaesthesia, Freeman Hospital, Newcastle Upon Tyne, UK

G. D. W. Curtis (21), Department of Microbiology and Regional Public Health Laboratory, John Radcliffe Hospital, Oxford, UK

C. Dahlgren (335), Department of Medical Microbiology, University of Linkoping, S-581 85 Linkoping, Sweden

P. De Baetselier (297), Algemene Biologie, Instituut voor Moleculaire Biologie, Vrije Universiteit Brussel, St-Genesius-Rode, Belgium

J. De Boever (163), Department of Obstetrics & Gynaecology, Academic Hospital, De Pintelaan 185, B-9000 Ghent, Belgium

M. DeLuca (111), Department of Chemistry, University of California, San Diego, La Jolla, California 92093, USA

C. De Simone (411), Clinica Malattie Infettive, Universita di Roma "La Sapienza" I. 00185 Roma, Italy

P. De Sole (389), Department of Clinical Chemistry, School of Medicine - Universita Cattolica del Sacro Cuore - Lgo A. Gemilli, 8 - 00168 Roma, Italy

D. Del Principe (339), Department of Paediatrics, University of Rome, Italy

A. M. Dhople (29), Medical Research Institute, 3325 West New Haven Avenue, Melbourne, Florida 32901, USA

J. Ennen (431), Borstel Research Institute, D-2061 Borstel, FRG

M. Ernst (315), Borstel Research Institute, D-2061 Borstel, FRG

A. Gadow (205, 209), Klinische Laboratorien der Klinik fur Innere Medizin Medizinische Hochschule Lubeck, D-2400 Lubeck 1, FRG

R. Galard (199), Department of Biochemistry, Hormone Laboratory, Vall Hebron General Hospital, Barcelona, Spain

G. H. Geelhaar (435), Department of Surgery, University of Heidelberg, D-6900 Heidelberg, FRG

M. L. Grayeski (565), Department of Chemistry, Seton Hall University, S Orange, New Jersey, USA

M. W. Griffiths (61), Hannah Research Institute, Ayr KA6 5HL, UK

M. Grob (415), Institute of Virology, University of Zurich, CH-8057 Zurich, Switzerland

M. B. Hallett (347), Department of Surgery, Welsh National School of Medicine, Heath Park, Cardiff, CF4 4XN, UK

M. J. Harber (3), Department of Renal Medicine, Welsh National School of Medicine, KRUF Institute, Royal Infirmary, Cardiff, UK

J. Haritz (215), Klinik fur Innere Medizin, Medizinische Hochschule Lubeck, D-2400 Lubeck, FRG

M. J. G. Hastings (439), Department of Medical Microbiology, University of Sheffield Medical School, Sheffield S10 2RX, UK

M. Heberer (357), Department of Surgery, University of Basel, Switzerland

T. Heinonen (525), Department of Biochemistry, Turku University, Turku, Finland

P. Hindocha (365, 443), Academic Department of Child Health, Queen Elizabeth Hospital for Sick Children, Hackney Road, London E2 8PS, UK

M. E. Holt (311), Department of Rheumatology, Welsh National School of Medicine, Cardiff CF4 4XW, UK

J. Hughes (269), Department of Chemical Pathology, St Mary's Hospital Medical School, Paddington, London W2 1PG, UK

D. E. Jenner (521), Department of Renal Medicine, Welsh National School of Medicine, KRUF Institute, Royal Infirmary, Cardiff CF2 1SZ, UK

W. Klingler (193), Institut fur Biochemische Endokrinologie Med Hochschule, 2400 Lubeck, FRG

F. Kohen (149), Department of Hormone Research, The Weizmann Institute of Science, Rehovot, Israel

H.-S. Krausz (219), Klinik fur Innere Medizin, Medizinische Hochschule Lubeck, D-2400 Lubeck 1, FRG

K. Kurkijarvi (125), Wallac Biochemical Lab, Box 10, Turku, Finland

E.-M. Lilius (393, 397, 401, 419, 423, 461), Department of Biochemistry, University of Turku, SF-20500 Turku, Finland

T. Lovgren (507), Wallac Biochemical Laboratory, Box 10, Turku, Finland

A. Lundin (71, 491, 545, 585), Research Centre and Department of Medicine, Karolinska Institute, Huddinge Hospital, S-141 86 Huddinge, Sweden

F.-E. Maly (331), Max-Planck-Institut fur Immunbiologie, Freiburg, FRG

P. A. Manger (57), PHLS, William Harvey Hospital, Ashford Kent TN24 OLZ, UK

B. J. McCarthy (49), Wira, Wira House, West Park Ring Road, Leeds LS16 6QL, UK

P. W. McWalter (17), Department of Bacteriology, Withington Hospital, Manchester M20 8LR, UK

G. Merenyi (569), Department of Physical Chemistry, Royal Institute of Technology, S-100 44 Stockholm, Sweden

G. Messeri (167), Clinical Chemistry Lab, USL 10D, Firenze, Italy

D. Morse (529), Department of Biochemistry, McGill University Montreal, Canada H3G 1Y6

A. Muller (343, 447), Department of Internal Medicine, Division of Haematology/Oncology, University Clinic Gottingen, Robert-Koch-Str 40, D-3400 Gottingen, FRG

J. Muller-Quernheim (307), Department of Pneumology, University of Mainz, Langenbeckstr, 65 Mainz, FRG

K. H. Nealson (87), Marine Biology Research Division, Scripps Institute of Oceanography, La Jolla, CA 92093, USA

J. C. Nicolas (537), INSERM Unite 58, 60 rue de Navacelles, 34100 Montpellier, France

L. Nilsson (25), Department of Clinical Bacteriology, University Hospital, 581 85 Linkoping, Sweden

T. E. T. Oikari (475), Wallac Oy, Box 10, Turku, Finland

C. Orlando (261), Endocrinology Unit, University of Florence, Italy

A. Patel (273), Bioluminescence Laboratory, Department of Biochemistry, University of Georgia, Athens, GA 30602, USA

M. Pazzagli (469), Endocrinology Unit, University of Florence, Italy

S. Peitzner (223), Klinik fur Innere Medizin, Medizinische Hochschule Lubeck, D-2400 Lubeck 1, FRG

E. Peterhans (323), Institute of Virology, University of Zurich, CH-8057 Zurich, Switzerland

P. N. Platt (319), University Department of Rheumatology, Royal Victoria Infirmary, Newcastle upon Tyne, UK

R. P. Prioli (81), Department of Medical Microbiology, St Mary's Hospital Medical School, Paddington, London W2 1PG, UK

S. Rajgopal (485), Laboratorie de Technologie de Separation, Universite de Technologie de Compiegne BP 233, 60206 Compiegne Cedex, France

A. Reinhertz (541), Department of Food Engineering & Biotechnology, Technion - Haifa, Israel

L. Reum (249), Mallinckrodt Diagnostica, von-Hevesy-Str 1-3, D-6057 Dietzenbach 2, FRG

P. A. Roberts (175), Department of Medical Biochemistry, Welsh National School of Medicine, Heath Park, Cardiff CF4 4XN, UK

A. Roda (129, 227), Istituto Scienze Chimiche, Universita di Bologna, 40127 Bologna, Italy

A. Rodriguez (105), Department of Biochemistry, McGill University, Montreal, Canada H3G 1Y6

R. Salerno (179), Endocrinology Unit, University of Florence, Italy

H. K. Schackert (351), Department of Surgery, University of Heidelberg, D-6900 Heidelberg, FRG

J. Scholmerich (141), Department of Medicine, University of California, San Diego, USA

R. E. Schopf (361), Department of Dermatology, Johannes Gutenberg University, D-6500 Mainz, FRG

E. Schram (289), Instituut voor Moleculaire Biologie, Vrije Universiteit Brussel, Brussels, Belgium

C. J. Stannard (53), Leatherhead Food RA, Randalls Road, Leatherhead, Surrey KT22 7RY, UK

L. Stohr (451), Department of Dermatology, J. W. Goethe-University, Frankfurt/Main, FRG

R. A. Stott (233), Department of Clinical Chemistry, Wolfson Research Laboratories, Queen Elizabeth Medical Centre, Birmingham B15 2TH, UK

C. J. Strasburger (239), Klinik fur Innere Medizin, Medizinische Hochschule, D-2400 Lubeck 1, FRG

A. Tanna (73), Department of Medical Microbiology, Wright-Fleming Institute, St Mary's Hospital Medical School, London W2 1PG, UK

G. H. G. Thorpe (243), Department of Clinical Chemistry, Wolfson Research Laboratories, Queen Elizabeth Medical Centre, Birmingham B15 2TH, UK

B. Tode (257), Klinik fur Innere Medizin, Medizinische Hochschule Lubeck, D-2400 Lubeck 1, FRG

N. Topley (409), Department of Renal Medicine, Welsh National School of Medicine, KRUF Institute, Royal Infirmary, Cardiff, UK

T. S. Tsai (75), United Technologies Packard, Downers Grove, IL 60515, USA

A. Tsuji (253), School of Pharmaceutical Sciences, Showa University, Hatanodai, Shinagawa-ku, Tokyo, Japan 142

S. Ulitzur (533, 591), Department of Food Engineering & Biotechnology, Technion - IIT, Haifa, Israel

F. Van Assche (511), Dept SBM, Limburgs Universitair Centrum, B-3610 Diepenbeek, Belgium

D. Vellom (133), Department of Chemistry, University of California, San Diego, La Jolla, California 92093, USA

W. Verstraete (33), Laboratory of Microbial Ecology, University of Ghent, Coupure L653, 9000 Ghent, Belgium

A. J. W. G. Visser (559), Department of Biochemistry, Agricultural University, De Dreijen 11, 6703 BC Wageningen, The Netherlands

I. Weeks (185), Department of Medical Biochemistry, Welsh National School of Medicine, Heath Park, Cardiff CF4 4XN, Wales, UK

R. Weinberger (555), Kratos Analytical Instruments, Ramsey, NJ 07446, USA

E. Wieland (503), Clinical Institute of Infarction Research of The Medical Clinic of The University of Heidelberg, Bergheimerstrasse 58, D-6900 Heidelberg, FRG

J. Willems (405), IRCKUL Campus Kortrijk; B-8500 Kortrijk, Belgium

W. G. Wood (189, 465), Klinik fur Innere Medizin, Medizinische Hochschule Lubeck, D-2400 Lubeck 1, FRG

L. F. J. Woods (581), Applied Microbiology Section, Leatherhead Food Research Association, Randalls Road, Leatherhead, Surrey KT22 7RY, UK

M. Wurl (427), Dept of Internal Medicine, Division of Haematology/Oncology, University Clinic of Gottingen, Robert-Koch-Str 40, D-3400 Gottingen, FRG

B. Yoda (587), Department of Medicine, Tohoku University School of Medicine, Sendai 980, Japan

Preface

This volume contains the proceedings of the Third International Symposium on Analytical Applications of Bioluminescence and Chemiluminescence which was held at the University of Birmingham, England, from 17-19 April 1984. It contains both the invited and contributed papers as well as those papers based on the posters presented at the symposium: the manuscripts were camera ready from the authors in order to speed publication.

We wish to thank the University of Birmingham for accomodation and conference facilities and staff of the Wolfson Research Laboratories at the Queen Elizabeth Medical Centre: Pat Armstrong, David Browning, Peter Broughton, Mary Edwards, Eileen Gillespie, Amanda Griffiths, Jacky Hogan, Fadia Homaidan, Angela O'Toole and Richard Stott, who ensured the smooth running of the symposium.

We gratefully acknowledge financial support from: Amersham International plc, Analytical Luminescence Laboratory, Bayer AG/Miles, BUPA Medical Centre Ltd, JS Pathology Services Ltd, Laboratorium Professor Dr. Berthold, Lumas/3M BV, Packard Instrument Ltd, Roche Diagnostic Systems, Sterilin Ltd, Technicon Instruments Corporation and Wellcome Diagnostics.

We are particularly grateful to the Plenary Lecturers: Drs. Harber, Verstraete, Nealson, DeLuca, Kohen, Allen and Schram and to the Chairmen of the Sessions: Drs. Vanstaen, Smith, Woodhead, Schroeder, Allen, Schram, Berthold and Brolin.

The Inaugural meeting of the Steering Committee for future Symposia on Bioluminescence and Chemiluminescence was held on 19 April 1984. The meeting was chaired by Professor

Tom Whitehead and was attended by 18 representatives from academia and industry who were from various countries. It was agreed that the next Symposium would be held in Germany (possibly Freiburg) in the summer of 1986. Dr. Marlene DeLuca, Dr. Larry Kricka and Dr. Phil Stanley were elected Chairman, Secretary and Treasurer, respectively, of the Steering Committee.

Contents

Contributors v

Preface xiii

I RAPID MICROBIOLOGY

Applications of Luminescence in Clinical Microbiology 3
M J Harber

Rapid Susceptibility Testing of Staphylococcus Aureus 17
Paul W McWalter

Bacterial ATP Content of Urines from Bacteriuric and Non-bacteriuric Patients 21
G D W Curtis, H H Johnston and W W Nichols

Rapid Detection of Bacteriemia by Bioluminescent Assay of Bacterial ATP 25
L Nilsson, O Molin and S Ansehn

Application of ATP Assays to Patient Care in Leprosy 29
A M Dhople

ATP Measurement by Bioluminescence: Environmental Applications 33
H Van de Werf and W Verstraete

Bioluminescent Determination of Microbial Activity on Textiles 49
B J McCarthy

ATP Assay as a Rapid Method for Estimating Microbial Growth in Foods 53
C J Stannard

Preliminary Evaluation of the Lumac Biocounter M2010 System, Measuring Bacterial ATP in Food, Using a Bioluminescent Technique 57
Pamela A Manger

A Rapid Method for Detecting Post-Pasteurization Contamination in Cream 61
M W Griffiths, J D Phillips and D D Muir

Effect of Imidazole Anti Fungals on Candida Albicans Demonstrated by Bioluminescent Assay of ATP 63
S Ansehn and L Nilsson

Determination of Intracellular ATP During Growth of Escherichia Coli in the Presence of Ampicillin 67
B Beckers, H R M Lang and A Beinke

Luminometric Detection of Bacteriuria in Primary Health Care 71
A Lundin, H Hallander, A Kallner, U Karnell Lundin and E Osterberg

Assay of Gentamicin and Chloramphenicol in Serum Using ATP Bioluminescence 73
A Tanna and C S F Easmon

Detection of Bacteriuria in Yeast Contaminated Specimen by Filtration and Bioluminescence Technique 75
TenLin S Tsai and Leroy J Everett

A Comparison of Methods for the Extraction of ATP From Mycobacteria 81
R P Prioli and I N Brown

II GENETIC ENGINEERING

Molecular Genetic Studies in Bioluminescence 87
Kenneth H Nealson and Richard Cassin

Cloning of the Bacterial Luciferase and Use of the Clone to Study the Enzyme and Reaction in vivo 101
Thomas O Baldwin, Tineke Berends and Mary L Treat

Aldehyde Biosynthesis in Luminescent Bacteria 105
A Rodriguez, L Wall, L M Carey, M Boylan, D Byers and E Meighen

III IMMOBILISED ENZYMES

Bioluminescent Assays Using Co-immobilised Enzymes 111
M DeLuca

Flow-injection Analysis with Immobilized Bacterial Bioluminescent Enzymes 125
K Kurkijarvi, T Heinonen, T Lovgren, J Lavi and R Raunio

Development of a Continuous-flow Analysis for Serum and Salivary Bile Acids Using Bacterial Bioluminescent Enzymes Immobilized on Nylon Coil 129
A Roda, S Girotti, S Ghini, B Grigolo, G Carrea and R Bovara

Continuous Flow Bioluminescent Assays for Femtomole Levels of NADH and TNT 133
D Vellom, J Hinkley, A Loucks, H Egghard and M DeLuca

Measurement of Bile Acid Patterns in Serum and Urine by Bioluminescence Assays Using Immobilized Hydroxysteroid dehydrogenases 141
Jurgen Scholmerich, Ian A Macdonald, Gerard van Berge Henegouwen, Alan F Hofmann and Marlene DeLuca

IV IMMUNOASSAY

Development of Luminescence-based Immunoassays for Haptens and for Peptide Hormones 149
F Kohen, E A Bayer, M Wilchek, G Barnard, J B Kim, W P Collins, I Beheshti, A Richardson and F McCapra

The Use of Isoluminol and Acridinium Labels in Immunoassay 159
G Barnard, J L Brockelbank, J B Kim and W P Collins

Direct Chemiluminescence Immunoassay for Progesterone 163
J De Boever, F Kohen and D Vandekerckhove

Homogeneous Luminescence Immunoassay: An Application for Urinary Estrogens Measurement 167
G Messeri, A L Caldini, F Franceschetti, R Salerno and M Serio

An Automated Luminescence Immunoassay for the Measurement of Estriol-16 alpha-glucuronide in Pregnancy Urine 171
A L Caldini, G Messeri, P Buzzoni and P Borri

Cyclic AMP in PMN Measured by Chemiluminescence Energy Transfer Assay 175
P A Roberts, A Patel, S C L Barrow, M B Hallett and A K Campbell

Evaluation of Luminescent Immunoassay Methods for Urinary Steroids 179
R Salerno, G Moneti, A Magini, A Tomasi and M Pazzagli

Chemiluminescent Acridinium Esters as Labels in Immunoassay 185
I Weeks and J S Woodhead

Routine Luminescence Immunoassays for Haptens and Proteins 189
W G Wood

Blood Spot 17-hydroxyprogesterone Luminescence Immunoassay 193
W Klingler, O Haupt and R Knuppen

Chemiluminescent Immunoassay of Salivary Testosterone in Normal Subjects and Hirsute Women 199
R Galard, R Catalan, A Lucas, S Schwartz and J Castellanos

Optimisation of Macro-solid-phase Luminescence Immunoassays Using Covalently Immobilised Antigens and Antibodies 205
A Gadow, C J Strasburger and W G Wood

Solid-phase Luminescence Immunoassays for Thyroid Parameters 209
A Gadow and W G Wood

Chemiluminescence Immunoassays for Alpha Fetoprotein for use in Early Pregnancy and Tumour Monitoring 215
J Haritz and W G Wood

Solid Phase Luminescence Immunoassays for Acute-phase Proteins Suitable for Routine Paediatric Application 219
Heidi-Susanne Krausz and W G Wood

A Two Hour Chemiluminescent Assay for Australia Antigen 223
S Peitzner and W G Wood

Development of a Sensitive Direct Solid-phase Competitive Enzyme Luminescent Assay for Serum and Salivary Cortisol 227
A Roda, S Girotti, S Lodi, S Preti and A Piacentini

Preparation of Microperoxidase and its use as a Catalytic Label in Luminescent Immunoassays 233
R A Stott, G H G Thorpe, L J Kricka and T P Whitehead

Solid Phase Antigen Luminescence Technique (SPALT) Assays for Cortisol and Gentamicin Determination in Serum 239
C J Strasburger, A Gadow and W G Wood

An Immunoassay for Serum Thyroxine Employing Enhanced Luminescent Quantitation of Horseradish Peroxidase Conjugates 243
Gary G H Thorpe, Eileen Gillespie, Robert Haggart, Larry J Kricka and Thomas P Whitehead

A New Immunoassay for Ecdysteroids Based on Chemiluminescence 249
Lutz Reum, Wolfgang Klingler and Jan Koolman

Enzyme Immunoassay Monitored by Chemiluminescence Reaction Using Bis(2,4,6-trichlorophenyl)oxalate-fluorescent Dye 253
A Tsuji, M Maeda and H Arakawa

The Routine Determination of Metalloproteins in Biological Fluids Using Chemiluminescence Immunoassays 257
Bettina Tode and W G Wood

Transferrin and Ceruloplasmin Measurement in Human Seminal Plasma by a Chemiluminescent Method 261
C Orlando, A L Caldini, T Barni, G Wood, B Tode, G Fiorelli, G Forti and M Serio

Chemiluminescence Immunoassay for the Demonstration of HBs AG Using a Fluorescein Isothiocyanate Labelled Anti HBs 265
H Berthold

Synthesis of a Novel Bioluminescent Conjugate of Progesterone for Immunoassay 269
J Hughes, Frances Short and V H T James

Calcium-sensitive Photoproteins as Bioluminescent Labels in Immunoassay 273
Ashok Patel and Milton J Cormier

V PHAGOCYTOSIS

Chemiluminescence as an Approach to the Study of Phagocyte Biochemistry and Humoral Immune Mechanisms 279
R C Allen

Luminescence Meter for the Kinetic Measurement of Phagocytic Activity and Other Luminescent Reactions 289
Eric Schram, Henri Roosens and Patrick de Baetselier

Generation of Macrophage-hybridomas for the Study of Macrophage Chemiluminescence 297
P de Baetselier, L Brys, E Vercauteren, L Mussche, R Hamers and E Schram

Selective Unresponsivness of Alveolar Macrophages (AM) in Pulmonary Alveolar Proteinosis (PAP) 307
Joachim Muller-Quernheim, Rudolf E Schopf, Peter Benes, Roman Rubin and Volker Schulz

Human Polymorphonuclear Leukocyte Luminol Chemiluminescence Identifies Albumin as an Oxygen Radical Scavenger 311
Mary E Holt and A K Campbell

Monocytes are the Effector Cells in Tumor Target Cell Induced Chemiluminescence: NK Cells do not Generate Activated Oxygen Species 315
M Ernst, A Lange, A Havel, J Ennen, A J Ulmer and H-D Flad

The Use of Chemiluminescence to Investigate Mechanisms of Activation of Polymorphonuclear Leucocytes by Monosodium Urate Crystals 319
P N Platt and I Bird

Cellular Chemiluminescence as a Tool in the Study of Surface Antigens, Antibody Specificities and FC Receptors 323
E Peterhans, T Arnold, M Grob and G Bertoni

Chemiluminescence Studies on the Adhesin-receptor Interaction of Different Strains of E Coli and Mouse Macrophages 327
K-H Buscher, V Klimetzek and W Opferkuch

Granulocyte Chemiluminescence Inducer (GCI): A Cytokine Produced by Mitogen-Stimulated Mononuclear Cells (MNC) 331
F-E Maly and A Kapp

Measurement of Luminol-Dependent Leukocyte Chemiluminescence Originated from Intra- and Extracellular Events 335
C Dahlgren, G Briheim and O Stendahl

Supernatant from Thrombin-activated Platelets Inhibits the Neutrophil Chemiluminescence 339
D Del Principe, A Menichelli, W De Matteis, R De Santis, A M Pentassuglio and A Finazzi-Agro

Reliability of the Detection of Human Monocyte Chemiluminescence in Mononuclear Cell Suspension 343
A Muller, P Schuff-Werner, M Wurl and G A Nagel

Dependence of Polymorphonuclear Leucocyte Luminol-dependent Chemiluminescence on Oxygen 347
M B Hallett, S W Edwards, S C L Barrow and A K Campbell

Perioperative Zymosan Induced Chemiluminescence of Polymorphonuclear Leukocytes in Surgical Cancer Patients 351
Hans Konrad Schackert, Michael Betzler, George F Zimmermann, Hans-Peter Geisen, Lutz Edler, G Harald Geelhaar and Christian Herfarth

Association of Chemiluminescence Induced by Pneumococcal Antigens in Whole Blood with Specific Antibody Titers 357
M Heberer, M Durig, M Ernst, J Sclenar, P Erb and F Harder

Agents Exerting Control Over Luminol-enhanced Phagocytic Chemiluminescence 361
Rudolf E Schopf

Opsono-phagocytic Variations of Serotypes Ia, Ib, II and III of Group B Streptococci as Observed by Neutrophil Chemiluminescence (CL). Role of the Alternative Pathway of Complement in Opsonization of Serotype III of Group B Streptococcus (GBS) 365
P Hindochta, Ruth Hill, C B S Wood, Urmila Patel and George Hunt

A Comparison of Whole Blood and Neutrophil Chemiluminescence in Patients Undergoing Cardio-pulmonary Bypass 369
P T Conroy, P N Platt and M J Elliott

Elucidation of the Phagocytosis Mechanism with the Aid of Luminous Bacteria 373
Mira Barak, S Ulitzur and D Merzbach

Opsonization - Characteristics and Mechanism with the Aid of Luminous Bacteria 377
Mira Barak, S Ulitzur and D Merzbach

Lectin Modified Chemiluminescence of Human Mononuclear Cells and Morphological Correlates in Electron Microscopy 381
J Barth, U Welch, U Schumacher and K G Ravens

A Chemiluminescent Assay for Mycoplasma in Cell Cultures 385
G Bertoni, R Keist, P Groschurth, R Wyler, R Keller and E Peterhans

Macrophage Chemiluminescence Activation by Bovine Gamma-globulins 389
P De Sole, S Lippa, F Meo, G De Sanctis, P Di Nardo, A Falera, C Giordano and P Massari

Kinetics of Lipoxygenase Reaction Measured by Chemiluminescence 393
Esa-Matti Lilius, Pekka Turunen and Simo Laakso

Chemiluminescence Emission from Enriched Fraction of Human Natural Killer Cells 397
Esa-Matti Lilius, Kaija Laurila, Jaakko Uksila and Olli Lassila

A Very Sensitive and Rapid Chemiluminescence Method for the Measurement of Phagocytosis 401
Esa-Matti Lilius and Matti Waris

Determination of Immune Complexes by Chemiluminescence of Stimulated Murine Phagocytes: Comparison with other Quantification Methods 405
J Willems and M Joniau

Influence of Bacterial Surface Components on the Stimulation of Phagocyte Chemiluminescence 409
N Topley, M J Harber and A W Asscher

B Lymphocyte Surface Bound ICs Stimulate PMNs to Produce Oxygen Derivatives which Impair T Cell Responses 411
C De Simone, M Ferrari, V Vullo, C Mastropietro and F Sorice

Lipoxygenase is not a Direct Source of Light in Sendai Virus-induced Chemiluminescence (CL) 415
Markus Grob and Ernst Peterhans

Diagnosis of Celiac Disease in Children Using Leukocyte-attached Immune Complex Luminescent Assay 419
Esa-Matti Lilius, Tuula Hamalainen and Marja-Riitta Stahlberg

Evaluation of Serum Opsonic Capacity Against Bordetella Pertussis by Luminol-dependent Phagocytosis-associated Chemilumienscence 423
Esa-Matti Lilius, Pekka Leivo, Matti Waris, Jussi Mertsola and Olli Ruuskanen

Luminol Dependent Chemiluminescence of Granulocytes of Tumor Patients 427
M Wurl, P Schuff-Werner, A Muller, K Gottsmann, S Krull and G A Nagel

Whole Blood Chemiluminescence: Influence of Cellular Constituents, Dilution and Stimulus on Luminol-and Lucigenin-dependent Chemilumienscence 431
J Ennen, M Ernst and H-D Flad

Chronobiological Investigation of Zymosan Induced and Luminol Amplified Chemiluminescence of PMN Leukocytes in Healthy Adults 435
G Harald Geelhaar, Michael Betzler, George F Zimmermann, Hans Konrad Schackert, Hans-Peter Geisen, Lutz Edler and Christian Herfarth

Studies of Group B Streptococcal (GBS) Opsonisation Using Luminol-dependent Chemiluminescence 439
M J G Hastings, J Deeley, K Oxley and C S F Easmon

Evaluation of Alternative Pathway of Complement for Opsonic Activity by Neutrophil Chemiluminescence. Its Correlation with the Particle Counter Method 443
P Hindocha, Ruth Hill, J D M Gould and C B S Wood

Augmentation of the Luminol-dependent Chemiluminescence Response of Purified Monocytes to Staph Aureus by Autologous Lymphocytes 447
A Muller, P Schuff-Werner, M Wurl and G A Nagel

Phagocytotic and Metabolic Activity of Granulocytes in AIDS-, Lymphadenopathy- and Hemophilia-Patients 451
L Stohr, M J Sessler, P Altmeyer, I Scharrer, H Holzmann, W Stille, E B Helm, M Elbert and J Gurenci

VI INSTRUMENTATION

Recent Developments in Automatic Luminescence Instrumentation 457
F Berthold

Automated Luminometer Setup for Continuous Measurement of 25 Samples 461
Esa-Matti Lilius, Matti Waris and Matti Lang

A Comparison of Two Semi-automatic Luminometers Suitable for Routine Luminescence Immunoassays 465
W G Wood, A Gadow and C J Strasburger

On-line Computer Analysis of Chemiluminescent Reactions and Application to Luminescent Immunoassays 469
M Pazzagli, A Tommasi, M Damiani, A Magini and M Serio

Stable Single Photon Calibration Standards at Different Wavelengths for Luminometry 475
T E T Oikari, I A Hemmila and E J Soini

Application of Light Guides for Enhancement of Signal to Noise Ratios at Low Levels of Luminescence Detectability 479
S E Brolin, P-O Berggren and P Naeser

VII BIOLUMINESCENCE AND BIOLUMINESCENCE ASSAYS

Binding Mechanism of Nucleotide Mimicking Dyes to Firefly Luciferase 485
Sunanda Rajgopal and M A Vijayalakshmi

Extraction and Automatic Luminometric Assay of ATP, ADP and AMP 491
A Lundin

Bioluminescent Determination of Free Fatty Acids: New Tool in Obesity Research 503
E Wieland and H Kather

Kinetic Analysis of Steroids with Bacterial Bioluminescence 507
T Lovgren, R Raunio, J Lavi and K Kurkijarvi

A Rapid Bioluminescence Assay for Non-cyclic Photophosphorylation of Isolated Chloroplasts 511
F Van Assche and H Clijsters

The Sorbitol Dehydrogenase Reaction in Bioluminescence Assay 515
S E Brolin, P Naeser and P-O Berggren

A Novel Assay for Agents Causing Membrane Perturbation 521
D E Jenner, M J Harber and A W Asscher

Measurement of Glutamate, Ammonia and Urea with Bacterial Bioluminescence 525
T Heinonen, R Raunio, J Lavi, T Lovgren and K Kurkijarvi

Bioluminescent Assays for Enzymes Metabolizing Long Chain Fatty Acid Derivatives 529
D Morse, R Szittner and E Meighen

Determination of 100 Chemicals by the Improved Bioluminescence Test for Mutagenic Agents 533
S Ulitzur, I Weiser, B Z Levi and M Barak

Bioluminescent Assays of Estrogens and Androgens 537
J C Nicolas, A M Boussioux, A M Boularan, B Colomb and A Crastes de Paulet

The Metabolic Activity of Bacteria Under Conditions of Low Water Activity - Luminous Bacteria as a Model 541
A Reinhertz, I J Kopelman and S Ulitzur

Automatic Luminometric Assay of Glycerol for Studies on Lipolysis 545
A Lundin, J Hellmer and P Arner

VIII CHEMILUMINESCENCE AND CHEMILUMINESCENCE ASSAYS

Review of Peroxyoxalate Chemiluminescence Detection for Liquid Chromatography 555
Robert Weinberger

A Luminol-mediated Assay for Oxidase Reactions in Reversed Micellar Systems 559
A J W G Visser and J S Santema

Enhancement of Lucigenin Chemiluminescence with Cyclodextrin 565
Mary Lynn Grayeski and Eric Woolf

In Situ Generation of Chemiluminescent Substances Through Hydroxylation 569
G Merenyi and J Lind

The Identification of Gamma Irradiated Foodstuffs by Chemiluminescence Measurements 573
W Bogl and L Heide

Chemiluminescence Method for Estimation of Autoxidation in Foods: Interfering Reactions 577
Bjorn E Bordalen

Determination of Choline and Acetylcholine by a Chemilumienscent Method 581
L F J Woods and P Neaves

Automatic Luminometric Assay of Choline and Acetylcholine 585
A Lundin, I Andersson, H Nilsson and A Sundwall

Spontaneous Chemiluminescence of Smoker's Blood 587
Binkoh Yoda, Ryuzo Abe, Yoshio Goto, Akio Saeki, Choichi Takyu and Humio Inaba

The Use of Luminous Bacteria to Study the Mode of Action and the Activity of Aminoglycoside Antibiotics 591
S Ulitzur and A Naveh

Index 595

I RAPID MICROBIOLOGY

APPLICATIONS OF LUMINESCENCE IN CLINICAL MICROBIOLOGY

M.J. Harber

Department of Renal Medicine, Welsh National School of Medicine, K.R.U.F. Institute, Royal Infirmary, Cardiff, U.K.

INTRODUCTION

Many diverse applications of luminescence technology have been described within the last decade but only one system, the firefly bioluminescence assay for adenosine triphosphate (ATP), has found regular use in the field of clinical microbiology. The basic principle behind this technique is that the ATP in microbial cells is present in a high and relatively constant amount under controlled conditions, and hence it may be used as an index of cell numbers. The firefly assay provides a simple, rapid and sensitive means of measuring microbial ATP, and several commercial kits are currently available for performing this method. A detailed discussion of the rationale for bioluminescence ATP analysis has been published elsewhere (1).

When any new technology is introduced it is necessary to explore all possible applications, but only a relatively small number of these are likely to become major innovations. Any proposed laboratory test should withstand close scrutiny. In particular, the following questions should be asked of any new technique:-
a) Is the test necessary? b) Does it provide the right information? c) Is it better than alternative methods?

The time is now right to ask these questions of luminescence methodology so as to maintain a balanced view of our achievements. I shall be examining the use of luminescence assays in the medical microbiology laboratory in this paper, and shall attempt to summarise achievements to date and prospects for the future.

ISBN 0-12-426290-2

BACTERIURIA DETECTION

The Clinical Problems

Urinary tract infections (UTI) occur with a particularly high prevalence in females and are also a frequent complication in patients with indwelling urinary catheters. The causative organisms, mostly Gram-negative bacilli, may possess particular virulence factors which determine their uropathogenicity (2). Three clinical types of UTI may be recognised - asymptomatic bacteriuria (ABU), symptomatic lower tract infection (cystitis) and symptomatic renal infection (acute pyelonephritis).

Because urine is often contaminated with skin commensals quantitative and qualitative bacteriology is required to detect UTI. The epidemiological studies of Kass (3) resulted in the definition of significant bacteriuria as a pure culture of $\geqslant 10^5$ colony-forming units (cfu)/ml urine, this being a statistical evaluation designed to distinguish infection from contamination. However, this definition is only applicable to ABU which is generally a benign condition for which treatment is not required in schoolchildren or non-pregnant adults (4). Recent studies have indicated that symptomatic UTI may be associated with as few as 10^2 coliforms/ml urine (5). Therefore interpretation of bacterial numbers alone is no easy matter, particularly in the range 10^2-10^5 cfu/ml. Failure to recognise urinary infection with low numbers of coliforms may account for many cases of the common syndrome of 'abacterial cystitis' (6).

Requirements of a Bacteriuria Test

Clinicians and laboratory staff have rather different requirements for a bacteriuria detection test. From a medical point of view the object is to identify patients who would benefit from antimicrobial therapy. Mass screening of healthy populations for ABU is not necessary (7), but there is a strong case for screening pregnant women and perhaps also neonates (if urine specimens can be obtained) and geriatric patients. All symptomatic infections require investigation.

From the laboratory point of view a rapid, automated preliminary test to identify non-infected urines, which may form a considerable part of the workload, will enable the staff to streamline their operations. To avoid the risk of overlooking infections any bacteriuria detection test should have high sensitivity (few false negatives) even at the expense of specificity if necessary. That is, some false positive results are acceptable as long as all false negatives are eliminated.

The most useful test of any is microscopic examination of urine for bacteria, leucocytes and epithelial cells, but this is impracticable with large numbers of specimens. Chemical bacteriuria detection tests (8) have never proved satisfactory, and more recent innovations have been based on microcalorimetry, photometry, electrochemical detection, electrical impedance measurement, colorimetric detection, radiometry, measurement of particle size distribution and bioluminescence (see ref.9 for review). The choice of method may depend on other requirements of a particular laboratory, but automation and costeffectiveness are major considerations.

Principle of the Bioluminescence Bacteriuria Test

The basic method for bacteriuria testing (10-13) involves extraction and measurement of bacterial ATP following specific extraction and hydrolysis of host cell ATP, techniques which have been made appreciably easier in recent years by the introduction of purpose-designed commercial reagents. In earlier studies the apyrase used for hydrolysis of non-microbial ATP was destroyed by heating before extraction of bacterial ATP, but this stage may be eliminated if the apyrase concentration is kept to a minimum and the bacterial ATP is measured immediately after its extraction. Inaccuracies can arise due to inhibitors of firefly luciferase in urine and variation in cellular ATP levels between different bacterial species (14), problems which are usually overlooked in commercial test protocols.

A threshold value of 10^5 or 10^4 cfu/ml urine is used to distinguish infected from noninfected samples, the latter value representing the approximate limit of sensitivity of the ATP assay. For reasons discussed above there is no evidence to support the use of 10^5 cfu/ml as a critical value in symptomatic UTI, and so the more sensitive value of 10^4 cfu/ml, although arbitary, would appear to be preferable. A recent report describing analysis of 34,600 routine urine specimens, using a threshold value of 10^4 cfu/ml with a commercial bioluminescence test system, revealed that 12.4% of these samples would have been discarded as uninfected if the higher threshold value of 10^5 cfu/ml had been used (15).

Commercial Developments

The gap between research developments and routine applications can only be bridged by commercial enterprise, new methodology being of wide appeal only if the instrumentation and reagents are packaged in kit form. A great deal of commercial interest has been shown in the bioluminescence bacteriuria test, and kits are now available in Europe and the U.S.A. from several companies. Those marketed by Lumac and A.L.L. are based on the principle outlined above, while the kit from Turner Designs differs radically in that it is based on the assay of total urinary ATP (bacterial plus somatic) before and after a 1h incubation in thioglycollate broth. This method appears to give an index of pyuria as well as bacteriuria, which would certainly be valuable additional information as pyuria usually accompanies symptomatic UTI (16), but the technique has yet to be formally validated.

Several studies have been described comparing commercial bacteriuria tests with other methods, and the Lumac kit has received particular scrutiny. This semi-automated method yields results within 1h and has been shown to have a level of sensitivity comparable to or even higher than that of commercial methods based on photometric growth detection (9,17). When compared with a reference cultural method, using a threshold value of 10^4 cfu/ml, the Lumac system was found to be accurate and reproducible giving no false negative results and only 13.6% false positives for general urine specimens (18). The major technical drawback is the amount of manual manipulation still required which reduces the cost-effectiveness of the method. The Berthold LB950 instrument is fully automated and allows processing of urine samples in only 8 minutes using the A.L.L. kit, but this combination is only in the preliminary stages of testing at present and may yet require some modification.

In conclusion, complete automation of the commercial bioluminescence tests will make this technique very attractive to routine hospital laboratories with a large workload. However, care needs to be exercised in the design of commercial kits to make sure that they fulfil the clinical requirements for a bacteriuria test as well as providing a convenience for the laboratory.

BACTERAEMIA DETECTION

The detection of bacteraemia poses far greater problems than detection of bacteriuria because the number of bacterial cells in question is considerably fewer. In fact, quantitation of absolute bacterial numbers in the blood stream is not possible, and bacteraemia can only be verified by use of an enrichment culture. Standard blood culture bottles may require an incubation period of a week or more, although the radiometric detection system currently available normally permits detection within a few days.

Preliminary studies have been undertaken using bioluminescence to detect bacterial growth following an enrichment culture to increase cell numbers to within the range of sensitivity of the ATP assay (19,20). The method used was analogous to the bacteriuria detection test in that non-microbial ATP was extracted and hydrolysed before extraction and measurement of bacterial ATP. The enormous ratio of host cell ATP: bacterial ATP was found to be a major problem, and this was amplified by binding of somatic ATP to cell debris whence it resisted hydrolysis by apyrase (20). However, Molin and colleagues (20) managed to separate bacteria from cell debris on a Percoll gradient which enabled them to detect bacterial ATP in simulated blood cultures, seeded with 10 cfu of various bacteria/ml, after only a 6-10h incubation period. This technique is an interesting approach to the problem of speeding-up the traditionally slow process of detecting bacteraemia.

ANTIMICROBIAL SUSCEPTIBILITY TESTING

The most commonly used method for antimicrobial susceptibility testing is the disc diffusion technique, but this usually requires an overnight incubation and it may give variable results depending on the concentration of antibiotic disc used, the choice of agar and the size of the inoculum. In recent years automated instruments allowing rapid reporting of antimicrobial susceptibility have become available commercially which are based on electrical impedance measurement or photometric detection of bacterial growth. However, introduction of new technology can create its own problems, and widespread concern has been expressed over the lack of standardization of methods for testing antimicrobial susceptibility. Despite this, the application of bioluminescence detection for bacterial growth to antibiotic sensitivity tests is a logical extension of the technique.

The first report on the use of bioluminescence for sensitivity testing showed that the method generally worked well but that there was a serious disagreement with an agar diffusion method in 7% of instances (21). This discrepancy appeared to be related to leakage of ATP into the supernatant caused by low concentrations of inhibitors of cell wall synthesis, a problem which was overcome by other workers by hydrolysing extracellular ATP with apyrase and measuring intracellular ATP only (22,23). Reliable results were then obtained for clinical isolates tested against a variety of antibiotics using a 2-3h incubation period (24). Results were expressed as the ATP level in broth cultures exposed to antibiotic as a percentage of the ATP level in unexposed cultures. This same method has also been used recently for determining the susceptibility of clinical isolates of *Staphylococcus aureus* to methicillin with a 2½h incubation (25), although in this instance total culture ATP was measured. Bacteria were classified as methicillin sensitive if the light emission from test cultures was $<40\%$ of that from control cultures, and results correlated exactly with those of disc diffusion and m.i.c. assays.

The bioluminescence ATP assay provides a more sensitive growth detection system than photometry or measurement of electrical impedance, albeit that the latter methods have the advantage of allowing continuous monitoring of bacterial growth. The bioluminescence system has been shown to be applicable to rapid antimicrobial susceptibility testing and it will be interesting to see if there is any commercial development of this technique.

ANTIBIOTIC AND VITAMIN ASSAYS

Rapid assays for antibiotics in clinical specimens are particularly important for the aminoglycoside group which are used to treat life-threatening infections, but which may create serious toxicity problems. Serum concentrations need to be monitored carefully to confirm that therapeutic peak levels are achieved while toxic trough levels are avoided.

Antibiotic assays incorporating bioluminescence ATP analysis were developed in this laboratory (26,27) and in Sweden (28,29). Both groups showed that the method compared favourably with results obtained using conventional agar diffusion assays. Accumulation of extracellular ATP was found to be a problem with some bacterial strains, and in one assay this ATP was hydrolysed with apyrase before extraction of intracellular ATP (28). Measurement of total culture ATP appears to be satisfactory in most cases, while measurement of extracellular ATP only has even been described (30). The method has been extended to provide rapid assays for a variety of antibiotics with different modes of action (1), and the ATP detection system has also been used to measure serum levels of folic acid (31).

In the original work on bioluminescence antibiotic assays impure reagents and insensitive apparatus limited the usefulness of the technique. However, modern commercial reagents for ATP extraction and measurement make the method considerably more attractive. These reagents may be injected directly into samples of broth cultures in instruments designed with multi-injection ports, so simplifying and speeding-up the procedure. Using Lumac reagents and instrumentation the assay of netilmicin in serum has been accomplished in this laboratory within a total time of 3h. Another time-saving measure is to reduce the concentration of serum in the assay medium to 1% of the final volume so that sera do not need to be heat-inactivated prior to assay, even with serum-sensitive test organisms (32). However, there may be a loss of sensitivity at low antibiotic concentrations, and up to 5% serum may be required for some assays when a 30 min pre-treatment at 56 °C might be necessary.

There is no doubt that for the assay of specific drugs modern EMIT or fluoroimmunoassays are superior to bioluminescence or any other microbiological technique in terms of accuracy, rapidity and convenience. However, these methods are not available to all routine hospital laboratories. The bioluminescence technique is as accurate as, and more rapid than, traditional microbiological assays and is potentially applicable to a wide range of antibiotics and vitamins. Furthermore, the method provides a measure of the biological activity of the analyte which could be an advantage in studies of drug interactions or metabolism. As with antimicrobial susceptibility testing there has not, as yet, been any commercial development of the method.

RESEARCH APPLICATIONS OF MICROBIAL ATP ANALYSIS

Adherence Tests

Quantitation of microbial mass in dental plaque can be a problem because of difficulty in dispersion of plaque samples. However, measurement of extractable ATP has been shown to provide a good index of viable cell mass, results being related to bacterial protein content (33) or to plaque wet weight (34). Determination of the adenylate energy charge in plaque following bioluminescence assay of ATP, ADP and AMP gives an even more precise estimate of microbial mass (35).

Bioluminescence ATP assay has also been used to measure the adherence of uropathogenic *Escherichia coli* to polystyrene (36). The method has recently been extended in this laboratory to test the adherence of catheter-associated urinary pathogens to segments of urinary catheters. Preliminary observations indicate that strains of *Pseudomonas* and *Proteus* isolated from catheterised patients with spinal injuries adhere more strongly to catheters than less common pathogens.

Hydrophobicity Test

Bacterial surface hydrophobicity is an important determinant of the interaction between bacteria and host epithelial or phagocytic cells. A common method for measuring hydrophobicity is to pass a bacterial suspension in phosphate buffer containing IM $(NH_4)_2SO_4$ through a mini-column containing octyl- or phenylsepharose. Organisms with a hydrophobic surface are retained on the gel (37). Analysis of bacterial numbers in the eluate by bioluminescence ATP assay has been used in this laboratory to give rapid and accurate results with this test. The concentration of $(NH_4)_2SO_4$ in the buffer was reduced to 0.25M because the very high osmolality in IM $(NH_4)_2SO_4$ adversely affected bacterial ATP levels.

LUMINESCENCE ENZYME IMMUNOASSAYS FOR BACTERIA AND VIRUSES

Luminescence systems other than the firefly ATP assay hardly seem to have been considered for application in medical microbiology. However, one very interesting development is the use of luminescence enzyme immunoassay to quantify bacterial cell numbers (38), bacterial toxins (39) and viral particles (40,41). In each case the analyte was labelled with horseradish peroxidase which was detected by measurement of the chemiluminescence emitted during the oxidation of pyrogallol, purpurogallin or luminol. The bacterial assay is the most sensitive technique yet described for rapid quantitation of cell numbers, as few as 30 cells of *Serratia marcescens* being detectable (38). There has been no recent development of this technique, but it is an inspiring approach to the problem of quantifying very low numbers of microorganisms.

CONCLUSIONS

The most important development in clinical microbiology since the previous International Luminescence Symposium has been the advent of several commercial systems for bioluminescence detection of bacteriuria. These methods have already entered routine use in many hospital laboratories in the U.S.A. and Europe, and further automation should ensure their success. However, it is hoped that the designers of commercial test kits will not lose sight of the clinical objectives of bacteriuria detection, and attention needs to be paid to the use of a threshold value for defining bacteriuria as well as to technical problems.

There has been very little development of assays other than those for bacteriuria detection in recent years. Commercial interest has been expressed in a bioluminescence test for bacteraemia, but nothing has yet materialised. It will be interesting to see what developments occur within the next few years, particularly with regard to microbiological applications of alternative systems such as luminescence immunoassays.

ACKNOWLEDGEMENTS

I am grateful to Mr. P.W. McWalter, Withington Hospital, Manchester, U.K., for supplying the proof of a paper prior to publication. Miss. J. Knowlden performed the netilmicin assays and Mr. C. Clayton the catheter adherence studies.

REFERENCES

1. Harber, M.J. (1982). Applications of luminescence in medical microbiology and hematology. *In* "Clinical and Biochemical Luminescence" (Eds. L.J.Kricka and T.J.N.Carter), pp. 189-218. Marcel Dekker, New York.
2. Harber, M.J., Mackenzie, R. and Asscher, A.W. (1984). Virulence factors of *Esch. coli*. *In* "Pyelonephritis" (Eds. H.Losse, A.W.Asscher, A.Lison and V.T.Andriole), vol.V, pp. 43-50. Thieme, Stuttgart.
3. Kass, E.H. (1956). Asymptomatic infections of the urinary tract. *Trans. Assoc. Am. Physicians* 69, 56-63.
4. Asscher, A.W. (1980). "The Challenge of Urinary Tract Infection". Academic Press, London.
5. Stamm, W.E., Counts, G.W. et al. (1982). Diagnosis of coliform infection in acutely dysuric women. *N. Engl. J. Med.* 307, 463-468.
6. Walpita, P. and Marsh, F.P. (1983). Do urethral *Escherichia coli* cause abacterial cystitis? *J. Clin. Pathol.* 36, 224-227.
7. Sussman, M., Asscher, A.W. et al.(1969). Asymptomatic significant bacteriuria in the non-pregnant woman. *Brit. J. Med.* 1, 799-803.
8. Mackinnon, A.E., Strachan, C.J.L. et al (1973). Detection of bacteriuria with a test for urinary glucose. *In* "Urinary Tract Infection" (Eds. W.Brumfitt and A.W.Asscher), pp. 11-15. Oxford University Press, London.
9. Pezzlo, M.T. (1983). Automated methods for detection of bacteriuria. *Am.J.Med.* 85, suppl. 1B, 71-78.
10. Thore, A., Ansehn,S. et al, (1975). Detection of bacteriuria by luciferase assay of ATP. *J. Clin. Microbiol.* 1, 1-8.
11. Picciolo, G.L., Chappelle, E.W. et al.(1975). Problem areas in the use of the firefly assay for bacterial detection. *In* "Analytical Applications of Bioluminescence and Chemiluminescence" (Eds. E.W.Chappelle and G.L.Picciolo), pp. 1-26, N.A.S.A., Washington.
12. Alexander, D.N., Ederer, G.M. and Matsen, J.M. (1976). Evaluation of an ATP assay as a screening method to detect significant bacteriuria. *J. Clin. Microbiol.* 3, 42-46.
13. Johnston, H.H., Mitchell, C.J. and Curtis, G.D.W. (1976). An automated test for the detection of significant bacteriuria. *Lancet* (ii), 400-402.

14. Nichols, W.W., Curtis, G.D.W. and Johnston, H.H. (1982). Analysis of the disagreement between automated bioluminescence-based and culture methods for detecting significant bacteriuria. *J. Clin. Microbiol.* 15, 802-809.
15. McWalter, P.W., Stern, G.S. et al.(1983). Routine bioluminescence urine screening. *Eur. J. Clin. Microbiol.* 2, 53.
16. Stamm, W.E. (1983). Measurement of pyuria and its relation to bacteriuria. *Amer. J. Med.* 85, (suppl. 1B), 53-58.
17. Szilagyi, G., Aning, V and Karmen, A. (1983). Comparative study of two methods for rapid detection of clinically significant bacteriuria. *J. Clin. Lab. Autom.* 3, 117-122.
18. McWalter, P.W. and Sharp, C.A. (1982). Evaluation of a commercially available semi-automated bioluminescence system for bacteriuria screening. *Eur. J. Clin. Microbiol.* 1, 223-227.
19. Beckers, B. and Lang, H.R.M. (1982). Bioluminescent measurement of ATP for the rapid detection of positive blood cultures. *Naturwissenshaften* 69, S 145-146.
20. Molin, O., Nilsson, L. and Ansehn, S. (1983). Rapid detection of bacterial growth in blood cultures by bioluminescent assay of bacterial ATP. *J. Clin. Microbiol.* 18, 521-525.
21. Vellend, H., Tuttle, S.A., et al.(1975). A rapid method for the determination of microbial susceptibility using the firefly luciferase assay for ATP. *In* "Analytical Applications of Bioluminescence and Chemiluminescence" (Eds. E.W.Chappelle and G.L. Picciolo), pp. 43-44. N.A.S.A., Washington.
22. Hojer, H., Nilsson, L. et al.(1976). *In vitro* effect of doxycycline on levels of ATP in bacterial cultures. *Scand. J. Infect. Dis. Suppl.* 9, 58-61.
23. Thore, A., Nilsson, L. et al.(1977). Effects of ampicillin on intracellular levels of ATP in bacterial cultures related to antibiotic susceptibility. *Acta. Pathol. Microbiol. Scand. Sect. B.* 85, 161-166.
24. Hojer, H., Nilsson, L. et al.(1979). Possible application of luciferase assay of ATP to antibiotic susceptibility testing. *In* "Proceedings of the 1st International Symposium on Analytical Applications of Bioluminescence and Chemiluminescence" (Eds. E.Schram and P.Stanley), pp.523-530. State Printing and Publishing, Westlake Village.

25. McWalter, P.W. (1984). Determination of susceptibility of *Staphylococcus aureus* to methicillin by luciferin-luciferase assay of bacterial ATP. *J. Appl. Bact.* 56, 145-150.
26. Harber, M.J. and Asscher, A.W. (1976). A new assay technique for antibiotics. *In* "Proceedings of the 9th International Congress of Chemotherapy" vol 11, "Chemotherapy" (Eds. J.D.Williams and A.M.Geddes), pp.125-131. Plenum, London.
27. Harber, M.J. and Asscher, A.W. (1977). A new method for antibiotic assay based on measurement of bacterial ATP using the firefly bioluminescence system. *J. Antimicrob. Chemother.* 3, 35-41.
28. Nilsson, L., Hojer, H. et al. (1977). A rapid semi-automated bioassay of gentamicin based on luciferase assay of bacterial ATP. *Scand. J. Infect. Dis.* 9, 232-236.
29. Hojer, H. and Nilsson, L. (1978). Rapid determination of doxycycline based on luciferase assay of bacterial ATP. *J. Antimicrob. Chemother.* 4, 503-508.
30. Nilsson, L. (1978). New rapid bioassay of gentamicin based on luciferase assay of extracellular ATP in bacterial cultures. *Antimicrob. Agents Chemother.* 14, 812-816.
31. Harber, M.J. and Asscher, A.W. (1979). Bioluminescence assay for antibiotics and vitamins. *In* "Proceedings of the 1st International Symposium on Analytical Applications of Bioluminescence and Chemiluminescence" (Eds. E.Schram and P.Stanley), pp. 531-542. State Printing and Publishing, Westlake Village.
32. Nilsson, L., Hojer, H. et al. (1979). A simplified luciferase assay of antibiotics in clinical serum specimens. *In* "Proceedings of the 1st International Symposium on Analytical Applications of Bioluminescence and Chemiluminescence" (Eds. E.Schram and P.Stanley), pp, 515-522. State Printing and Publishing, Westlake Village.
33. Robrish, S.A., Kemp, C.W. and Bowen, W.H. (1978). Use of extractable ATP to estimate the viable cell mass in dental plaque samples obtained from monkeys. *Appl. Environ. Microbiol.* 35, 743-749.
34. Distler, W., Kroncke, A. and Maurer, G. (1980). ATP content of human dental plaque as a measure of viable cell mass. *Caries Res.* 14, 265-268.
35. Kemp, C.W. (1979). Adenylate energy charge: a method for the determination of viable cell mass in dental plaque samples. *J. Dent. Res.* 68, (D), 2192-2197.

36. Harber, M.J., Mackenzie, R. and Asscher, A.W. (1983). A rapid bioluminescence method for quantifying bacterial adhesion to polystyrene. *J. Gen. Microbiol.* 129, 621-632.
37. Magnusson, K-E. (1980). The hydrophobic effect and how it can be measured with relevance to cell-cell interactions. *Scand. J. Infect. Dis.Suppl.* 24, 131-134.
38. Halmann, M., Velan, B. and Sery, T. (1977). Rapid identification and quantitation of small numbers of microorganisms by a chemiluminescent immunoreaction. *Appl. Environ. Microbiol.* 34, 473-477.
39. Velan, B. and Halmann, H. (1978). Chemiluminescence immunoassay; a new sensitive method for determination of antigens. *Immunochemistry* 15, 331-333.
40. Velan, B., Schupper, H. et al.(1979). Solid phase chemiluminescent immunoassay. *In* "Proceedings of the 1st International Symposium on Analytical Applications of Bioluminescence and Chemiluminescence" (Eds. E.Schram and P.Stanley), pp. 431-437. State Printing and Publishing, Westlake Village.
41. Miller, R.J. and Reichard, D.W. (1981). Chemiluminescent immunoassay for the detection of virus/antibody aggregates. *In* "Bioluminescence and Chemiluminescence" (Eds. M.A.DeLuca and W.D.McElroy), pp.645-649. Academic Press, New York.

RAPID SUSCEPTIBILITY TESTING OF *STAPHYLOCOCCUS AUREUS*

Paul W. McWalter

Department of Bacteriology, Withington Hospital, Manchester, M20 8LR, England

INTRODUCTION

Rapid determination of antibiotic susceptibility is important in the management of serious *Staphylococcus aureus* infection. Conventional disc diffusion methods for testing susceptibility may involve 18 hours delay before an antibiogram can be obtained. Luciferin-Luciferase assay of bacterial adenosine triphosphate (ATP) has been described as a possible rapid alternative to disc diffusion for determining antibiotic susceptibility (1,2,3). The application of this technique for testing the susceptibility of *Staphylococcus aureus* to methicillin has been recently reported (4). Due to the use of overnight broth cultures as inocula for these Luciferin-Luciferase assay systems no overall time reduction has been achieved.

Preliminary investigations carried out in this laboratory demonstrated that modification of the Luciferin-Luciferase methicillin testing system would allow the susceptibility of *Staphylococcus aureus* to be determined using organisms taken directly from solid growth media. In this study the modified Luciferin-Luciferase bacterial ATP assay method was used to determine the susceptibility of strains of *Staphylococcus aureus* directly from primary inoculation media.

MATERIALS AND METHODS

Test strains

100 strains of *Staphylococcus aureus* (identified on primary inoculation media by colonial appearance and a positive slide coagulase test) were examined in this study. The susceptibility of each strain to methicillin, gentamicin,

ISBN 0-12-426290-2

clindamycin, erythromycin and fusidic acid was determined by disc diffusion, minimum inhibitory concentration (m.i.c.) and Luciferin-Luciferase assay of bacterial ATP.

Disc diffusion

The technique of Stokes and Ridgway (5) was used to carry out disc diffusion. Sensitivity discs (Oxoid) were used at the following concentrations; methicillin 10μg, gentamicin 10μg, clindamycin 2μg, erythromycin 5μg and fusidic acid 10μg. Methicillin testing was carried out on Oxoid DST agar supplemented with 5% sodium chloride. The other antibiotics were tested on DST agar containing 5% lysed horse blood. Test plates were incubated overnight, methicillin at 30°C, the remainder at 37°C.

Minimum inhibitory concentration

The agar incorporation and multiple replication technique used to calculate m.i.c. values has been previously reported (4). Methicillin plates were incubated overnight at 30°C, the remainder at 37°C overnight. Antibiotics were tested at the following concentrations: methicillin 1,2,4,6,8,10,12 and 16μg/ml; gentamicin 0.1,0.5,1,2,4,6,8,10 and 12μg/ml; clindamycin and erythromycin 0.05,0.1,0.25,0.5,1,2 and 4μg/ml; fusidic acid 0.5,1,2,4,6,8,10 and 12μg/ml.

Luciferin-Luciferase assay

The following Lumac/3M bv equipment and reagents were used; M2010 multijet biocounter, NRB, Lumit PM and Lumit buffer. Six colonies of *Staphylococcus aureus* were touched with a wire loop and suspended in 1ml of distilled water. Suspensions were adjusted to McFarlands No1 standard. Standardised suspension (0.1ml) was added to each of 7 one dram vials containing 1.9ml $\frac{1}{4}$ strength Brain Heart Infusion broth (Oxoid). Five vials contained one of the following antibiotics; methicillin (4μg/ml), gentamicin (1μg/ml), clindamycin (0.25μg/ml), erythromycin (0.25μg/ml), fusidic acid (2μg/ml), the remaining two being growth controls. The methicillin vial and a control vial were incubated at 30°C for three hours, the remainder at 37°C for 2 hours. After incubation Luciferin-Luciferase ATP assay was carried out as previously reported (4). Test broth RLU values were expressed as a percentage of their growth controls (1,4). Values of 40% or less were taken as indicating susceptibility (4). All assays were carried out in duplicate.

RESULTS

The correlation between the three testing systems is shown in Table 1. Two strains of *Staphylococcus aureus* tested

TABLE 1

Percentage correlation between testing systems

Antibiotic	m.i.c./ATP	disc diff./m.i.c.	ATP/disc diff.
methicillin	98	100	98
gentamicin	98	100	98
clindamycin	98	99	97
erythromycin	98	98	95
fusidic acid	98	99	97

directly by Luciferin-Luciferase ATP assay were shown to be contaminated with other organisms, explaining the 2% discrepancy noted with all antibiotics when compared to the other two systems. Luciferin-Luciferase assay of pure cultures of these strains correlated with disc diffusion and m.i.c. On one occasion with clindamycin and fusidic acid, and twice with erythromycin, Luciferin-Luciferase results failed to correlate with disc diffusion. All these strains were sensitive by disc diffusion and resistant by m.i.c./ Luciferin-Luciferase.

All susceptible strains gave RLU percentage values of <40 when tested by Luciferin-Luciferase ATP assay, resistant strains having values of $>78.5\%$. Maximum variation in RLU values between duplicates was 2.3%.

DISCUSSION

Agreement between the three testing systems was excellent. The Luciferin-Luciferase antibiotic concentrations were those commonly used as systemic break-point values. The results using these values correlated well with m.i.c. findings. The m.i.c./Luciferin-Luciferase resistant, disc diffusion sensitive strains had m.i.c. values which would be reported as 'intermediately sensitive' using a two concentration break-point technique.

This investigation supports the findings previously reported (4) that expression of results as percentage RLU values provides a simple method of demonstrating antibiotic susceptibility.

CONCLUSIONS

Results indicate that a single point Luciferin-Luciferase assay of bacterial ATP using organisms taken directly from primary inoculation media is an accurate, rapid alternative to disc diffusion for determining the susceptibility of *Staphylococcus aureus*.

ACKNOWLEDGEMENTS

The author wishes to thank Lumac/3M bv for support with equipment and reagents.

REFERENCES

1. Hojer, H.,Nilsson, L.,Ansehn, S. and Thore, A. (1976). In-vitro effect of doxycycline on levels of adenosine triphosphate in bacterial cultures, *Scandinavian Journal of Infectious Diseases, Supplement*, 9, 58-61.
2. Thore, A.,Nilsson, L.,Hojer, H.,Ansehn, S. and Brote, L. (1977). Effects of ampicillin on intracellular levels of adenosine triphosphate in bacterial cultures related to antibiotic susceptibility, *Acta Pathologica et Microbiologica Scandinavica, Section B*, 85, 161-166.
3. Hojer, H.,Nilsson, L.,Ansehn, S. and Thore, A. (1979). Possible application of luciferase assay of ATP to antibiotic susceptibility, *In* "Proceedings of the International Symposium on Analytical Applications of Bioluminescence and Chemiluminescence" (Eds. E. Schram and P.E. Stanley). pp523-530. State Printing and Publishing Inc., Westlake Village, California.
4. McWalter, P.W. (1984). Determination of susceptibility of Staphylococcus aureus to methicillin by luciferin-luciferase assay of bacterial adenosine triphosphate, *Journal of Applied Bacteriology*, 56, 145-150.
5. Stokes, E.J. and Ridgway, G.L. (1980). "Clinical Bacteriology", 5th Edition, Arnold, London.

BACTERIAL ATP CONTENT OF URINES FROM BACTERIURIC AND NON-BACTERIURIC PATIENTS

G.D.W. Curtis, H.H. Johnston and W.W. Nichols

Department of Microbiology and
Regional Public Health Laboratory,
John Radcliffe Hospital, Oxford, UK.

INTRODUCTION

Of the many methods advocated for screening for bacteriuria in recent years only bioluminescence is both rapid (less than 30 min. to obtain results) and capable of automation (1). It is therefore not surprising that many workers have reported experiences with both locally developed and commercial systems (1-4). Bioluminescence based methods must be compared with a reference method to assess their accuracy but the use of a variety of reference methods has led to conflicting results (3-5). We have suggested that the use of a single criterion of a quantitative microbial count is an unsuitable reference for the heterogeneous group of specimens received by most laboratories (6).

We have attempted to overcome this problem by using information available in the laboratory to decide whether each urine came from a patient with bladder bacteriuria. The results are reported in the present communication.

RESULTS AND DISCUSSION

Urine samples (512) were placed into one of four categories (Infected, Uninfected, Contaminated, Query) by reference to information available from laboratory investigations (quantitative culture and identification of isolates, somatic cell count) and details provided on the request card (symptoms, therapy, type of specimen, time of collection). The presence of squamous epithelial cells, diphtheroid organisms or mixed bacterial flora was regarded as evidence of contamination (45 samples). 34 samples which we could not categorise were placed in the 'Query' group. There

ISBN 0-12-426290-2

were 113 'Infected' and 320 'Uninfected' samples remaining. The bioluminescence method was calibrated empirically against the subjective assessment. The threshold of positivity for the bioluminescence method was defined as that value (1.68nM ATP) which gave 21% falsely positive results (Table 1)- an acceptable level for a screening test.

TABLE 1

Comparison between Bioluminescence and Subjective Assessment

		Bioluminescence		
		-	+	
Subjective assessment	-	97	16	113
	+	67	253	320
		164	269	433

Sensitivity 97/113 x 100 = 86% (14% falsely negative)
Specificity 253/320 x 100 = 79% (21% falsely positive)

The results of Table 1 represent a discrete evaluation.

The usefulness of a diagnostic test can also be gauged from the degree of separation of the two populations (infected and uninfected patients in this case) which is observed when the levels of the tested analyte in each population are plotted as frequency distributions (7).

Frequency distributions of apparent bacterial ATP concentrations in 'Infected' and 'Uninfected' samples (Fig. 1) were each approximately normal with means of 1.21 and -0.74 and SDs of 1.09 and 1.04 respectively. 'Contaminated' and 'Query' specimens were also approximately normally distributed with means of 0.53 and 0.14 and SDs of 0.93 and 1.16 respectively (units as Fig. 1; data not shown).

By attempting to determine the presence or absence of bacteria in the bladder we are approaching a true measurement of the sensitivity and specificity of the bioluminescence method. More rigorous searching for clinical information than we were able to carry out in this preliminary investigation should reduce the number in the 'Query' category.

We propose that a bioluminescence-based or indeed any other automated method for the detection of 'significant bacteriuria' will only achieve general acceptance if its

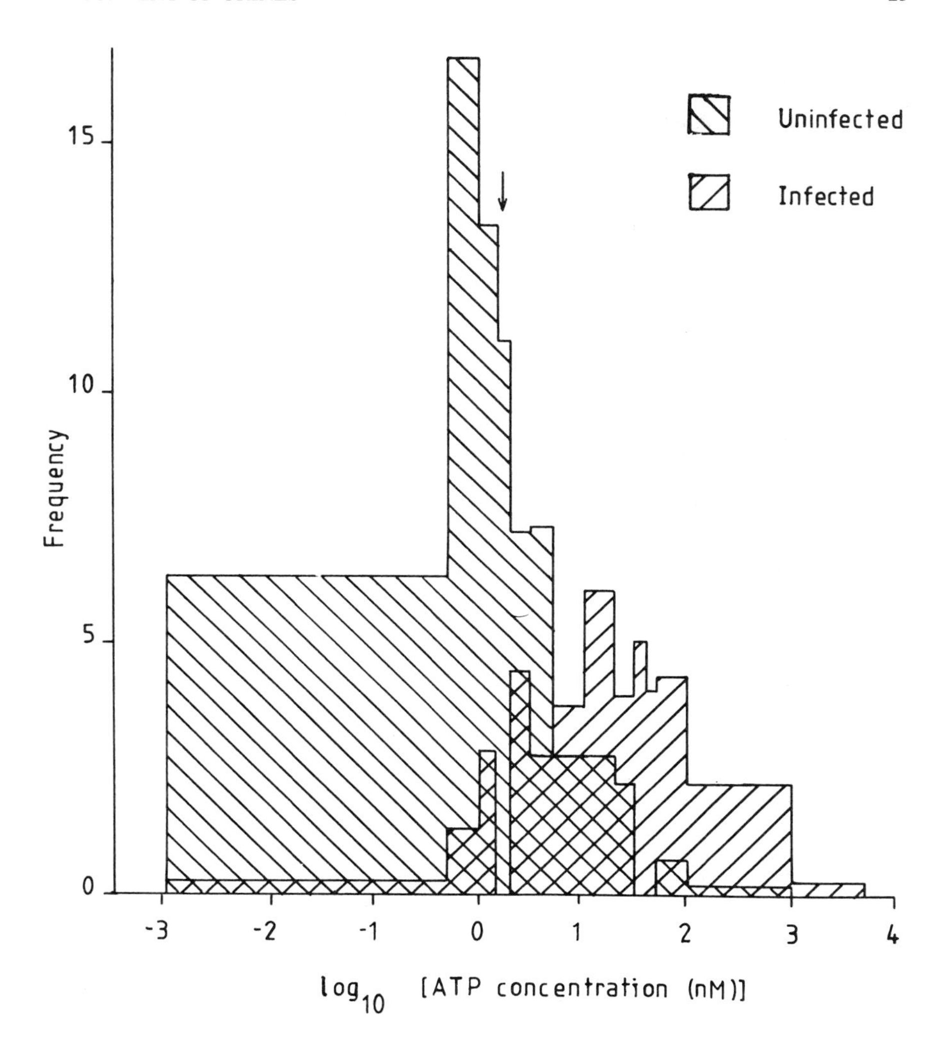

Fig. 1 Frequency distributions of apparent bacterial ATP concentrations in 'Infected' and 'Uninfected' samples. Urine samples were centrifuged (1600g, 40 min, 4°C) and the sediment resuspended in 1/10 volume distilled water. Analysis was performed in the flow system of Nichols et al. *(5) without probe dilution. The measurements of bacterial ATP in urine were those which were obtained in the third study of reference 6.*

ability to detect the clinical condition of interest (urinary tract infection) is proven by the type of frequency distribution plot shown in Fig. 1.

ACKNOWLEDGEMENT

This work was supported by an Equipment Research and Development grant (55/79, R/E 1048/240) from the Department of Health and Social Security.

REFERENCES

1. Johnston, H.H., Mitchell, C.J. and Curtis, G.D.W. (1976). An automated test for the detection of significant bacteriuria. *Lancet* ii, 400-402.

2. Thore, A., Lundin, A. and Ansehn, S. (1983). Firefly luciferase ATP assay as a screening method for bacteriuria. *Journal of Clinical Microbiology* 17, 218-224.

3. McWalter, P.W. and Sharp, C.A. (1982). Evaluation of a commercially available semi-automated bioluminescence system for bacteriuria screening. *European Journal of Clinical Microbiology* 1, 223-227.

4. Mackett, D., Kessock-Philip, S., Bascombe, S. and Easmon, C.S.F. (1982). Evaluation of the Lumac kit for the detection of bacteriuria by bioluminescence. *Journal of Clinical Pathology* 35, 107-110.

5. Nichols, W.W., Curtis, G.D.W. and Johnston, H.H. (1982). Analysis of the disagreement between automated bioluminescence-based and culture methods for detecting significant bacteriuria, with proposals for standardizing evaluations of bacteriuria detection methods. *Journal of Clinical Microbiology* 15, 802-809.

6. Nichols, W.W., Curtis, G.D.W. and Johnston, H.H. (1984). Detection of bacteriuria by bioluminescence : effect of pre-analysis centrifugation of specimens. *Journal of Applied Bacteriology* 56, 247-257.

7. Wald, N.J. and Cuckle, H.S. (1980). Alpha fetoprotein in the antenatal diagnosis of open neural tube defects. *British Journal of Hospital Medicine* 23, 473-489.

RAPID DETECTION OF BACTERIEMIA BY BIOLUMINESCENT ASSAY OF BACTERIAL ATP

L. Nilsson, Ö. Molin and S. Ånséhn

Department of Clinical Bacteriology
University Hospital, 581 85 Linköping, Sweden

INTRODUCTION

The occurrence of large amounts of nonbacterial ATP from somatic cells is a major problem in the assay of bacterial ATP in clinical specimens.

A selective method for distinguishing bacterial and non-bacterial ATP in urine specimens by incubation of urine samples with detergent and ATPase has been presented (1). The usefulness of this procedure has been shown in several bioluminescent assay procedures for detection of bacteriuria (1-3).

When applied to blood cultures the detergent ATPase treatment did not sufficiently eliminate blood cell ATP (4).

This study deals with possible ways to reduce blood cell ATP and to develop a method for rapid detection of bacteriemia.

MATERIALS AND METHODS

Blood cultures

A two-phase blood culture system was used (4). The blood cultures were inoculated with 10% (vol/vol) blood.

ISBN 0-12-426290-2

Elimination of blood cell ATP

To reduce blood cell ATP a sample from a simulated or a clinical blood culture was incubated for 10 min at 37°C with a triton apyrase solution (4). Separation of blood cell debris from bacteria was performed on a Percoll gradient (Pharmacia Fine Chemicals, Uppsala, Sweden) (4). The position of the bacterial layer can be determined by density maker beads (Pharmacia Fine Chemicals) or by gradient centrifugation of a bacterial culture with a bacterial count exceeding 10^6 CFU/ml, which gives a visible layer at a specific region of the gradient.

To eliminate remaining blood cell ATP, a sample from the bacterial layer was incubated with an apyrase solution for 10 min at 37°C.

Extraction of intracellular bacterial ATP

After the elimination of blood cell ATP, bacterial ATP was extracted with boiling Tris/EDTA buffer (4). The extraction inhibits the action of the apyrase and disrupts the bacteria thus releasing their ATP content.

Luciferase assay of ATP

Luciferase reagent (ATP monitoring reagent, LKB-Wallac, Turku, Finland) was added to each extract and the light intensity was measured in a 1250 Luminometer (LKB-Wallac) and registered on a 1250 Display (LKB-Wallac).

Calculation of assay results

Sample ATP levels were calculated by using assays of standard amounts of ATP as reference, and correction for background luminescence was made. An internal standard technique was used to correct for inhibition of the luciferase reaction in the extracts, i.e. the addition of known amounts of ATP to the extracts.

RESULTS AND DISCUSSION

Blood cultures inoculated with 10% (vol/vol) fresh blood contained 10^{-4}M blood cell ATP after extraction with boiling Tris buffer. Detergent ATPase treatment before extraction reduced this level to 10^{-6}M. Fractionated centrifugation to settle most of the blood cells into a pellet followed by treatment of the supernatant with detergent and

ATPase gave a further decrease of blood cell ATP. This treatment did not release all ATP since a background of ATP (10^{-8}) reappeared after the extraction. This background makes it impossible to detect less than 10^6 CFU of bacteria per ml.

A better separation was necessary to overcome this background of blood cell ATP (10^{-8}M). To achieve this, samples from the blood cultures were subjected to density gradient centrifugation. The small difference between densities of blood cells and bacteria necessitated lysing of blood cells with detergent before gradient centrifugation. The released ATP was eliminated by apyrase. Blood cell ATP which was bound to cell debris and escaped the ATPase activity was separated on the gradient. By this procedure the detection limit was decreased to 10^{-10}M, which is due to the background luminescence from the luciferase reagent and the sensitivity of the equipment. The yields of bacteria in samples from the gradient was equal to or greater than in the blood culture bottles due to the zone sharpening effect of the centrifugation.

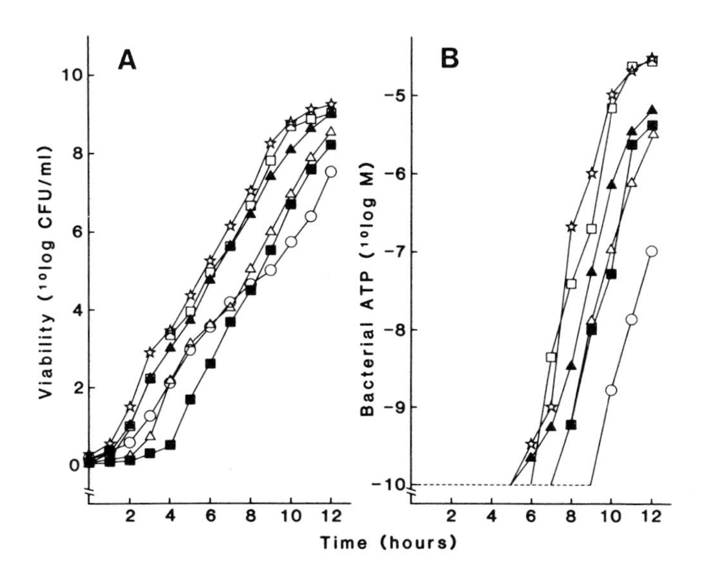

Fig. 1 Growth of Escherichia coli (□), *Klebsiella pneumonia* (☆), *Pseudomonas aeruginosa* (○), *Proteus mirabilis* (△), *Staphylococcus aureus* (■) *and Streptococcus faecalis* (▲) *monitored by viability (CFU/ml) (A) and bacterial ATP (M) (B). The cultures contained 10% (vol/vol) blood.*

In simulated blood cultures inoculated with 10 CFU of bacteria per ml blood bacterial ATP levels exceeded the detection limit (10^{-10}M) after 6 to 10 h of incubation (Fig. 1). At the time of detection the amount of bacteria just exceeded 10^4 CFU/ml (Fig. 1).

The bioluminescent assay was evaluated on 512 clinical blood cultures. The results of macroscopic examination and subculture were the basis for establishing true positive and negative results.

At the detection limit (10^{-10}M) the bioluminescent assay detected 94.2 % of the positive blood cultures within 6 to 12 h incubation but the specificity was only 85.8 %. The predictive values for positive and negative results were 29.5 and 80.3 % respectively. By increasing the detection limit the sensitivity and specificity increased. The maximum correctly classified blood cultures were reached at a detection limit of 2 x 10^{-9}M ATP. At this limit of detection the bioluminescent assay detected 82.9 % of the positive blood cultures and 99.6 % of the negative blood cultures were correctly classified. The predictive values for positive and negative results also reached optimal figures, 93.6 and 99.6 % respectively. Furthermore the bioluminescent assay detected 52.9 % of the positive blood cultures one day earlier than with macroscopic examination and subculture.

We conclude that the bioluminescent assay of bacterial ATP in blood cultures is an objective and sensitive method for rapid detection of bacteriemia.

REFERENCES

1. Thore, A., Ånséhn, S., Lundin, A. and Bergman, S. (1975). Detection of bacteriuria by luciferase assay of adenosine triphosphate, *J. Clin. Microbiol.* 1, 1-8.
2. Nichols, W.W., Curtis, G.D.W. and Johnston, H.H. (1982). Analysis of the disagreement between automated bioluminescence-based and culture methods for detecting significant bacteriuria, with proposals for standardizing evaluations of bacteriuria detection methods, *J. Clin. Microbiol.* 15, 802-809.
3. Thore, A., Lundin, A. and Ånséhn, S. (1983). Firefly luciferase ATP assay as a screening method for bacteriuria, *J. Clin. Microbiol.* 17, 218-224.
4. Molin, Ö., Nilsson, L. and Ånséhn, S. (1983). Rapid detection of bacterial growth in blood cultures by bioluminescent assay of bacterial ATP, *J. Clin. Microbiol.* 18, 521-525.

APPLICATION OF ATP ASSAYS TO PATIENT CARE IN LEPROSY

A. M. Dhople

Medical Research Institute
3325 West New Haven Avenue
Melbourne, Florida 32901

For 110 years (1873 - 1983) inability to cultivate *Mycobacterium leprae* in vitro has been a major bottleneck in leprosy research. The most serious problem is to obtain useful information on the state of bacterial load during chemotherapy of leprosy. The occurance of drug resistance, one of the anticipated outcomes of chemotherapy, has now become well established in lepromatous leprosy on dapsone therapy since the first report of its kind in 1966 (1). This is of particular importance since it is quite clear that transmission of resistant bacilli takes place leading to primary sulfone resistant disease in contacts (2). Therefore, in any leprosy control program the determination on the response of a patient to a given therapy and the detection and quantitation of drug-resistant disease should be given the highest priority.

The only method presently available to determine viability of *M. leprae* is its transmission to normal or immunosuppressed mice or rats (3, 4). Eventhough the mouse foot pad inoculation gives the final, medically scientific proof of drug resistance and undoubtedly is of great value in distinguishing between partially resistant patients and fully susceptible patients, it is most unlikely to be a standard method because (a) it is expensive and somewhat difficult technique to be available in endemic areas, and (b) the time interval to obtain any valid information is extremely long. Therefore, we proposed to employ adenosine triphosphate (ATP) as a biochemical indicator in chemotherapeutic studies on leprosy patients in order to obtain fast and reliable information on the status of *M. leprae* harvested from these subjects during the course of anti-leprosy therapy. The

ISBN 0-12-426290-2

ubiquitous distribution and metabolic importance of ATP make precise, quantitative measurement of this compound a particularly promising method for detecting and quantitating viable microorganisms.

The method for the assay of ATP is based on the bioluminescent reaction between ATP, luciferin and luciferase. The suspensions from skin biopsy specimens of leprosy patients are first decontaminated with NaOH; they are then treated with a mixture of trypsin, chymotrypsin and collagenase. These two steps remove almost all host tissue material. Host ATP adsorbed on purified *M. leprae* is removed by means of Triton X-100 and ATPase, and bacterial ATP is extracted by means of chloroform and heat (5, 6). For the assays Chem-Glo photometer of Aminco was used, while the substrate and enzyme were obtained from DuPont company.

Biopsy specimens were obtained from 14 randomly selected leprosy patients, both untreated and treated with dapsone for various periods, from Argentina, Brazil, Surinam and Thailand. *M. leprae* suspensions from these specimens were used for ATP assays and also for mouse foot pad inoculations to determine viability and dapsone sensitivity. Nine of the fourteen patients had no previous anti-leprosy treatment. ATP levels of *M. leprae* from these nine patients varied from 1.09 to 1.38 picograms per one million cells; futhermore, *M. leprae* from all these patients gave standard growth curves in the foot pads of mice eight to eleven months post inoculations. This suggests that *M. leprae* from all nine patients were viable.

Of these nine patients, biopsy specimens were obtained again from three patients four to seven months after initiating dapsone treatment. *M. leprae* from two patients did not show presence of any ATP, thus, suggesting the loss of viability of organisms following dapsone treatment; this was subsequently confirmed by the mouse foot pad assay technique. *M. leprae* from the third patient still contained the same level of ATP as before initiating treatment, indicating possible resistance to dapsone. On subsequent analysis of mouse foot pad assay results, it was found to be true. On the basis of ATP assay results, and even before obtaining mouse foot pad assay data, the clinician was informed of possible dapsone resistance in the patient. The treatment of the patient was changed to rifampin and within three months there was complete loss of ATP from *M. leprae* of this patient and also failure of *M. leprae* to multiply in foot pads of mice. Thus, here is an example where ATP data gave instantaneous information that was advantageous to both the patient and clinician.

Of the original 14 patients, five were under dapsone treatment for varying periods. *M. leprae* from two of these patients contained ATP within normal range, gave standard growth curves in the foot pads of mice and also were found to be dapsone resistant as determined by mouse foot pad assays. *M. leprae* from the remaining three patients did not contain any ATP and also failed to multiply in the foot pads of mice. Again, soon after obtaining ATP data on two dapsone resistant patients, clinicians were asked to treat the patients with rifampin. Three months of rifampin treatment resulted in total loss of viability of *M. leprae* both by ATP and mouse foot pad techniques. These results have been very encouraging and suggest strongly that ATP determinations of *M. leprae* give information on the status of a patient under chemotherapy and can identify drug-resistant cases instantaneously at very early stages of therapy and thus avoid further complications due to the disease.

A pilot project has been set up in collaboration with leprologists from Argentina, Brazil, India and Surinam to evaluate ATP assay technique in the prognosis of leprosy patients under therapy. These were all previously untreated patients. Initial biopsy was taken for ATP assay and mouse foot pad inoculations. The patients were then given standard dapsone treatment. The biopsy specimens from these patients were taken at an interval of four to six weeks for up to six months and the results on ATP assays (obtained instantaneously) were compared with the mouse foot pad inoculation data, obtained eight to ten months later. The interpretation of ATP data is simple. If ATP per aliquot of *M. leprae* cells decreases during treatment, it will be interpreted as positive response to the treatment since the drug therapy has inhibited the biosynthetic capability of *M. leprae* cells and their ability to grow and multiply, thus, rendering them non-viable. On the other hand, if ATP per aliquot remains steady or increases over the original level, the conclusion will be that the treatment has no effect in inhibiting the metabolism of *M. leprae* and thus, are still able to grow and multiply in host.

The results of the first twenty patients have been encouraging. Based on ATP assay results, six of the twenty patients have been found to be dapsone resistant. Even after six months of dapsone therapy, ATP levels of *M. leprae* from these patients remained unchanged. Subsequent mouse foot pad assay results confirmed these findings - i.e., *M. leprae* from these patients gave standard growth curves in the foot pads of mice fed with dapsone mixed in their feed. In the case of the other fourteen patients, the ATP levels of *M. leprae* started declining within one month after initiating the dapsone treatment and within four to six months lost all their intracellular ATP. *M. leprae* harvested from these patients failed to multiply in the foot pads of mice indicating that they became nonviable after the dapsone treatment. Thus, it has been possible to determine the response of a patient to a given therapy within the first two months of the treatment using ATP assay technique.

In conclusion, therefore, it can be stated that on the basis of the results obtained so far, the ATP assay technique seems to be promising in obtaining the information on the status of *M. leprae* from leprosy patients under chemotherapy, especially their viability and drug sensitivity. This method gives instantaneous information and is far more cheaper to adopt widely in endemic areas.

REFERENCES

1. Pettit, J.H.S., Rees, R.J.W. and Ridley, D.S. (1966). Studies on sulfone resistance in leprosy, *Internatl. J. Leprosy* 34, 375-390.
2. Pearson, J.M.H. (1978). Epidemiology and some implications of sulfone resistance to leprosy. *In* Proceedings of XI International Leprosy Congress, Mexico.
3. Shepard, C.C. (1960). The experimental disease that follows the injection of human leprosy bacilli into foot pads of mice, *J. Expt. Med.* 112, 445-454.
4. Fieldsteel, A.H. and McIntosh, A.H. (1971). Effect of neonatal thymectomy and antithymocyte serum on susceptibility of rats to *M. leprae* infection, *Proc. Soc. Expt. Biol. Med.* 138, 408-413.
5. Dhople, A.M. and Hanks, J.H. (1973). Quantitative extraction of ATP from cultivable and host-grown microbes, *Appl. Microbiol.* 26, 399-403.
6. Dhople, A.M. and Storrs, E.E. (1982). Adenosine triphosphate content of *M. leprae*: Effect of purification procedures, *Internatl. J. Leprosy* 50, 83-89.

This work has been supported through the grants from National Institutes of Health and German Leprosy Relief Association.

ATP MEASUREMENT BY BIOLUMINESCENCE : ENVIRONMENTAL APPLICATIONS

H. Van de Werf and W. Verstraete

Laboratory of Microbial Ecology, University of Ghent, Coupure L 653, 9000 Ghent, Belgium

INTRODUCTION

To assess and monitor the number of microorganisms and their activity in samples of waters, soils, sediments or waste materials by classical microbiological methods requires considerable labor and time. The development of a reliable rapid technique which can be used unequivocally to measure microbial biomass is therefore of general interest.

Recently, Karl (1) has written a comprehensive review on the cellular nucleotide measurements and their applications in microbial ecology. The present effort focusses on the ATP measurement as an analytical tool and a control parameter for the applied environmental microbiologist.

DEFINITION OF TERMS

ATP measurements are performed in a multitude of ways and unfortunately also expressed in a variety of forms. To avoid confusion, the terms which are generally of use in environmental microbiology are briefly defined.

Total ATP

A variety of reagents are known to extract efficiently ATP from all types of cells (pro- and eucaryotic) such as e.g. boiling Tris, boiling citrate-phosphate, dilute H_2SO_4, trichloroacetic acid, chloroform, dimethylsulfoxide, etc. The ATP thus extracted represents the total ATP and is expressed as µg/kg sample dry weight.

ISBN 0-12-426290-2

Microbial ATP

In contrast to the reagents mentioned above, NRB (a quaternary detergent) was found to be capable to release ATP selectively from microbial cells (bacteria, fungi, protozoa, algae; in other words from unicellular organisms). We have performed a considerable amount of work with this reagent to verify this critical joint, particularly in the context of soils. Table 1 illustrates the recovery of various forms of ATP from different soils. Clearly the combination of a Tris-EDTA-Azide buffer with NRB is specific for microbial ATP.

TABLE 1

Detection Efficiency of ATP Added in the Form of Biomass to Soils

Soil	Biomass added	% recovered NRB	TCA
sand	*E. coli*	100	96
	Aspergillus niger	95	73
	Oryza sativa	0	96
sandy loam	*E. coli*	89	74
	Aspergillus niger	95	81
	Oryza sativa	0	96
sandy clay loam	*E. coli*	84	87
	Aspergillus niger	99	83
	Oryza sativa	0	94
sandy loam	nematode mixture	0	87

Specific ATP content

It is well known that microbial biomass can vary considerably in ATP content depending on its age and physiological status. Often, this is expressed as a C/ATP ratio, varying from 50 to 10 000 with an average of 250 (1). Recently, the specific ATP content is preferably expressed in a direct way, i.e. mg ATP/g biomass. This value ranges from 0.1 to 10 mg ATP/g biomass, with an average of about 2 for normal viable cells. When expressed, as is often done in the form of mg ATP/g biomass-C, the range becomes 0.2 - 20 and the average 4 (indeed, cells contain ca. 50 % C on the dry weight).

Table 2 illustrates that for axenic cultures, the specific ATP content not only varies with the age of the cells, but also with the extraction reagent used. It can be seen that for the young cultures, NRB is a better reagent than TCA. For the old cells, NRB can compete with TCA for all cultures except for the *Pseudomonas aeruginosa* and the *Mucor sp.* The reasons for this apparent lower extraction efficiency for older cells, are unknown. Note that for logarithmically growing cultures the specific ATP content indeed falls in the range 2 - 12 (average 4.8) mg ATP/g biomass-C, while for cells in the stationary phase, relying on maintenance metabolism, this parameter drops to 0.1 - 1.6 (average 0.5). In view of the fact that in most environments, microorganisms mostly "survive" rather than explosively grow, the latter value is of major importance.

TABLE 2

ATP Levels in Bacterial and Fungal Biomass (dry wt) in Relation to Culture Age as Determined by Two Methods

Species	Culture age (days)	Biomass density mg g^{-1}	Oxygen uptake rate mg O_2 g^{-1} h^{-1}	mg ATP g^{-1} biomass-C TEA NRB	TCA
Micrococcus luteus	1	0.22	ND*	0.2	0.2
	3	1.34	104.3	12.9	2.2
	9	1.33	10.6	1.6	0.2
Bacillus subtilis	1	0.37	37.4	3.7	3.7
	3	0.52	60.5	5.9	7.2
	13	0.48	ND	0.2	0.1
Escherichia coli	1	0.47	62.4	11.7	7.3
	3	0.67	85.9	2.8	3.7
	10	1.09	23.4	0.6	0.6
Pseudomonas aeruginosa	1	0.48	430.4	6.2	10.5
	3	0.73	68.0	4.0	2.2
	10	0.89	9.0	0.1	0.9
Mucor sp.	1	1.11	8.5	1.0	0.4
	3	6.38	14.1	8.8	2.2
	10	4.13	2.0	0.1	1.0
Aspergillus niger	1	2.18	37.1	1.7	3.4
	3	4.03	11.0	0.6	0.6
	10	15.46	1.0	0.2	0.03
Pseudomonas fluorescens	3	—	—	8.4	—
	10	—	—	0.8	—
Streptomyces albus	3	—	—	3.1	—
Saccharomyces sp.	3	—	—	3.4	—
	10	—	—	0.4	—
Penicillium sp.	3	—	—	3.5	—
	11	—	—	0.5	—
Tetrahymena pyriformis	3	—	—	1.3	—

*ND=not detectable

Adenylate Energy Charge

The cumulative effects of the adenosine phosphates on the rate of cellular metabolism is commonly expressed by the so-called adenylate energy charge parameter (1). To calculate this parameter, the determination of the ADP and AMP pools in addition to the ATP pool, is necessary. This of course severely complicates the methodology. Furthermore, most microorganisms have control systems to maintain the energy charge within the 0.5 - 0.9 range. Only strongly stressed cells yield values below this range. In view of the methodological problems and the restricted significance of the energy charge, this parameter is subsequently not further discussed.

Methodology

The light emission reaction catalyzed by the firefly luciferase is sensitive to a considerable number of interferences. For environmental samples, it is of major importance that the quenching of the light emitted by the enzyme is counteracted. Verstraete et al. (2) therefore advocate dilution down to 10^{-2} and even 10^{-3} levels for materials such as soils. Furthermore, to convert the relative light units to ATP, standard addition for each sample appears necessary.

The necessity to work with purified luciferase enzyme preparations has been stressed by Tate and Jenkinson (3). Indeed, the authors found that the ATP content was, on average, 24 % lower when measured using the purified preparation of luciferase than when measured using the crude preparation, presumably because the crude preparation contained enzymes that converted GTP to ATP.
When for each sample one dilution series is prepared, and the appropriate dilution is analyzed for ATP twice as such and once more with standard addition, the percent variability on the analyses was found by us to fall in the 5 - 10 % range.

APPLIED ATP MEASUREMENTS

Soils

Soil microbiologists are faced with the problem that they have no unequivocal methods to quantify the microbial biomass present in soils. The method developed by Van de Werf and Verstraete (4) has been applied to a large series of soils and evaluated with regard to a variety of soil characteristics and treatments. It was found that the TEA/NRB ATP values correlate very significantly with the total soil microbial biomass as determined by the indirect fumigation-reinoculation technique (Fig. 1).

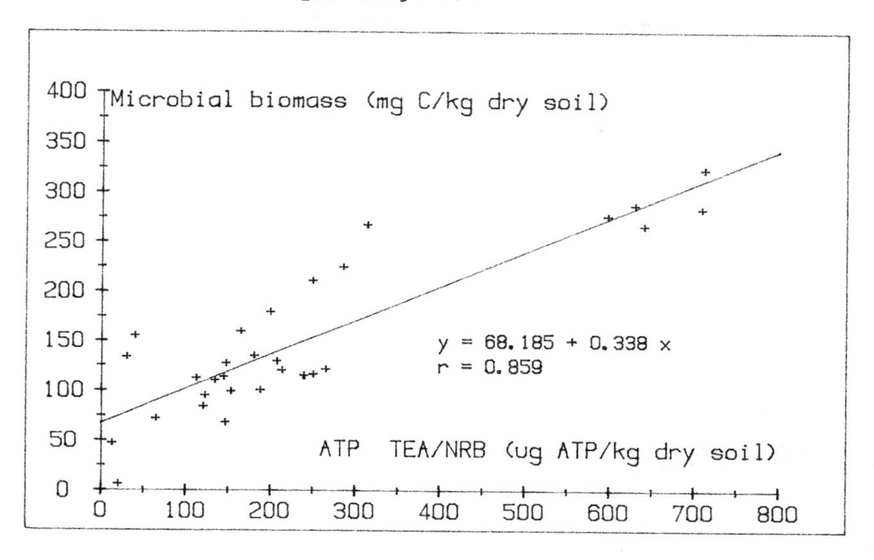

Fig. 1. Relationship between ATP determined by the TEA/NRB procedure and the microbial biomass determined by the fumigation reinoculation method (for details see 5)

It therefore appears that the latter time consuming (20 days) method can be replaced by the rapid ATP method (10 minutes). It was also found that combination of both the TEA/NRB and the TCA extraction procedures, yielded interesting information about the overall biological status of the soil. Soil ATP appeared to be majorly of microbial origin and only to a minor extent due to the mesobiota. An exception to this were soils supporting, in the present or the past, high densities of plants and hence containing a lot of roots and rootlets. For these soils, total ATP largely surpassed microbial ATP. This permitted the following conclusions : soils for which ATP-TEA/NRB > ATP-TCA, are clearly populated with actively growing microorganisms and are not in a steady state condition. Soils for which (ATP-TEA/NRB) x 2.4 ≅ ATP/TCA can be interpreted as normal and equilibrated. Finally, soils for which ATP-TCA surpasses ATP-TEA/NRB with a factor of 4.0 or more must be considered as special and most probably rich in plant material.

The absolute values of either total or microbial ATP in soils vary considerably. Verstraeten et al. (5) report a range of 20 to 712 (mean 309) µg ATP-TEA/NRB/kg soil DW and 59 to 4719 (mean 1038) µg ATP-TCA/kg soil DW respectively. This and further work (6) permitted to deduce the following tendencies :

1. Clayey and calcareous soils have considerably higher microbial ATP values in comparison to loamy and sandy loam soils. Generally sandy soils show the lowest ATP figures.
2. Differences in landuse are also reflected : microbial ATP is always higher under grassland than under arable land.

In view of the latter findings, it is clear that the measurement of soil ATP (total and microbial) offers perspectives to soil taxonomy and soil fertility. Incidently, the soil fertility index, as developed by Sys for land evaluation was applied by Angerosa (7) to 3 different soils and found to correlate significantly with the soil microbial ATP in two out of the three cases. For total soil ATP, the correlation was significant only for 1 out of three soils.

The fumigation-reinoculation method developed by Jenkinson & Powlson (8) allows to approximate the amount viable microbial biomass in the soil. Hence, it is possible to calculate the specific ATP content of the soil microorganisms. The values found range from 1.0 to 3.4 mg ATP/g biomass-C with an average of 1.8. The latter value is in accordance with the concept that most of the biomass under normal soil conditions, is relying on maintenance metabolism.

Further interesting uses for microbial ATP measurements in soils are at present the area of rhizosphere microbiology and the area of rapid detection of soil pollution. Indeed, when studying the effect of the plant on the rhizosphere, the microbiologist must partition plant cells from microbial cells. The specific NRB reagent offers a lot of prospects in this context. As to the detection of impacts of pollutants, the ATP analysis is capable to reveal these phenomena rapidly and effectively. Table 3 illustrates the effect of a number of chemicals applied to soil, upon the microbial and total soil ATP. It is clear that the base, the acid, the heavy metal and finally the pesticide all have a considerable impact on both types of ATP. Hence, these effects can easily be detected. In our experience, no other microbial analysis can compare with ATP in terms of rapidity and sensitivity for revealing ecotoxicological stress of the microbial community.

Waters and Sediments

ATP measurements have been used to estimate biomass in the water column and the sediments of marine and fresh water

TABLE 3

Evolution of ATP Levels (µg ATP/kg soil dry weight) in Gistel Soil After Treatment with Several Pollutants

a. ATP(TEA/NRB)

	Time (weeks) 0	2	4	12	24	t-test to control
Control	598	600	539	504	457	
$Ca(OH)_2$ pH +1	598	638	701	378	460	- 0.33
H_2SO_4 pH -1	598	433	380	222	290	3.45*
Pb^{++} 50 ppm	598	504	410	238	247	3.04*
Pb^{++} 500 ppm	598	462	373	234	240	3.47*
Pb^{++} 1000 ppm	598	440	357	225	232	3.61*
Pb^{++} 5000 ppm	598	324	299	108	81	3.64*
Pyrazon	598	586	450	293	462	1.51
Pyrazon	598	528	426	274	353	2.78

b. ATP(TCA)

	Time (weeks) 0	2	4	12	24	t-test to control
Control	4719	3035	2058	1191	1017	
$Ca(OH)_2$ pH +1	4719	2092	1947	1826	1803	- 0.24
H_2SO_4 pH -1	4719	1852	1762	369	345	2.89*
Pb^{++} 50 ppm	4719	1543	1437	1383	1055	1.21
Pb^{++} 500 ppm	4719	1487	1218	1185	854	1.69
Pb^{++} 1000 ppm	4719	1212	1032	962	803	1.94
Pb^{++} 5000 ppm	4719	1084	880	398	273	2.94*
Pyrazon 0,6 ppm	4719	2184	1913	1414	1150	0.67
Pyrazon 6 ppm	4719	2398	1639	1639	1043	0.62

t-test on paired observations $t_{4;0.05} < 2.776$

systems. For marine waters, surface ATP concentrations range from 0.5 µg or more of ATP per liter for eutrophic waters to 0.1 to 0.5 µg of ATP per liter for regions with moderate productivity to less than 0.1 µg of ATP per liter for oligotrophic parts of the ocean (1). For inland waters, the absolute concentration of ATP appears also to be significantly correlated with the nutrient status and the rate of primary production (9). Stephens and Shultz (10) evaluated periphyton-ATP as a rapid response parameter to nutrient additions in an oligotrophic creek. The method appeared to be very rapid and reliable, but it responded more to variations in light intensity than to those of nutrient levels.

Aquatic sediments integrate the chemistry and biology of the overlaying layer of water and are therefore of major interest to aquatic ecologists. Generally, levels of total ATP are quite high in sediments (100 - 10 000 µg ATP/kg DW) (1, 11). However, in contrast to soils, sediments can contain considerable levels (up to 50 %) of meiofaunal ATP (12). Sofar, such partitioning is based on the determination of the numerical abundance of the various taxa, and obviously requires a lot of work. Since in ecological studies, the subdivision of the community into the most important trophic parts is of major importance, it can be expected that the current availability of ATP extractants selective for the microbial compartment will encourage more extensive work along these lines. The specific ATP content of total and microbial biomass in waters and soils has received insufficient attention sofar. Several authors have suggested that on a mg ATP/g C basis, the ATP content of various meiofaunal taxa are very similar (12, 13). The data tabulated by Karl (1) for nematodes and copepods suggest that on the average, these organisms contain levels which are in the range 5 - 10 mg ATP/g biomass-C, which is considerably higher than bacteria under non-growth conditions.

Capone et al. (14) recently used ATP to assay the effect of metals on anoxic salt marsh sediments. They found that levels of 1000 ppm of $HgCl_2$, $PbCl_2$ and $NiCl_2$ did not strongly affect the ATP-level while the same concentration of $FeCl_2$, $K_2Cr_2O_7$, $CdCl_2$, $CaCl_2$ and $ZnSO_4$ decreased the ATP level with 40 to 80 %.

Sanitary Engineering

During the preparation and distribution of drinking water, the sanitary engineer has to constantly monitor the level of microbial contamination. Unfortunately, no methods are currently available which permit to instantaneously detect low levels ($\leqslant$ 1 fecal coliform/100 ml) water (15). In view of the high sensitivity of the ATP luminescence method, attempts have been made to use ATP for rapid detection of fecal contamination of waters. Picciolo et al. (16) described an apparatus for in-line monitoring the hygienic quality of wastewater effluents and drinking water supplies. The processed sample is mixed with the luciferase-luciferin mixture as it flows into a coiled cell adjacent to the photomultiplier tube. Processing steps include addition of an ATP-ase for hydrolysis of non bacterial ATP, addition of HNO_3 as an extracting agent and subsequent dilution of the acid in the processed sample. The assay is obviously very rapid, but its sensitivity appeared to be restricted to 10^5 - 10^6 *Escherichia coli*/ml.

Hence, the first major problem to be tackled is the sensitivity. For routine ATP analysis, 10^2 - 10^3 bacterial cells per ml constitute the lower limit. Of course, concentration of the cells e.g. by membrane filtration can be invoked to lower this limit with a factor 10. However, the second problem is that the method should differentiate between plain and fecal bacteria (or any other hygienic indicator). The ubiquitous occurrence of ATP in all living organisms poses in this context a serious limitation on the ATP assay.

Recently, the Arcat procedure has been proposed as a method to rapidly and selectively detect fecal coliforms (17). The procedure is based on the addition of a phage sensitive host *Escherichia coli ATCC 13706* to samples and, after plating and incubating for 6 hours, counting the number of coliphages. The authors claim a detection limit of 6 FC/100 ml for a 6 hour total testrun. Obviously, the addition of the host cells is an elegant selection criterium. We therefore asked the question if this approach could be usefull in a modified way. The procedure developped consists out of 3 phases. Firstly, the specific host cells are added. Secondly, the mixture is incubated for a short period to allow cell infection and lysis. Finally, cellular ATP levels are measured. Hence the plating and plaque enumeration phase of the Arcat procedure is replaced by an ATP determination. If the sample is contaminated with coliphages (and hence has, according to Wentsel et al. (17) a history of *Escherichia coli* contamination), then ATP levels should decrease compared to those of control samples.

A series of experiments were set up to test this hypothesis (18). Ten ml host culture *Escherichia coli ATCC 13702* grown 18 hours in nutrient broth at 37°C, were added to 100 ml sample or to decimal dilution with sterile saline of the sample. To each beaker, 5 ml of nutrient broth 20-fold concentrated was furthermore added. The samples were subsequently incubated in a water bath at 37°C. At regular intervals, aliquots were withdrawn and the ATP content was determined by the TEA/NRB procedure (2). A typical example of the results thus obtained is given in Table 4.

TABLE 4

Evolution of the ATP Content (µg/ℓ) of Lake Water Samples upon Incubation according to the ARCAT Procedure. The Fecal Coliform Level in the Initial Sample, as Determined by Plate Counting, Amounted to 5600/100 mℓ

Dilution of the sample	Time of incubation				t-test to control
	0	3	6	8	
10^{0}	32	372	521	481	1.32
10^{-1}	41	364	518	497	1.35
10^{-2}	38	363	537	493	1.26
10^{-3}	38	382	562	677	1.06
control	37	386	562	743	

t-test on paired observations for $x \leqslant 0$ $t_{3;0.05} \leqslant 3.182$

The overall results of this study can be summarized as follows :

* In samples containing less than 10 000 fecal coliforms per 100 ml, the procedure resulted in a noticeable (but unfortunately not statistically significant) drop of ATP in concordance with the principle underlying the ARCAT procedure. In samples with higher coli densities, the growth of the phage-insensitive coli and other bacteria, masked the lysis of the host cells.
* In samples containing less than 10 000 fecal coliforms per ml, the presence of these indicator organisms could be detected by the ATP measurements in 6 - 8 hours and with a minimum sensitivity of 4 - 8 FC per 100 ml.

These results show that the method permits to reveal low levels of fecal coli in a selective and relatively rapid way. Its overall performance is comparable with the current radiometric, electrochemical, enzymatic or chromatographic "rapid" methods described by Cundell (15). Yet, some drawbacks must also be recognized. Indeed, the method reflects in an indirect way the fecal coli and due to labor and chemicals, is fairly expensive for routine analysis.

Several authors have proposed ATP as a control parameter for activated sludge systems. Kucnerowicz & Verstraete (19) studied for a series of laboratory units the influence of various operational parameters on the ATP content of the sludge. They found, as illustrated in Fig. 2, very nice linear relationships between the specific ATP content of the sludge and the cellular residence time, the sludge loading rate, the specific substrate removal rate and the oxygen uptake rate. Several other authors examined the ATP content of activated sludges and also reported specific levels in the order of 500 - 1500 µg ATP/g MLVSS (Mixed Liquor Volatile Suspended Solids) (10, 21). There is general agreement that ATP offers the best estimate of sludge activity while plate counts, oxygen uptake and enzyme activities (such as dehydrogenase) are not so reliable. All this indicates that the ATP measurement could be a useful method of monitoring and controlling activated sludge water treatment systems. In practice however, cost and labor implications have prohibited this parameter to compete with the oxygen based control devices. The same implies to the development of on-line microbiological toxicity monitoring devices for wastewater streams. ATP is generally equally or more sensitive than the oxygen utilization rate. This is e.g. illustrated in Table 5. To activated sludge samples (sludge age 7.5 d, sludge concentration 2 g/l) slug doses of different toxic chemicals were added and the subsequent decrease of activity of the sludges was monitored as a function of time. It was found that indeed the ATP measurement qualifies as a sensitive and reliable indicator of intoxication (22). Similar results have recently been reported by Parker (23). However, here too cost aspects currently tend to prohibit further development of the ATP techniques in this respect.

Finally, a few words about composting of solid wastes. Recently, composting has received considerable attention and major advances in process control have been made (24). A major question remains the evaluation of the maturity (stability) of the endproduct. Few data for total ATP in compost are available. Collin (25) reported from 100 to 5800 µg ATP/kg dry matter while Anid (26) found values between 3 000 and 22 000. Some results from our laboratory are summarized in Table 6. We consider a compost stable and suitable for plant growth promoting when its microbial ATP level does not surpass 10 000 µg/kg dry weight. This level is then ca. 50 times the level of a normal soil. Hence, when applied at the normal dose of 50 tons compost/ha (or 25 tons DW per 2 500 tons DW of top soil), this amount will raise the soil ATP with a factor of 1.5 - 2.0. At higher microbial levels, the microorganisms respire too intensively thus inducing anoxic sites which are deleterious for plant growth.

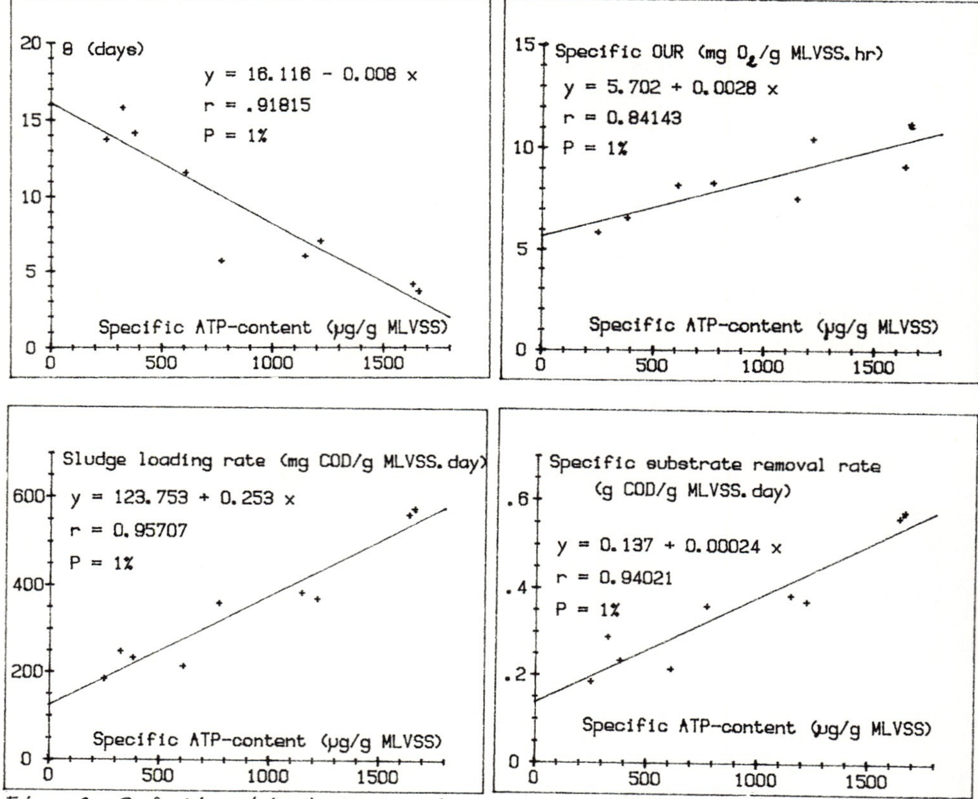

Fig. 2. Relationship between the specific ATP content of activated sludge and operational parameters

TABLE 5

Procentual ATP Levels and Oxygen Uptake Rate (OUR) of Activated Sludge Samples Treated with Toxicants

Time after intoxication (hr)	TEGO 1000 ppm ATP	OUR	TEGO 100 ppm ATP	OUR	MBT 4500 ppm ATP	OUR	Cr^{6+} 500 ppm ATP	OUR	NaOCl 10 dpm ATP	OUR
0	100	100	100	100	100	100	100	100	100	100
1	11	38	52	109	56	28	43	41	93	99
2	4	35	44	94	45	0	48	35	79	95
6	0	0	45	96	63	0	46	46	68	69

TEGO : quaternary detergent N-dodecyldi(amino-ethyl)-glycine
MBT : 2-mercaptobenzothiazol
Cr^{6+} : dosed as CrO_4^{2-}
NaOCl: dosed as bleaching water

TABLE 6

Evolution of ATP during Composting of Household Waste (µg/kg dry weight)

Time days	Composting Plant: Gent ATP	Bilzen ATP	Dendermonde ATP	Tenneville ATP
40	48592	22872	81424	26357
60	65183	7170	49853	10549
80	9273	10468	10267	6877
140	12183	10581	12024	8801

ATP in µg/kg dry material

CONCLUSIONS

1. Measurement of ATP by bioluminescence in a sample of water, soil, sediment or waste material can be performed rapidly and reliably. Qualibration by standard addition and use of high quality enzyme preparations is necessary to

counteract interferences.
2. A most interesting potential of the current ATP measurement technology is the possibility to partition the biomass. Indeed, by use of the NRB extractant the microbial ATP can be differentiated from that of all higher forms of life.
3. ATP can be expressed in a variety of ways. Particularly the concepts total ATP (µg/kg sample DW), microbial ATP (µg/kg sample DW) and the specific ATP (µg/g biomass carbon or dry weight) appear usefull in environmental microbiology.
4. The ATP measurement is a most powerful analytical tool and control parameter for the environmental microbiologist. For the soil microbiologist, it provides a unique way to rapidly and selectively approximate soil microbial biomass. The same is true for the aquatic microbiologist. For the sanitary microbiologist, ATP permits to rapidly detect fecal contamination of potable water, to monitor and control biological wastewater treatment systems, to assess compost maturity, to detect toxicants, etc... However, in all these areas, alternative techniques are available which particularly in terms of economics, compete with the ATP bioluminescence analyses.

REFERENCES

1. Karl, D. (1980). Cellular nucleotide measurements and applications in microbial ecology. *Microbiological Reviews* 44, 739-796.
2. Verstraete, W., Van de Werf, H., Kucnerowicz, F., Ilaiwi, M., Verstraeten, L.M.J. & Vlassak, K. (1983). Specific measurement of soil microbial ATP. *Soil Biol. Biochem.* 15, 391-396.
3. Tate, K.R. & Jenkinson, D.S. (1982). Adenosine triphosphate measurement in soil : improved method. *Soil Biol. Biochem.* 14, 331-335.
4. Van de Werf, R. & Verstraete, W. (1979). Direct measurement of microbiota in soils. *In* "Proceedings Int. Symp. Analytical Applications of Bioluminescence and Chemiluminescence" (Eds E. Schram et al.). p. 333-338. State Printing Publ. Inc., Westlake Village, Californie.
5. Verstraeten, L.M.J., De Coninck, K., Vlassak, K., Verstraete, W., Van de Werf, H. & Ilaiwi, M. (1983). ATP content of soils estimated by two contrasting extraction methods. *Soil Biol. Biochem.* 15, 396-402.
6. De Coninck, K., Verstraeten, L.M.J. & Vlassak, K. (1983). The ATP-content of Belgian soils. Tagung der Deutschen Bodenkuntlichen Gesellschaft. Trier 04.09.83-10.09.83.
7. Angerosa, M.O. (1983). Evaluation of microbial activity in soils with special reference to adenosine triphos-

phate. Ph.D.Thesis. State University of Ghent. Coupure L 653, 9000 Gent.

8. Jenkinson, D.S. & Powlson, D.S. (1976). The effects of biocidal treatments on metabolism in soil V. A method for measuring soil biomass. *Soil Biol.Biochem.* 11, 193-199.
9. Cavari, B. (1976). ATP in Lake Kinnevet : indicator of microbial biomass or of phosphorus deficiency ? *Limnol. Oceanography* 21, 231-236.
10. Stephens, D.W. & Schultz, D.J. (1982). Extraction of periphyton adenosine triphosphate and evaluation of its use as a rapid response bioassay parameter. *Arch. Hydrobiol.* 93, 327-340.
11. Simmons, G.M., Wharton, R.A., Paker, B.C. & Anderson, D. (1983). Chlorophyll a and adenosine triphosphate levels in antarctic and temperate lake sediments. *Microb. Ecol.* 9, 123-135.
12. Yingst, J.Y. (1978). Patterns of micro- and meiofaunal abundance in marine sediments, measured with the adenosine triphosphate assay. *Marine Biology* 47, 41-54.
13. Goerke, H. & Ernst, W. (1975). ATP content of estuarine nematodes : contribution to the determination of neiofauna biomass by ATP measurements. *In* "Proceedings of the Ninth European Marine Biology Symposium" (Ed H. Barnes). pp. 683-691. Aberdeen, Scotland, Aberdeen University Press.
14. Capone, D.G., Reese, D.D. & Kiene, R.P. (1983). Effects of metals on methanogenesis, sulfate reduction, carbon dioxide evolution, and microbial biomass in salt marsh sediments. *Applied Environmental Microbiology* 45, 1486-1591.
15. Cundell, A.M. (1981). Rapid counting methods for coliform bacteria. *Adv. Appl. Microbiol.* 27, 169-183.
16. Picciolo, G.L., Thomas, R.R., Deming, J.W. & Chapelle, E.W. (1977). Environmental applications of the firefly luciferase ATP assay : flow techniques for monitoring of waste water effluent and of drinking water supplies. *In* "Second bi-annual ATP methodology symposium" (Ed. Borun, G.A.). pp. 547-567. S.A.I., Technology Co-CA. March 22-24, San Diego, Californie.
17. Wentsel, R.S., O'Neill, P.E. & Kitchens, J.F. (1982). Evaluation of coliphage detection as a rapid indicator of water quality. *Applied Environmental Microbiology* 43, 430-434.
18. Van Damme, G. (1983). Snelle detectie van fecale contaminatie in water. Masters Thesis, State University of Ghent, Coupure L 653, 9000 Gent.

19. Kucnerowicz, F. & Verstraete, W. (1979). Direct measurement of microbial ATP in activated sludge samples. *J. Chem. Tech. Biotechnol.* 29, 707-712.
20. Droste, R.L. & Sanchez, W.A. (1983). Microbial activity in aerobic sludge digestion. *Water Res.* 17, 975-983.
21. Williamson, K.J. & Nelson, P.O. (1981). Influence of dissolved oxygen on activated sludge viability. *J. Water Pollut. Control. Fed.* 53, 1533-1540.
22. Kucnerowicz, F. (1978). ATP als parameter voor aktief slib systemen. M. Sci. Thesis. State University of Ghent, Coupure L 653, 9000 Gent.
23. Parker, C. (1982). Surrogate parameter analysis for organic priority pollutants. *J. Water Pollut. Control. Fed.* 54, 77-86.
24. Finstein, M.S., Miller, F.C., Strom, P.F., MacGregor, S.T. & Psarinos, K.M. (1983). Composting ecosystem management for waste treatment. *Bio/Technology* June, 347-353.
25. Collin, F. (1977). Mise au point d'une méthode de détermination de l'ATP dans les composts. *Sols et déchets solides* 11, 355-360. Actes du 1er Symposium sur la recherche en matières de sol et déchets solides. Ministère de la culture et de l'environnement. Paris.
26. Anid, P.J. (1982). Caractérisation de l'état de maturation du compost. *Annales de Gembloux* 88, 119-131.

BIOLUMINESCENT DETERMINATION OF MICROBIAL ACTIVITY ON TEXTILES

B. J. McCarthy

Wira, Wira House, West Park Ring Road, Leeds, LS16 6QL, England

INTRODUCTION

Test methods for assessing the susceptibility of textile materials to microbial degradation, or assessing the efficacy of biocides, generally involve a method for deliberately infecting the substrate with suitable bio-deteriogens, combined with a method of testing (usually tensile strength or visual assessment) for measuring the extent of the resultant degradation. Microbial growth on textile materials may result in a variety of interrelated spoilage phenomena (eg odour production, etc), all of which may contribute to significant financial and material losses during manufacture.

A method (1) providing for the detection and estimation of microbial growth on textiles by the firefly bioluminescent assay of adenosine triphosphate (ATP) has been described. The objective of the present study is to compare the stages in the development of substrate spoilage assessed visually, with microbial activity on the textile substrate (expressed in terms of ATP concentrations) using various standard natural and synthetic materials.

MATERIALS AND METHODS

(a) Organisms

A mixed spore suspension was prepared as specified in BS 6085:1981 (2) using 14-day old cultures of standard test organisms (Table 1) obtained from the Commonwealth

ISBN 0-12-426290-2

Mycological Institute, Kew, UK.

TABLE 1

Cultures specified for testing in accordance with BS6085:1981

Aspergillus niger V. Teigham
Aspergillus terreus Thom.
Aureobasidium pullulans (De Barry) Arnaud
Paecilomyces variotii Bainer
Penicillium funiculosum Thom.
Penicillium ochro-chloron Biourge
Scopulariopsis brevicaulis (Sacc.) Bain Var Glabra Thom.
Trichoderma viride Pers. Ex. Fr.
Chaetomium globosum

(b) Textile substrates

100% Worsted flannel (Style 526) - Wool
100% Dacron (Style 767) - Polyethylene terephthalate
100% Orlon (Style 864) - Polyacrylonitrile
100% Nylon (Style 301) - Polyamide
100% Acetate taffeta (Style 111) - di-acetate

Samples of the standard test materials were supplied by Testfabrics Inc, Middlesex, NJ, USA.

(c) Test procedure

All fabric specimens were tested in accordance with Section Three (Agar Plate test) of BS 6085:1981 using mineral salts medium without sucrose nutrient. All specimens were incubated at 28°C for up to 28 days. Day 0 samples were tested immediately following addition of the mixed spore suspension.

(d) Visual assessment

The extent of growth was assessed visually and recorded in accordance with the scheme specified in the standard (ie no growth - 0; heavy growth - 5).

(e) Quantitative assessment

Details of sample preparation and ultrasonication are provided elsewhere (1). ATP determinations were performed using Lumac reagents and a Biocounter M2010 (Lumac B.V.). A mean ATP value (from three assays) was devised for each textile specimen. The results were weight corrected and expressed as picograms per 0.1 gram sample (pg/0.1gm).

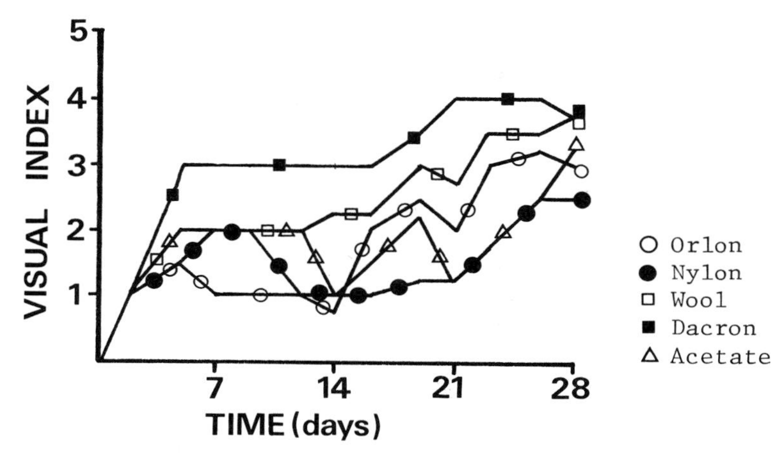

Fig. 1 Growth of inoculum on substrate-visual assessment

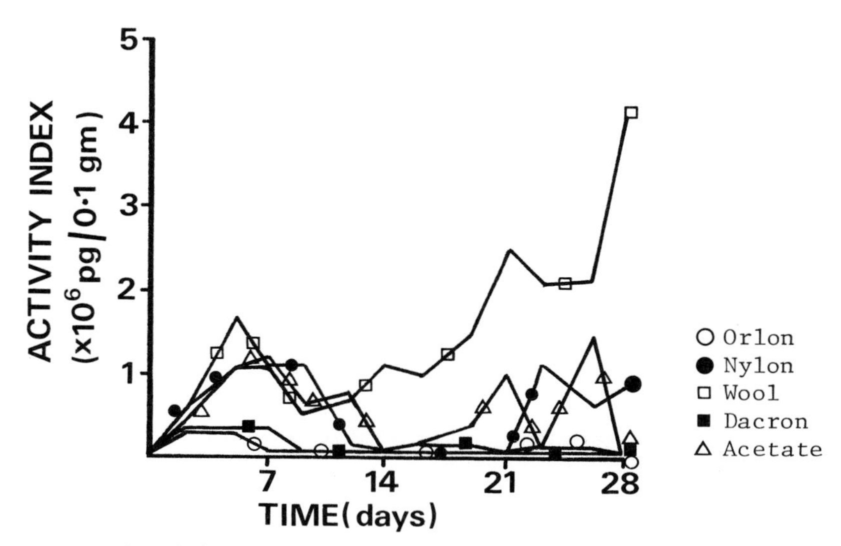

Fig. 2 Growth of inoculum on substrate-activity assessment

RESULTS AND DISCUSSION

Standard test substrates were inoculated with a pooled mixed spore suspension and incubated under standard conditions to assess the development of microbial spoilage. The increasing grades of observed growth (normally assessed only on Day 28) are shown in Figure 1. Both natural and synthetic fabrics were subject to microbial deterioration; between 5 and 16 days (mean values) were required to observe visible growth on the various materials. A cotton control cloth exhibited heavy growth (grade 5) after 14 days confirming inoculum viability. Visual assessment would indicate the susceptibility of materials to spoilage as follows: Dacron > Wool > Acetate > Orlon > Nylon.

The test specimens were subjected to an ultrasonication process to remove the resultant biomass from the fabric allowing the release of ATP. Initial germination and outgrowth, not detected by visual assessment, are indicated by rapidly increasing ATP levels (Figure 2). Residual traces of nutrients contained in the spore suspension may provide the initial impetus for growth on otherwise inert substrates. The synthetic materials Orlon and Dacron, although exhibiting visual staining, show relatively low levels of initial activity followed by cryptic growth, indicating little or no utilisation of the substrate fibres. ATP levels indicate extensive utilisation of Wool and Acetate (natural) fibres, with partial utilisation of Nylon.

ATP determinations may be used to demonstrate the extent of growth of biodeteriogens on a textile material, reflecting the extent to which the substrate may be utilised as a food source, and indicating its susceptibility to biodeterioration: ie Wool > Acetate > Nylon > Dacron > Orlon. Significant increases in ATP levels (within 48 hours) suggest a rapid method for recording susceptibility of textile materials to spoilage and assessing biocide efficacy.

REFERENCES

1. McCarthy, B. J. (1983)
Bioluminescent assay of microbial contamination on Textile Materials. International Biodeterioration Bulletin 19 (2): 53-57
2. BS6085:1981
Methods of test for determination of the resistance of textiles to microbiological determination. British Standards Institution, London

ATP ASSAY AS A RAPID METHOD FOR ESTIMATING MICROBIAL GROWTH IN FOODS

C.J. Stannard

*Leatherhead Food R.A., Randalls Road,
Leatherhead, Surrey KT22 7RY, England*

INTRODUCTION

ATP assay has been used successfully to estimate bacteria and yeasts in pure cultures. However, early workers attempting to apply the technique to foods experienced problems due to the large quantities of non-microbial ATP present (1, 2). There are two possible approaches to this problem. Firstly, organisms may be separated from the food before their ATP is extracted and assayed. The other approach is to destroy interfering non-microbial ATP enzymically. The objective of this paper is to demonstrate the development of appropriate sample preparation techniques for two foods (raw meat and fruit juices) to allow the estimation of the microbial flora by ATP measurement.

MATERIALS AND METHODS

Meat Samples

Samples of raw beef, pork and lamb were stomached 1:10 in 0.05M phosphate/citrate buffer, pH 5.8. The homogenate was then subjected to a three-stage separation process. A brief centrifugation (2000 *g*, 10 s) was followed by stirring with cation-exchange resin (Bio-Rex 70, 100–200 mesh, Bio-Rad Laboratories) which had previously been equilibrated to pH 5.8. After the resin had settled, the supernatant was passed through a membrane filter (0.22 μm pore size, 25 mm diameter, Millipore). ATP from the bacteria isolated on the filter was extracted and assayed using an LKB 1250 Luminometer and reagents (LKB Instruments Ltd, Croydon, Surrey). ATP estimates were compared with colony count of the homogenate (30°C, 24 h, Plate Count Agar).

 ISBN 0-12-426290-2

Fruit Juice Samples

Fruit juices (orange, grapefruit and pineapple flavours) were artificially contaminated with a spoilage yeast. Juice samples were centrifuged (1200 *g*, 10 min) and the pellet resuspended. The suspension was treated with ATP extractant for somatic cells (F-NRS®, Lumac) and apyrase (Somase®, Lumac). After 45 minutes' incubation at room temperature, the sample was washed by centrifugation to remove apyrase. Yeast ATP was then extracted and assayed as above. Yeast ATP estimates were compared with colony count (30°C, 48 h, Malt Extract Agar).

On some occasions, when low numbers of yeast were to be assayed, 24 hours' incubation of the juice was necessary to increase numbers to the level detectable by the instrument.

A commercially available kit, the Lumac Fruit Juice Test Kit, was also evaluated during this study.

RESULTS

The three-stage separation system removed most of the interfering non-microbial ATP from raw meat samples. Large particles of meat were deposited during the brief centrifugation; smaller, colloidal particles were adsorbed on to the cation-exchange resin, and soluble ATP passed through the membrane filter. ATP extracted from the bacteria isolated on the filter compared well with colony count (Fig. 1), over the range $10^5 - 10^9$ cfu/g meat.

For fruit juice samples, both a partial separation of yeasts and a selective destruction of non-microbial ATP were found to be necessary. The Lumac Fruit Juice Test Kit, which relies solely upon a selective destruction technique, was found to give unreliable and irreproducible results. By incorporating centrifugations into the method, a relationship between yeast ATP and colony count was obtained over the range $10^4 - 10^7$ cfu/ml juice (Fig. 2). Below this range, apparent yeast ATP remained constant.

When low numbers of yeast in juice were incubated to increase the level to at least 10^4 cfu/ml, a reduction in ATP/cfu values was observed on some occasions.

CONCLUSIONS

ATP assay was shown to provide an adequate estimate of colony count in two foods, meat and fruit juices. Preparation of the sample is necessary to remove non-microbial ATP from the food prior to assay of microbial ATP. The techniques involved in sample preparation are dependent on the type of food and the organisms likely to be present.

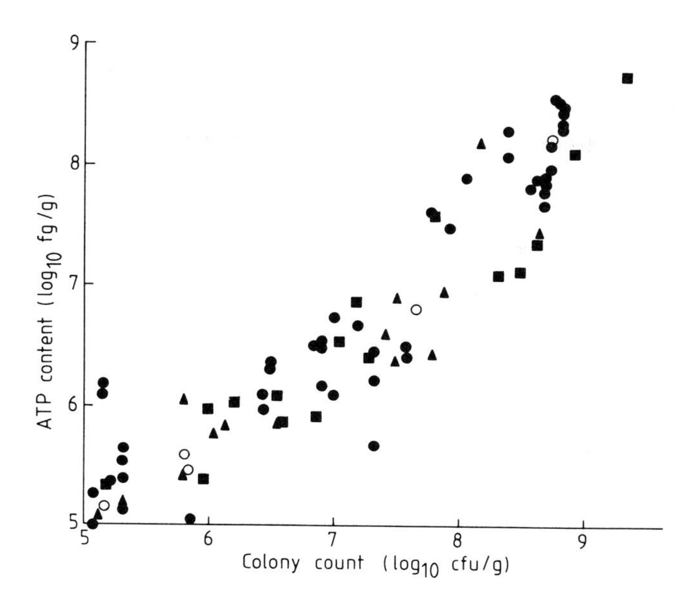

Fig. 1 Relationship between microbial ATP and colony count for raw meats. ●, Beef (○, coincident points); ▲, Pork; ■, Lamb.

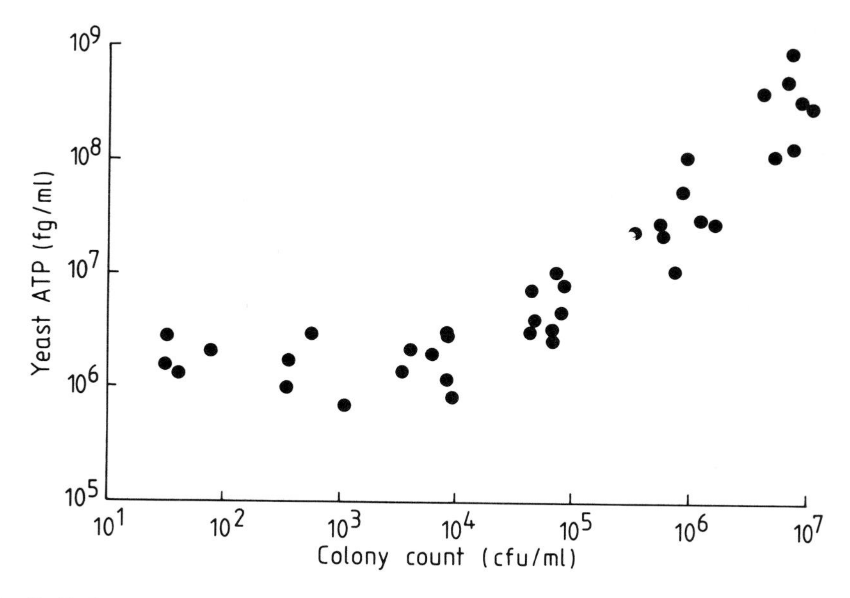

Fig. 2 Relationship between yeast ATP and colony count for artificially contaminated, pasteurised orange juice.

ACKNOWLEDGEMENTS

This work was supported by the Ministry of Agriculture, Fisheries and Food.

REFERENCES

1. Sharpe, A.N., Woodrow, M.N. and Jackson, A.K. (1970). Adenosine triphosphate levels in foods contaminated by bacteria, *Journal of Applied Bacteriology* 33, 758–767.

2. Williams, M.L.B. (1971). Limitations of the DuPont Luminescence Biometer in the microbiological analysis of foods, *Canadian Institute of Food Technology Journal* 4, 187–189.

PRELIMINARY EVALUATION OF THE LUMAC BIOCOUNTER M2010 SYSTEM, MEASURING BACTERIAL ATP IN FOOD, USING A BIOLUMINESCENT TECHNIQUE

Pamela A Manger

PHLS, William Harvey Hospital, Ashford, Kent, TN24 OLZ, England

INTRODUCTION

The semi-automation of microbiological sciences has been a slow process, with bacteriologists reluctant to leave their microscope and petri dishes. Despite this, one of the systems to enter the field is that of bioluminescence. Food bacteriology is one of the areas where a rapid method for assessing total bacterial numbers could be advantageous. Perishable foods, especially those that are imported, are checked for total bacterial numbers as part of a more extensive examination for the presence of potential food poisoning organisms. The work is routine and sometimes tedious, often yielding very low bacterial counts on large batches of a single product. The Lumac Biocounter M2010 Multijet was used in conjunction with Lumac Industrial Microbial Control Kit (IMC) to assess its usefulness in a routine food laboratory.

Method

All the foods were received by the laboratory, as samples for testing. They were steamed lightly to remove all organisms without damaging the food, and then they were artificially inoculated with dilutions of pure cultures of food associated organisms. Suspensions of the food were prepared by making 1:10 dilutions in Maximum Recovery Diluent, emulsifying in a Colworth stomacher, and dispensing in appropriate volumes. The organisms used included laboratory isolates of *Escherichia coli* *Staphylococcus aureus* and *Bacillus cereus*. Dilutions

ISBN 0-12-426290-2

were made of overnight broths, of the order 10^{-1} to 10^{-8} These were added to the food suspensions, giving a further dilution factor of 1:10. The bacterial content of every suspension was then assessed by two methods. The cultural method used was a modified Miles & Misra (1938) using Columbia blood agar incubated aerobically at 37°C; the Lumac method involved measuring the bacterial ATP by a bioluminescent technique, with the results being in the form of relative light units (RLUs) - these determinations were performed in triplicate with uninoculated food as a control. The IMC kit included somatic nucleotide releasing and destroying enzymes, bacterial ATP releasing enzyme and luciferin-luciferase as light source.

Result The initial experiment used turkey meat and *E coli* and when RLUs were plotted against Miles & Misra count, a straight line graph was produced. Below 10^4 orgs/gram the graph curved towards the RLU axis, showing a lack of sensitivity of the Lumac system, at lower bacterial numbers.

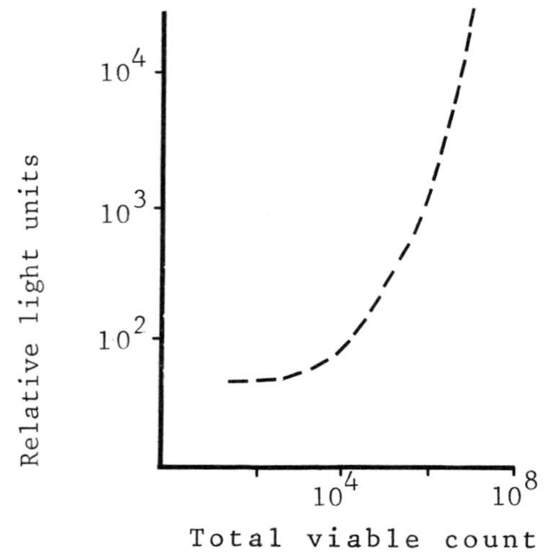

Fig. 1 Turkey with E coli

Other foods were tested against *E coli* and these included raw clams, fresh cream gateaux, boiled rice, pasteurised cream, cooked prawns, paté and tinned meat. Figs. 2 and 3 show the results of these and it is interesting to note the high level of RLUs obtained with the uninoculated cream, and consequently the high count at which the graph begins to curve towards the RLU axis.

Food was tested against the organism with which it was commonly associated in food poisoning outbreaks and Fig. 4 shows graphs of the results using cooked meat with *Staphylococcus aureus* and boiled rice with *Bacillus cereus*.

The final experiment involved the turkey meat used initially; this had been shown to contain large numbers of

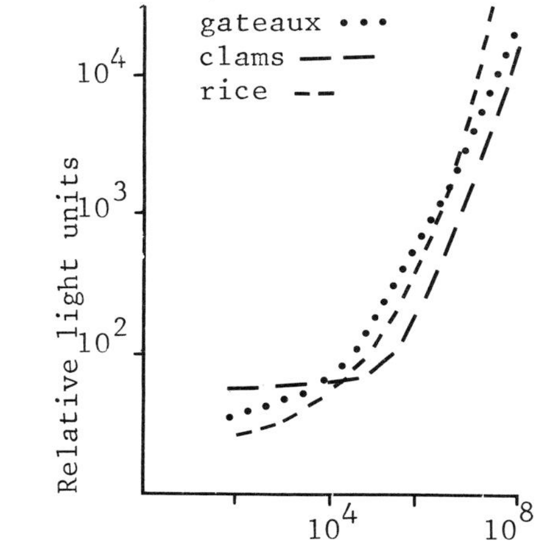

Fig. 2 Clams, boiled rice and gateau with E. coli

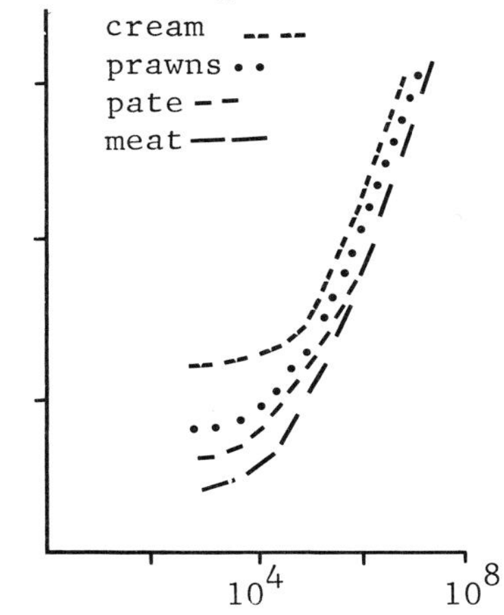

Fig. 3 Cream, paté, prawns & tinned meat with E. coli

(controls (rlu); gateaux, 35; prawns, 27; cream, 127; clams, 19; rice, 29; pate, 24; meat, 21)

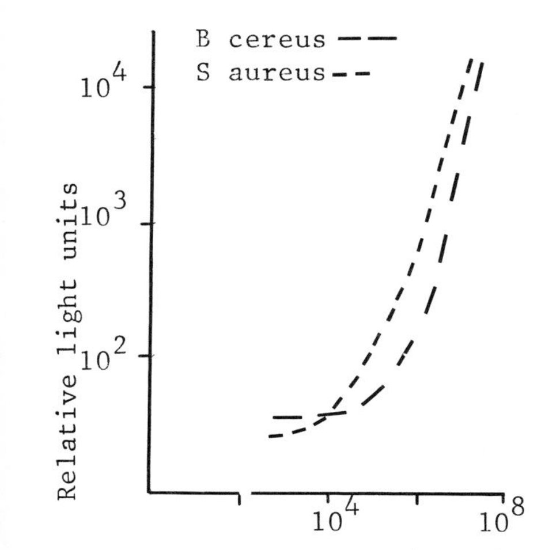

Fig. 4 B. cereus with rice, and S. aureus with meat

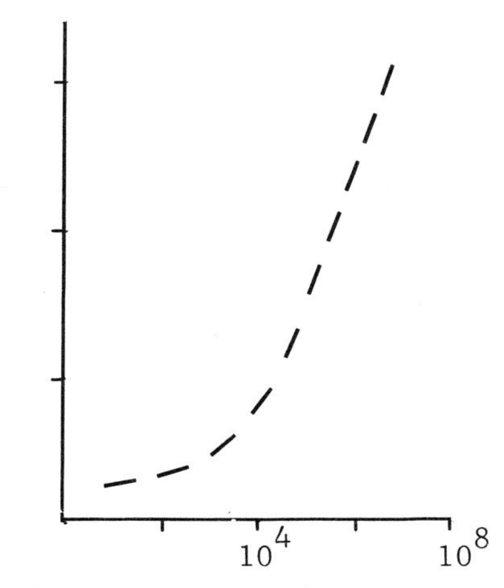

Fig. 5 Turkey with a streptococcus

streptococci when first received, and these had been counted using both the Miles & Misra and Lumac methods. The streptococcus was then added to the sterilised turkey in serial dilutions, and its numbers assessed by the two methods. Fig. 5 shows that the number of streptococci originally present in the meat can be accurately calculated from the graph when its ATP concentration is known in RLUs.

Conclusion The bacterial numbers determined by cultural methods correlated well with those obtained by measuring bacterial ATP using the Lumac biocounter. For food containing 10^3 orgs/gram or more, this was found to be an accurate method of assessment, which had the advantage of speed, the test taking approximately 2 hours. For foods containing less than 10^3 orgs/gram, the method was less sensitive. The ATP measurements were all carried out in triplicate, and errors in reproducibility amounted to less than 0.2%.

There seems to be little doubt about the accuracy of the Lumac bioluminescence method, where more than 10^3 orgs/gram of food are present. Where the food contains less than that number, the sensitivity of the test is lost, but in many instances bacterial numbers of less than 10^3 orgs/gram would be considered insignificant in food samples, so it could also be a useful screening test. The technique is much faster than the usual counting methods, taking only 2 hours to test a batch of samples. This has several advantages in the foods laboratory; (a) the laboratory worker is left free to do other, possibly more interesting, work, (b) a saving in time is a saving in money, (c) on routine food samples a search for pathogens need only be carried out on samples with a higher total count, thus saving time and materials (d) a rapid result means that food with a "holding order" on it can be released more quickly. Set against the saving in time and materials is the initial outlay on the equipment. This need only include the biocounter, but if it is interfaced with a microcomputer, the adding of reagents is precisely and automatically timed, thus ensuring reproducibility.

One of the disadvantages of the system is that it cannot identify the organisms that are present, and until some highly selective liquid culture media are developed, it cannot be used to detect specific organisms. This means that a mixed system of automated and cultural methods has to be employed to analyse fully the bacterial content of a food sample. Despite this, the Lumac system takes much of the monotony out of the routine testing of food samples, and gives accurate and reproducible results.

A RAPID METHOD FOR DETECTING POST-PASTEURIZATION CONTAMINATION IN CREAM

M.W. Griffiths, J.D. Phillips and D.D. Muir

Hannah Research Institute, Ayr KA6 5HL, Scotland

INTRODUCTION

In a survey of the factors affecting the shelf-life of cream in Scotland, 80% of retail pots showed evidence of post-pasteurization contamination (1), mainly by Gram-negative, psychrotrophic bacteria. Detecting the small number of organisms present involves incubating the cream at 6-7°C for periods up to 14 d, followed by a plate count. This procedure is obviously unsuitable for predicting the shelf-life of a product such as cream.

It has been shown, however, that incubation at 21°C for 25 h allows psychrotrophs to grow (2). Gram-positive bacteria also proliferate under these conditions, but by including specific inhibitors in the medium their growth may be suppressed whilst allowing Gram-negative organisms to multiply (3). This pre-incubation procedure, coupled with a rapid method of enumerating bacteria, *e.g.* ATP assay, enables early detection of post-pasteurization contamination (4).

METHOD

A sample of cream (10 g) was added to a sterile container followed by crystal violet (20 μg/g), penicillin (200 U/g) and nisin (400 U/g). This mixture was incubated at 21°C for 25 h and then diluted with sterile 2% w/v citrate solution (90 ml). The bacterial ATP was determined by the method of Bossuyt (5) using the Milk Bacteria Kit supplied by Lumac B.V.

ISBN 0-12-426290-2

RESULTS

Pre-incubation of creams at 21°C for 25 h in the presence of crystal violet-penicillin-nisin to prevent the growth of Gram-positive organisms (P-INC test) followed by enumeration of bacteria by plate count gave a good estimate of post-heat treatment contamination. Selective cut-offs of 3.2×10^7 CFU/g were used as this was the level of bacteria above which organoleptic defects became apparent. Using this technique, 97.4% of samples fell into the two quadrants which identify acceptable and unacceptable creams.

A good correlation exists between the P-INC plate count and the P-INC count determined by ATP assay. Using 3.5 log units (RLU x multiplication factor) as a cut-off for the ATP method, 95.4% of samples could be assigned to the appropriate quadrants.

The comparison between the P-INC test count obtained by ATP assay and the count obtained after the shelf-life test (7 d storage of cream at 6°C) showed that 94.5% of samples could be assigned to the two selective quadrants.

The P-INC test combined with bacterial enumeration by ATP assay gives a reliable method of assessing post-pasteurization contamination of short shelf-life dairy products within 26 h of production.

REFERENCES

1. Phillips, J.D., Griffiths, M.W. and Muir, D.D. (1981). Factors affecting the shelf life of pasteurized double cream. *Journal of the Society of Dairy Technology* 34, 109-113.
2. Griffiths, M.W., Phillips, J.D. and Muir, D.D. (1980). Rapid plate counting techniques for enumeration of psychrotrophic bacteria in pasteurized double cream. *Journal of the Society of Dairy Technology* 33, 8-10.
3. Phillips, J.D., Griffiths, M.W. and Muir, D.D. (1983). Accelerated detection of post-heat-treatment contamination in pasteurized double cream. *Journal of the Society of Dairy Technology* 36, 41-43.
4. Griffiths, M.W., Phillips, J.D. and Muir, D.D. (1984). Methods for rapid detection of post-pasteurization contamination in cream. *Journal of the Society of Dairy Technology* 37, 22-26.
5. Bossuyt, R. (1981). Determination of bacteriological quality of raw milk by an ATP assay technique. *Milchwissenschaft* 36, 257-260.

EFFECT OF IMIDAZOLE ANTIFUNGALS ON CANDIDA ALBICANS DEMONSTRATED BY BIOLUMINESCENT ASSAY OF ATP

S. Ånsēhn and L. Nilsson

Department of Clinical Bacteriology
University Hospital, 581 85 Linköping, Sweden

INTRODUCTION

Some imidazole antimycotics have been suggested to exert a fungistatic action at low concentrations due to blocked ergosterol synthesis, and a fungicidal action at high concentrations due to a direct membrane damage (DMD) (1). Whether ketoconazole (KCZ) causes a DMD or not is a question under discussion (1,2,3). Since intracellular (IC) ATP parallels viability of *C. albicans* exposed to econazole (4), and leakage of ATP (extracellular (EC) ATP) occurs when bacteria are exposed to membrane active agents (5,6), we exposed *C. albicans* to KCZ and tioconazole (TCZ) and correlated IC- and EC ATP with viability in an effort to further elucidate the mode of action of the drugs.

MATERIAL AND METHODS

Determination of IC ATP in C. albicans cultures

KCZ and TCZ, dissolved in dimethylsulphoxide, were diluted in MOPS buffered YNB broth (pH 7.0) and inoculated with *C. albicans*. After incubation at 37°C EC ATP was eliminated with apyrase and IC ATP was extracted with boiling Tris/EDTA buffer (5). ATP monitoring reagent (LKB-Wallac) was added and the light emission was measured in a 1250 Luminometer (LKB-Products) (5).

ISBN 0-12-426290-2

Determination of EC ATP

ATP monitoring reagent buffered with Tris containing EDTA was added to the cultures (5). The purified and buffered reagent provided for a constant light intensity for several min. Kinetics of ATP leakage was studied in MOPS buffer (pH 7.0) at 22°C.

RESULTS AND DISCUSSION

The relationship between IC- and EC ATP and viability in *C. albicans* cultures exposed to TCZ is shown in Fig. 1. Viability parallels IC ATP indicating growth inhibition at $\leq$ 31.2 µg/ml. At higher concentration a fungicidal activity is indicated by a dramatic decrease in IC ATP, a loss of viability and a concomitant increase of EC ATP.

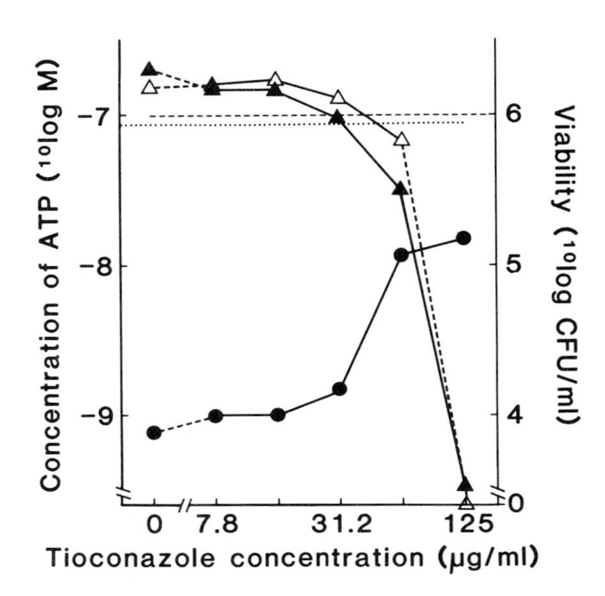

Fig. 1 IC (▲) and EC (●) ATP and viability (△) in C. albicans cultures exposed to TCZ for 2 hours. IC ATP (········) and CFU/ml (-----) at the start of the experiment.

TABLE I

Relationship between viability, IC- and EC ATP in C. albicans cultures exposed to TCZ and KCZ

Conc. (μg/ml)	Viability (CFU/ml)	% stained cells	IC ATP (M)	EC ATP (M)
TCZ 0	$6.6\ 10^{7}$	0	$6.2\ 10^{-6}$	$6.0\ 10^{-10}$
50	$2.0\ 10^{4}$	100	$3.0\ 10^{-9}$	$7.5\ 10^{-9}$
100	-	100	-	$1.2\ 10^{-8}$
200	-	100	-	$1.4\ 10^{-8}$
400	-	100	-	$1.4\ 10^{-8}$
KCZ				
50	$7.0\ 10^{6}$	6	$7.7\ 10^{-7}$	$4.2\ 10^{-10}$
100	$2.5\ 10^{6}$	20	$3.1\ 10^{-7}$	$6.9\ 10^{-9}$
200	$1.5\ 10^{5}$	48	$7.0\ 10^{-8}$	$1.0\ 10^{-8}$
400	-	50	$1.2\ 10^{-8}$	$1.3\ 10^{-8}$

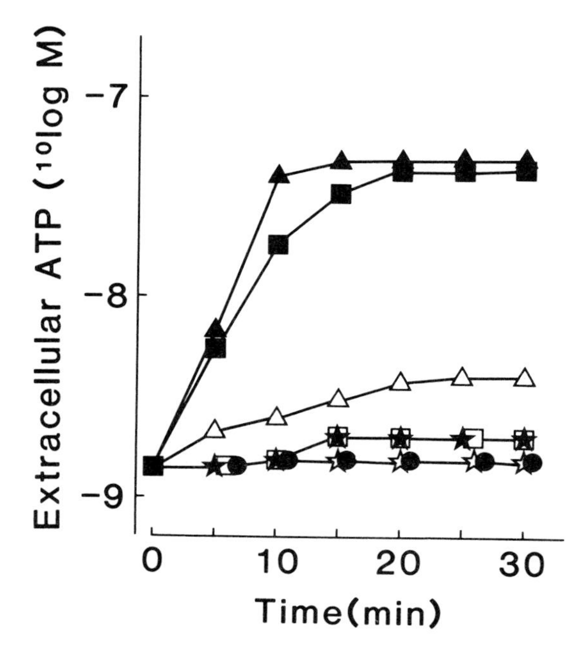

Fig. 2 Kinetics of ATP leakage from C. albicans cells in MOPS buffer exposed to TCZ and KCZ respectively: 75 μg/ml (▲, △), 37.5 μg/ml (■, □) and 18.75 μg/ml (★, ☆). Control without antifungal drug (●).

Leakage of ATP occurred within min after exposure to high concentrations of TCZ and maximal levels were reached within 20 min (Fig. 2). KCZ caused a less extensive leakage indicating a less pronounced membrane damage, which corresponds with methylene blue uptake (1) (% stained cells) in another experiment (Table I).

Bioluminescent assay of ATP in C. *albicans* cultures exposed to high concentrations of TCZ and KCZ demonstrates a rapid and heavy leakage of ATP accompanied by a simultaneous decrease in IC ATP and viability (Table I). The leakage is due to a direct membrane damage, which is less pronounced for KCZ than for TCZ.
Lower imidazole concentrations exert a growth inhibiting effect without significant leakage of ATP.
These findings support the hypothesis that imidazole antifungals basically act in the same manner bvt at different concentrations.

REFERENCES

1. Sud, I.J. and Feingold, D.S. (1981). Heterogenity of action mechanisms among antimycotic imidazoles, *Antimicrob. Agents Chemother.* 20, 71-74.
2. Beggs, W.H. (1983). Comparison of miconazole- and ketoconazole induced K^+ release from *Candida* species, *J. Antimicrob. Chemother.* 11, 381-383.
3. Uno, J., Shigematsu, M.L. and Arai, T. (1982). Primary site of action of ketoconazole on C. *albicans*, *Antimicrob. Agents Chemother.* 21, 912-918.
4. Ånsēhn, S.(1977). In vitro synergistic action of antimycotics and antibiotics on C. *albicans*, *Curr. Ther. Res.* 22, 92-99.
5. Nilsson, L. (1981). New rapid bioassay of gentamicin based on luciferase assay of extracellular ATP in bacterial cultures. *Antimicrob. Agents Chemother.* 14, 812-816.
6. Thore, A., Nilsson, L., Höjer, H., Ånsēhn, S. and Bröte, L. (1977) Effects of ampicillin on intracellular levels of adenosine triphosphate in bacterial cultures related to antibiotic susceptibility, *Acta path. microbiol. scand. Sect. B* 85, 161-166.

DETERMINATION OF INTRACELLULAR ATP DURING GROWTH OF ESCHERICHIA COLI IN THE PRESENCE OF AMPICILLIN

B. Beckers, H.R.M. Lang[x] and A. Beinke

Department of Medical Microbiology, Medical Faculty University of Aachen, Aachen, Fed. Rep. Germany.

[x]*Department of Medical Chemistry, University of Vienna, Vienna, Austria*

INTRODUCTION

The ATP determination with firefly bioluminescence is a sensitive indicator for the presence and growth of bacteria in biological material. We investigated the applicability of the ATP assay for rapid determination of antibiotic susceptibility of Escherichia coli ATCC 25923 against ampicillin. Our aim was to obtain a reliable determination of minimal inhibitory concentration (MIC) after a short period of incubation.

MATERIAL AND METHODS

For determination of MIC the broth dilution test was used.

Dilutions of ampicillin : the broth dilution test was performed with a final concentration of 8, 4, 2 and 1 mcg/ml of ampicillin. Tubes containing medium alone served as growth controls.

Inoculum : bacteria cultivated on DST-agar (Oxoid) were harvested after 18 h of cultivation and cell density adjusted to 10^6/ml by optical turbidity. Each tube was inoculated with 1 ml of this suspension and then incubated at 37^oC.

Determination of colony forming units (CFU) : im-

ISBN 0-12-426290-2

mediately after inoculation and after 3 6 and 24 hours an aliquot was removed and of this 10-fold serial dilutions were dropped on DST-agar. The colony counts was evaluated after an incubation of 24 hours at 37°C.

Determination of intracellular ATP : at the same intervals the amount of intracellular ATP was determined. The difference between total ATP and extracellular ATP was taken as value for intracellular ATP. Total ATP was measured after mixing 50 µl sample with 100 µl NRB (a nucleotide releasing reagent for bacterial cells,Lumac/3M) and addition of 100 µl of firefly lantern extract(Sigma). Light emission of luciferine-luciferase reaction was recorded during 10 sec in a Biocounter M 2000 (Lumac/3M) and results were expressed in relative light units (RLU). For determination of extracellular ATP buffer was used instead of NRB.

RESULTS

Fig. 1 shows the effect of ampicillin on the growth of E.coli. In comparison to controls without drug, the addition of 1mcg/ml ampicillin did not influence the increasing number of viable bacteria during incubation whereas 4 and 8 mcg/ml lead to a rapid and continuous decrease of CFU. At a concentration of 2 mcg/ml a delayed growth is observable. But at the end of incubation no difference

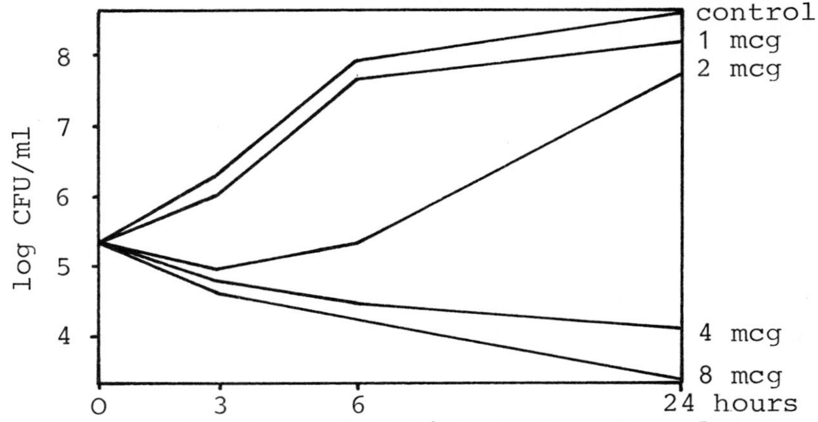

Fig. 1 Determination of CFU/ml in E.coli cultures containing different concentrations of ampicillin

was found between 1 and 2 mcg/ml. The MIC of ampicillin for E.coli ATCC 25923 is 4 mcg/ml (1). The corresponding ATP curves in Fig. 2 run in a quite different manner.

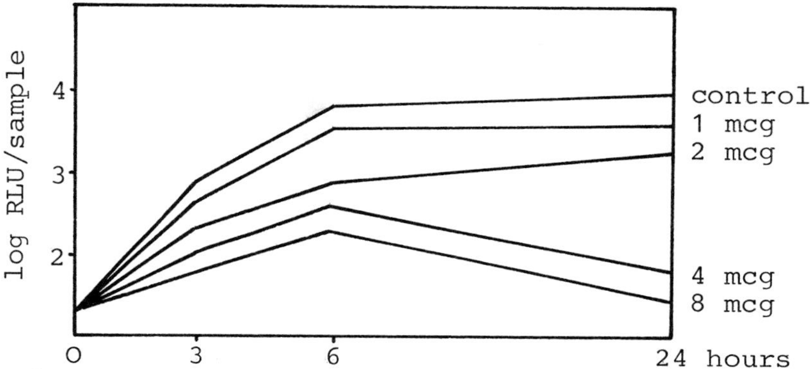

Fig. 2 Determination of intracellular ATP in E.coli cultures containing different concentrations of ampicillin

The expected early decrease of intracellular ATP in the tubes with 4 and 8 mcg/ml as found by determination of CFU was not notified, but the ATP concentrations increase in all tubes during the first 6 hours of incubation. This increase depended from the used concentration of drug. Using low concentrations of ampicillin the increase is more than 100-fold compared to values immediately after inoculation. In the case of high amounts of ampicillin the increase was only 10-fold.
A clear distinction between inhibition with 4 mcg/ml and growth with 2mcg/ml can be observed after an incubation period of 24 hours. This finding corresponds well with the MIC of the tested strain.

DISCUSSION

If cells of Escherichia coli are incubated in the presence of ampicillin, there is no correlation between the concentrations of intracellular ATP and the CFU values. At the inhibitory concentration of the drug,CFU were decreased,whereas there was an increased ATP concentration during the first 6 hours of incubation. Similar results were observed

when incubating Staphylococcus aureus in the presence of penicillin.
Several authors (2,3,4) reported about reliable determination of antimicrobial susceptibility with bioluminescence after 2 to 4 hours of incubation. The drugs investigated were gentamicin, chloramphenicol, polymyxin B and tetracycline.Gentamicin produced a rapid fall in bacterial ATP and a bacteriostatic drug such as tetracycline inhibited the ATP synthesis, the ATP content remained at a constant low level.
Our results presented may be explained by the mode of action of ampicillin.ß-lactam antibiotics induce a variety of morphological changes on bacterial cells,mainly in subinhibitory concentrations.These forms may contain elevated ATP concentrations, at the same time the cell division is inhibited.
Determination of intracellular ATP in cultures of Escherichia coli exposed to ampicillin seems not to be suitable for reliable determination of MIC within 6 hours of incubation.

REFERENCES

1) Washington,J.A. and Sutter,V.L.(1980).Dilution susceptibility test:agar and macro-broth dilution procedures.In"Manual of Clinical Microbiology"(Eds.E.H.Lennette et al.)3rd edn.pp.453-462.American Society for Microbiology,Washington,D.C.
2) Picciolo,G.L.and Chappelle,E.W.(1977).Applcation of firefly luciferase assay for ATP to antimicrobial drug sensitivity testing,NASA Technical Note D-8439,pp.1-116.
3) Fauchère,J.L.et al.(1982).Détermination rapide de la concentration minimale inhibitrice d'un antibiotique par dosage de l'ATP bactérien. Ann.Microbiol.(Inst.Pasteur)133A,293-299
4) Thore,A.et al.(1977).Effects of ampicillin on intracellular levels of ATP in bacterial cultures related to antibiotic susceptibility, Acta path. microbiol. scand. Sect. B, 85,161-166.

LUMINOMETRIC DETECTION OF BACTERIURIA IN PRIMARY HEALTH CARE

A. Lundin[1], H. Hallander[2], A. Kallner[3], U. Karnell Lundin[1] and E. Österberg[4]

[1]*Research Centre and Department of Medicine, Karolinska Institute, Huddinge Hospital, S-141 86 Huddinge, Sweden and LKB-Wallac, SF 201 01 Turku 10, Finland*
[2]*National Bacteriological Laboratory, S-105 21 Stockholm, Sweden*
[3]*Karolinska Hospital, S-104 01 Stockholm, Sweden*
[4]*Sollentuna Hospital, S-191 23 Sollentuna, Sweden*

Urine contains three ATP pools (i) non-cellularly associated, and associated with (ii) somatic cells and (iii) bacterial cells. In preliminary experiments the first two pools were estimated by adding an aliquot of urine directly to a firefly reagent measuring the stable light emission subsequently, releasing ATP from somatic cells and measuring the increase of the light emission. In other aliquots non-bacterial ATP was degraded with Triton X-100 and apyrase under various conditions, bacterial ATP extracted (and apyrase inactivated) with TCA and estimated by the firefly assay. Total ATP in urine was determined by extraction of untreated urine with TCA. Using conventional bacteriological culture as reference, a method for determination of bacterial ATP was selected as a candidate for luminometric detection of bacteriuria. It could be performed in a single cuvette in 15 min. The diagnostic performance was studied as a function of the ATP limit used for designating a positive test. In 781 urine specimens obtained in primary health care (prevalence 17%) its performance was compared to a sediment test counting bacterial and white cells, a granulocyte esterase test, a nitrite test and a dip-slide test. With the exception of the tests for white cells and granulocyte esterase an acceptable number of false results (7-8%) was obtained by all tests. With the ATP test it was possible to reduce the number of false results to 2% by follow-up testing of specimens with intermediate ATP levels (3-25 nmol/l). Thus the ATP test may become an attractive alternative for detection of bacteriuria in primary health care.

ISBN 0-12-426290-2

ASSAY OF GENTAMICIN AND CHLORAMPHENICOL IN SERUM USING ATP BIOLUMINESCENCE

A. Tanna and C.S.F. Easmon

Department of Medical Microbiology,
Wright-Fleming Institute,
St. Mary's Hospital Medical School,
London W2 1PG

The measurement of bacterial growth by adenosine triphosphate (ATP) estimation using firefly luciferin luciferase reaction is rapid and simple. We have applied this technique to the rapid assay of gentamicin and chloramphenicol in serum.

The effectiveness of a variety of extraction procedures including those using commercial extractants and luciferin luciferase reagents was compared in terms of extraction, efficiency, reproducibility and cost, using standard antibiotic assay organisms. The lumac-nucleotide releasing agent with L.K.B. luciferin luciferase reagent met with these specifications.

Growth curves using ATP measurement were found to correlate well with viable counts for Staphylococcus aureus, Klebsiella edwardsii and Sarcina lutea. Standard curves were then prepared for the antibiotics to be assayed. ATP in mV was plotted against the antibiotic concentrations.

ATP assay for both gentamicin and chloramphenicol gave results comparable with EMIT and HPLC assays respectively using over 50 serum samples for each drug. The correlation coefficient for gentamicin was 0.99 and for chloramphenicol 0.96. The assays took no longer than 2 hr and the cost per estimation was 20 p or £1 for serum level with four standards.

ISBN 0-12-426290-2

DETECTION OF BACTERIURIA IN YEAST CONTAMINATED SPECIMEN BY FILTRATION AND BIOLUMINESCENCE TECHNIQUE

TenLin S. Tsai and Leroy J. Everett

United Technologies Packard, Downers Grove, IL 60515 U.S.A.

INTRODUCTION

The applications of bioluminescent (B) assay of microbial ATP (1) in clinical specimens such as the determination of significant bacteriuria have often been limited by the interference of nonbacterial ATP, or other substances which inhibit the ATP-firefly luciferase-luciferin reaction (2,3). False positive results arise from the former and false negative results from the latter. We report here a rapid procedure combining double membrane filter technique with bioluminescent assay. This new approach provides in one passage the concentration of bacteria and the removal of nonbacterial (such as yeast and red cells) ATP and other interfering substances.

MATERIALS AND METHODS

All media were obtained from BBL (Cockeysville, MD). A 24 hr culture (37°C) of *E. coli* (ATCC® 25922) in trypticase soy broth and a 48 hr culture (25°C) of *Saccharomyces cerevisiae (variety ellipsoides)* in Sabouraud dextrose broth were used to inoculate the zero culture urine specimens from healthy adults. Spiked samples were plated on trypticase soy agar (37°C, 18 hrs) for enumeration of *E. coli* colony forming units (CFU). Yeast CFU were counted on potato dextrose agar plates (pH5.6, supplemented with 0.01% chloramphenicol and 0.007% chlorotetracycline) after 2 days incubation at 25°C.

Bioluminescent reagents obtained from Packard Instrument Co. (Downers Grove, IL) were prepared and used per manufacturer's direction. The Lumac/3M Bacteriuria Screening Kit was purchased from Lumac Medical Products Division/3M (St. Paul, MN). Packard's luminometer, PICOLITE® 6500 in conjunction with HP-85 was used for the B assay and interpretation of results. The 17 x 60 mm glass shell vials from Brockway, Inc. (Parkersburg, WV) were treated to eliminate contaminating ATP residues. Filter holders used in this study are Swinney (Gelman Sciences, Inc., Ann Arbor, MI), Swin-Lok (Nuclepore, Pleasanton, CA) and Swinnex (Millipore Corp., Bedford, MA). Polycarbonate membrane filters were obtained from Nuclepore.

ANALYTICAL APPLICATIONS OF
BIOLUMINESCENCE AND CHEMILUMINESCENCE

ISBN 0-12-426290-2

Vials containing 25 μl of sample or blank and 50 μl of NRS/Somase reagent (from Lumac) were incubated at 37°C for 25-40 min. The B assay was then carried out in the 6500 with the automatic injection of 100 μl of extracting agent to lyse bacteria for 2 min and 100 μl of the ATP-firefly luciferase-luciferin reagent to start reaction. The light response was corrected by internal standarization. Standard curves for *E. coli* and *S. cerevisiae* were established as log CFU/ sample vs. log [ATP]/ sample. These curves were then used to interpret the deduced log [ATP] results from light response of unknown samples.

Ten milliliters of the diluted urine (1:50 with 0.0003 M Phosphate buffer, pH 7.2) in a sterile disposable syringe was filtered through two membrane filters by an infuser with a constant upward speed (2.5 ml/min). This represents an assay of 200 μl of urine. The 3 μm filter, being upstream of the 0.4 μm filter, retains cell debris and somatic cells. The bacteria in the filtering sample are concentrated on the second stage 0.4 μm filter, leaving soluble ATP and other interfering substances voided in the end filtrate. The filters were retrieved from the filter holders and placed on the bottom of the vial to be assayed.

RESULTS

Removing Nonbacterial ATP by Two Bioluminescence Systems.

E. coli spiked urine samples were analyzed with the soluble B assay receiving NRS/Somase or just buffer during incubation. Results illustrated in Fig. 1 indicate that when nonbacterial ATP is not reduced a high ATP background is seen. The lowest detectability of bacteria is about 10^5/ml urine and cannot be used dependably to screen bacteriuria. The resolution of bacteria at higher levels is poor. With a 5 log range of increasing *E. coli* CFU there is merely a 2 log increase in the predicted CFU. Good CFU correlation is achieved with B assay and aerobic plate counts (APC) using reagents from either source as long as somatic ATP was previously removed from samples.

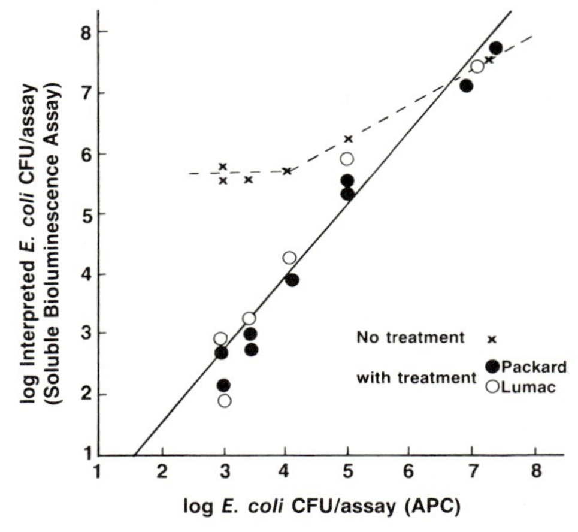

Fig. 1 *Removing nonbacterial ATP from urine sample is a prerequisite for detecting bacteriuria.*

Urine samples spiked with increasing levels of *E. coli* were analyzed with Lumac's Kit where somatic ATP were minimized by NRS/somase treatment or Packard's DF/B system where only bacteria retaining 0.4 μm filters were assayed. Results obtained from the two procedures are compared in Table 1. Slightly better accuracy and precision values are seen with DF/B system because it separates and removes somatic ATP from bacterial ATP, whereas Lumac method only reduces the nonbacterial ATP. The slope of the correlation curve from DF/B results is closer to 1 indicating a closer prediction of bacterial levels in urine samples.

TABLE 1

Comparison of Two Bioluminescence Systems for Detecting Bacteriuria in Noncontaminated Samples

Parameters	DF/B	NRS/Somase
CFU prediction curve		
Correlation coefficient, r	0.983	0.979
Slope	1.190	1.372
Accuracy[1]	1.05	1.06
Precision[2]	7.56	9.00
Assay running time	10 min	30 min (25 min incubation)
Assay range	$5 \cdot 10^3$-10^8 CFU/ml	10^4 CFU/ml, or 10^5 CFU/ml[3]

[1] Log CFU by bioluminescence/log CFU by aerobic plate count
[2] Coefficient of variation (%)
[3] Urine specimens may be screened with Lumac™ Bacteriuria Screening Kit at either 10^4 CFU/ml (200 relative light units) or at 10^5 CFU/ml (500 relative light units).

Detecting Bacteriuria in Yeast Contaminated Urine Specimens

Since yeast cells contain approximately one hundredfold ATP per cell as that of *E. coli* small contamination of yeast in urine would lead to erroneous enumeration of bacteria. The efficacy of the two B systems is tested in urine samples which are spiked with a constant level of yeast and an increasing level of *E. coli*. Both 3 μm and 0.4 μm filters were assayed for ATP, respectively, for yeast and bacteria. It is clearly shown in Fig. 2 that the Lumac kit fails to distinguish lower bacterial levels due to the interference of yeast ATP. The DF/B method, however, predicts bacterial levels accurately ($r=0.977$) in yeast containing samples and in addition, quantitates yeast and bacteria separately by assaying their respective retaining filters (Fig. 3). The original innoculation in the samples were accurately predicted.

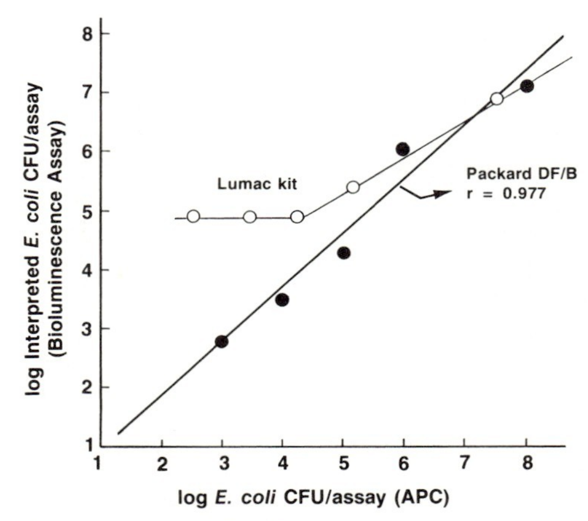

Fig. 2 *Comparison of two bioluminescence systems for detecting bacteriuria in yeast contaminated samples.*

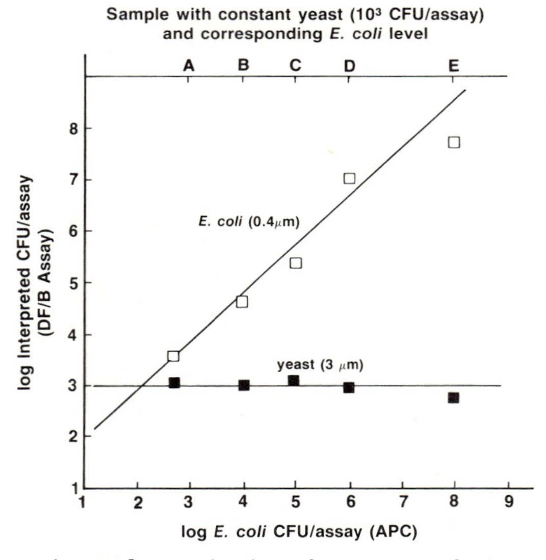

Fig. 3 *Separation and quantitation of two types of microorganisms by DF/B method in samples spiked with increasing levels of E. coli and constant level of yeast.*

Twenty two urine samples were assayed with the DF/B method. The results agreed closely with conventional plate count (r=0.972), and the following predictive values were derived from the data: sensitivity = 100%, specificity = 75% (3/4, 3 negative by DF/B out of 4 negative by APC), positive predictive value = 95%, negative predictive value = 100%, false positive rate = 5% and false negative rate = 0%.

DISCUSSION AND CONCLUSION

Previous studies have optimized the B assay of the membrance filter (4) and have established the separation and quantitation of a variety of microorganisms by the DF/B technique (5). The commercial development of a preassembled double filtration unit will enhance the utility of this technique.

The DF/B system provides: 1) a shorter actual assay time since no incubation is needed; 2) precise report of the bacterial level in a range of five orders of magnitude instead of just an arbitrary cutoff, 3) an automated system where automatic injections, sample changing, and data management are included; 4) separation of yeast and other somatic cells from bacteria in the sample thereby reducing false positive results from urine containing large amounts of yeast and red cells, and 5) the capability of measuring more than one type of cells or microorganisms. Consequently, the double filtration/bioluminescence system compared to other systems (6) offers a rapid and high resolution tool for bacteriuria screening, especially in contaminated urine specimens.

REFERENCES

1. Chappelle, W. E., and Levin, G. V. (1968). Use of the firefly bioluminescent reaction for rapid detection and counting of bacteria, *Biochem. Med* *2, 41-51.*
2. Conn, Rex B., Charache, P., Chappelle, W. E. (1974). Limits of applicability of the firefly luminescence ATP for the detection of bacteria in clinical specimens, *Amer. J. Clin. Pathol. 63, 493-501.*
3. Deming, J. W., McGarry, M. W., Chappelle, E. W., Picciolo, G. L. (1977). Procedures for the quantitation of bacteria in fluids: Firefly luciferase ATP assay used with concentration procedures. *In* "2nd Bi-Annual ATP Methodology Symposium" (Ed. G. A. Borun) pp. 465-490, SAI Technology Co., San Diego, Calif.
4. Tsai, T. L. S. (1982). Method of Concentrating and Measuring Unicellular Organisms. *U.S. Patent 400556 pending.*
5. Everett, L. J., Tsai, T. L. S., Ramey, N. J. (1983). High Resolution Method of Measuring ATP and Concentrating and Measuring Unicellular Organisms. *U.S. Patent 495206 pending.*
6. Pezzlo, M. T. (1983). Automated methods for detection of bacteriuria, *Amer. J. Med. 7, 71-78.*

A COMPARISON OF METHODS FOR THE EXTRACTION OF ATP FROM MYCOBACTERIA

R.P. Prioli and I.N. Brown

Department of Medical Microbiology, St. Mary's Hospital Medical School, Paddington, London W2 1PG, England.

INTRODUCTION

Colony counts on suspensions of mycobacteria take a long time and require special media. They are also liable to error because of clumping and carry over (1). In the case of the non-cultivable mycobacteria, *Mycobacterium leprae* and *M.lepraemurium*, viability is usually assessed by staining reaction, animal inoculation, or by the uptake and metabolism of a suitable substrate (2). These methods are indirect or, again, time consuming. A rapid method for measuring the viability of these organisms is urgently needed for both clinical and experimental work.

A possible approach is to measure mycobacterial ATP using the firefly bioluminescent reaction. We have compared five extraction procedures using *M.bovis*, BCG as a model. This organism grows readily in culture and colony counts can be made in 14-21 days.

MATERIALS AND METHODS

Bacteria

The Glaxo substrain of BCG was used. Organisms were grown as a dispersed growth in Middlebrook 7H9 broth (Difco) containing Tween 80. Estimates of total count were made by measuring opacity and colony counts by plating out suitable dilutions of culture on Middlebrook 7H10 agar. Plates were incubated in plastic bags at 37°C for 2-3 weeks. Details of the methods are given in Brown *et al* (3).

ISBN 0-12-426290-2

Extraction procedures

1. *Boiling tris EDTA buffer* 0.1 ml bacterial suspension was added to 0.2 ml buffer (0.1M, pH 7.75) and the mixture heated in a boiling water bath for 5 minutes.

2. *Heat and chloroform* 0.03 ml 23% chloroform was added to 0.1 ml bacterial suspension which was then heated in a boiling water bath for 5 minutes. The pellet was rehydrated with 0.4 ml tris-EDTA buffer (4).

3. *Dimethylsulphoxide* 1.0 ml bacterial suspension was filtrated by vacuum suction onto a 45μ, 25 mm filter pad. The bacteria were washed with 1.0 ml tris-EDTA buffer and then using a fresh collection tube, treated with 0.2 ml 90% DMSO for 30 seconds and washed with two lots of 1.0 ml buffer. The ATP content of the filtrate was measured.

4. *NRB Lumac* Equal volumes of NRB reagent and bacterial suspension were mixed and left at room temperature for 10 minutes before sampling. For some experiments the reaction mixture was heated to 60°C.

5. *Extralight* Equal volumes of reagent and bacterial suspension were mixed and allowed to react at room temperature for 60 seconds before sampling.

Measurement of ATP

ATP monitoring reagent and ATP standard were purchased from LKB Wallac. 0.2 ml ATP standard or bacterial extract was added to 0.2 ml monitoring reagent in a polystyrene measuring cuvette. Peak light output was recorded in mV on a LKB Wallac luminometer 1250.

RESULTS

Comparison of the five extraction procedures

Details of the methods chosen for study are summarized in table 1. The methods were compared on 4 separate occasions using broth cultures containing approximately 10 million BCG/ ml. Five samples of culture were extracted using each procedure. The results are given in table 2.

TABLE 1

Details of the extraction methods

Method	Volume of culture (ml)	Extraction time	Temperature (C°)
Boiling buffer	0.1	5 min	100
Heat & Chloroform	0.1	5 min	98
NRB	0.25	10 min	20
DMSO	1.0	30 sec	20
Extralight	0.25	60 sec	20

TABLE 2

ATP extraction from BCG cultures

BCG	Extraction procedures (mean mV/ml)				
CFU/ml	Boiling buffer	Heat & chloroform	NRB	DMSO	Extralight
5×10^7	3765	3483	667	not done	194
2×10^7	1575	1593	328	not done	106
1×10^7	833	730	124	31	27
6×10^6	392	380	59	19	13
Efficiency of method %	99-114	100	16-21	4-7	5-10

All the methods were reproducible and produced a ranking which agreed with that subsequently obtained by colony count. The hot methods were more efficient than the cold methods. Taking the heat and chloroform as the point of comparison, of the two hot methods boiling buffer proved the best and was therefore chosen for further study.

Further studies using the boiling buffer method

ATP was extracted from 22 samples of neat or diluted BCG culture and colony counts were made on the same suspension. A linear correlation was obtained between light output and colony counts (correlation coefficient $r^2 = 0.97933$). Previous experiments had confirmed the linear relationship between ATP concentration and mV/ml light output over the range 10^{-5} - 10^{-10}M ATP standard.

Suspensions prepared from clinical isolates of tubercle bacilli were extracted with boiling buffer and NRB reagent at 60C. Both methods were applicable but the boiling buffer method consistently released about 4 times the amount of ATP detectable after NRB extraction. The boiling buffer method also killed the tubercle bacilli.

CONCLUSIONS

We have found that treatment of mycobacterial suspensions with boiling tris-EDTA buffer is the best method for extracting ATP. It has been applied successfully to BCG and human tubercle. It is a simple, rapid and safe technique because it kills the mycobacteria.

REFERENCES

1. Fenner, F., Martin, S.P. and Pierce, C.H. (1949). The enumeration of viable tubercle bacilli in cultures and infected tissues, *Annals of the New York Academy of Sciences* 52, 751-764.
2. Brown, I.N. (1983). Animal models and immune mechanims in mycobacterial infection. *In* "The Biology of the Mycobacteria" (Eds C. Ratledge and J. Stanford). Vol. 2, pp. 173-234. Academic Press, London.
3. Brown, C.A., Brown, I.N. and Sljivic, V.S. (1980). Active suppression masks an underlying enhancement of antibody production *in vitro* by spleen cells from BCG infected mice, *Immunology* 40, 303-309.
4. Dhople, A.M. and Hanks, J.H. (1973). Quantitative extraction of ATP from cultivable and host grown microbes - calculation of ATP pools, *Applied Microbiology* 26, 399-403.

II GENETIC ENGINEERING

MOLECULAR GENETIC STUDIES IN BIOLUMINESCENCE

Kenneth H. Nealson* and Richard Cassin[+]

Marine Biology Research Division, Scripps Inst. of Oceanography, La Jolla, CA 92093

[+] *Dept. of Biology, Stanford University Palo Alto, CA. 94305*

Although the luminous bacteria are the only light-emitting organisms to have been examined on the molecular genetic level, modern techniques of nucleic acid chemistry and recombinant DNA methods offer substantial new opportunities for the study of bioluminescence in many systems. This article will review the published literature on the molecular genetics of luminous bacteria as well as work currently in progress on this system, and will speculate on possible future opportunities to extend the study of bioluminescence to the molecular genetic level in eukaryotes, with particular reference to dinoflagellates.

LUMINOUS BACTERIA

Bioluminescent bacteria are ubiquitous in the marine environment and are occasionally found also in terrestrial and aquatic habitats. The marine forms can be planktonic, gut symbionts, saprophytic, parasitic, or in specialized light organs of fish and squid. Although there is considerable taxonomic and ecological diversity in marine luminescent bacteria, they all appear to share a common mechanism (referred to as the "*lux*" system) for light emission (Fig. 1).

Light production is catalyzed by bacterial luciferase, a dimeric mixed-function oxidase composed of two different (alpha and beta) subunits of approximately 40K MW each (1). At full induction luciferase can constitute 5% or more of total cellular protein (2). Luciferase catalyzes the

ISBN 0-12-426290-2

oxidation of reduced flavin mononucleotide (FMNH2) and a long-chain aliphatic aldehyde to produce oxidized FMN, the corresponding long chain fatty acid . . . and light. The precise number of specific proteins required for light emission is not yet clear, but it is known that in addition to luciferase, a fatty acid reductase is required to recycle fatty acid back to aldehyde. Isolated from *Photobacterium phosphoreum*, this enzyme exhibits two activities: a 51K ATP-dependent acyl protein synthetase, and a 58K NADPH-dependent acyl CoA reductase (3,4). Reducing power is provided by an NADPH-FMN oxidoreductase (5). It is also possible that other enzymes catalyzing the de novo synthesis of specific aldehydes or fatty acids may be necessary. Dark mutants have been isolated that become bright upon exogenous addition of either long-chain fatty acid or aliphatic aldehydes, suggesting that if any of the genes required for aldehyde synthesis or acid reduction is disabled, light emission is impaired.

$$RCHO + FMNH_2 + O_2 \xrightarrow{luciferase} RCOOH + FMN + H_2O + h\nu$$

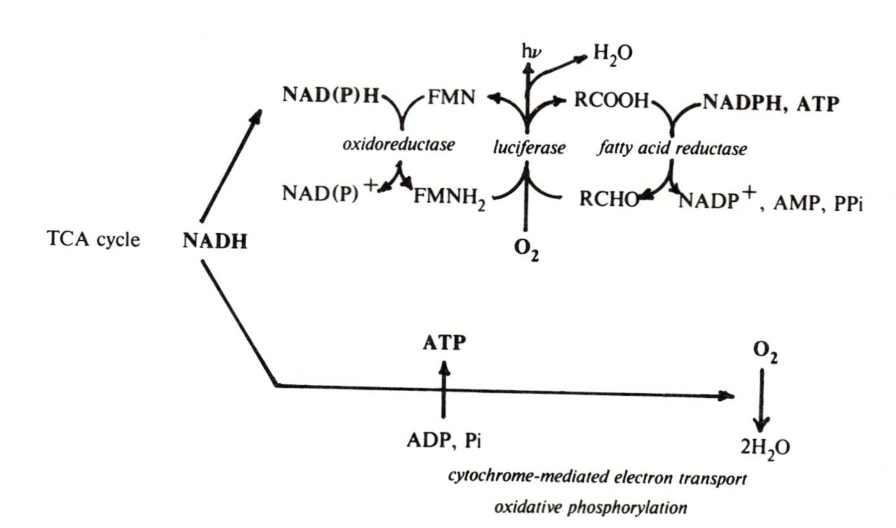

Fig. 1 Schematic representation of the bacterial bioluminescence reaction in relation to other reactions in the cell.

Luminous bacteria can expend 10% or more of their total cellular energy producing light (6). One might thus expect that the system would be tightly regulated, and indeed, in most luminous bacteria, the synthesis of the light-emitting system is regulated via a complex control mechanism called autoinduction (7). This process is perhaps best elaborated in *Vibrio fischeri* (8), which when grown in batch culture exhibits a lag in the appearance of bioluminescence relative to growth, followed by a rapid increase in luminescence at high cell density. This observation ultimately led Eberhard, et al. (9,10) to identify a small molecule produced by the cells (the "autoinducer") which accumulates in the growth medium, and which at a critical concentration results in a 1000-fold increase in light production. When cultures of *V. fischeri* are maintained at low density in a carbon-limited chemostat, they are continuously dark due to dilution of the autoinducer, but when an appropriate amount of autoinducer is added to the chemostat growth chamber, the lux system is induced and the culture begins to luminesce (11).

The autoinducer isolated from the growth medium of *V. fischeri* was found to be N-(beta-ketocaproyl)homoserine lactone ([N-(3-oxohexanoyl)-3-aminohydro-2(3H)-furanone]) (9,10). Synthetic autoinducer was prepared and shown to mimic the activity of the natural inducer in all *V. fischeri* strains and in the closely related *V. logei*, but in no others, indicating that some specificity is involved in the autoinduction process. Studies using inhibitors of protein and mRNA synthesis suggested that control of light production by the autoinducer is at the level of gene transcription (7). Mutants deficient in the production of autoinducer can be made to luminesce upon addition of the purified or synthesized autoinducer to the medium. In addition, specific mutants of *V. fischeri* have been selected that produce normal, functional autoinducer, but which are unable to respond to it. These experiments suggest that a specific receptor or cofactor is required to interact with the autoinducer in stimulating light production. Interestingly, some non-luminous *Vibrios* have been found to produce an activity that induces luminescence in *V. harveyi* (12).

THE MOLECULAR GENETICS OF BACTERIAL BIOLUMINESCENCE

The initial molecular genetic experiments on the system were performed by Cohn, et al. (13). Using available protein sequence data for the alpha subunit of luciferase (*lux*-A) from *V. harveyi* (B392), they synthesized a mixture of eight oligonucleotide probes, each 17 nucleotides long, which was then used as a probe to identify sequences in a lambda charon 13 genomic library carrying the entire alpha subunit of the lux gene and the 5' end of the beta subunit. The 1.85-kb EcoR1 fragment carrying the lux sequences was further subcloned into M13 mp7 and mp8. Partial nucleotide sequencing of these subclones and comparison with the known amino acid sequence revealed the orientation of the *lux*-A gene and the location of the *lux*-B gene immediately downstream.

Belas, et al. (14) then cloned both lux-A and lux-B genes from a seawater isolate of *V. harveyi* (BB7) in order to reconstruct the bioluminescent system in *E. coli*, therein to study the mechanisms of its regulation. *E. coli* containing the plasmid with the lux genes did luminesce constitutively, but (considering the difference in gene dosage) almost three orders of magnitude less than wild-type *V. harveyi* at the same cell density. This low-level constitutive activity could not be increased by adding exogenous aldehyde, and appeared to be due to the presence of a low-level promoter on the cloning vector. Apparently, the essential regulatory elements either had not been cloned with the lux genes or, if cloned, were not functional.

To increase the transcriptional efficiency of the cloned *lux* genes in *E. coli* they were recloned into pGMC12, downstream from a lambda promoter which is regulated by a thermosensitive repressor. Thus in the presence of exogenous aldehyde at 30°, *E. coli* carrying the plasmid luminesced at the same low level as the first construction, but when the temperature of the medium was increased to 39° the culture luminesced at nearly the same level as wild-type *V. harveyi*. Polyacrylamide gel electrophoresis of 35S-methionine-labeled proteins revealed three bands which were not prominent in the 30° cultures. Two of the three bands were the same molecular weights as the alpha and beta subunits of luciferase. The third was not identified. These experiments indicated that indeed, expression of bioluminescence in *E. coli* was limited by reduced transcription of the cloned *lux* genes, and that appropriate regulatory sequences (such as those coding for the autoinducer and its putative 'receptor') were either not present or not functional on the *E. coli* plasmid. In addition, the requirement of exogenous aldehyde even with the strong lambda promoters suggested that again, genes coding for the appropriate substrates were either not present or not functional in the *E. coli* clones.

Successful cloning of a functional bioluminescence system was recently accomplished by Engebrecht, et al. (15). We used a different luminous species, *V. fischeri* (MJ-1) which is a light organ symbiont of the fish *Monocentris japonicus* (16). The autoinducer molecule from *V. fischeri* had previously been isolated, characterized, and synthesized, rendering this species particularly desirable for studying genetic regulation of the light-producing system.

Bam H1 restriction fragments from *V. fischeri* DNA were cloned into pACYC184 to generate a recombinant library, and of the 10,000 resulting *E. coli* transformants screened, three were luminescent without the addition of exogenous aldehyde (tetradecanal) or other reagents. A 9-kb Sal 1 fragment internal to the Bam H1 sites was subcloned into pBR322 and also produced transformants which luminesced without addition of aldehyde. The pattern of light emission as a function of cell density closely mimicked the bioluminescent behavior of wild type *V. fischeri*, indicating that many of the genes for normally-regulated expression of the *lux* system were present in the recombinants and functional in *E. coli*.

Employing Tn5 and mini-Mu insertional mutagenesis, the functional regions of the recombinant plasmids were defined (15). If a transposable element inserts in an operon, transcription of downstream genes is impaired and genetic function is blocked. More than 200 such mutations were generated and mapped, identifying both 5' and 3' boundaries of the complete cloned unit as well as the approximate locations of certain functions (see Fig. 2). Two transcriptional units were identified: The downstream or right operon (R) of about 7.5-kb contains a regulatory region coding for autoinducer production (and responding to autoinducer) as well as the regions encoding the structural genes for aldehyde synthesis and luciferase (Fig. 2). The upstream (left) operon L of approximately 1-kb is regulatory, and codes for the putative receptor or cofactor that somehow interacts with the autoinducer to begin transcription of operon R.

Transcription assays were performed with transposon mini-Mu containing the *E. coli lac*Z gene at one end. When the transposon is inserted in the target operon in one orientation, the *lac*Z gene comes under transcriptional control of that operon and β-galactosidase is synthesized at the same rate as other gene products encoded in that operon. If, however, the transposon is inserted in the other orientation, no β-galactosidase is synthesized (17). These assays revealed that operon R is transcribed to the right, and operon L to the left. Furthermore, upon induction, operon R increases its rate of transcription approximately 10-fold, while operon L shows no increase.

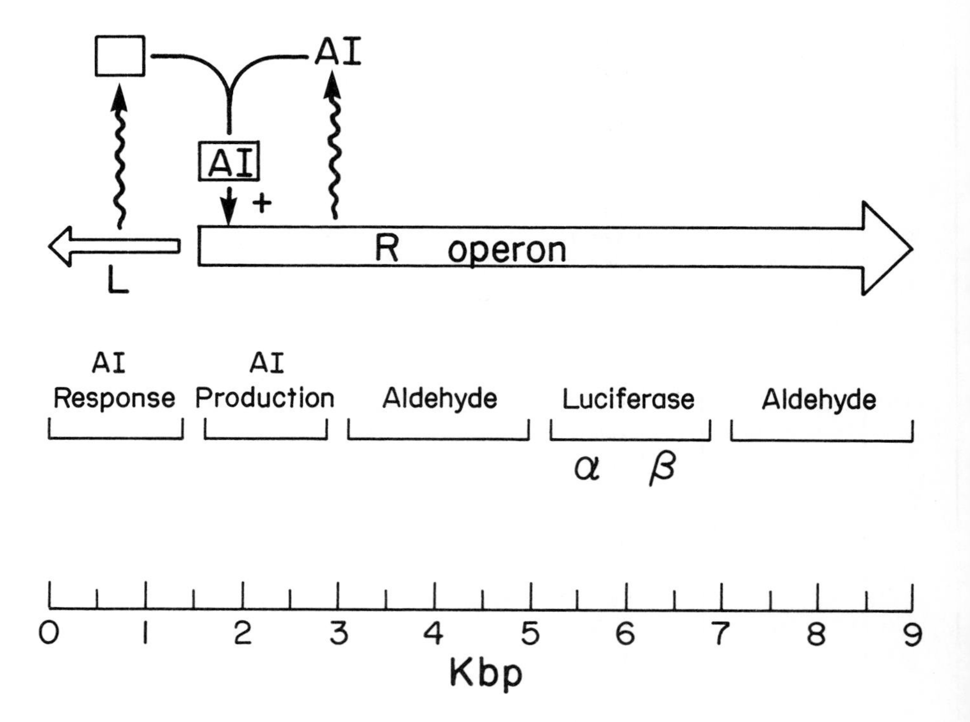

Fig. 2. Organization and hypothesized operation of lux operons from V. fischeri (after Engebrecht et al. 1984). The direction of transcription of each is indicated by the open arrows. The regulatory regions are involved with autoinducer production and recognition; the R operon is induced by the action of AI plus the gene product of the L operon.

Although the structure of the autoinducer is known, and also that the level of bioluminescence is a function of its concentration in the medium, the actual mechanism of action of this small molecule remains obscure. The transposon mutagenesis experiments of Engebrecht and her colleagues revealed that mutations in operon L produced two phenotypes, depending upon the precise location of the transposon insertion in operon L: dim mutants with less than 10% light production was less than 1% of wild-type.

The dark mutants produced no detectable autoinducer, and the dim mutants produced less than 10% of that found in the wild type. Addition of exogenous autoinducer did not induce light production in the dark mutants, but did increase light production in the dim mutants by more than 10 fold. Luciferase (*lux* A & B) activity in these strains was also reduced proportionally to the amount of light production. Clearly, sequences in operon L mediate either the synthesis of autoinducer, response to it, or both. Perhaps more interesting still is the observation that synthesis of autoinducer from operon R is regulated by the presence of autoinducer synthesized from the same operon. Thus, the system is truly "autoregulated", and led Engebrecht and her colleagues to formulate a model of positive control as shown in Fig. 2.

These functions are reasonably well-mapped, but it is possible that the recombinants carry sequences which code for bioluminescent functions not yet clearly characterized. For instance, three aldehyde-related genes have recently been defined in the two aldehyde regions, using complementation of point mutants (18). It is not known whether sequences coding for FMN synthesis are contained in the plasmid, or whether *E. coli* provides the reduced flavin substrate. An oxidoreductase function has not yet been identified on the recombinant plasmid. The possible redundancy of this and other functions in both *E. coli* and on the plasmid has thus far rendered this question refractory to analysis. Now that a generalized transducing bacteriophage has been isolated for *V. fischeri* (19) it will be possible to study the molecular genetics and regulation of the *lux* system in the wild-type bacteria using mutants and the cloned genes.

In addition to the regulatory features that have been revealed using molecular techniques, a variety of other areas are also being exploited. Hybridization probes made by nick translation of cloned *lux* genes have been used to demonstrate that all luminous bacteria contain sequences that hybridize with the *lux* genes from *V. harveyi* and *V. fischeri*. The functional similarities noted between luciferases are thus echoed at the structural (nucleotide) level, and from such studies it seems likely that bacterial luciferase has evolved only once, and not yet diverged enough to mask the similarities at the molecular level. These data suggest that hybridization probes might be effectively used to identify virtually any bacterial *lux* system for molecular manipulations or cloning purposes.

In other hybridization experiments, Potrikus et al. (1984) reported cross reaction of several non-luminous bacteria with hybridization probes from the *lux* genes of *V. harveyi* (21). The *lux* gene may thus be more widely distributed then was previously appreciated.

Lux probes have also been used by Cohn (22) and Haygood (23) in studies of symbiosis in *Anomalopid* fishes which have large light organs containing non-cultural bacteriods. When the DNA from these bacteriods was hybridized to probes made against the 1.85 kb *V. harveyi lux* genes, a positive reaction was seen, indicating the presence of the *lux* genes in the bacteroids (22). Furthermore, some non-luminous bacteria that were morphologically similar to the light organ bacteroids were cultured; DNA from these strains did not hybridize with the *lux*, suggesting that they were non-luminous contaminants, and not light organ symbionts (23).

One of the fundamental questions in bioluminescence concerns the origin and evolution of the capacity to emit light. Hastings has recently proposed that bioluminescence may have arisen 30 times or more in the course of evolution (24). A valuable index of molecular relatedness and evolution is found in comparative amino acid sequence data. Such data are now possible to obtain routinely for the luminous bacteria; after the 1.85 kb piece of DNA was cloned from *V. harveyi*, the nucleotide sequence of the genes coding for the entire alpha subunit and the N-terminus of the beta subunit were obtained, yielding the corresponding amino acid sequences (22). The sequence homologies support the previous proposal (1) that the luciferase subunits arose by a tandem gene duplication.

MOLECULAR GENETICS OF EUKARYOTIC BIOLUMINESCENCE: REGULATION OF DINOFLAGELLATE LIGHT-PRODUCING GENES

The major "black box" in modern biology is regulation of gene expression during the development of higher organisms, and although the basic pattern of gene regulation appears reasonably well-understood in bacteria, it is now generally accepted that these mechanisms bear only superficial resemblance to parallel processes in eukaryotic organisms. Our combined interest in bioluminescence and eukaryotic molecular genetics has prompted us to begin a long-term research effort which addresses the questions of eukaryotic bioluminescence on the molecular level, and hopefully contributes more generally to the understanding of gene regulation in eukaryotes.

The effort currently underway in our laboratory is focused on elaborating the mechanisms which regulate the expression of genes coding for bioluminescence in dinoflagellates. This group was selected because of the extensive published literature on the biochemistry of their light-producing systems (25), and because of their ability to grow in unialgal culture (26).

In *Gonyaulax*, a substituted open-chain tetrapyrole (the 'luciferin' substrate) -possibly derived from chlorophyll - is oxidized by molecular oxygen catalyzed by a specific 'luciferase' (27). Luminescence is expressed both as a low-intensity glow (without stimulation) and as luminous flashes of about 100 ms. A luciferin binding protein (LBP) is involved in the control of the flashes. Both the flash and glow are controlled over the diurnal cycle, with about 10 fold more luciferase and LBP being present in the night phase cells (29,30).

Our major effort is centered on isolation and cloning of the luciferase and LBP genes, and requires construction of *Gonyaulax* cDNA libraries from day-and night-phase cells (31). The cDNA libraries are then screened to identify recombinant clones carrying the luciferase and the LBP coding sequences. There are numerous methods of screening such libraries, but we are beginning with one which appears to be particularly appropriate for identifying sequences which are expressed differentially at night or during the day. cDNA made from night-phase mRNA is hybridized with an excess of day-phase mRNA. The only sequences not hybridizing are those that are preferentially expressed at night or during the day. The reaction mixture is then passed over a hydroxyapatite column, wherein the double-stranded cDNA-mRNA hybrids are bound. The single-stranded molecules are then eluted, including highly-enriched luciferase and LBP cDNA. Differential screening of the cDNA libraries should permit specific identification of recombinant clones carrying the genes of interest. The isolation and cloning of these sequences paves the way for the molecular genetic analysis of the two genes and their regulatory mechanisms. This work will include a structural analysis of luciferase and LBP genomic DNA by restriction mapping and nucleotide sequencing. Comparison of cDNA and genomic sequences will reveal the location and structure of any possible intervening sequences, and pave the way for the study of mRNA processing in these organisms.

If our method of hybridizing cDNA made from night-phase cells against an excess of day-phase mRNA is successful, the gene isolation experiments themselves should verify whether or not the day-night difference in luciferase and LBP protein levels is due to different mRNA levels. This same method can then be used to identify mRNA expressed only during certain parts of the cell cycle. Using 3H-thymidine incorporation experiments, we will extend the cell-cycle experiments further by determining when in the cycle the luciferase and LBP genes are replicated, and if their expression is cell-cycle regulated. Essential to these regulatory studies will be nuclear run-off experiments to assay rates of luciferase and LBP transcription in day-and night-phase cells and at different points in the cell-cycle, and assays of the relative stabilities of luciferase and LBP mRNA's at different points in the cell and diurnal cycles.

We are particularly interested in examining diurnal or cell-cycle specific "orchestration" of arrays of genes, and hope to elaborate a putative regulatory interaction between the dinoflagellate bioluminescent and photosynthetic systems. The "subtractive probe" method which we are using to isolate luciferase and LBP cDNA's should also be very useful in identifying such coordinately-expressed gene arrays. We are hopeful that this aspect of the work will generate additional research questions leading ultimately to a more detailed understanding of the cellular and molecular genetic mechanisms regulating both light production and photosynthesis in these marine dinoflagellates.

CONCLUSION

The examination of bioluminescence on the molecular genetic level, whether in bacteria or eukaryotes, may contribute not only to the elaboration of molecular mechanisms of light emission by living organisms and to an understanding of the molecular genetics underlying circadian rhythms generally, but also of basic genetic processes themselves. The work described herein represents only a beginning of what we see as an entirely new phase of bioluminescence research. With more than 70 years of accumulated work since E.N. Harvey's earliest efforts in 1913 (32) as a base, and with the tools of genetic engineering readily available, we expect to see increasing applications of molecular genetics technology to problems of bioluminescence which are otherwise intractable with traditional methods of cell physiology and biochemistry.

REFERENCES

1. Ziegler, M.M., and Baldwin, T.O. (1981). Biochemistry of bacterial bioluminescence, *Curr.Top.Bioenerg.* 12, 65-113.
2. Hastings, J.W., Riley, W.H. and Massa, J. (1965). Purification, properties, and chemiluminescent quantum yield of bacterial luciferase, *J.Biol.Chem.* 240, 1473-1481.
3. Riendeau, D., Rodriguez, A. and Meighen, E. (1982). Resolution of the fatty acid reductase from *Photobacterium phosphoreum* into acyl-protein synthetase and acyl-CoA reductase activities, *J.Biol.Chem.*, 257, 6908-6914.
4. Rodriguez, A., Riendeau, D., and Meighen, E. (1983). Purification of the acyl-CoA reductase component from a complex responsible for the reduction of fatty acids in bioluminescent bacteria, *J.Biol.Chem.* 258, 5233-5237.
5. Jablonski, E., and DeLuca, M. (1978). Studies of the control of luminescence in *Beneckea harveyi*. Properties of the NADH and NADH-FMN oxidoreductases, *Biochemistry* 17, 672-678.
6. Karl, D.M., and Nealson, K.H. (1980). Regulation of cellular metabolism during synthesis and expression of the luminous system in *Beneckea* and in *Photobacterium*, *J.Gen.Micro.* 117, 357-368.
7. Nealson, K.H., Platt, T., and Hastings, J.W. (1970). The cellular control of the synthesis and activity of the bacterial bioluminescent system, *J.Bacteriol.* 104, 313-322.
8. Nealson, K.H. (1977). Autoinduction of bacterial luciferase: Occurrence, mechanism, and significance, *Arch.Microbiol.* 112, 73-79.
9. Eberhard, A., Eberhard, C., Burlingame, A.L., Kenyon, G.L., Oppenheimer, N.J., and Nealson, K.H. (1981). Purification, identification and synthesis of *Photobacterium fischeri* autoinducer. *In* "A Symposium on bioluminescence and chemiluminescence: Basic Chemistry and Analytical Applications." (Eds. M. DeLuca and W. McElroy). pp. 113-120. Academic Press, New York.
10. Eberhard, N., Burlingame, A.L., Eberhard, C., Kenyon, G.L., Nealson, K.H., and Oppenheimer, N.J. (1981a). Structural identification of autoinducer of *Photobacterium fischeri* luciferase, *Biochemistry* 120, 2444-2449.

11. Rosson, R. and Nealson, K.H., (1981). Autoinduction of bacterial bioluminescence in a carbon-limited chemostat. *Arch.Microbiol.* 129, 299-304.
12. Greenberg, E.P., Hastings, J.W., and Ulitzur S. (1979). Induction of luciferase synthesis in *Beneckea harveyi* by other marine bacteria, *Arch.Microbiol.* 120, 87-91.
13. Cohn, D.H., Ogden, R.C., Abelson, J.N., Baldwin, T.O., Nealson, K.H., Simon, M.I., and Mileham, A.J. (1983). Cloning of the *Vibrio harveyi* luciferase gene: Use of a synthetic oligonucleotide probe, *Proc.Nat.Acad.Sci. USA* 80, 120-123.
14. Belas, R., Mileham, A., Cohn, D., Hilmen, M., Simon, M., and Silverman, M. (1982). Bacterial bioluminescence: Isolation and expression of the luciferase genes from *Vibrio harveyi*, *Science* 218, 791-793.
15. Engebrecht, J., Nealson, K.H., and Silverman, M. (1983). Bacterial bioluminescence: Isolation and genetic analysis of functions from *Vibrio fischeri*, *Cell* 32, 773-781.
16. Tebo, B.M., Linthicum, D.S., and Nealson, K.H. (1979). Luminous bacteria and light-emitting fish: The ultrastructure of the symbiosis, *Biosystems* 11, 269-280.
17. Casadaban, M.J., and Cohen, S.N. (1979). Lactose genes fused to exogenous promoters in one step using Mu-lac bacteriophage: In vivo probe for transcriptional control sequences, *Proc.Nat.Acad. Sci. USA* 76, 4530-4533.
18. Engebrecht, J. (1984). Personal communication.
19. Levisohn, R. and Nealson, K.H. (1984). Isolation and characterization of a generalized transducing phage of *V. fischeri* (MJ-1), (MS submitted).
20. Lucas, S. and Nealson, K.H. (unpublished data).
21. Potrikus, C.J., Greenberg, E.P., Hamlett, N.V., Gupta, S., and Hastings, J.W. (1984). Hybridization of *V. harveyi* luciferase genes to non-luminous marine bacteria, *Abstracts of Amer.Soc.Microbiol.* 84A-1.
22. Cohn, D.H. (1983). Isolation, organization and expression of the luciferase genes from *Vibrio harveyi* Ph.D thesis, Univ.Cal. San Diego.
23. Haygood, M.G. (1984). Iron regulation of luminescence: Implications for the ecology and symbiotic associations of the luminous bacteria. Ph.D thesis, Univ. Cal. San Diego.
24. Hastings, W.J. (1983). Biological diversity, chemical mechanisms, and the evolutionary origins of bioluminescent systems, *J.Mol.Evol.* 19, 309-321.
25. Hastings, J.W. (1978). Bacterial and dinoflagellate

luminescent systems. *In* "Bioluminescence in Action" (Ed. Peter J. Herring). pp. 129-170. Academic Press. New York.
26. Haxo, F.T. and Sweeney, B.M. (1955). Bioluminescence in *Gonyaulaz polyedra*. *In* "The luminescence of biological systems" (Ed. F.H. Johnson). *Amer.Assoc. Adv.Sci.*, Washington, D.C.
27. Dunlap, J.C., and Hastings, J.W. (1981). The biological clock in *Gonyaulax* controls luciferase activity by regulating turnover, *J.Biol.Chem.* 256, 10509-10514.
28. Krasnow, R., Dunlap, J.C., Taylor, W., Hastings, J.W., Vetterling, W., and Gooch, V. (1980). Circadian spontaneous bioluminescent glow and flashing of *Gonyaulax polyedra*, *J.Comp.Physiol.* 138, 19-26.
29. Dunlap, J.C., Taylor, W., and Hastings, J.W. (1981). The control and expression of bioluminescence in dinoflagellates. *In* "Bioluminescence: Current Perspectives" (Ed K.H. Nealson). Burgess Publishing Company, Minneapolis.
30. Sulzman, F.M., Krieger, N.R., Gooch, V.D., and Hastings, J.W. (1978). A cricadian rhythmn of the luciferin binding protein from *Gonyaulax polyedra*, *J.Comp. Physiol.* 128, 251-257.
31. Maniatis, T., Fritsch E.F., and Sambrook, J. (1982). Molecular cloning: A laboratory manual. Cold Spring Harbor Laboratory.
32. Harvey, E.M. (1913). The temperature limits of phosphorescence of luminous bacteria, *Biochem.Bull.* 2, 456-457.

CLONING OF THE BACTERIAL LUCIFERASE AND USE OF THE CLONE TO STUDY THE ENZYME AND REACTION IN VIVO

Thomas O. Baldwin, Tineke Berends and
Mary L. Treat

*Department of Biochemistry and Biophysics and
The Texas Agricultural Experiment Station,
Texas A&M University,
College Station, Texas 77843*

INTRODUCTION

The genes encoding the two subunits of luciferase of *Vibrio harveyi* are contained on a single 4.0 kb Hind III fragment of DNA. We have cloned this fragment into pBR322 and demonstrated expression of the luciferase genes in *E. coli* (1-3). By studying the bioluminescence of the parent clone, pTB7, and a series of subclones, we have gained some understanding of the properties of the enzyme and the reaction *in vivo*. Specifically,

1.) Proper folding of the α subunit into a stable conformation in the cell requires the presence of the β subunit.

2.) There is at least one enzyme in *E. coli* capable of supplying luciferase with reduced flavin.

3.) The bioluminescence reaction in *V. harveyi* does not require the participation of an energy transfer system, as has been suggested.

4.) The luciferase clone and several of the subclones are readily applicable to studies of regulatory genetic elements in *E. coli* and numerous other organisms.

METHODS

Vibrio harveyi, strain 392 was grown on NaCl complete medium as described. *E. coli*, strain RR1, was used for the initial cloning experiments; subcloning experiments used TB1, an r^-,m^+ strain of *E. coli* developed in this laboratory. Genomic DNA was purified from late log phase

ANALYTICAL APPLICATIONS OF
BIOLUMINESCENCE AND CHEMILUMINESCENCE

ISBN 0-12-426290-2

V. harveyi cells by standard techniques, digested with Hind III, and the resulting fragments ligated into pBR322. *E. coli*, strain RR1, was transformed with the resulting recombinant plasmids to ampicilin resistance. Cells were spread onto plates, incubated overnight at 37°, transferred to a photographic darkroom and observed for luminescence; none was apparent. A small amount of n-decyl aldehyde was then spread under the lid of each plate and the plates observed again for luminescence. From a total of about 6,000 colonies carrying recombinant plasmids, 12 were found to express luminescence when supplied with exogenous aldehyde. One of these plasmids, pTB7, was studied in detail. The restriction map of the plasmid is given in Fig. 1.

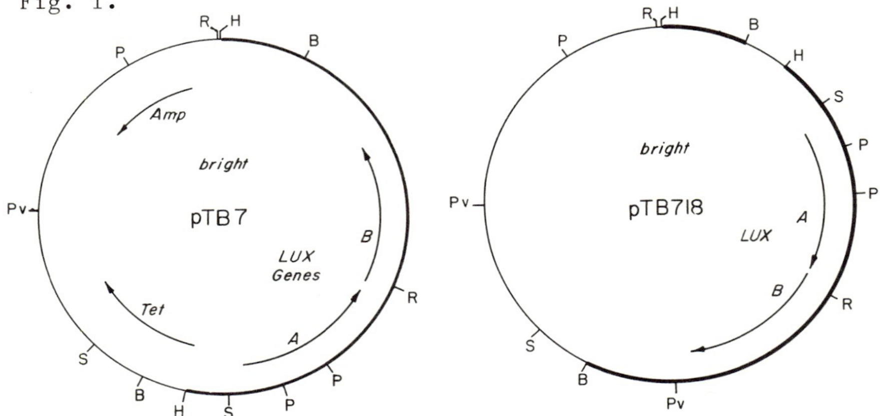

Fig. 1 Plasmids pTB7 and the derivative pTB718.

RESULTS AND DISCUSSION

It was obvious from the requirement for aldehyde for bioluminescence that pTB7 did not express the structural genes for the aldehyde-biosynthetic apparatus; however, it was apparent that the luciferase in *E. coli* was being supplied with reduced flavin. To determine whether the gene encoding a flavin reductase activity was contained between the end of the lux B gene and the Hind III site, we digested pTB7 with BamHI and religated to obtain the reverse alignment pTB718 (Fig. 1). The two plasmids, pTB7 and pTB718, were indistinguishable in terms of luminescence intensity following aldehyde stimulation. These observations suggested that the flavin reductase activity most likely was not encoded on the plasmid, but was more likely a host activity. To further test the possibility that the activity was encoded on the plasmid, we constructed an additional subclone, pMT1 by digesting pTB718 with PvuII and religating to delete the fragment from close to the 3' end of lux B to the PvuII site on pBR322 (Fig. 2). The plasmid pMT1 expressed luciferase at the same level as pTB7 and pTB718, indicating that the DNA that had been deleted contained no information regarding expression of bioluminescence in *E. coli*.

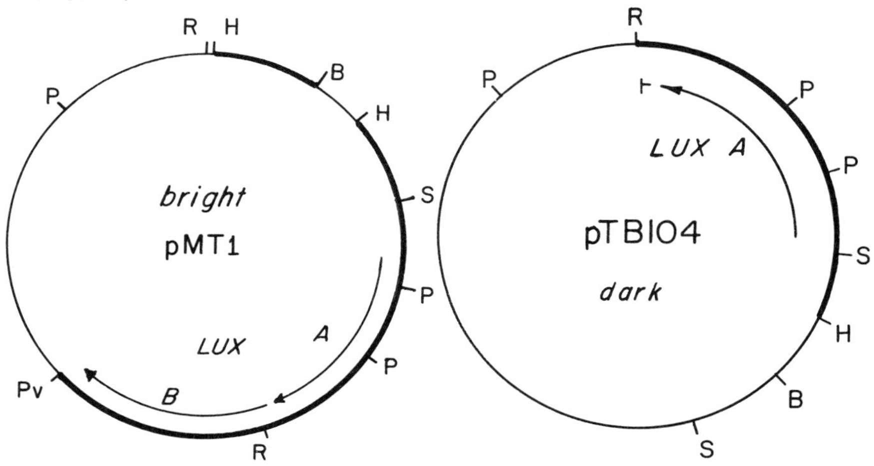

Fig. 2 The plasmids pMT1 and pTB104.

Another question of interest that we have investigated pertains to the ability of the individual luciferase subunits to fold properly into a stable conformation in the living cell. Plasmid pTB7 was digested with EcoRI and religated to delete the fragment containing all but the 5' 38 bases of the β subunit. The resulting plasmid, pTB104 (Fig. 2), contained the same 5' regulatory region as pTB7, but no α subunit could be detected in lysates of cells carrying pTB70 under conditions that luciferase was readily detected in lysates of pTB7 carrying cells. These observations suggest that the α subunit is not stable in cells without folding together with the β subunit.

Two deletion subclones were constructed to determine the location of the promoter sequence that regulated the transcription of the lux genes. The subclone pTB718 suggested that the promoter sequence was actually within the coding region of the tetracycline resistance gene of pBR322 but oriented in the opposite direction. To be certain that the promoter was within the HindIII to BamHI fragment of pBR322, we digested pTB718 with HindIII and religated to delete the HindIII fragment and also digested pTB7 with SalI and religated to obtain a second deletion of the suspected region.

The resulting plasmids, pTB70 and pAB12, showed indistinguishable levels of luciferase synthesis that were 100-fold lower than for pTB7, pTB718, or pMT1.

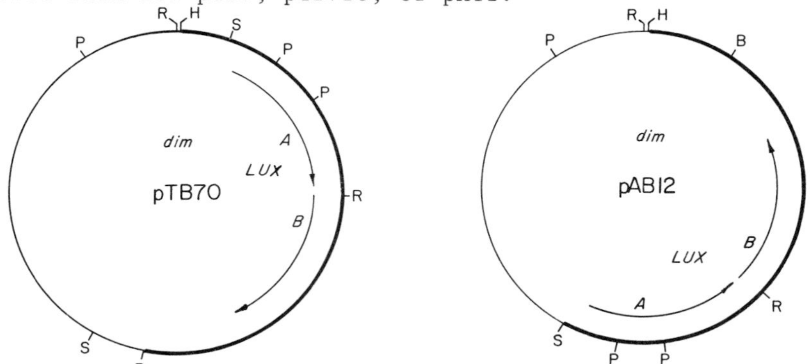

Fig. 3 Plasmids pTB70 and pAB12.

Bioluminescence emission spectra were measured using a fluorimeter from SLM Instruments. The emission spectra of cells carrying pTB7 and of wild-type *V. harveyi* were indistinguishable. The bioluminescence emission spectra of the reactions *in vitro* were also measured and found to be identical. If there were an energy transfer system in *V. harveyi* (see ref. 4 for discussion), one should find some spectral shift if the luciferase were required to function alone. None was found. (Supported by NSF & NIH).

REFERENCES

1. Baldwin, T.O. (1982) Cloning of the luciferase structural genes and the expression of bioluminescence in *E. coli*. Presented at the 38th Southwest and 6th Rocky Mountain Combined Regional Meeting, El Paso, Texas.
2. Baldwin, T.O., Berends, T., Bunch, T.A., Holzman, T.F., Rausch, S.K., Shamansky, L., Treat, M.L. and Ziegler, M.M. (1984) The cloning of the luciferase structural genes from *Vibrio harveyi* and the expression of bioluminescence in *Escherichia coli*. *Biochemistry*, (in press).
3. Cohn, D., Ogden, R.C., Abelson, J., Baldwin, T.O., Nealson, K.H., Simon, M.I. and Mileham, A.J. (1983) Cloning of the luciferase genes from *Vibrio harveyi*. *Proc. Natl. Acad. Sci. USA* 80, 120-123.
4. Ziegler, M.M. and Baldwin, T.O. (1981) Biochemistry of bacterial bioluminescence. *In* "Current Topics in Bioenergetics" (Ed. D. Rao Sanadi). Vol. 12, pp. 65-113. Academic Press, New York.

ALDEHYDE BIOSYNTHESIS IN LUMINESCENT BACTERIA

A. Rodriguez, L. Wall, L.M. Carey, M. Boylan,
D. Byers and E. Meighen

Department of Biochemistry, McGill University
Montreal, Canada H3G 1Y6

INTRODUCTION

The biosynthesis of long chain aliphatic aldehydes for light emission in the luminescent bacterium, *Photobacterium phosphoreum*, is catalyzed by a fatty acid reductase enzyme complex in a reaction consisting of (a) ATP dependent activation of fatty acid and (b) NADPH dependent reduction of an acyl enzyme intermediate. The enzyme complex contains three polypeptides, 34K, 50K, and 58K, with only the latter two polypeptides being essential for fatty acid reduction (1,2).

RESULTS

The 50K polypeptide is responsible for fatty acid activation resulting in the formation of the acyl-enzyme intermediate (acyl-50K) with the concomitant hydrolysis of ATP to AMP. The level of acylation of the 50K protein with [^{3}H]-fatty acid (+ATP) is increased remarkably by the presence of either the 34K or 58K polypeptides (Fig. 1A). The acyl-50K has been isolated and shown to be reduced to aldehyde on addition of NADPH and the 58K protein (2).

The 58K polypeptide catalyzes the NADPH dependent reduction of the acylated 50K polypeptide and other activated acyl derivatives such as acyl-CoA. In the absence of NADPH, this protein catalyzes acyl transfer from acyl-50K or acyl-CoA to thiol acceptors (Fig. 2). Protein acylation with [^{3}H] acyl-CoA results in the labelling of the 58K and 34K proteins, but not the 50K protein (Fig. 1B).

The purified 34K protein has also been shown to catalyze acyl transfer to thiol groups and to oxygen acceptors such as water and glycerol (Fig. 2). In the presence of the 50K

ANALYTICAL APPLICATIONS OF
BIOLUMINESCENCE AND CHEMILUMINESCENCE

ISBN 0-12-426290-2

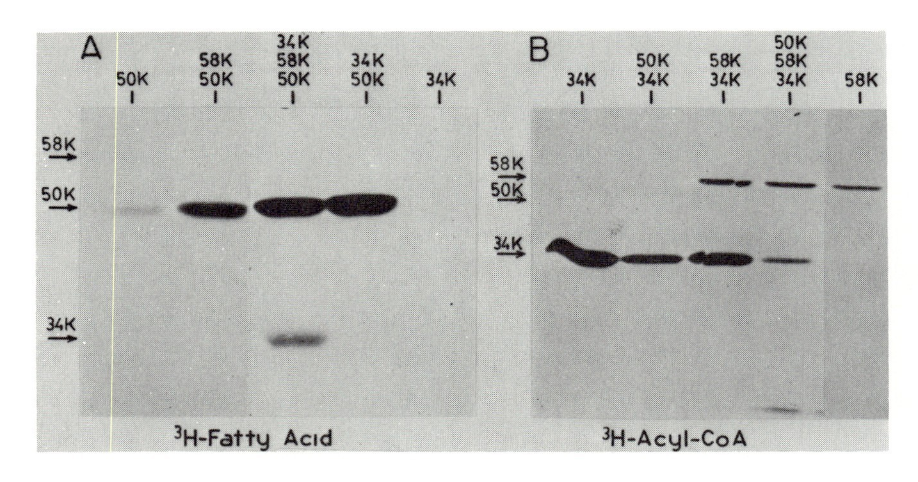

Fig. 1 SDS-PAGE and fluorography of acylated polypeptides.

polypeptide, the specificity is changed so that acyl-CoA hydrolase activity is primarily observed. Hydrolysis of both sulfur and oxygen esters has been observed suggesting that the 34K protein can be involved in supplying fatty acids for the luminescent system directly from the fatty acid synthetase system and/or membrane lipids. Incubation of the 34K protein with acyl-CoA results in its acylation (Fig. 1B) indicating that the enzyme bound fatty acid is an intermediate in the acyl-CoA cleavage reaction.

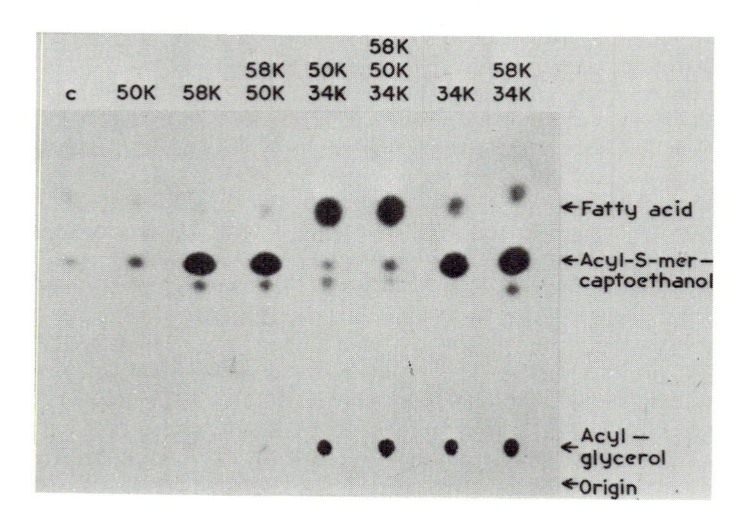

Fig. 2 TLC of ^{3}H-acyl-CoA cleavage products (c, control).

The different enzyme activities associated with the three polypeptide components of the fatty acid reductase complex, as well as their acylation properties, are summarized in the following model. Although acyl-transfer to thiols catalyzed

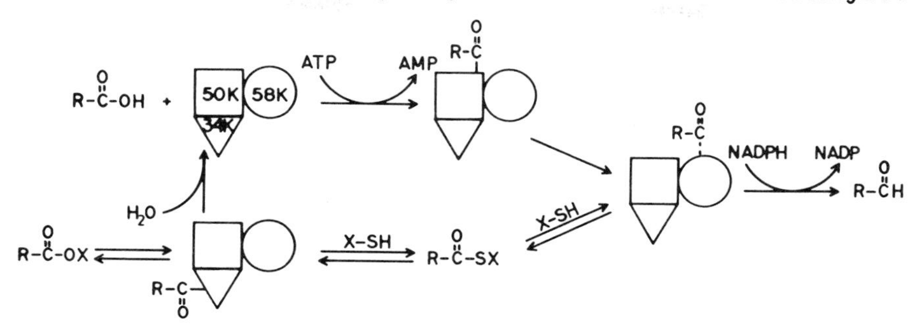

by the 34K and 58K proteins is unlikely to occur *in vivo*, the esterase activity of 34K in the complex could provide an important regulatory control in supplying fatty acids to the luminescent system only when needed.

Protein acylation with either ^{3}H-fatty acid and ATP or ^{3}H-acyl-CoA can be used to identify the peptides involved in fatty aldehyde biosynthesis in different luminescent strains and clones (Fig. 3). The pattern on SDS gels of acylated

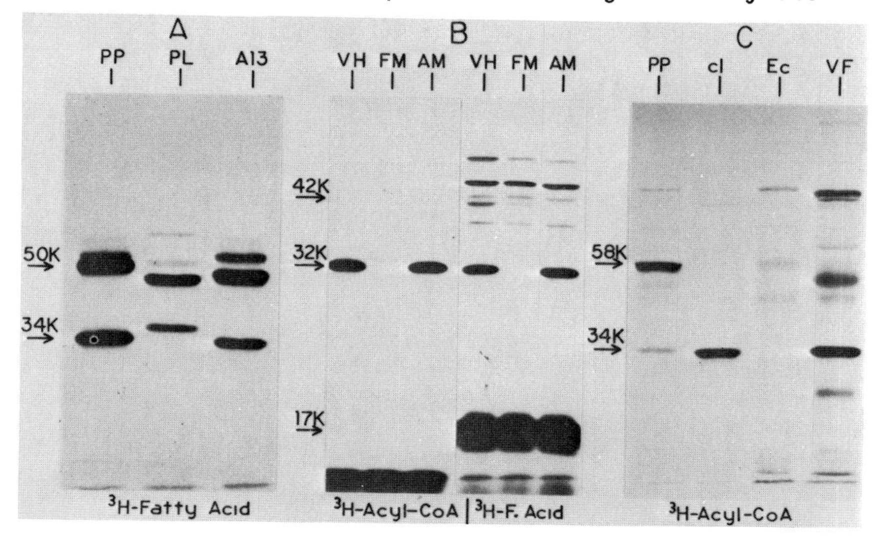

Fig. 3 Acylation of protein extracts from P. phosphoreum (PP and A13), P. leiognathi (PL), V. harveyi (VH), fatty acid (FM) and aldehyde (AM) mutants, E. coli (Ec), V. fischeri (VF) and Ec clone with VF DNA (cl).

polypeptides in crude extracts serves as a fingerprint for a particular bacterial strain. A close similarity exists between the acylation patterns with ^{3}H-fatty acid and ATP for three *Photobacterium* strains (Fig. 3A).

Soluble extracts from the wild type and two dark mutants of *Vibrio harveyi* show different fatty acid (+ATP) and acyl-CoA patterns for the acylated polypeptides (Fig. 3B). A 32kDa polypeptide is not acylated in the fatty acid stimulatable dark mutant. A different dark mutant, stimulated only by aldehyde, lacks the acylated 42kDa-band present in the wild type after incubation with ^{3}H-fatty acid and ATP. These results along with recent induction studies suggest that the 32kDa polypeptide supplies free fatty acid whereas the 42kDa protein may be responsible for activation of fatty acids for their subsequent reduction to aldehyde.

An *Escherichia coli* clone with a plasmid containing DNA encoding the luminescent system of *V. fischeri* contains a 34kDa polypeptide specifically labelled with ^{3}H-acyl-CoA (Fig. 3C). The acylated polypeptide is present in *V. fischeri* but absent in *E. coli*.

CONCLUSIONS AND RECOMMENDATIONS

Three polypeptides are involved in aldehyde biosynthesis in *P. phosphoreum*; two are required for fatty acid reduction and the third polypeptide is necessary for supplying fatty acids. Although only acyl-CoA reductase activity has been detected in *V. harveyi* (3) whereas fatty acid reductase has been measured in *Photobacterium* strains, differential acylation of polypeptides has implicated the presence of similar systems in other luminescent bacteria.

REFERENCES AND ACKNOWLEDGEMENTS

Supported by the Medical Research Council of Canada.

1. D. Riendeau, A. Rodriguez and E. Meighen (1982). Resolution of the fatty acid reductase from *Photobacterium phosphoreum* into acyl protein synthetase and acyl-CoA reductase activities, *J. Biol. Chem.* 257, 6908-6915.
2. A. Rodriguez, L. Wall, D. Riendeau and E. Meighen (1983). Fatty acid acylation of proteins in bioluminescent bacteria, *Biochemistry* 22, 5604-5611.
3. D. Byers and E. Meighen (1984). *Vibrio harveyi* aldehyde dehydrogenase: partial reversal of aldehyde oxidation and its possible role in the reduction of fatty acids for the bioluminescence reaction, *J. Biol. Chem.* (in press).

III IMMOBILISED ENZYMES

BIOLUMINESCENT ASSAYS USING CO-IMMOBILIZED ENZYMES

M. DeLuca

Department of Chemistry
University of California, San Diego
La Jolla, California 92093, USA

For some years there has been a continuing interest in immobilized enzymes. These preparations have been useful for producing various chemicals, they have been used to assay compounds that are of research or clinical interest, and they can be considered as models for membrane bound enzymes in the cell.

Why should immobilized enzymes be preferred over the soluble forms? In many cases the immobilized enzymes are more stable than the soluble enzymes. They can be packed into columns through which a sample is continuously pumped and therefore these enzymes can be re-used for multiple assays. Also such flow systems can be conveniently automated. Finally, when one is using co-immobilized multi-enzyme systems, these co-immobilized enzymes have been shown to be much more efficient in catalyzing the coupled reactions than the comparable soluble enzymes (1-4). This increase in efficiency has been attributed to a buildup of intermediates in the micro-environment as well as locally high concentrations of the enzymes.

There are many different ways of immobilizing enzymes and many different supports have been used. These procedures have been reviewed extensively (5). It may be necessary to use different methods for immobilizing different enzymes and this is mostly a process of trial and error. The experiments reported here have all been done with enzymes immobilized onto cyanogen bromide activated Sepharose 4-B. This procedure was first described by Axen et al. (6).

The two bioluminescent systems which we have used as detection systems are the firefly luciferase and the bacterial oxidoreductase/luciferase. Firefly luciferase catalyzes the ATP dependent oxidative decarboxylation of luciferin with the

ISBN 0-12-426290-2

emission of light, equation (I).

$$\text{(I)} \quad LH_2 + ATP + O_2 \xrightarrow{M^{++}} L{=}O + CO_2 + AMP + PPi + h\nu$$

The usefulness of this reaction is due to the specificity of the enzyme for ATP. None of the other nucleotide triphosphates are active as substrates and therefore ATP can be measured in biological samples without interference from other naturally occurring compounds.

The bacterial luminescent system consists of two enzymes which catalyze the following reactions:

$$\text{(II)} \quad NAD(P)H + FMN + H^+ \longrightarrow NAD(P)^+ + FMNH_2$$

$$\text{(III)} \quad FMNH_2 + RCHO + O_2 \longrightarrow FMN + RCOOH + H_2O + h\nu$$

The first reaction is catalyzed by an NAD(P)H:FMN oxidoreductase and the $FMNH_2$ produced is used by luciferase, reaction III, in the presence of a long chain aldehyde and oxygen to produce an acid, FMN and light. Luciferase has been purified from a variety of salt water bacteria (7). The enzymes we use are obtained from *Vibrio harveyi* and the oxidoreductases are purified as described by Jablonski and DeLuca (8). There are two oxidoreductases in this bacteria, one specific for NADH and another which utilizes NADPH. Using these two enzymes it is possible to measure either NAD^+ or $NADP^+$ linked substrates. If such specificity is not required, a commercially available diaphorase can be used (9).

With these two systems, firefly and bacterial, it is possible to measure any compound or enzyme which produces ATP or NAD(P)H. There are over 250 enzymes known that produce NADH or NADPH. So the total number of enzymes and substrates that can be assayed is quite large. The rationale for developing the various assays has been to co-immobilize a specific enzyme or enzymes which will act on the metabolite to be measured, along with either the firefly luciferase or the bacterial oxidoreductase/luciferase. There are three advantages of having the luciferase co-immobilized with the coupling enzymes. These are: (1) It is possible to measure ATP or NAD(P)H in the micro-environment where the concentration is much greater than it would be in the macro-environment. (2) Photon emission is not subject to diffusional restrictions as are chemical entities, and light can be measured rapidly and sensitively. (3) Both the firefly and bacterial luciferase reactions are irreversible and therefore these enzymes will pull any reactions that might be thermodynamically unfavorable.

When considering what matrix to use for immobilization,

it is possible to immobilize either on a surface or in pores of a Sepharose bead. Fig. 1 shows schematically the difference between surface or porous immobilization. By co-immobilizing onto the porous Sepharose 4-B, we estimate that

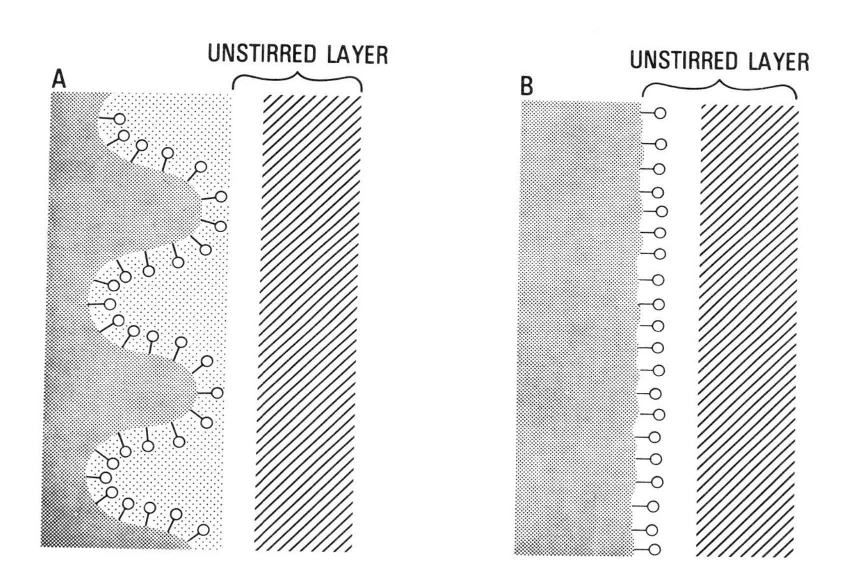

Fig. 1. Schematic representation of an enzyme immobilized in a porous matrix, A; or on the surface of a solid support, B.

most of the enzymes will be in the pores and one can approach local concentrations of enzymes as high as 10^{-5} Molar. These are concentrations which would be completely impractical to achieve with the soluble enzymes. Fig. 2 shows a comparison of a two-enzyme system, oxidoreductase/luciferase co-immobilized relative to the soluble enzymes. At various concentrations of NADH as the limiting substrate the co-immobilized enzymes produce about 100 times more light than the soluble forms. Fig. 3 shows data obtained with a four enzyme system: hexokinase, glucose-6-phosphate dehydrogenase, oxidoreductase, luciferase. Using this preparation it is possible to assay NADPH, Glucose-6-phosphate or glucose. In general, the more enzymes involved in the reactions the less sensitive the assay.

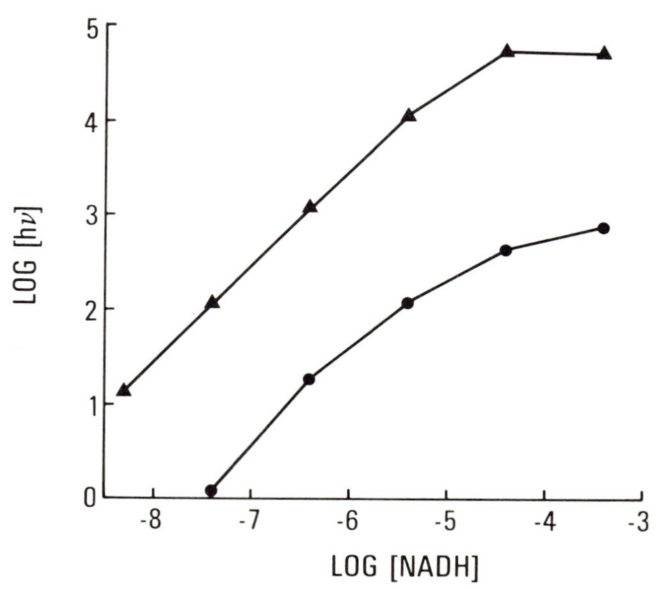

Fig. 2. Light obtained vs. NADH concentration with ●—● soluble oxidoreductase-luciferase; or ▲—▲ the co-immobilized enzymes.

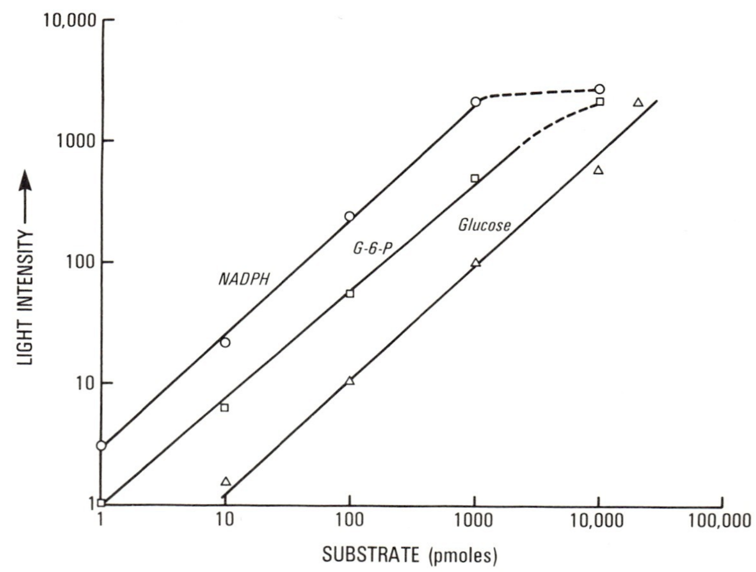

Fig. 3. Light obtained with varying concentrations of substrates using co-immobilized hexokinase, glucose-6-phosphate dehydrogenase, oxidoreductase and bacterial luciferase.

We have immobilized approximately 25 different enzymes in various combinations. Table 1 shows data obtained for a typical group of enzymes.

TABLE 1

Enzyme coupled	% Activity recovered on Sepharose
NADH:FMN oxidoreductase	90
Creatine kinase	84
Firefly luciferase	80
Hexokinase	60
Malate dehydrogenase	45
Lactate dehydrogenase	25
7α hydroxysteroid dehydrogenase	20
Bacterial luciferase	15

Recoveries range from almost 100% to a low of about 10-15%. The bacterial luciferase always exhibits a low activity on immobilization. Two important factors in obtaining active immobilized enzymes are the protein concentration and the pH of the solution during immobilization. The protein concentration is brought to 8 mgs per ml by the addition of BSA to the enzymes before coupling. More protein than this appears to interfere with the binding, while less protein produces inactive enzyme probably due to multiple links between the enzyme and the Sepharose. The best coupling we have obtained is in 0.1 M pyrophosphate buffer, pH 8.0.

Table 2 shows some of the metabolites and enzymes which have been assayed using firefly luciferase co-immobilized with other enzymes.

A summary of some metabolites and enzymes that have been assayed, using the bacterial enzymes, and the range in which the assay is linear is shown in Table 3 (11,12).

The lower limit of detection will depend on how much of the active enzymes have been immobilized as well as the equilibrium constants for the various enzymes. The greatest sensitivity is obtained when the equilibrium of the reactions catalyzed lies in the direction of NADH production. For example, if malate is assayed with malate dehydrogenase which has an equilibrium constant of 6×10^{-13} M, the lowest detectable concentration of malate is 100 pmoles. If the malic enzyme is used, 10 pmoles are measurable and this enzyme has an equilibrium constant of 5×10^{-5} M (11). In general, assays for substrates are in the pmole range while enzymes can be assayed in the lower femtomole range due to their turnover.

TABLE 2

Metabolite	Linear range (pmoles)
PEP	150 - 15,000
ADP	5 - 15,000
PCr	6 - 12,000
Enzyme	**Amount detectable (femtomoles)**
Creatine kinase	1
Pyruvate kinase	8
Adenylate kinase	20

TABLE 3

Metabolite	Linear range (pmoles)
NAD^+, $NADP^+$	0.2 - 2,000
D-Glucose	10 - 1,500
L-Malate	10 - 10,000
3α Hydroxysteroid	0.5 - 1,000
L-Lactate	100 - 100,000
Enzyme	**Linear range (femtomoles)**
LDH	3 - 700
ADH	15 - 3,000
G-6-PDH	1.5 - 100

The kinetics of light emission are also variable with the different combinations of enzymes. With the four enzyme system used to measure glucose, the slope of increasing light is measured as shown in Fig. 4. This initial slope is proportional to the concentration of glucose if it is the limiting substrate in the assay. The initial decrease of light observed is due to a light stimulated luminescence in the enzyme. Fig. 5 shows the time course of light emission

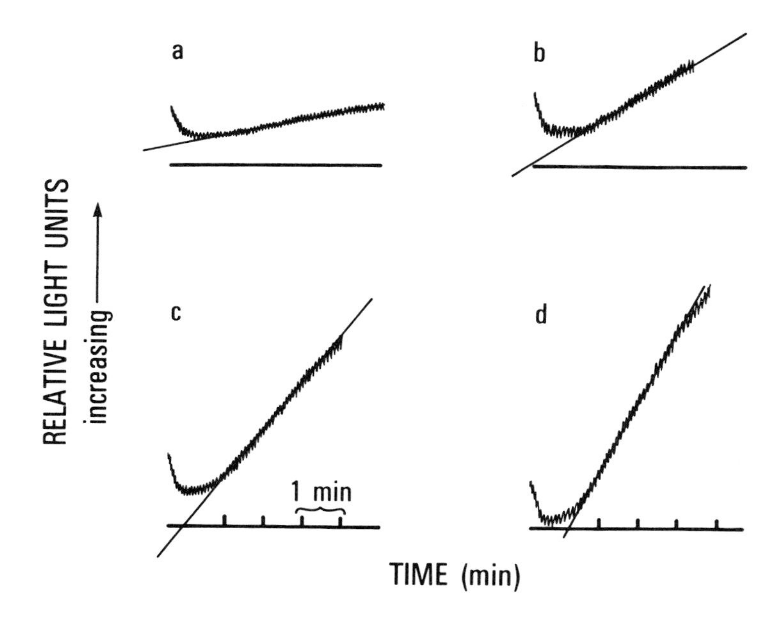

Fig. 4. Time course of light output in the assay of D-glucose: a) background; b) 2, c) 5, and d) 10 pmols of glucose.

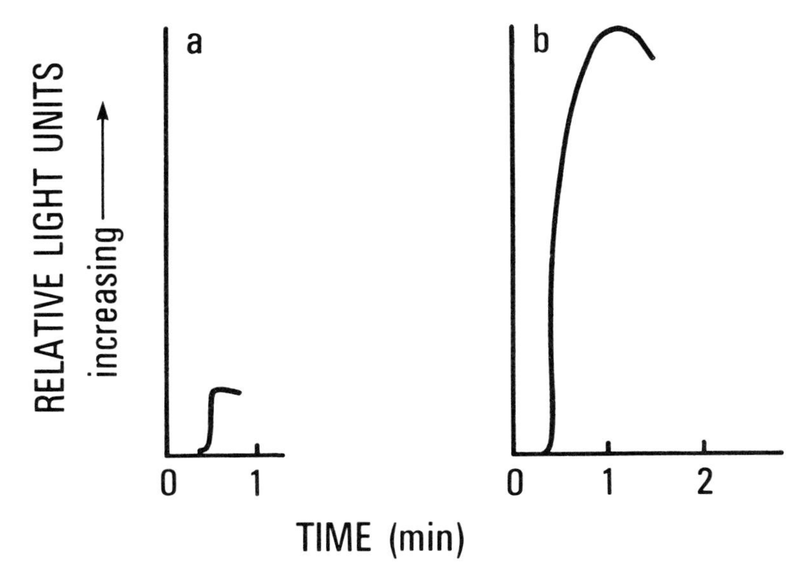

Fig. 5. Time course of light output in the assay of $NADP^+$. (a) background; (b) 2 pmols of $NADP^+$.

obtained when measuring $NADP^+$. This is a three-enzyme system consisting of 6-phosphogluconic dehydrogenase, NADPH:FMN oxidoreductase and luciferase. In this case, a peak of light is obtained in less than one minute and this peak is proportional to the $NADP^+$ concentration in the presence of excess 6-phosphogluconic acid (11).

Recently, Drs. Roda and Schoelmerich have developed a series of assays which are specific for various bile acids. The general reactions catalyzed are:

(IV) OH-bile acid + $NAD^+ \longrightarrow$ O=bile acid + NADH + H^+

(V) H^+ + NADH + FMN $\longrightarrow NAD^+ + FMNH_2$

(VI) $FMNH_2 + RCHO + O_2 \longrightarrow FMN + RCOOH + H_2O + h\nu$

The first reaction is catalyzed by a specific hydroxysteroid dehydrogenase and produces one NADH for every hydroxysteroid oxidized. The NADH is used by the oxidoreductase to form $FMNH_2$ and this is oxidized by luciferase with the production of light. Table 4 shows the bile acids for which bioluminescent assays have been developed.

TABLE 4

Bile Acid	Lower Limit of Detection	Ref.
7α hydroxysteroid	5 pmoles	(13)
12α hydroxysteroid	4 pmoles	(9)
3α hydroxysteroid	1 pmole	(12)

The detection limit is defined as the concentration of bile acid which produces a peak light intensity twice that of background. Fig. 6 shows the time course of light emission from a chenodeoxycholyltaurine standard (A) and a serum sample (B) (13), using immobilized 7α hydroxysteroid dehydrogenase and the bacterial system. Fig. 7 shows the correlation obtained using the bioluminescent assay compared with gas-liquid chromatography on serum samples from patients. The serum samples were heated at 70° C for 15 minutes to inactivate any dehydrogenases which might interfere with the assay. Fig. 8 shows results obtained with the bioluminescence and radioimmunoassay. The agreement is excellent in both cases. The sensitivity of the bioluminescent assay is comparable to RIA and the method is much more

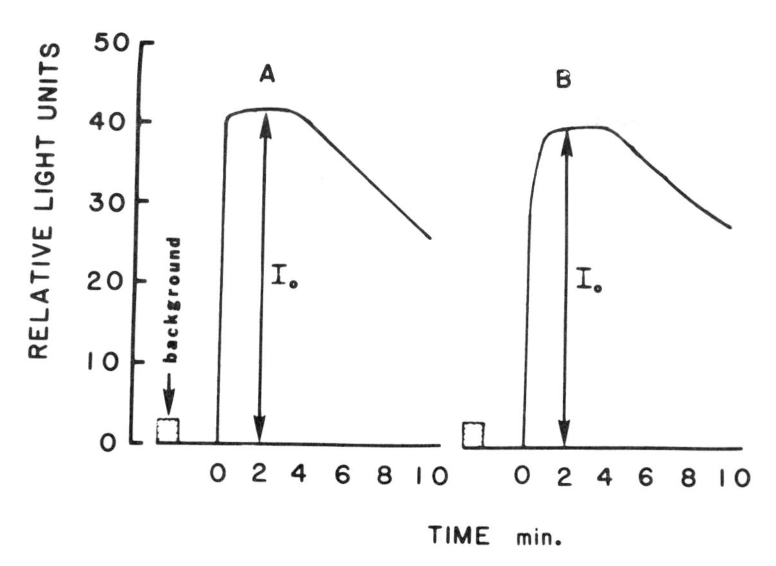

Fig. 6. Time course of light emission for (A) a 5 μM chenodeoxycholyltaurine standard and (B) a serum sample.

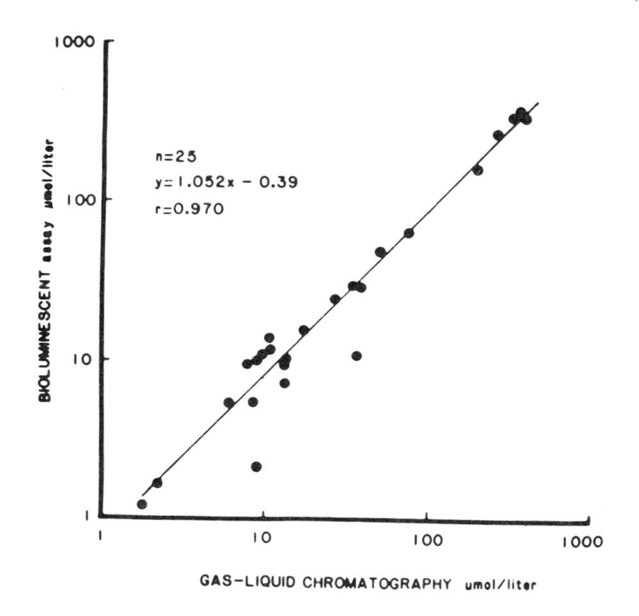

Fig. 7. Comparison of results obtained from bioluminescent assay with gas-liquid chromatography on serum samples

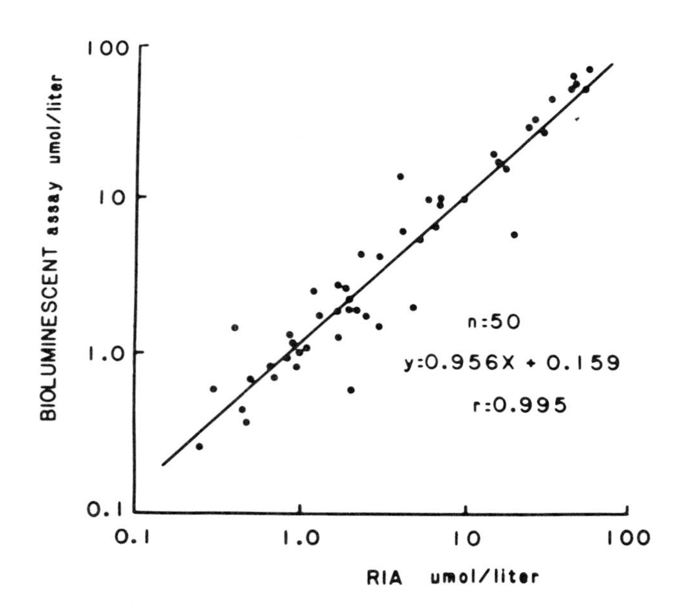

Fig. 8. Results obtained by bioluminescent assay compared with those obtained by radio-immunoassay.

rapid requiring approximately one minute per assay.

Schoelmerich et al. (9) developed a similar assay for 12α hydroxy bile acids and 3α hydroxy bile acids. For these assays a commercially available diaphorase was used rather than the NADH:FMN oxidoreductase from V. harveyi. Comparable results were obtained with either enzyme.

With the 3α hydroxysteroid assay the light output was linear from 1 pmole - 20,000 pmoles. Comparison of the bioluminescent assay with gas-liquid chromatography gave an excellent correlation: $r = 0.99$; $n = 31$. The 3α hydroxysteriod dehydrogenase had different affinities for the various bile acids. These differences could be eliminated by preparing the standards in buffer containing 3% BSA. The most reproducible and rapid assay was obtained when the reaction was initiated by the injection of $NADP^+$ into a solution containing all of the other substrates and immobilized enzymes.

All of the assays discussed previously have been performed with a suspension of the Sepharose-enzymes in a final volume of 0.5 ml. Kricka et al. designed an automated flow system using these immobilized enzymes in a flow cell (14). The immobilized enzymes are packed into a small piece of tygon tubing which is mounted in front of the phototube. A typical time course of light emission for repetitive samples of glucose-6-phosphate is shown in Fig. 9. This assay uses the

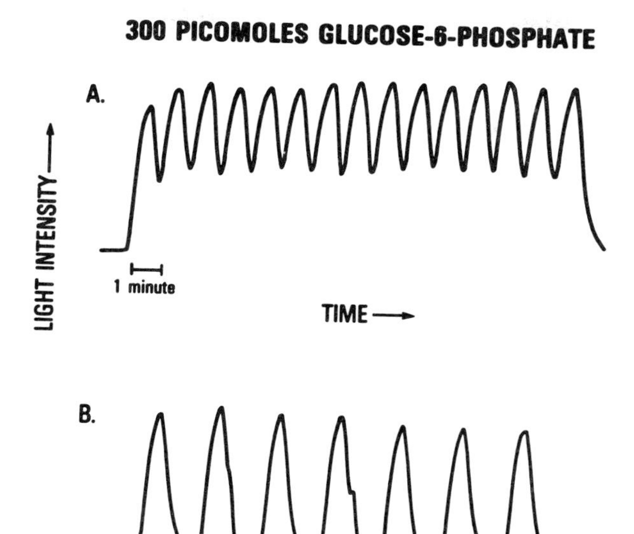

Fig. 9. Typical response for repeated analyses of glucose-6-phosphate: (A) 20 second wash; (B) 80 second wash between samples.

three-enzyme system glucose-6-phosphate dehydrogenase, NADH: FMN oxidoreductase and bacterial luciferase. The lower limit of detection is 1.5 pmoles of glucose-6-phosphate. A single flow cell was used for up to 700 assays. Similar cells were constructed for use in analysis of 7α hydroxysteroids, NADH and ATP. The reusability of these enzymes makes this a very attractive method of assay.

Further development of the automated flow system has been carried out by Mr. Dan Vellom in our laboratory. He has redesigned the flow cell and modified the instrumentation so that it is possible to measure 6 femtomoles of NAD(P)H. This is approximately 1000x more sensitive than the previous detection limits. We believe that some other changes in the system will lead to even further increases in sensitivity.

In summary, these bioluminescent assays using co-immobilized enzymes are specific, rapid, sensitive and the enzymes are reusable. Ultimately, we believe they will be used routinely in the clinical as well as the research laboratory.

ACKNOWLEDGEMENTS

This work was supported by grants from the National Institutes of Health and the National Science Foundation.

REFERENCES

1. Mosbach, K. and Mattiasson, B. (1970). Matrix bound enzymes, Part II: Studies on a matrix-bound two-enzyme system, Acta Chem. Scand. 24, 2093.
2. Mattiasson, B. and Mosbach, K. (1971). Studies on a matrix-bound three-enzyme system, Biochim. Biophys. Acta 235, 253.
3. Srere, P., Mattiasson, B., and Mosbach, K. (1973). An immobilized three-enzyme system: A model for micro-environmental compartmentation in mitochondria, Proc. Natl. Acad. Sci. USA 70, 2534.
4. Siegbahn, N. and Mosbach, K. (1982). An immobilized cyclic multi-step enzyme system - the urea cycle, FEBS Letts. 137, 6.
5. "Immobilized Enzymes. Methods in Enzymology." (1976). (Ed. K. Mosbach). Vol. 44. Academic Press, New York.
6. Axen, R., Porath, J., and Ernback, S. (1967). Chemical coupling of peptides and proteins to polysaccharides by means of cyanogen halides, Nature (London) 214, 1302.
7. Hastings, J.W., Baldwin, T.O., Nicoli, M.Z. (1978). Bacterial luciferase: Assay, purification and properties. In "Methods in Enzymology" (Ed. M. DeLuca). Vol. LVII, p. 135. Academic Press, New York.
8. Jablonski, E. and DeLuca, M. (1977). Purification and properties of the NADH and NADPH specific FMN oxidoreductases from *Beneckea harveyi*, Biochemistry 16, 2932.
9. Schoelmerich, J., Hinkley, J.E., Macdonald, I.A., Hofmann, A.F., and DeLuca, M. (1983). A bioluminescent assay for 12α hydroxy bile acids using immobilized enzymes, Anal. Biochem. 133, 244.
10. Wienhausen, G.K., Kricka, L.J., Hinkley, J.E., and DeLuca, M. (1982). Properties of bacterial luciferase/NADH:FMN oxidoreductase and firefly luciferase immobilized onto sepharose, Appl. Biochem. Biotech. 7, 463.
11. Wienhausen, G., and DeLuca, M. (1982). Bioluminescent assays of picomole levels of various metabolites using immobilized enzymes, Anal. Biochem. 127, 380.
12. Schoelmerich, J., van Berge Henegouwen, G.P., Hofmann, A.F., and DeLuca, M. (1984). A bioluminescent assay for total 3α-hydroxy bile acids in serum using immobilized enzymes, Clin. Chim. Acta 137, 21.
13. Roda, A., Kricka, L.J., DeLuca, M., and Hofmann, A.F.

(1982). Bioluminescence measurements of primary bile acids using immobilized 7α-hydroxysteroid dehydrogenase: Application to serum bile acids, J. Lipid Res. 23, 1354.
14. Kricka, L.J., Wienhausen, G.K., Hinkley, J.E. and DeLuca, M. (1983). Automated bioluminescent assays for NADH, glucose-6-phosphate, primary bile acids, and ATP, Anal. Biochem. 129, 392.

FLOW-INJECTION ANALYSIS WITH IMMOBILIZED BACTERIAL BIOLUMINESCENCE ENZYMES

K. Kurkijärvi *, T. Heinonen +, T. Lövgren *,
J. Lavi *, R. Raunio +

*\) *Wallac Biochemical Lab., Box 10, Turku, Finland*
+) *Dept. of Biochemistry, Turku University, Turku, Finland*

INTRODUCTION

A flow injection method for the analysis of glutamate, ethanol, glucose, oxalacetate, and aspartate has been developed, using immobilized bacterial bioluminescence enzymes coupled to the proper additional enzyme(s) (GlDH, ADH, GlucDH, MDH; ASAT).

The use of bacterial luciferase (BL) and NADH:FMN oxido-reductase (OR) in NADH monitoring is well documented. Any reaction producing or consuming NADH can be measured with the BL-OR system (1,2).

Recently bioluminescent flow methods for NADH, glucose-6-phosphate, and primary bile acids using immobilized enzymes have been published (3,4,5).

EXPERIMENTAL

Bacterial luciferase and oxidoreductase (LKB-Wallac, Turku, Finland) were immobilized on CNBr-activated Sepharose CL 4B (Pharmacia Fine Chemicals, Uppsala, Sweden) with the proper coupling enzyme(s) (Sigma, St. Louis, Missouri, USA), either separately (MDH, ASAT) or on the same particles (GlDH, ADH, GlucDH). All other reagents used were analytical grade.

The construction of the plug flow column(s) was as described earlier with minor modifications (3). The light emission was measured with LKB-Wallac 1250 Luminometer and registered on a Model 2210 chart recorder (see Fig. 1).

ISBN 0-12-426290-2

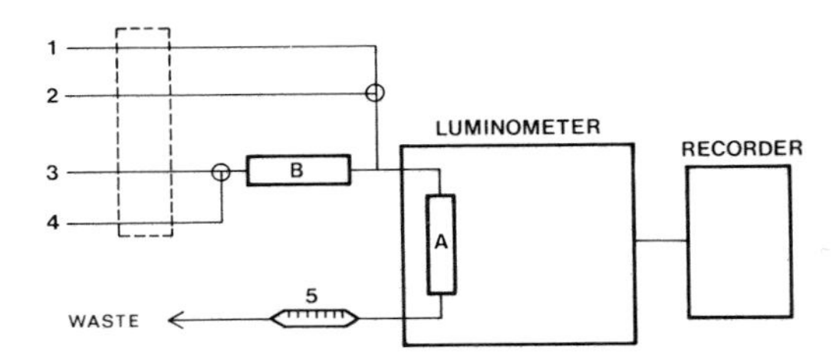

Fig. 1 Manifold of the flow injection system

A = enzyme column consisting of Sepharose-immobilized bioluminescence enzymes and the proper dehydrogenase (GlDH, ADH or GlucDH)

B = ASAT-MDH column

1 = 10 μM FMN, 0.001 % decanal, 0.1 mM DTT, and 0.1 M KCl in 0.05 M potassium phosphate buffer, pH 7.5

2 = NAD^+ (10 mg/ml) in 0.05 M potassium phosphate, pH 7.5 + glutamate, ethanol or glucose injection

3 = 0.05 M potassium phosphate, 0.1 M KCl, pH 7.5 + oxalacetate injection (containing 5 nmoles NADH)

4 = 0.05 potassium phosphate, 0.1 M KCl, 0.05 M α-ketoglutarate + aspartate injection (containing 5 nmoles NADH)

5 = flow meter

RESULTS

The recovery of light emitting activity (BL and OR) was above 60 % as compared to the corresponding soluble system. The additional enzymes (GlDH, ADH, GlucDH, MDH, and ASAT) retained their activities also quite well upon immobilization (30-50 %).

The linear ranges of the flow-injection analysis were compared to the soluble and immobilized batch-systems.

TABLE 1
Linear ranges of bioluminescent substrate assays

	Linear range (nmoles/assay)		
Analyte	Soluble	Immobilized batch	Flow-injection
Glutamate	0.002-10	0.01-40	0.01-600
Ethanol	10-500	10-1000	10-1000
Glucose	0.01-150	0.03-800	0.05-200
Oxalacetate	0.01-1	0.01-5	0.05-5
Aspartate	0.01-1	0.01-5	0.05-5

The flow rate affects significantly the response of the flow-injection system (Fig. 2). Total flow rate of 0.60 ml/min was chosen because it was found optimal for the sampling time, sensitivity and reagent consumption.

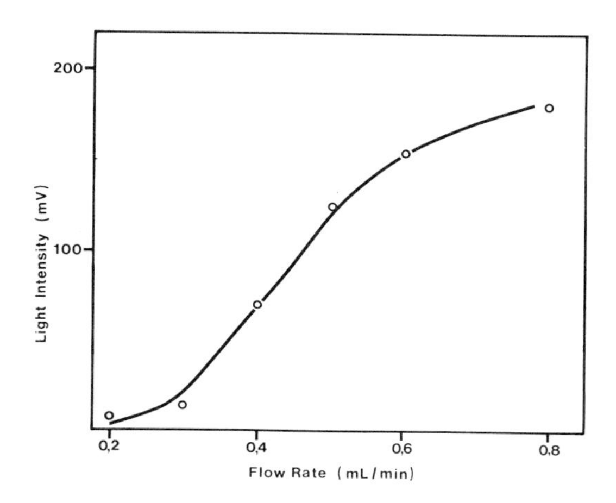

Fig. 2 The effect of total flow rate on peak height. 5 nmoles of glucose were injected in aliquots of 30 µl.

The precision of the flow-injection method was excellent (CV<5 %, n = 10-30, see Fig. 3). No significant sample dispersion was found when injection volumes from 5 to 50 µl were used. With higher volumes the peak heights were not linear to the amount of analyte. More than 30 samples per hour can be analyzed without any significant carry-over.

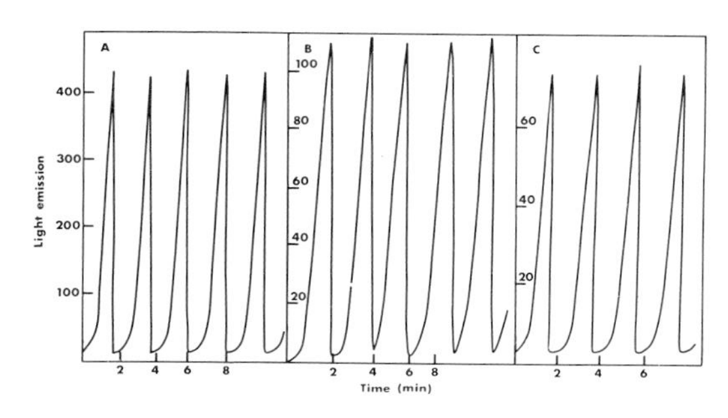

Fig. 3 Time course of light peaks
A = 35 nmoles of glutamate
B = 200 nmoles of ethanol
C = 0.9 nmoles of glucose

The stability of immobilized enzymes was followed systematically during storage at 4°C for 4 months. No decrease in activities was found. Preparations containing the BL-OR system need fresh DTT to be present. The operational stability of the immobilized columns was good and more than 1000 samples can be analyzed during several weeks with no significant decrease in the response.

CONCLUSIONS

The flow-injection analysis using immobilized bio-luminescence enzymes as a detector is quick, sensitive, precise, stable, and easy to automate for routine analysis of metabolites.

REFERENCES

1. Lövgren, T., Thore, A., Lavi, J., Styrelius, J., Raunio, R. (1982). Continuous monitoring of NADH-converting reactions by bacterial bioluminescence, *J. Appl. Biochem.* 4, 103-108
2. Kurkijärvi, K., Raunio, R., Lavi, J., Lövgren T. (1984). Stable light emitting bacterial bioluminescence reagents, *In* "Bioluminescence and Chemiluminescence: Instrum. & Appl." (in press) CRC Press, Boca Raton, U.S.A.
3. Kurkijärvi, K., Raunio R., Korpela, T. (1982). Packed-bed reactor of immobilized bacterial bioluminescence enzymes, *Anal. Biochem.* 5, 415-418
4. Kricka, L., Wienhausch, G., Hinkley, J., De-Luca, M. (1983). Automated bioluminescent assays for NADH, glucose-G-phosphate, primary bile acids, and ATP, *Dual. Biochem.* 129, 392-397
5. Roda A., Girotti, S., Ghini, S., Grigolo, B., Carrea, G., Bovara, R. (1984). Continuous-flow determination of primary bile acids, by bioluminescence, with use of nylon-immobilized bacterial enzymes, *Clin. Chem.* 30, 206-210

ACKNOWLEDGEMENTS

This work was kindly supported by the Finnish Academy of Sciences.

DEVELOPMENT OF A CONTINOUS-FLOW ANALYSIS FOR SERUM AND SALIVARY BILE ACIDS USING BACTERIAL BIOLUMINESCENT ENZYMES IMMOBILIZED ON NYLON COIL

A. Roda*, S. Girotti*, S. Ghini*, B. Grigolo*,
G. Carrea** and R. Bovara**

**Istituto Scienze Chimiche, Università di Bologna*
40127 Bologna, Italia
***Istituto di Chimica degli Ormoni, C.N.R.,*
20100 Milano, Italia

INTRODUCTION

It is well established that serum bile acid (BA) is a useful test for the diagnosis of liver and intestinal disease (1). The high cost or the insensitivity of the existing methods allowed us to develop a new ultrasensitive bioluminescent assay for BA (2).The overall scheme of the bioluminescent assay is:

$$\text{Bile acid-OH} + NAD^{+} \xrightleftharpoons{\text{HSD}} \text{Bile acid-OXO} + NADH + H^{+} \quad (1)$$

$$H^{+} + NADH + FMN \xrightleftharpoons{\text{OXRED}} NAD^{+} + FMNH_2 \quad (2)$$

$$FMNH_2 + \text{decanal} + O_2 \xrightarrow{\text{LUCIFERASE}} FMN + \text{decanoic acid} + H_2O + \text{light} \quad (3)$$

where HSD is 3α-or 7α-or 12α-hydroxysteroid dehydrogenase, separately or co-immobilized with bioluminescent enzymes on nylon tubes.

ISBN 0-12-426290-2

MATERIAL AND METHODS

Immobilisation of enzymes on nylon 6 tubes has previously been described (2).

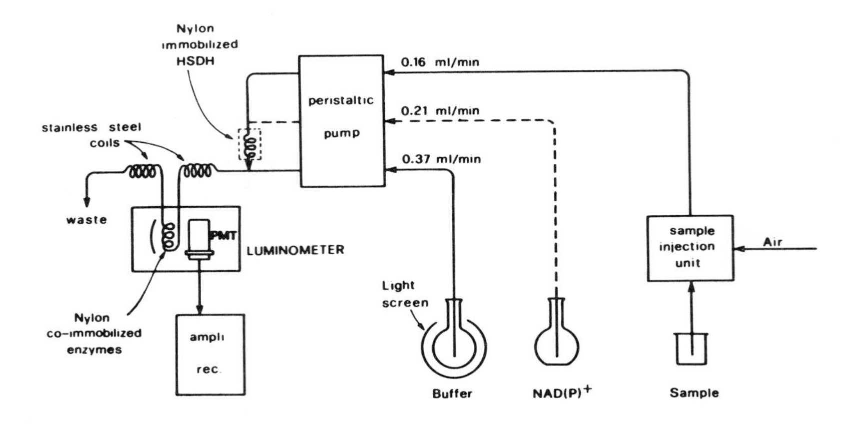

Fig. 1. Continuous-flow system for bioluminescent assays with coimmobilized enzymes. Buffer: 0.1 M of KH_2PO_4, pH 7.0; 0.5 mM DTT; 1 mM NAD^+; 10 μM FMN; and 27 μM decanal. Dotted lines indicate the apparatus used with separately-immobilised HSD.

The flow system (Fig.1) involves two streams: the primary one is continuous and contains the working bioluminescent solution; the other stream is a continuous flow of air into which a known volume of sample is intermittently inserted.

When separately-immobilised HSD was used (Fig.1,dotted lines), BA oxidation was carried out at pH 9.2 in 0.01 M phosphate. With 12 α-HSD the 1 mM NAD^+ was substituted by 1mM $NADP^+$, since this enzyme is $NADP^+$ dependent.

RESULTS AND DISCUSSION

The K_m values of nylon-immobilised enzymes are higher than those of the free ones, except for NAD(P)H:FMN oxidoreductase, where the values are similar. The increased values are probably due to reduced substrate diffusion, usually present in such flow reactions (3). The activity recoveries were between 2 and 9% in spite the fact that all the added enzymes was covalently linked to the nylon matrix. However activities were adequate to make BA assays still sensible. The half-life time of immobilised enzyme was between twenty and fourty days.

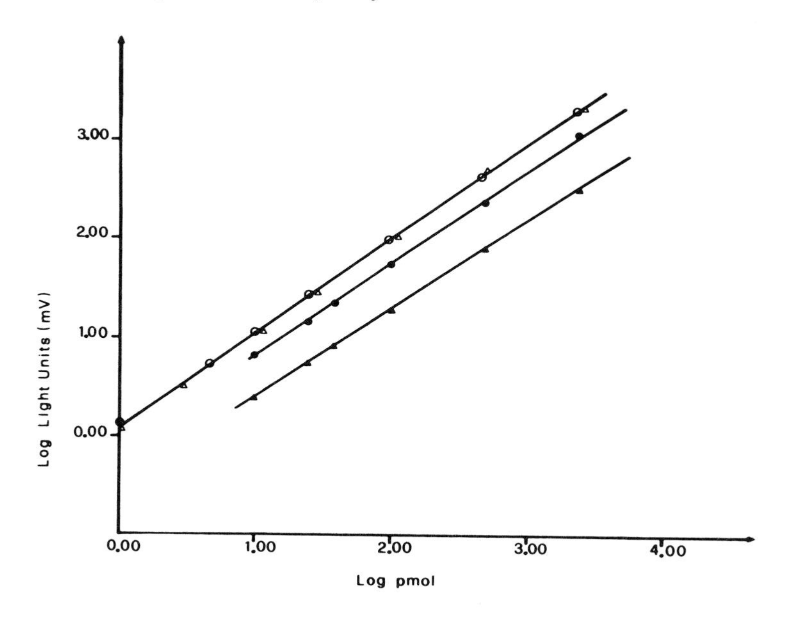

Fig. 2 Standard curves for cholic acid when use 7α-HSD (○-○), 3α-HSD (△-△) and 12α-HSD (▲-▲) separately-immobilized with bioluminescent enzymes. For comparison 7α -HSD (●-●) co-immobilized is shown.

Fig. 2 shows standard curves obtained for cholic acid determination with separately-immobilised enzymes. The lower sensitivity shown by 12α-HSD reflect the lower activity of NAD(P)H:FMN oxidoreductase toward NADPH. When HSD was coimmobilized the sensitivity was markedly reduced, since the bile acids was not completely transformed in the assay.

Similar results were obtained with the other conjugated or unconjugated bile acids and more than 20 samples per hour were analyzed without carry-over. Serum samples, which were directly analyzed after a 10 fold dilution with 0.1 M phosphate buffer pH 7.0, gave results in good agreement with those obtained by RIA, EIA and HPLC. Both intra-assay (CV: 6-10%) and inter-assay (CV: 7-10%) precision was satisfactory for serum samples at low and high concentrations. BA can be also analyzed directly on 50 µl of saliva sample and results agree well with those obtained by immunological methods.

The method here described is simple, reliable, cheap and the sensitivity is very high, particularly when separately-immobilised HSD are used (detection limit 1 pmol). The response is scarcely influenced by variations of flow rate and injection volume, and thousands of assays can be carried out with a simple reactor coil. Potentially, the bioluminescent reactor (oxidoreductase and luciferase) can be used for the analysis of a variety of other NAD(P)H-generating metabolites.

REFERENCES

1. Festi, D., Morselli Labate, A.M., Roda, A., Bazzoli, F., Frabboni, R., Rucci, P., Taroni, F., Aldini, R., Roda, E. and Barbara, L. (1983).Diagnostic effectiveness of serum bile acids in liver diseases, as evaluated by multivariate statistical methods, *Hepathology* 3,707-713
2. Roda, A., Girotti, S.,Ghini,S., Grigolo, B., Carrea, G. and Bovara, R.(1984). Continuous-flow determination of primary bile acids, by bioluminescence, with use of nylon-immobilized bacterial enzymes, *Clinical Chemistry* 30, 206-210.
3. Carrea, G., Bovara, R. and Cremonesi, P. (1984). Continuous-flow automated assay of steroids with nylon-tube-immobilized hydroxysteroid dehydrogenases. *Analytical Biochemistry* 136, 328-335.

CONTINUOUS FLOW BIOLUMINESCENT ASSAYS FOR FEMTOMOLE LEVELS OF NADH AND TNT

D. Vellom[+], J. Hinkley[+], A. Loucks[#], H. Egghart*, M. DeLuca[+]

Departments of [+]Chemistry and [#]Reproductive Medicine
University of California, San Diego
La Jolla, California 92093, USA

*U.S. Army Belvoir Research and Development Center
Fort Belvoir, Virginia 22060, USA

INTRODUCTION

Bioluminescence has recently been exploited as a useful analytical tool for measuring low levels of a variety of biologically important compounds (1-3). The light producing enzymes from luminous marine bacteria have proven to be particularly useful when coupled to specific dehydrogenases. The reactions catalyzed by these enzymes are the following:

(I) $NAD(P)H + H^+ + FMN \longrightarrow NAD(P)^+ + FMNH_2$

(II) $FMNH_2 + RCHO + O_2 \longrightarrow FMN + RCOOH + FMN + h\nu + H_2O$

The first reaction is the oxidation of a reduced pyridine nucleotide with the formation of $FMNH_2$. This reaction is catalyzed either by an NADH or NADPH:FMN oxidoreductase, both of which have been purified from *Vibrio harveyi*. Alternatively, a commercially available diaphorase which will utilize either nucleotide may be substituted (4). The bacterial luciferase catalyzes the reaction of $FMNH_2$ with decanal and oxygen to produce FMN, decanoic acid and light. By using the specific oxidoreductase with luciferase it is possible to measure any substance which can be coupled to the production or disappearance of NADH or NADPH.

It has been demonstrated with several systems that co-immobilization of the luciferase and oxidoreductase along with a specific NAD or NADP dependent dehydrogenase greatly

ISBN 0-12-426290-2

enhances light production relative to the same amount of the soluble enzymes (5,6). Using these coimmobilized enzymes, bioluminescent assays have been developed for NADH and NADPH, various metabolites, steroid hormones and bile acids (4,7-9). These assays were initially developed using a suspension of the Sepharose bound enzymes. Sensitivity was typically in the range of 10-1000 pmoles. Kricka et al. (10) developed an automated system utilizing these immobilized enzymes in a flow cell. With this system it was possible to reproducibly measure 6 pmoles of NADH. We have modified the earlier flow system and are now able to detect 6 fmoles of NADH. We have used this modified system for the measurement of NADH, NADPH, and 6-phosphogluconic acid.

MATERIALS AND METHODS

Reagent grade FMN, decanal, and β-mercaptoethanol were obtained from Sigma Chemical Co. NADH, NAD, NADP, and NADPH were the highest purity available from Boehringer-Mannheim. Cyanogen Bromide and Trinitrotoluene were purchased from Eastman Kodak. Sepharose 4B and CL-6B were products of Pharmacia.

Diaphorase "from Microorganism" was purchased from Boehringer-Mannheim. Bacterial luciferase was isolated from a frozen cell paste of *V. harveyi* as described by Hastings et al. (11). The NADPH and NADH:FMN oxidoreductase was purified according to the method of Jablonski and DeLuca (12). TNT reductase was purified from a soil bacteria as described in Beckman Technical Report #DAAK70-77-C-0129.

Sepharose 4B or CL-6B was activated by cyanogen bromide according to the method of Axen (13). The enzymes were immobilized as described previously (7) with the exception that BSA was omitted. The immobilized enzymes were stored frozen (-20° C) at a concentration of 1 g Sepharose/10 ml 50 mM Tris, pH 7.0 containing 2 mM β-mercaptoethanol, 1 mM EDTA, and .02% NaN_3.

All light measurements were done in an LKB Model 1250 luminometer.

NADH Assay. Decanal (0.15% in ethanol) was diluted in buffer (50 mM sodium phosphate, pH 6.0 or 50 mM Tris·HCl, pH 7.0) to 0.0006% final concentration. 2 mM β-mercaptoethanol and 2 mM EDTA were added to stabilize the immobilized enzymes. 0.1 mM FMN was dissolved in distilled water. Either phosphate or Tris buffer was used to prepare the samples and to wash the flow cell between successive samples. All buffers and H_2O were thoroughly degassed under vacuum before use each day.

6PGA Assays. The enzymes coimmobilized were 6-phosphogluconic acid dehydrogenase, NADPH:FMN oxidoreductase, and luciferase. Assays were performed in 50 mM Tris·HCl, pH 7.9. Reagents were prepared as for the NADH assay except that the FMN reagent also contained 1.6 mM NADP, 0.4 mM Mg-acetate and 36 mM NH_4-acetate.

TNT Assay. TNT reductase catalyzes the following reaction:

$$\text{(III) } NADH + H^+ + TNT \longrightarrow NAD^+ + TNT_{red}$$

Therefore, the enzymatic assay for TNT involves measurement of a decrease in NADH concentration. TNT reductase was used in soluble form in 50 mM sodium phosphate, pH 6.0. 10 µl samples with or without TNT were added to 50 µl 100 nM NADH at room temperature. 50 µl TNT reductase (0.002 units) were added to start the reaction. The mixture was incubated for 10 min at room temperature followed by 30 sec at 100° C to stop the reaction. Samples were then assayed for remaining NADH by adding 300 µl decanal-phosphate buffer pH 7.0, 10 µl Sepharose immobilized NADH:FMN oxidoreductase/luciferase, and 10 µl 130 µM FMN. Peak light intensity was read on the luminometer. Stock solutions of TNT at 0.5 M concentration were made up in acetone, diluted 1:100 in ethanol and then serially diluted in phosphate buffer pH 6.0.

RESULTS AND DISCUSSION

Flow cells were constructed from 20 µl glass capillary pipets (Drummond Microcap) as shown in Fig. 1. The lower end of the capillary tube is first constricted by very brief heating with a flame. The Sepharose-enzyme bed is supported on a short bed of washed glass beads (Supabrite glass beads, Type 100-5005, 3M Co.). A small reservoir constructed from a pasteur pipet and tygon tubing is used to introduce a measured amount, 50 µl of a 1 g/10 ml suspension of Sepharose enzymes into the capillary tube. Suction from a peristaltic pump connected downstream from the flow cell is used to pack the bed. Silicone rubber tubing, 0.25 mm I.D., is used to connect the flow cell to 22 ga. stainless steel tubing which passes through the rubber septum on the injection port on an LKB 1250 luminometer.

The manifold is shown in Fig. 1. An AAI autoanalyzer pump and sampler (Technicon Instruments Corp.) were used to deliver substrates and cofactors to the flow cell. Pump tubing was silicone rubber with 0.25 mm I.D. All other tubing was 0.3 mm I.D. teflon or glass. Three separate channels were used to feed reagents to the flow cell with

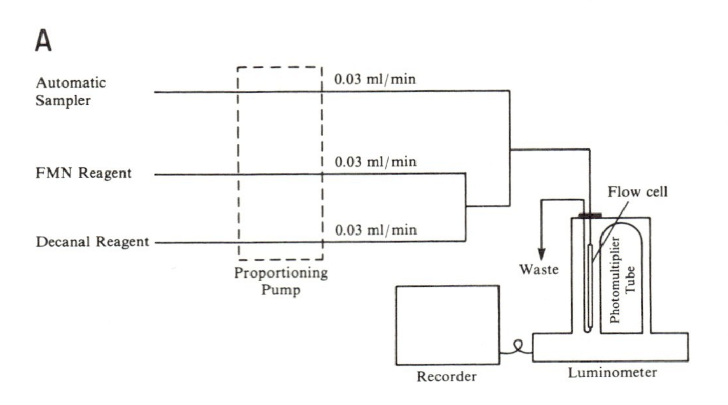

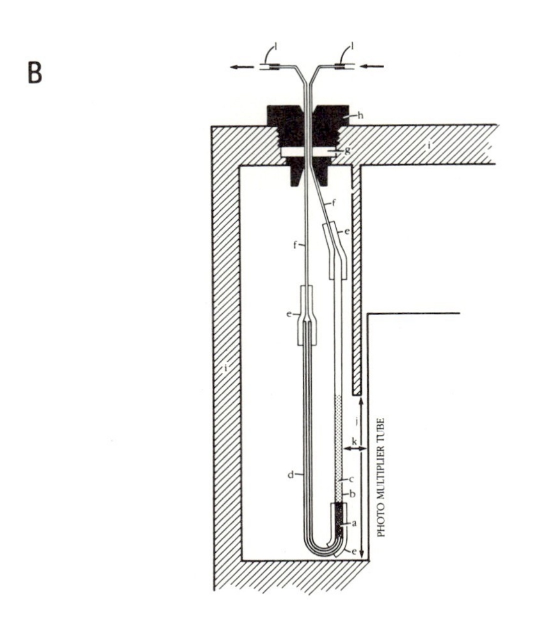

Fig. 1. (A) The flow manifold. (B) Design of immobilized enzyme packed capillary flow cell and its orientation in the luminometer: (a) glass beads bed support; (b) capillary tube flow cell (20 μl Drummond Microcap); (c) Sepharose-enzyme bed; (d) thick wall capillary for outflow; (e) silicone rubber tubing; (f) 22 ga. stainless steel hypodermic needle; (g) rubber septum; (h) knurled screw port; (i) body of sample chamber: 38 mm; (k) distance from flow cell to PMT ∿ 5 mm: (l) teflon tubing.

microbore "T" fittings as mixing chambers. Each channel had a flow rate of 30 µl/min, giving a combined rate of 90 µl/min. 5 ml reservoirs of FMN and decanal were used. A controller cam for the sampler was designed to provide a 2 min sampling period followed by a 4 min wash.

The amount of light obtained with known concentrations of NADH is shown in Fig. 2. The linear range of the assay is from 100 fmoles/ml to 1 nmole/ml. Since the sample volume is 60 µl the lower limit of detection is 6 fmoles of NADH.

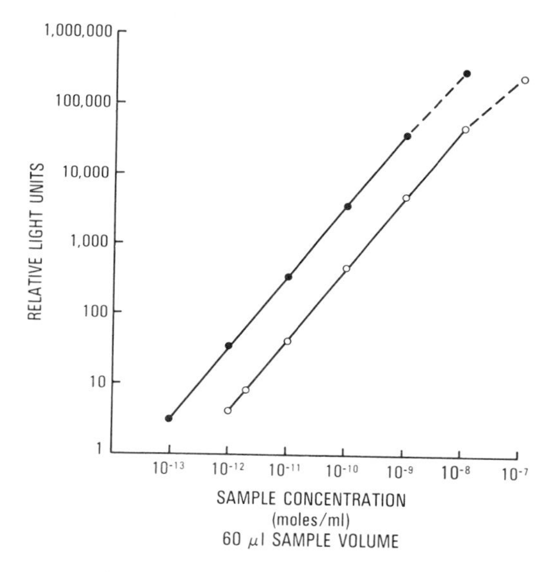

Fig. 2. Peak light intensity as a function of the amount of NADH (●) and 6PGA (O). 2 min sampling time = 60 µl, 4 min wash time.

This is a significant improvement over an earlier flow method where the lower detectable amount of an NADH was 6 pmoles (10). Five factors contribute to this increased sensitivity: 1) a reduced bed volume in the flow cell has decreased the background; 2) a reduced flow rate allowed the sample to remain in contact with the reactive bed for a longer period of time; 3) the capillary tube's smaller diameter allowed a greater proportion of the emitted light to reach the PMT; 4) the flow cell's location, closer to the PMT, increased the measured light intensity; 5) the smaller manifold tubing reduced the dead volume relative to the earlier system. Thus we have increased the measurable light output of a given NADH concentration by about 1000-fold over the previous flow system. Our sensitivity is limited by a small relatively constant baseline light emission and by the sensitivity of the luminometer. Fig. 2 contains data for light output vs. concentration of 6PGA. The lower limit of detection of 6PGA was 60 fmoles.

An analytical trace for repetitive NADH samples is shown in Fig. 3. Intra-assay coefficients of variation were from 2-5%.

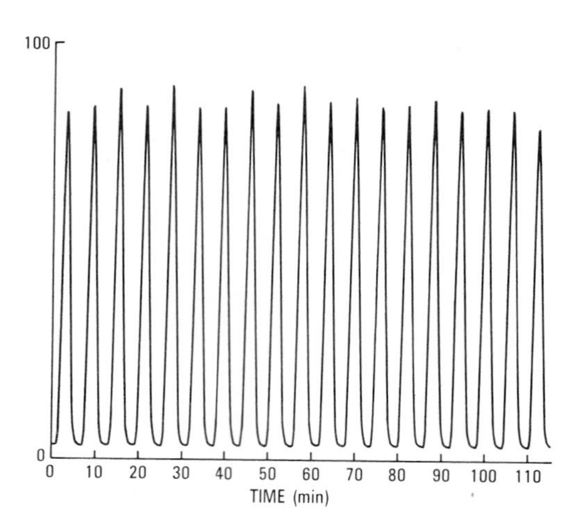

Fig. 3. Analytical traces obtained for repeated analysis of 100 pmole/ml of NADH. 2 min sampling time, 4 min wash time.

The enzymatic assay for TNT we currently employ utilizes the suspension type bioluminescent detection system. Since the presence of TNT in a sample would be detected by a decrease in NADH concentration, it is desirable to lower the NADH concentration utilized in the reaction such that a small amount of TNT would give a detectable decrease in light intensity. The TNT reductase enzyme we are working with contains an associated NADH oxidase activity. This limits our sensitivity for detecting TNT by oxidizing NADH in a reaction not dependent on TNT. Using our present assay conditions we are able to reproducibly detect 400 fmoles of TNT in 10 µl of aqueous solution, i.e. 40 nM. Experiments using the flow cell detection system indicate approximately the same level of detection. Experiments are currently underway to integrate the TNT assay into the continuous flow system using immobilized TNT reductase.

ACKNOWLEDGEMENTS

This work was supported by a Contract from the U.S. Army, (RFQ DAAK70-49-A-0019).

REFERENCES

1. "Methods in Enzymology." (1978). (Ed. M. DeLuca). Vol. LVII. Academic Press, New York.
2. "Luminescent Assays." (1982). (Eds. M. Serio and M. Pazzagli). Raven Press, New York.
3. "Clinical and Biochemical Luminescence." (1982). (Eds. L.J. Kricka and T.N.M. Carter). Marcel Dekker, New York.
4. Schoelmerich, J., Hinkley, J.E., Macdonald, I.A., Hofmann, A.F. and DeLuca, M. (1983). A bioluminescent assay for 12α hydroxy bile acids using immobilized enzymes, Anal. Biochem. 133, 244.
5. Ford, J. and DeLuca, M. (1981). A new assay for picomole levels of androsterone and testosterone using co-immobilized luciferase, oxidoreductase, and steroid dehydrogenase, Anal. Biochem. 110, 43.
6. DeLuca, M. and Kricka, L.J. (1983). Coimmobilized multi-enzymes: An *in vitro* model for cellular processes, Arch. Biochem. Biophys. 226, 285.
7. Wienhausen, G. and DeLuca, M. (1982). Bioluminescent assays of picomole levels of various metabolites using immobilized enzymes, Anal. Biochem. 127, 380.
8. Schoelmerich, J., van Berge Henegouwen, G.P., Hofmann, A.F. and DeLuca, M. (1984). A bioluminescent assay for total 3α-hydroxy bile acids in serum using immobilized enzymes, Clin. Chim. Acta 137, 21.
9. Roda, A., Kricka, L.J., DeLuca, M. and Hofmann, A.F. (1982). Bioluminescence measurements of primary bile acids using immobilized 7α-hydroxysteroid dehydrogenase: Application to serum bile acids, J. Lipid Res. 23, 1354.
10. Kricka, L.J., Wienhausen, G.K., Hinkley, J.E. and DeLuca, M. (1983). Automated bioluminescent assays for NADH, glucose-6-phosphate, primary bile acids, and ATP, Anal. Biochem. 129, 392.
11. Hastings, J.W., Baldwin, T.O. and Nicoli, M.Z. (1978). Bacterial luciferase: Assay, purification and properties. *In* "Methods in Enzymology" (Ed. M. DeLuca). Vol. LVII, p. 135. Academic Press, New York.
12. Jablonski, E. and DeLuca, M. (1977). Purification and properties of the NADH and NADPH specific FMN oxidoreductases from *Beneckea harveyi*, Biochemistry 16, 2932.
13. Axen, R., Porath, J. and Ernback, S. (1967). Chemical coupling of peptides and proteins to polysaccharides by means of cyanogen halides, Nature (London) 214, 1302.

MEASUREMENT OF BILE ACID PATTERNS IN SERUM AND URINE BY BIOLUMINESCENCE ASSAYS USING IMMOBILIZED HYDROXYSTEROID DEHYDROGENASES

Jürgen Schölmerich, Ian A.Macdonald, Gerard van Berge Henegouwen, Alan F.Hofmann, and Marlene DeLuca

Departments of Chemistry and Medecine, University of California, San Diego, USA

INTRODUCTION

Bile acid concentrations in serum or biological fluids might be useful in the detection or assessment of liver or intestinal disease (1). The methods used so far for measurement of bile acids either lack specifity for the single bile acids or sensitivity or simplicity.

MATERIALS AND METHODS

Luciferase from Beneckea Harveyi, strain B 392, was prepared as described (2). Bacterial diaphorase was purchased from Boehringer, Mannheim, FRG; 3α-hydroxysteroiddehydrogenase (HSD) from Sigma, St.Louis, USA and from Nygaard, Oslo, Norway, 7α-HSD from Sigma. 12α-HSD was prepared as described (3). Bile acids used were from a varity of sources and had a purity of $\geq$98%.

Coimmobilization of luciferase, diaphorase and the appropriate HSD on activated Sepharose 4B was performed as described (4,5). Principle and assay procedure were as described for 3α-OH bile acids (5).

Using three equations (6) the concentration of the main bile acids in serum and urine were calculated. Serum samples were diluted 1:4 - 1:20 and heated (68°C) for 15 min. Urine samples were used after centrifugation. Solvolysis of bile

ISBN 0-12-426290-2

acid sulfates was performed with arylsulfatase. Results of the assays were validated by comparison with results obtained by GLC.

RESULTS

A linear relationsship between bile acid concentration and light emission was found over a concentration range of 0.4 - 1000 µmoles/l for all three systems.

Table 1

Intraassay precision of the test system

	serum		urine	
	conc (µM)	CV (%)*	conc (µM)	CV (%)**
3α	4.3	8.1	9.6	5.9
	111.2	6.2	499.2	7.1
7α	18.9	7.4	18.9	7.4
	121.9	5.4	-	-
12α	4.3	7.8	9.0	7.6
	38.8	8.2	-	-

* 10 determinations ** 6 determinations

Intraassay precision (table 1) and recovery of added standards (93 - 102% for serum, 84 - 95% for urine) were satisfactory.

Application of all three assays on a mixture of standards showed a good agreement between calculated and measured values (table 2).

Table 2

Application to mixed standards

concentration	3α	7α	12α
(µM)	µM (%)*	µM (%)*	µM (%)*
1000	902.5(90.2)	600.0(90.9)	571.0(86.3)
100	101.1(100.1)	59.0(89.4)	68.1(102.2)
10	8.8(88.2)	6.3(95.4)	7.1(107.0)

* % of calculated value

The comparison of values obtained by the bioluminescence assays and those measured by GLC revealed an excellent agreement. Correlation coefficients were similar for all assay systems either in serum or in urine as shown for total urinary bile acids after solvolysis (figure 1).

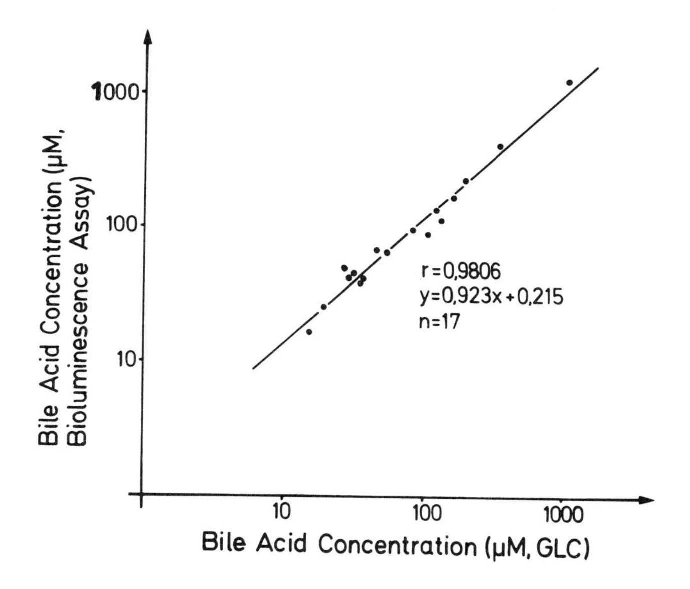

Fig.1 Comparison of urinary bile acid concentration (3α) after solvolysis obtained by bioluminescence and GLC.

Correlation for the main single bile acids was good in serum and urine ($r>0.89$). There was no correlation between the difference of the values obtained with both methods and the total concentration (figure 2).

DISCUSSION

The assays described appear to be an ideal method for measuring bile acid concentrations in serum and urine. Sensitivity is comparable to competetive binding assays and better than that of other routine methods (7). Use of the equations for

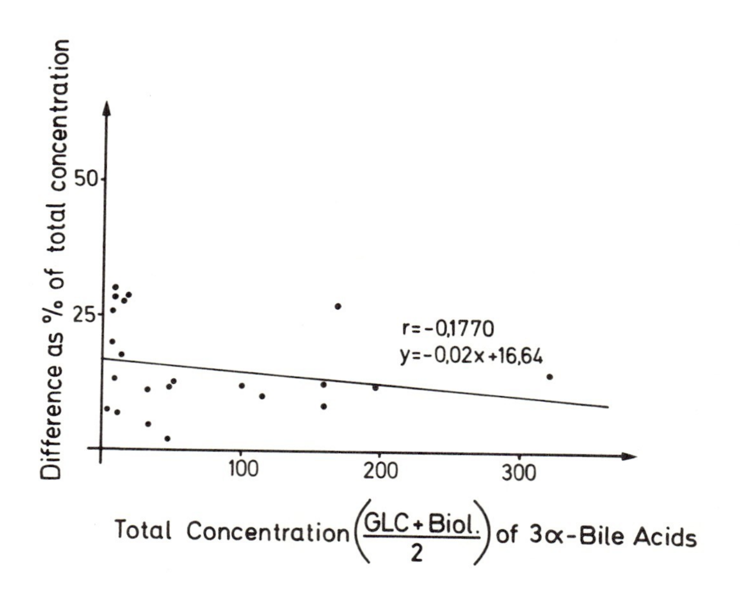

Fig. 2 Relationsship between the difference of values obtained by bioluminescence and GLC and the total serum bile acid concentration.

calculation of single bile acids increases the possible error. However, results are still accurate enough for clinical or experimental use.

The rapidity and simplicity of the assay may facilitate the assessment of the clinical utility of serum bile acid measurement (1). In addition, the assays can be used to study bile acid metabolism and transport in experimental systems (8). Use of a flow cell (9) permits automation and thus, application to the routine clinical laboratory.

REFERENCES

1. Hofmann, A.F. (1982). The aminopyrine demethylation breath test and the serum bile acid level: nominated but not yet elected to join the common liver tests, Hepatology 2, 512-517.
2. Gunsalus-Miguel, A., Meighen, E.A., Nicoli, M.

Z., Nealson, K.H., and Hastings, J.W. (1972). Purification and properties of bacterial luciferase, J.Biol.Chem. 247, 398-404.

3. Macdonald, I.A., Meier, E.C., Mahony, D.E., and Costain, G.A. (1976). 3α-, 7α-, and 12α-hydroxysteroid dehydrogenase activities from clostridium perfringens, Biochim.Biophys.Acta 450, 142-152.
4. Roda, A., Kricka, L.J., DeLuca, M., and Hofmann, A.F. (1982). Bioluminescence measurement of primary bile acids using immobilized 7α-hydroxysteroid dehydrogenase: application to serum bile acids, J.Lipid Res. 23, 1354-1361.
5. Schölmerich, J., van Berge Henegouwen, G.P., Hofmann, A.F., and DeLuca, M. (1984). A bioluminescence assay for total 3α-hydroxy bile acids in serum using immobilized enzymes, Clin.Chim.Acta 137, 21-32.
6. Macdonald, I.A., Williams, C.N., and Musial, B.C. (1980). 3α-, 7α-, and 12α-OH group specific enzymic analysis of biliary bile acids: comparison with gas-liquid chromatography, J. Lipid Res. 21, 381-385.
7. Street, J.M., Trafford, D.J.H., and Makin, H. L.J. (1983). The quantitative estimation of bile acids and their conjugates in human biological fluids, J.Lipid Res. 24, 491-511.
8. Schölmerich, J., Kitamura, S., and Miyai, K. (1983). Changes of the pattern of biliary bile acids during isolated rat liver perfusion, Biochem.Biophys.Res.Commun. 115, 518-524.
9. Kricka, L.J., Wienhausen, G., Hinkley, J.E., and DeLuca, M. (1983). Automated bioluminescent assays for NADH, glucose-6-phosphate, primary bile acids, and ATP, Anal.Biochem. 129, 392-397.

IV IMMUNOASSAY

DEVELOPMENT OF LUMINESCENCE-BASED IMMUNOASSAYS FOR HAPTENS AND FOR PEPTIDE HORMONES

F. Kohen, E.A. Bayer*, M. Wilchek*,
G. Barnard+, J.B. Kim+, W.P. Collins+,
I. Beheshti#, A. Richardson# and F. McCapra#

*Departments of Hormone Research and *Biophysics,
The Weizmann Institute of Science, Rehovot, Israel.
+King's College Medical School, Denmark Hill, London, and
#Department of Molecular Sciences, University of Sussex,
United Kingdom*

INTRODUCTION

Existing radioimmunoassay (RIA) procedures for haptens and for peptide hormones have the advantage of high sensitivity and specificity. However, RIA involves the use of an isotopically tagged antigen or antibody which may pose problems associated with radioactive waste disposal, short half-life and radiolysis of the labelled marker. To avoid these drawbacks while retaining the specificity of an immunoassay, we explored the use of luminescence as an end-point in the assay. Two different approaches were investigated. In the direct approach, haptens (e.g. steroids, drugs or prostaglandins), peptide hormones (e.g. hCG) or antibodies (e.g. monoclonal anti β-hCG) are covalently linked to chemiluminescent molecules (e.g. isoluminol or acridinium ester derivatives), and the resulting conjugates serve as labels in immunological reactions. At the end of the reaction the light production of the oxidized labels is measured by chemiluminescence. In the indirect approach, advantage is taken of the high affinity of avidin ($K_D = 10^{-15}$M) for biotin for amplifying the sensitivity of the assay.

In this system, a biotinylated antigen or specific antibody serves as the primary probe. After the immunological reaction with immobilized antibody and sample, a secondary probe consisting of excess avidin and biotinylated enzyme

ISBN 0-12-426290-2

are added. The biotinylated enzymes which we have examined include horseradish peroxidase, NAD^+-dependent enzymes (glucose-6-phosphate dehydrogenase) and ATP-dependent enzymes (e.g. myokinase and pyruvate kinase). Depending on the particular enzyme used, the end point is determined by bioluminescence using the bacterial or firefly luciferase system or by chemiluminescence with luminol/H_2O_2 system. The results of solid-phase immunoassays for haptens and peptide hormones based on luminescence monitoring are reported here.

MATERIALS AND METHODS

Carboxy derivatives of steroids and drugs, urinary steroid metabolites, and prostaglandins were linked covalently to amino alkyl derivatives of isoluminol (6-amino-2,3-dihydrophthalazine- 1,4-dione) to yield the corresponding hapten-aminohexyl ethyl isoluminol (AHEI), hapten-aminobutyl ethyl isoluminol (ABEI) (see Fig. 1 for a representative steroid-isoluminol conjugate), and hapten-aminoethyl ethyl isoluminol (AEEI) conjugates, essentially as previously described (1,2). The synthesis of progesterone-11 α-hemisuccinate acridinium ester conjugate (see Fig. 1) will be reported elsewhere.

a. ESTRADIOL- ISOLUMINOL

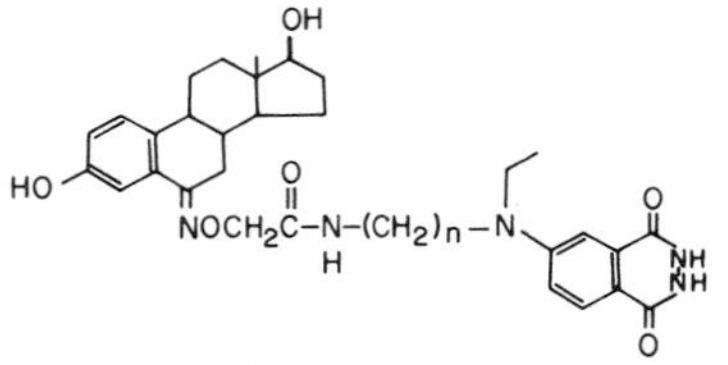

n=2 to 6

b. PROGESTERONE - ACRIDINIUM ESTER

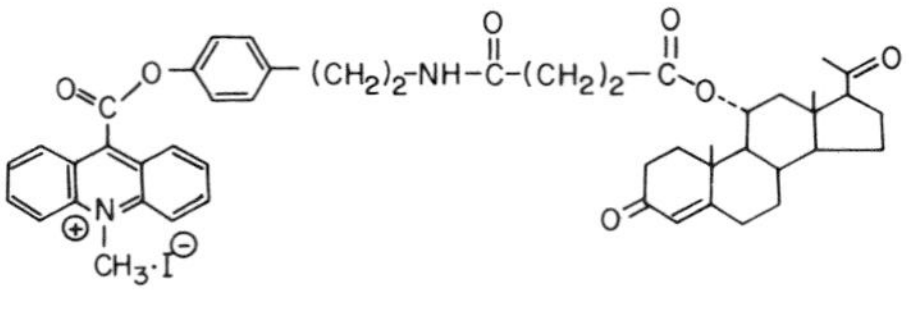

Fig. 1. Representative structures of steroid-chemiluminescent marker conjugates: (a) estradiol-6-[0]-carboxymethyl oxime ABEI, and (b) progesterone-11α-hemisuccinate acridinium ester.

Lysine residues of peptide hormones and of purified immunoglobulin fractions (IgG) were reacted with the N-hydroxysuccinimide ester derivatives of either aminobutyl ethyl isoluminol, acridinium ester, or biotin to yield the corresponding conjugate [i.e., protein-isoluminol (3), protein-acridinium ester (4) or biotinylated antigen or biotinylated antibody conjugates (5)]. Purified IgG fractions of monoclonal antibodies to hormones were purified by chromatography on Sepharose Protein-A and were conjugated covalently to immunobeads from Bio-Rad (1). All other reagents and assay procedures were as previously described (1,2), unless specified otherwise.

Light emission was measured with a Luminometer Model 2080 (Lumac Systems, Basel) or with an LKB Luminometer 1250 (Wallac, Finland).

RESULTS

Chemiluminescence-based immunoassays for haptens

Using the haptens conjugated to the chemiluminescent markers described in Materials and Methods, two types of formats were developed for chemiluminescence-monitored immunoassays for haptens: (a) assays that do not require physical separation of bound and free hormone, so-called "homogeneous" assays; (b) assays that require a separation step, so-called "heteregeneous" assays. The homogeneous immunoassays, based on antibody-enhanced chemiluminescence, proved satisfactory with respect to sensitivity and precision (2). However, they were affected by nonspecific interference from luminescent compounds present in extracts of plasma (2) or diluted urine samples (2). To overcome these problems, assays requiring separation of bound and free ligand were developed.

Separation of bound and free antibody fractions was achieved either in the liquid or solid phase. Dextran-coated charcoal was adopted for separation of bound and free forms in immunological reactions carried out in the liquid phase. Sensitivity and precision similar to those of RIA were achieved for plasma steroids such as progesterone, cortisol and testosterone. However, these assays were also affected by luminescent compounds present in biological extracts or in diluted urine. Moreover, dextran-coated charcoal caused stripping of the bound fraction.

The problems encountered using liquid-phase techniques could be overcome using solid-phase methods. In this case, the specific antibodies (monoclonal or polyclonal) were either adsorbed onto the surface of polystyrene tubes or balls (7). Alternatively, the antibodies were covalently linked to polyacrylamide beads (1). In the immunoadsorption technique, aspiration of the reaction mixture and subsequent washing with buffer removed all potentially interfering substances with the concomitant reduction of background chemiluminescence. The light yield of the label that was bound onto the walls of the antibody-coated tube was then measured at alkaline pH with a microperoxidase/H_2O_2 system. This approach is simple and has enabled the development of techniques for the measurement of urinary steroid metabolites [e.g. pregnanediol-3α-glucuronide (8), estrone-3-glucuronide (7), and estriol-16α-glucuronide], of plasma steroids (7) [progesterone, estradiol and testosterone] and of prostaglandins (9). The precision of steroid assays regarding interassay variation using the antibody-coated tube method is shown in Table 1.

TABLE 1

Interassay Variations of Urinary Steroid Measurements by Chemiluminescence Immunoassay Using the Antibody-Coated Tube Method

Urinary Steroid	n	Pool 1 Mean	Pool 1 (SD)	Pool 1 CV,%	Pool 2 Mean	Pool 2 (SD)	Pool 2 CV,%
Pregnanediol-3α-glucurone (μmol/l)	6	4.6	(0.3)	6.5	9.95	(0.91)	9.2
Estrone-3-glucuronide (nmol/l)	6	23.2	(3.55)	15.3	117.5	(10.0)	8.5

Although the antibody-coated tube method can be considered a single-tube assay, it has certain drawbacks, such as: (a) possible loss of immunoreactivity of the adsorbed antibodies upon storage; and (b) considerable batch-to-batch variation in the polystyrene tubes with respect to behaviour of adsorbed antibody. On the other hand, the use of a primary or a secondary antibody covalently coupled to polymer beads as the solid phase reagent adds to the stability and speed of the assay and avoids problems observed with the antibody-coated tube method (1).

Using this approach, sensitive dose response curves for haptens were obtained, with a linear segment between 5-100 pg hapten/tube. Representative dose response curves for progesterone utilizing two different marker conjugates (progesterone-11α-aminobutyl ethyl isoluminol or progesterone-11α-acridinium ester conjugate) are shown in Fig. 2.

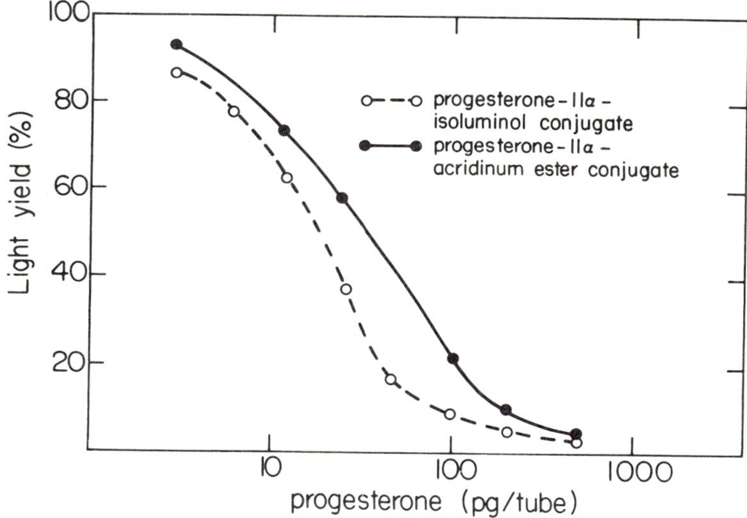

Fig. 2. Dose-response curves for progesterone as measured by chemiluminescence immunoassay, utilizing two different progesterone-chemiluminescent marker conjugates.

The above method was found to be suitable for clinical application. For instance, plasma levels of estradiol-17 could be monitored during gonadotropin therapy for ovulation induction (10). In addition, urinary levels of pregnanediol-3α-glucuronide (1) and estrone-3-glucuronide (1) could be measured in serial samples of early morning urine to assess follicular development and luteal function in human ovarian cycles. These results suggest that chemiluminescence based-immunoassays for haptens can provide satisfactory non-isotopic alternatives to RIA.

Luminescence-based solid-phase immunoassay for peptide hormones

Solid-phase immunoassays monitored by luminescence have been developed for the measurement of human chorionic gonadotropin (hCG) and luteinizing hormone (LH) from biological fluids. The methods involved the use of two types of formats: (a) Competitive immunoassays, and (b) two-site immunoluminometric immunoassays. Each format will be discussed separately below.

Competitive Immunoassays: These techniques utilize two types of labelled probes: (a) a direct probe [e.g. hCG-isoluminol (3) or acridinium ester conjugate (see Fig. 3)], or (b) an indirect probe [e.g. biotinyl hCG conjugate (see Fig. 3)]. In each case a solid-phase antibody (either the primary or a secondary antibody directed against the first antibody) is utilized to separate bound and free hormone. After the respective binding reactions, the label that is bound is measured by chemiluminescence (e.g. hormone-isoluminol or hormone-acridinium ester conjugate) or by bioluminescence (hormone-biotin conjugate) utilizing the avidin:biotinyl NAD^+- or ATP-dependent enzyme complex. Schematic diagrams of the steps involved in the end-point determinations using the different probes are shown in Fig. 4.

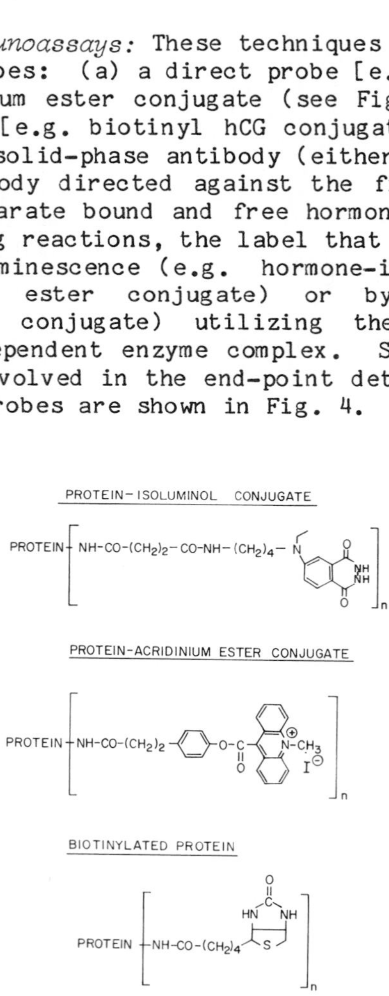

PROTEIN-ISOLUMINOL CONJUGATE

PROTEIN-ACRIDINIUM ESTER CONJUGATE

BIOTINYLATED PROTEIN

Fig. 3. Structure of the various probes used in the development of luminescence-based assays for peptide hormones.

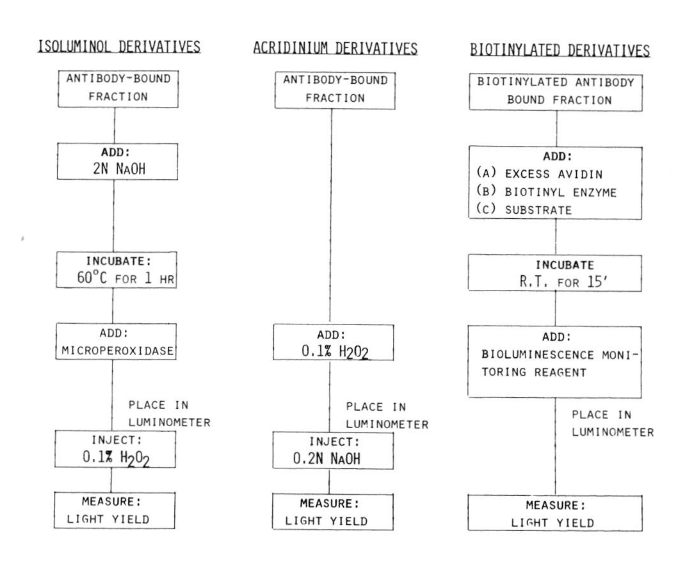

Fig. 4. Luminometric end-point determinations.

Sensitive dose-response curves for hCG (3) and for LH (11) were obtained using either the direct or indirect probe. Fig. 5 shows the results.

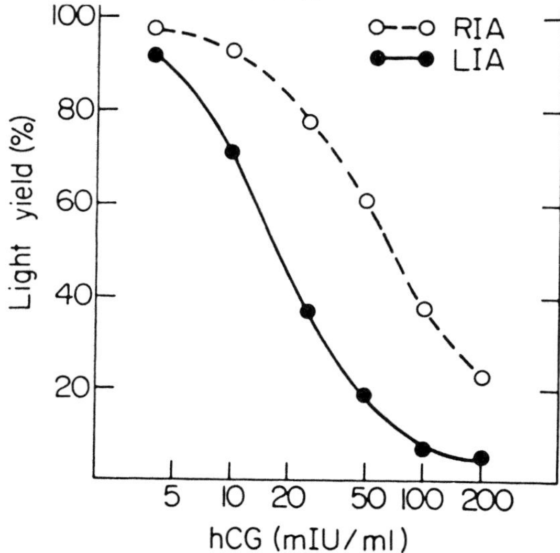

Fig. 5. Dose-response curves for hCG using biotinyl hCG conjugate and bioluminescence as an end-point (●) or I^{125}hCG as the label (O).

Two-site Immunoluminometric Assays: These sandwich methods were developed using two different types of labelled-antibody probes: (a) a direct probe (e.g. hormone-specific monoclonal antibody labelled with isoluminol or an acridinium ester derivative); or (b) an indirect probe (e.g. biotinyl monoclonal antibody). In each case an excess concentration of hormone-specific polyclonal antibody adsorbed onto the walls of polystyrene tubes or coupled to polymer beads serves as the solid-phase reagent. After the respective binding reactions, the end point is determined by bioluminescence using the bacterial or firefly luciferase system or by chemiluminescence. A typical dose-response curve for hCG utilizing the indirect probe (biotinyl antibody) is shown in Fig. 6.

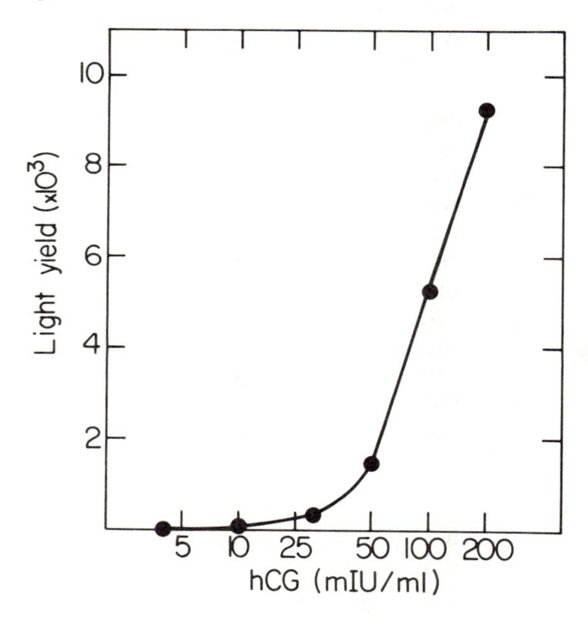

Fig. 6. An immunobioluminometric [IBMA] dose-response curve for hCG. In this system excess polyclonal anti-hCG IgG adsorbed to tubes served as the solid-phase, and biotinylated monoclonal anti-β-hCG served as the labelled primary probe. After the binding reaction, excess acetylated avidin, and biotinylated glucose-6-phosphate dehydrogenase are added. After 30 min incubation, the solutions are removed by aspiration. The substrates (glucose-6-phosphate and NAD^+) are added. After a short incubation (15'), the bioluminescence monitoring reagent (luciferase/FMN/decanal) is added, and the light that is generated is integrated for 10 sec.

CONCLUSIONS

In this review we have described various direct and indirect probes used in the development of luminescence-based immunoassays of haptens and peptide hormones. Utilizing the direct probes chemiluminescence-monitored assays were established for the measurement of steroids (1,2), urinary steroid metabolites (1,2,7), drugs (7), prostaglandins (9) and peptide hormones (3,11) from biological fluids. On the other hand, the indirect probes, mediated via the avidin-biotin system, were utilized at this stage to develop bioluminescence-based assays for peptide hormones. The use of the avidin-biotin complex (5,12) served to amplify the sensitivity of the assay since the tetrameric structure of avidin can accommodate the biotinyl residues on the primary probe (e.g. biotinyl antibody, see Fig. 3) as well as the biotinyl residues present on the enzyme.

Biotinylation of antibodies, enzymes, or peptide hormones can be achieved under very mild conditions and the resulting conjugates show very little loss of specific activity (12). Another advantage of the biotinylated probes is that various options exist for the end-point determination (e.g. luminescence, fluorescence, spectrometry). At this stage, the described format requires the stepwise addition of the various probes. We are currently investigating the application of premade complexes which comprise the appropriate probes (i.e., biotinyl antibody, avidin and biotinyl enzyme) in order to minimize the number of steps necessary for assay as well as to further increase the sensitivity of the assay.

ACKNOWLEDGEMENTS

We are grateful to Mrs. M. Kopelowitz for secretarial assistance and to the Bosch Foundation and the U.S. Binational Science Foundation (Jerusalem) for financial support.

REFERENCES

1. Kohen, F., Lindner, H.R. and Gilad, S. (1983). Development of chemiluminescence monitored immunoassays for steroid hormones, *J. Steroid Biochem.* 19, 414-418.
2. Kohen, F., Pazzagli, M., Serio, M., de Boever, J. and Vanderckhove, D. (1984) Chemiluminescence and bioluminescence immunoassays. *In:* "Alternative Immunoassays" (Ed. W.P. Collins). John Wiley and Sons, in press.

3. Barnard, G.J., Kim, J.B., Brockelbank, J.L. and Collins, W.P. (1984). The measurement of human chorionic gonadotrophin by chemiluminescence immunoassay and immunochemiluminometric assay, *Clin. Chem.* in press.
4. Weeks, I., Beheshti, I., McCapra, F., Campbell, A.K. and Woodhead, J.S. (1983). Acridinium esters as high-specific-activity labels in immunoassay, *Clin. Chem.* 29/8, 1471-1479.
5. Wilchek, M. and Bayer, E.A. (1984). The avidin-biotin complex in immunology, *Immunology Today* 5, 39-45.
6. Pazzagli, M., Messeri, G., Caldini, A.L., Monetic, G., Martinazzo, G. and Serio, M. (1983). Preparation and evaluation of steroid chemiluminescent tracers, *J. Steroid Biochem.* 19, 407-412.
7. Collins, W.P., Barnard, G.J., Kim, J.B., Weerasekera, D.A., Kohen, F., Eshhar, Z. and Lindner, H.R. (1983). Chemiluminescence immunoassays for plasma steroids and urinary steroid metabolites. *In:* "Immunoassays for Clinical Chemistry: A Workshop Meeting", Edinburgh, 1982, (Eds. W.M. Hunter and J.E.T. Corrie), pp. 373-397, Churchill Livingstone, Edinburgh.
8. Barnard, G., Collins, W.P., Kohen, F. and Lindner, H.R. (1981). A preliminary study of the measurement of urinary pregnanediol-3 α-glucuronide by a solid-phase chemiluminescence immunoassay, *In:* "Bioluminescence and Chemiluminescence", (Eds. W.D. McElroy and M.A. DeLuca), pp. 311-317, Academic Press, New York.
9. Weerasekera, D.A., Koullapis, E.N., Kim, J.B., Barnard, G.J., Collins, W.P., Kohen, F. and Lindner, H.R. (1983). Chemiluminescence immunoassay of thromboxane B_2. *In:* "Advances of Prostaglandin, Thromboxane and Leukotriene Research", Vol. 12 (Eds. B. Samuelsson, R. Paoletti and P. Ramwell), pp. 285-190, Raven Press, New York.
10. De Boever, J., Kohen, F. and Vanderckhove, D. (1983). A solid phase chemiluminescence immunoassay for plasma estradiol-17 β for gonadotropin therapy compared with two different radioimmunoassays, *Clin. Chem.* 29, 2068-2072.
11. Brockelbank, J.L., Kim, J.B., Barnard, G.J.,Collins, W.P., Gaier, B. and Kohen, F. (1984). The measurement of urinary LH by a solid-phase chemiluminescence immunoassay, *Annals of Clin. Biochem.*, in press.
12. Bayer, E.A. and Wilchek, M. (1980). The use of the avidin-biotin complex as a tool in molecular biology, *Methods Biochem. Anal.* 26, 1-45.

THE USE OF ISOLUMINOL AND ACRIDINIUM LABELS IN IMMUNOASSAY

G. Barnard, J.L. Brockelbank, J.B. Kim and W.P. Collins

Department of Obstetrics and Gynaecology
King's College School of Medicine
Denmark Hill, London SE5 8RX, England

INTRODUCTION

An immunoassay may be defined as an analytical procedure involving the competitive reaction between a limiting concentration of specific antibody and two populations of antigen, one of which is labelled or immobilised. The signal is inversely proportional to the concentration of analyte. This distinguishes an immunoassay (e.g. radioimmunoassay; RIA or chemiluminescence immunoassay; CIA) from an immunometric assay (e.g. immunochemiluminometric assay; ICMA) in which antibodies are used in excess. In this method, an antibody for a known antigenic determinant is immobilised on a solid-phase and binds the analyte. The second antibody is specific for an alternative epitope of the antigen and is labelled with a chemiluminescent marker. The signal is proportional to the concentration of analyte.

ISOLUMINOL DERIVATIVES

At the 2nd International Symposium on Applications of Bioluminescence and Chemiluminescence held in San Diego during August 1980, Barnard et al. reported a separation CIA for the measurement of pregnanediol-3α-glucuronide in diluted urine (1). Subsequently, several solid-phase and liquid-phase CIAs using derivatives of isoluminol have been developed for the measurement of haptens in extracts of peripheral plasma and in diluted urine and serum (2,3). Quality control data has been accumulated for the CIA of three steroids in plasma extracts using Dextran-coated charcoal or Sac-Cel (Wellcome Diagnostics Ltd) as the separation reagent and some aspects of the between batch variation are shown in Table 1. In addition, we have developed and evaluated CIAs for the direct measurement of

ISBN 0-12-426290-2

progesterone and oestradiol in aliquots of saliva using sheep anti-rabbit IgG covalently linked to polyacrylamide beads (3). The concentration (mean ± S.D.) of progesterone in daily samples of early morning saliva throughout 6 menstrual cycles are shown in Fig. 1.

TABLE 1

CIAs for Haptens: Between Batch variation

Hormone	units	Dextran-coated Charcoal separation				
		No. of	QC1		QC2	
		batches	Mean	CV%	Mean	CV%
Oestradiol	(pmol/l)	16	111	14.4	1826	9.0
Testosterone	(nmol/l)	10	2.3	13.0	11.8	10.2
Progesterone	(nmol/l)	30	10.8	17.6	16.7	6.0
			Sac-Cel separation			
			QC3		QC4	
Oestradiol	(pmol/l)	4	540	10.0	1190	3.9
Testosterone	(nmol/l)	5	2.5	12.0	22.2	6.8

Recently, it has been shown that protein-isoluminol conjugates (e.g. hCG-ABEI, a non-specific anti-hCG-ABEI and a sheep anti-rabbit IgG-ABEI) may be prepared with a hemisuccinamide derivative of ABEI (ABEI-H) which is available commercially. These conjugates have acceptable quantum yields upon oxidation at high pH (3) and the specific activity of each labelled protein has not changed over 18 months. The incubation with sodium hydroxide at high temperature, however, is an undesirable feature of the assay and alternative methods to enhance sensitivity are being investigated.

ACRIDINIUM ESTERS

In 1981, Simpson et al. reported the use of acridinium esters as labels in immunoassays (4). The use of these compounds may increase quantum efficiency, avoid serious quenching when associated with proteins or haptens, and allow simpler oxidation systems involving alkaline peroxide alone. We have developed and are evaluating a CIA for the measurement of plasma progesterone which correlates well with the results from a reference RIA ($r = 0.93$; $n=30$) and gives the same clinical information regarding the presence of a corpus luteum. In addition, we have developed a CIA and ICMA for the measurement of hCG using acridinium ester labels. Both methods give similar results for the early detection of pregnancy to those obtained from a conventional RIA. An analogue presentation of an ICMA calibration curve for hCG and the analysis of 3 plasma samples is shown in Fig. 2.

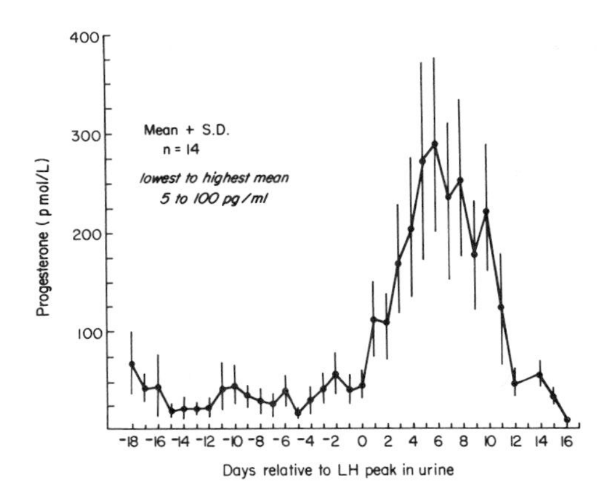

Fig. 1. The measurement of salivary progesterone throughout 6 menstrual cycles by direct CIA

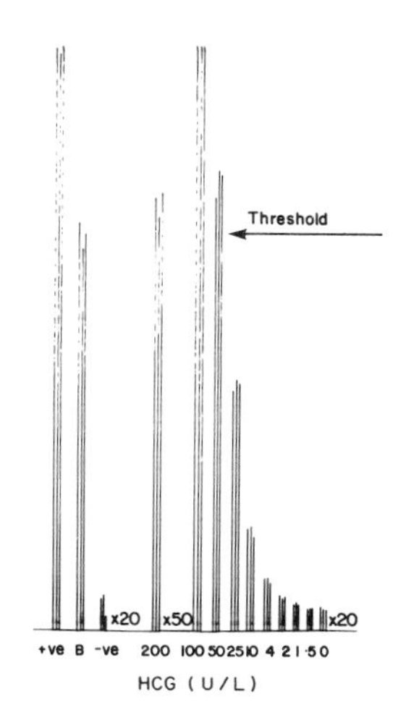

Fig. 2. Analogue ICMA calibration curve for hCG using an acridinium-labelled antibody (B = borderline result; threshold for pregnancy set at 48 U/L)

DISCUSSION

The results to date show that derivatives of isoluminol are

useful alternatives to isotopic labels for the measurement of haptens in diluted biological fluids or in concentrated extracts. The use of activators in place of sodium hydroxide will make the methods easier and quicker to perform. Acridinium esters are preferable labels for the measurement of proteins because the activated moiety dissociates from the carrier molecule immediately prior to light emission. Both methods involve rapid kinetics and necessitate the precise injection of oxidant into the tube which is positioned in front of the photo-detector. More recently, we have developed a bioluminescence immunoassay and an immunobioluminometric assay for the measurement of hCG using glucose-6-phosphate dehydrogenase and bacterial bioluminescence monitoring reagents (2). The continuous light emission obtained from these sensitive immunoassays has obviated the need for an injection system. Light detection may be performed in existing luminometers or liquid scintillation counters. Alternatively, we envisage that these systems may be used in association with dipstick or microtitre plate luminescence photometers.

REFERENCES

1. Barnard, G., Collins, W.P., Kohen, F. and Lindner, H.R. (1981). A preliminary study of the measurement of urinary pregnanediol-3α-glucuronide by a solid-phase chemiluminescence immunoassay. *In* "Bioluminescence and Chemiluminescence. Basic Chemistry and Analytical Applications" (Eds M. DeLuca and W.D. McElroy) pp 311-317. Academic Press, New York.

2. Kohen, F., Bayer, E., Wilchek, M., Barnard, G., Kim, J.B., Collins, W.P., Beheshti, I., Richardson, A. and McCapra, F. (1984). Development of luminescence based immunoassays for haptens and for peptide hormones. This volume

3. Barnard, G. (1984). Chemiluminescence immuno- and immunochemiluminometric assay. *In* "Alternative Immunoassays" (Ed W.P. Collins). John Wiley and Sons, Chichester, in press.

4. Simpson, J.S.A., Campbell, A.K., Woodhead, J.S., Richardson, A., Hart, R. and McCapra, F. (1981). Chemiluminescent labels in Immunoassay. *In* "Bioluminescence and Chemiluminescence. Basic Chemistry and Analytical Applications" (Eds M. DeLuca and W.D. McElroy) pp 673-679. Academic Press, New York.

DIRECT CHEMILUMINESCENCE IMMUNOASSAY FOR PROGESTERONE

J. De Boever*, F. Kohen+ and D. Vandekerckhove*

*Department of Obstetrics & Gynaecology, Academic Hospital, De Pintelaan 185, B-9000 Ghent, Belgium
+ Department of Hormone Research, The Weizmann Institute of Science, Rehovot 76100, Israel

INTRODUCTION

Radioimmunoassay (RIA) is widely used in many clinical laboratories to determine progesterone concentrations in extracts of serum or plasma. Recently direct RIA methods for progesterone, omitting extraction, have been described. RIA is however associated with problems such as health hazards, radiolysis of labeled reagents and radioactive waste disposal. Recent studies, reviewed by Kohen *et al.* (1) have shown that luminescent labels can substitute for radioisotopes, potentiating the development of chemiluminescence and bioluminescence immunoassays.

We have developed a solid phase chemiluminescence immunoassay (CIA) for the direct measurement of progesterone in serum. Preliminary results are presented.

MATERIALS

Reagents were as described (2); [1,2,6,7 - ^{3}H] progesterone (specific activity 104 kCi/mol) was purchased from Amersham Int. Ltd., Amersham, Bucks., U.K.; Danazol (17α-pregna-2,4-dien-20-yno [2,3-d] isoxazol-17-ol) was a gift from Winthrop Labs, Brussels, Belgium.
Reagent Solutions including buffer and oxidizing system were described previously (2).

METHODS

Antibodies to progesterone-11-hemisuccinate coupled to BSA

ANALYTICAL APPLICATIONS OF
BIOLUMINESCENCE AND CHEMILUMINESCENCE

ISBN 0-12-426290-2

were raised in rabbits as described (3) and characterized in terms of titer, affinity and specificity by RIA procedures (4,5).

Chemiluminescent marker conjugate. Progesterone-11-hemisuccinate-aminobutylethyl-isoluminol (P11-ABEI) was synthesised as described (6). A stock 5 µg/mL solution in ethanol was stored at 4°C.

Preparation of immunoadsorbant. A purified IgG fraction of polyclonal antiprogesterone antibody was covalently coupled to immunobead matrix (2) and treated as directed by the supplier (Bio-Rad). A stock of 10 mg immunoadsorbant per mL was kept at 4°C. Before immunoassay a fraction of the immunoadsorbant was diluted 100 fold with a suspension of immunobead matrix containing 1 mg matrix beads per mL (diluted solid-phase antibody).

Sample collection and extraction. Blood samples were obtained by venipuncture; serum was stored frozen at -20°C. 0.3 mL serum was extracted with 3 mL petroleum ether (P.E.). The mean efficiency of extraction as determined by measuring the analytical recovery of tritiated progesterone added to the serum samples was 87% (SD 4.7).

Immunoassay methods. For chemiluminescence immunoassay 0.05 mL of serum sample was diluted with 0.2 mL assay buffer containing 500 ng of Danazol. Standards contained per mL : 0.2 mL male serum, 2.0 µg Danazol and from 62.5 pg to 8 ng progesterone. Duplicate 0.05 mL aliquots of diluted serum samples or of standards were added to lumacuvettes (Lumac Systems, Basel, Switzerland) containing 0.1 mL diluted solid-phase antibody, and incubated at 4°C for 1 hour. 50 pg of P11-ABEI in 0.1 mL of assay buffer were added to all tubes. The contents were mixed, incubated overnight at 4°C, 0.9 mL of wash solution was added, and the tubes were centrifuged (10 min, 2000 x g, room temperature). The wash procedure was repeated once. The supernatant fluid was decanted and the luminescent marker bound to solid-phase antibody in the precipitate was assayed as described (2). Light emission was integrated during 10 s with a Biocounter M2000 of Lumac.

RIA was performed as described previously (4,5).

RESULTS

Validation of the Chemiluminescence Immunoassay.

Accuracy. Analytical recovery of progesterone added to normal male serum was 99.3% over a range of 1.25 to 40 ng progesterone per mL (RIA : 98.7%).

Sensitivity. The least amount of progesterone that could be distinguished from zero (mean -2 SD) was calculated from calibration curves prepared in duplicate. The mean value was 2 pg (RIA 4.3 pg).
Precision. Intra-assay CV was 9% (RIA 10%). Inter-assay CV was 10.6% (RIA 12%).
Correlation with RIA. The concentration of progesterone in 53 serum samples as determined by RIA (x) and by CIA (y) gave a linear regression equation of $y = 1.036\ x - 0.086$ ($r = 0.92$) (Fig. 1).

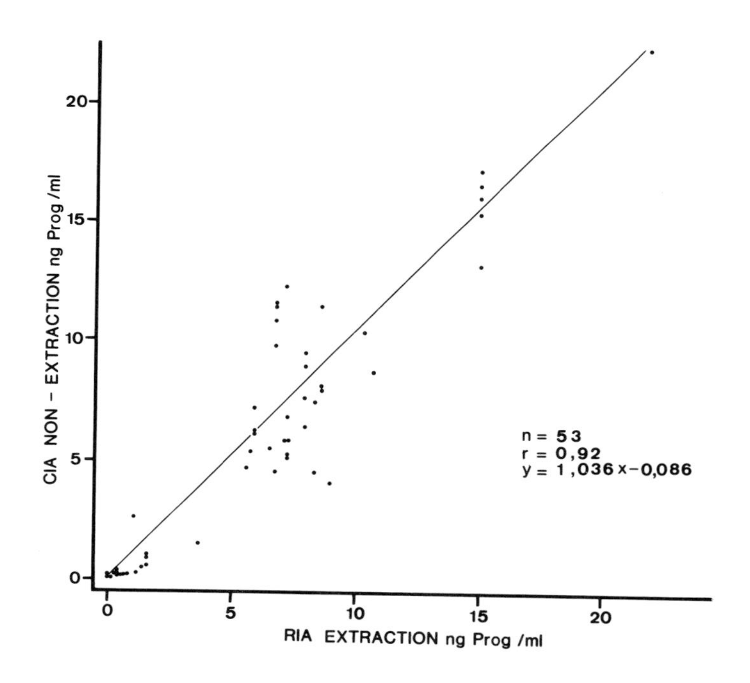

Fig. 1. Correlation between progesterone concentrations determined by RIA and direct CIA methods.

CONCLUSIONS

A direct CIA for progesterone, i.e. without prior extraction of serum, has been validated. The assay can be performed with acceptable accuracy, sensitivity and precision and may offer a valuable alternative method to radioimmunoassay methods, including direct RIA for progesterone. Interference of serum compounds with the measurement of chemiluminescence is not encountered. This may be ascribed to the small amount

(10 μl) of serum sample required for the assay and to the use of a suitable solid-phase antibody.

ACKNOWLEDGEMENTS

We are grateful to Dr. P . Stanley and Mr J. Vossen of Lumac/3M, Schaesberg, The Netherlands, for the loan of a Biocounter M2000; to Mrs D. Leyseele, Miss M. Eeckhout and Mrs J. Osher for technical assistance; to Mr H. Chretien and Mrs S. Hoste for graphical and secretarial assistance. This work was supported by grants to F.K. form the Bosch Foundation and the Binational Science Foundation (Jerusalem). F.K. is the Fanny and Samuel Kay Research Fellow at the Weizmann Institute.

REFERENCES

1. Kohen, F., Pazzagli, M., Serio, M., De Boever, J. and Vandekerckhove, D. (1984). Chemiluminescence and bioluminescence immunoassays. *In* "Alternative Immunoassays" (Ed W.P. Collins) In press. John Wiley and Sons, London.
2. De Boever, J., Kohen, F., Vandekerckhove, D. (1983). Solid-phase chemiluminescence immunoassay for plasma estradiol-17β during gonadotropin therapy compared with two radioimmunoassays. *Clin Chem* 29, 2068-2072.
3. Kohen, F., Bauminger, S. and Lindner, H.R. (1975). Steroid immunoassay. *In* "Proc. Fifth Tenovus Workshop" (Eds E. Cameron, S. Hillier and K. Griffiths). pp 11-32. Alpha Omega Publishing Ltd., Cardiff.
4. Bauminger, S., Kohen, F. and Lindner, H.R. (1974). Steroids as haptens : optimal design of antigens for the formation of antibodies to steroid hormones. *J. Steroid Biochem* 5, 739-747.
5. De Boever, J., Vandekerckhove, D. and Van Maele, G. (1979). Specific radioimmunoassays for 17β-estradiol and progesterone in serum. *Arch Int Physiol Biochim* 87, 790-797.
6. Pazzagli, M., Kim, J.B., Messeri, G., Martinazzo, G., Kohen, F., Prancheschetti, F., Tommasi, A., Salerno, R., and Serio, M. (1981). Luminescent immunoassay (LIA) for progesterone in a heterogeneous system. *Clin Chim Acta* 115, 287-296.

HOMOGENEOUS LUMINESCENCE IMMUNOASSAY: AN APPLICATION FOR URINARY ESTROGENS MEASUREMENT

G.Messeri*, A.L. Caldini**, F.Franceschetti***
R.Salerno** and M.Serio**

*Clinical Chemistry Lab., USL 10D, Firenze, Italy
**Endocrinology Unit, University of Firenze, Italy
***Physiopathology of Reproduction Service
University of Bologna, Italy

INTRODUCTION

When the binding to the specific antiserum induces a substantial modification of the light emission features of a chemiluminescence (CL) labelled steroid (enhancement, quenching, kinetic shift etc.) an homogeneous immunoassay requiring no bound/free separation step can be developed.

Estradiol-17- beta-hemisuccinate-ABEI (E2-hs-ABEI) conjugate showed a five fold increase of the light efficiency following the binding to anti-estriol-16,17-hemisuccinate-BSA-serum. This opportunity was exploited for the development of an homogeneous immunoassay for the measurement of "total" estrogens in non-pregnancy urine.

MATERIALS AND METHODS

The synthesis of ABEI, of E2-hs-ABEI and the reagents for immunoassay were as previously described (1). The antiserum was raised in rabbits by using estriol-16,17-dihemisuccinate-BSA conjugate as the immunogen. Measurement of light emission were made with an automatic luminometer (Picolite Model 6500; Packard Instrument Co.) .

Urine (10 uL) were incubated for 1h at 50 °C with 0.2 mL of acetate buffer (0.1 mol/L,pH 5.2) containing 1000 Fishmann U. of beta-glucuronidase and 15000 Roy U. of sulpha-

ISBN 0-12-426290-2

tase. The hydrolysed samples (10 uL) or the standard solutions (3.5-500 pg/tube) were transferred to polystyrene assay tubes. With the aid of the luminometer injection system 0.2 ml of the tracer (0.1 pmoles) and 0.1 mL of the antiserum (titre: 1/20.000) were added to each tube. After a one hour incubation at r.t., 0.1 mL of microperoxidase (5 umol/L) were added and , after a 10 seconds delay, the light emission was started by the injection of 0.1 mL of H_2O_2 (1.5mL/L). The light output was recorded for 10 sec.

RESULTS

The antiserum specificity was tested by the conventional competition experiments. Cross-reactivity (%) was respectively: Estriol (100), E2 (161), Estrone (193), 16beta-OH-E2 (87) but it was less than 0.01% for Testosterone, Pregnandiol, Pregnantriol, Cortisol,DHA, Androsterone, Ethiocolanolone and Cholesterol.

When using E2 as the competitor, the sensitivity of the calibration curve was 2pg/tube (4 ug/L urine).

Recovery tests, performed by adding increasing amounts of authentic E2 (50,100 and 200 ug/L) to 15 urines from different subjects resulted respectively (%mean, range): 92.1 (82.3-116.2), 95.7 (79.4-116.0) and 92.3 (84.5-113.5).

The possible interference of urine was tested by measuring the light emission of tubes containing respectively buffer, and buffer plus increasing amount of tracer in the presence and in the absence of 0.5 uL of urine from 30 different subjects. The urine resulted to rise significantly the blanck value, but this increase was negligible if compared to the elevated output of the tracer troughout all the points of the calibration curve. No significant difference was found when measuring the light emission of the tracer alone or in the presence of the urine aliquot (0.5 uL).

Precision, as evaluated by testing two pooled urine samples at low and high estrogen content resulted respectively (mean+/-SD,CV%): 40.5+/-2.6,6.5 and 138.5, 9.2,6.7 (within run), 41.6+/-4.1,9.8 and 145.0+/-12.5,8.6 (between run).

When comparing the concentration of "total" urinary estrogens as measured by the luminescence immunoassay (y) and a conventional fluorometric method (x) the regression equation was y=2.506x - 2.833 (r=.958, n=50).

The concentration of "total" estrogens in early morning urine samples, daily collected by 20 healthy women is shown in fig.1.

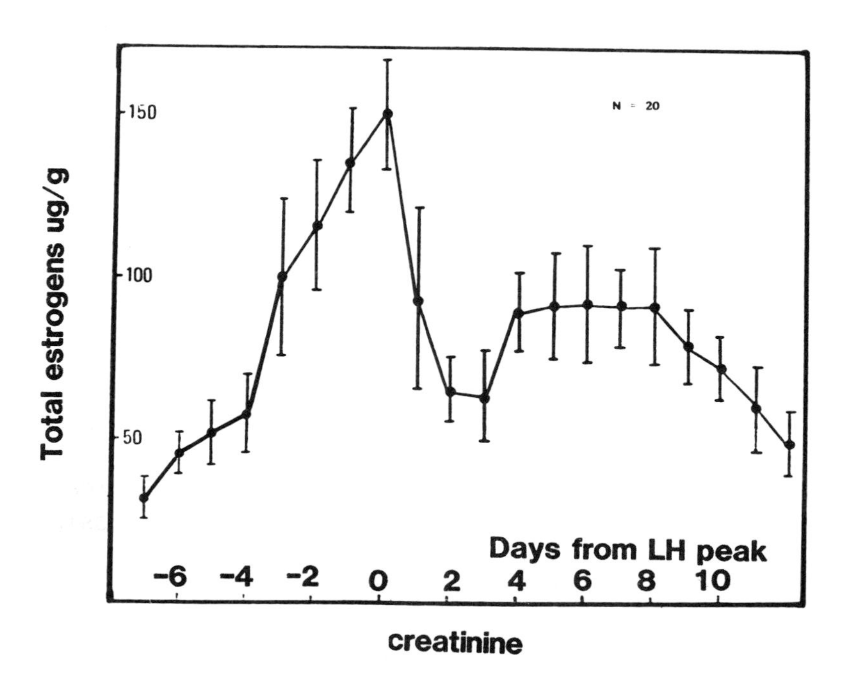

Fig.1 Reference values throughout the menstrual cycle (mean+/- SE).

DISCUSSION

The affinity of the antiserum we used appeared higher towards estrone and E2 than towards estriol. As a consequence, when using E2 as the reference compound, the method should underestimate Estriol (-40%) and overestimate Estrone (-10%). This drawback is anycase common to the chemical methods, as different recoveries were reported for the different estrogens (2).

The amount of urine we used in the test was so reduced that no interference on the CL reaction was observed.

In terms of accuracy and precision the method appears fully comparable to the most widely used RIA methods.

When compared to the fluorometric method, immunologic values appeared markedly higher than the fluorometric ones. Such discrepancies between immunological and chemical estrogens assay were already reported (2) and were mainly attributed to procedural losses associated with the solvent extraction and the acid hydrolysis of the chemical method.

The method we described allows the direct measurement of "total" estrogens on enzymically hydrolysed urines in the range 5 - 500 ug/L. With the aid of an automatic luminometer, the assay can be accomplished semi-automatically within 3 h and it appears a typical example of a possible application of the CL homogeneous immunoassay.

ACKNOWLEDGEMENTS

This work was supported by a grant of the University of Firenze. R.Salerno is a fellow of the Lega Italiana per la Lotta contro i Tumori.

REFERENCES

1. Messeri,G. et al (1984). Homogeneous Luminescence Immunoassay for Total Estrogens in Urine, Clin.Chem. (in press).
2. Baker,T.S. et al (1979). The direct radioimmunoassay of estrogen glucuronides in human female urine,Biochem.J. 177, 729-738.

AN AUTOMATED LUMINESCENCE IMMUNOASSAY FOR THE MEASUREMENT OF ESTRIOL-16 alpha-GLUCURONIDE IN PREGNANCY URINE

A.L.Caldini*, G.Messeri**, P.Buzzoni*** and P.Borri***

**Endocrinology Unit, University of Florence, Italy*
***Clinical Chemistry Lab., USL 10/D, Florence, Italy*
****Dept. of Obstetrics and Gynecology, University of Florence, Italy*

INTRODUCTION

In pregnancy estriol-16alpha-glucuronide (Eg) represents 67.1% of total urinary estriol (1) and it was reported to be an useful test in feta well-being evaluation.

A chemiluminescent (CL) immunoassay for the measurement of this steroid was developed. The light emission of the steroid - CL tracer conjugate increases about four fold following its binding to the specific antibody:thus an homogeneous immunoassay,not requiring bound /free separation step,could be developed for the direct measurement of Eg in crude pregnancy urine samples.

MATERIALS AND METHODS

The antiserum to Eg was raised in rabbits by using estriol - 16,17-dihemisuccinate-BSA conjugate as the immunogen (2). The antiserum titre and specificity were assessed as previously described (3).

6(N-(6-aminobuthyl)-N-ethyl)-amino-2,3-dihydrophthalazine -1,4-dione (ABEI) was covalently linked to estriol-16alpha-hemisuccinate-BSA to yield estriol-16alpha-hemisuccinate-ABEI (E3-ABEI). The procedures for the synthesis, the purification and the characterization of the CL tracer were described elsewhere (4).

ISBN 0-12-426290-2

The assay buffer was borate buffer, 0.1 mol/L, pH 8.6, containing 9 g of NaCl, 80 mg of BSA and 80 mg of human IgG per liter.

The assay procedure was as following: dilute the urine samples 1:10000 with assay buffer. Dilute the standard solution (Eg) to the appropriate concentration (80 - 10000 ng/L) with assay buffer. Transfer 0.1 mL of diluted standard or sample to the assay tubes and place the tubes in the luminometer (Picolite 6500, Packard Instrument Co.) where 0.1 mL of E3-ABEI (100 fmoles), 0.1 mL of antiserum (titre 1:20000) and 0.1 mL of microperoxidase (5 umol/L) are automatically added. After 1 hour of incubation at room temperature, the automatic injection of 0.1 mL of hydrogen peroxide (0.15%) initiates the luminescence reaction. The light emission was measured using the integration mode over a 10 sec interval after the injection of the starter reagent.

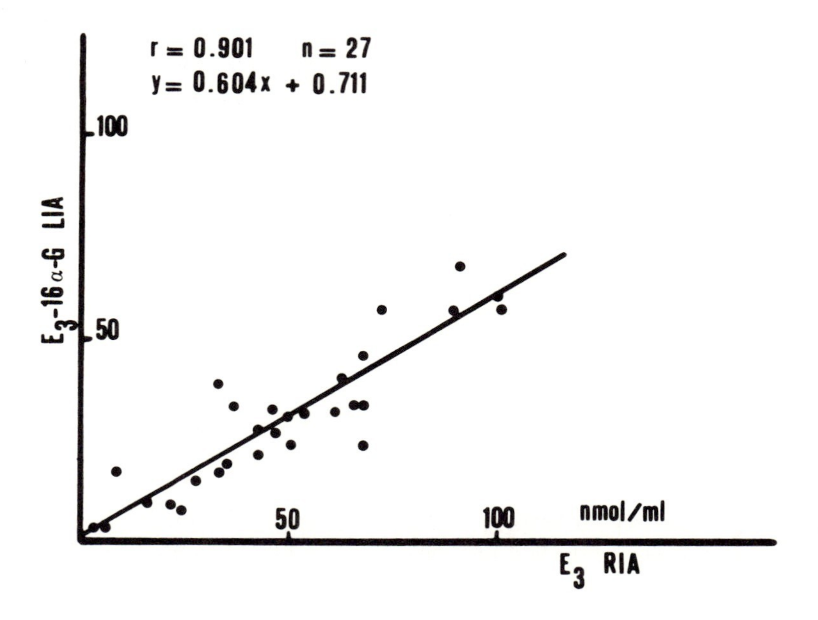

Fig.1 Correlation between the luminescence immunoassay for EG and a RIA method for total urinary estriol.

RESULTS

The effect of urine on the CL reaction was found to be negligible at the low urine concentration used in the assay (0.01 uL/tube) (data not shown).

The sensitivity of the method, calculated from 10 different calibration curves , was 10 pg per tube.

The within assay precision (n=15), evaluated using two samples of pooled urine with low and high Eg content,was respectively (CV) 7.2% and 5.1%. The between assay precision (n=12) was evaluated using in the same pooled samples and resulted 10.4% and 8.5% respectively.

Results by the proposed method correlate well with those by a conventional RIA method for total urinary estriol (Fig.1).

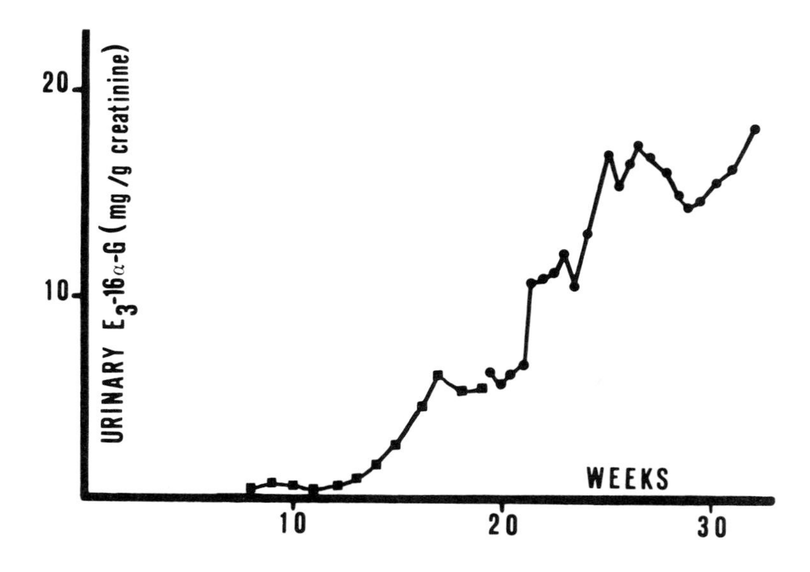

Fig.2 Concentrations of EG (ug/g creatinine) in early morning urine of two healthy women in relation to gestational age.

CONCLUSIONS

The method described here allows the direct measurement of Eg in diluted pregnancy urine and gives reliable results for Eg concentrations between 1 and 100 mg/L.The sensitivity and precision of this method are comparable to those of conventional RIA methods. Since no separation step is required the method is automated and it can be accomplished within 90 min.

An efficient follow-up of high risk pregnancy requires the repetition of the test during gestation. The measurement of Eg in early morning urine gives useful information about fetal well being with considerable psychological and practical advantages over blood drawing or 24 h urine collection. In Fig.2 are shown the concentrations of Eg in early morning urine of two healthy women in relation to gestational age.

ACKNOWLEDGEMENTS

This paper has been supported by a grant from the Italian Research Council (Progetto Finalizzato Tecnologie Biomediche). We thank Bouty S.p.A., Milan, Italy for technical assistance.

REFERENCES

1. Alexander,S. et al (1979). Renal clearance of estriol and its conjugates in normal and abnormal pregnancies, J. of Clin. End. and Metab.49,588-593.
2. Roda,A. et al (1980). Production of a high titer antibody to bile acids, J.steroid Biochem.13,449-454.
3. Messeri,G. et al (1984). Homogeneous luminescence immunoassay for total estrogens in urine, Clin. Chem. (in press).
4. Pazzagli, M. et al (1983). Preparation and evaluation of steroid chemiluminescent tracers, J.steroid Biochem.19, 407-412.

CYCLIC AMP IN PMN MEASURED BY CHEMILUMINESCENCE ENERGY TRANSFER ASSAY

P.A. Roberts, A. Patel, S.C.L. Barrow, M.B. Hallett* and A.K. Campbell

*Departments of Medical Biochemistry and *Surgery, Welsh National School of Medicine, Heath Park, Cardiff, CF4 4XN, U.K.*

INTRODUCTION

Oxygen radical production by PMN can be activated by primary phagocytic and chemotactic stimuli and can be modified by a variety of secondary regulators which may act through changes in intracellular calcium and cyclic nucleotides (1). The aim of the work reported here is to characterise adenosine as a secondary regulator and the role played by cyclic AMP in PMN oxygen radical production.

The philosophy underlying our experiments is ultimately to measure both cyclic AMP and Ca^{2+} in intact cells (2). Accordingly a homogeneous chemiluminescence energy transfer immunoassay for cyclic AMP has been developed (3) which has the necessary properties to be used intracellularly. The stability of the chemiluminescent - labelled cyclic AMP (ABEIscAMP) used in this assay, has therefore been investigated both in the presence of phosphodiesterase(PDE) and in human erythrocyte ghosts.

METHODS

Human PMN were separated from whole heparinised blood by centrifugation through a stepwise Percoll gradient. Rat PMN isolation (4), monitoring of oxygen radical production, ghost preparation (1) and radioimmunoassay of cyclic AMP (5) are described elsewhere. The energy transfer assay for cyclic AMP (3) was able to measure standards over the range of 2,500 to 0.025 pmoles.

RESULTS

Adenosine and forskolin inhibited oxygen radical production

ISBN 0-12-426290-2

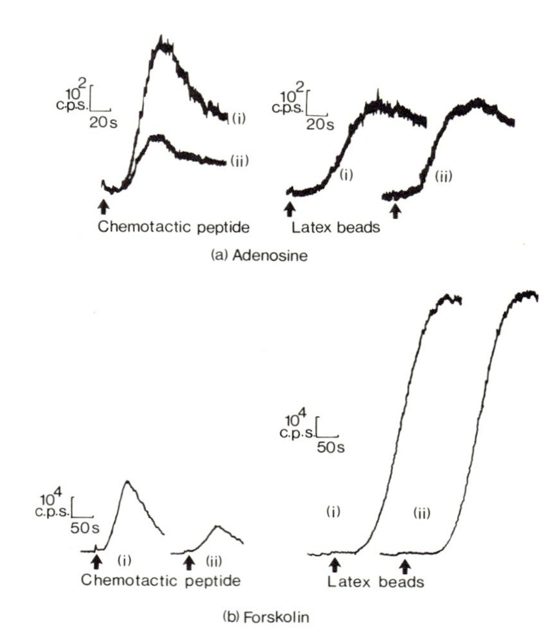

Fig. 1. Chemotactic peptide-stimulated oxygen radical production of a) human PMN after preinubcation for 2 min. with (i) Krebs and (ii) adenosine (10 μM), b) Rat PMN after preincubation with (i) ethanol in Krebs (0.1%) and (ii) forskolin (10 μM).

induced by chemotactic peptide, but not by latex beads (Fig. 1). 8-phenyltheophylline, an adenosine antagonist overcame this inhibition indicating that adenosine was not acting as an oxygen radical scavenger.

Both adenosine and forskolin elevated cyclic AMP levels in PMN, in both cases magnifying the elevation caused by chemotactic peptide: for instance addition of chemotactic peptide to human PMN elevated cyclic AMP levels from 2.6±.04 to 4.6±3.2 pmoles/10^6 cells in the absence of adenosine (10 μM), but from 1±0.4 to 9.45±2.54 pmoles/10^6 cells in its presence. Likewise addition of chemotactic peptide to rat PMN in the absence of forskolin elevated cyclic AMP levels from 6.77±0.86 to 15.79±0.5 pmoles/10^6 cells, whereas in its presence cyclic AMP was elevated from 7.99±0.74 to 33 ± 8.7 pmoles/10^6 cells.

ABEIscAMP was degraded by PDE *in vitro* and in ghosts (Table 1). Addition of IBMX (0.5 mM) reduced this. Similar losses in immunological cross-reactivity were also seen with ^{3}H-cAMP.

TABLE I.
Stability of ABEIscAMP and ^{3}H-CAMP in vitro and in human erythrocyte ghosts
% Unbound label

	ABEIscAMP		^{3}H-cAMP	
	-IBMX	+IBMX	-IBMX	+IBMX
PDE (1 mU)	92	-	80	-
PDE (0.3 mU)	76	-	60	-
Buffer	9	-	3	-
Ghosts prepn.				
Addition	15	5	25	22
Entrapment	21	15	35	40
Resealing	60	37	73	56
+ 30 min.	99	58	92	62

In order to establish a more sensitive chemiluminescent indicator of oxygen radicals a variety of bioluminescent substances have been compared with luminol. Pholasin from the mollusc *Pholas dactylus* appeared to provide a signal capable of detecting activation of single cells (Fig. 2).

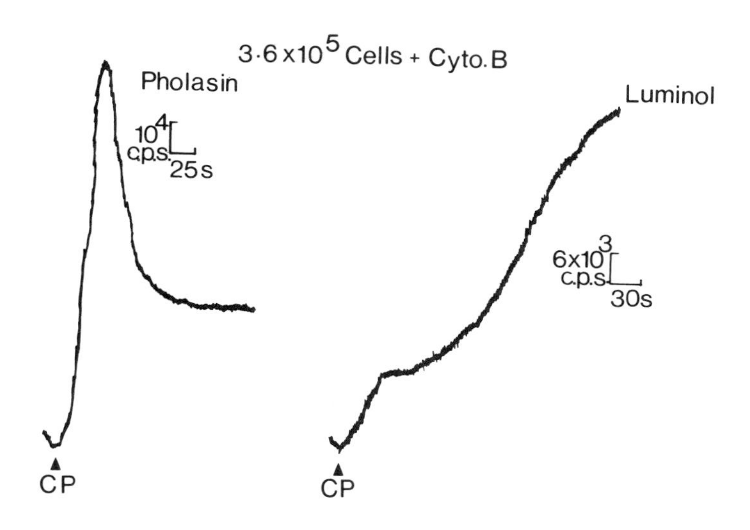

Fig. 2. *Chemotactic peptide-stimulated oxygen-radical production of human PMN monitored by Pholasin.*

CONCLUSIONS AND DISCUSSION

Adenosine and forskolin inhibited the response of PMN to chemotactic peptide, but not latex beads, concomitantly with an elevation in cyclic AMP. This suggests that the inhibition is specific for stimuli which require an increase in intracellular Ca^{2+} to activate PMN. The elevation in cyclic AMP seen with adenosine suggested that it was acting via an A_2 receptor. In addition, the use of forskolin has demonstrated the importance of this elevation in this response.

The observed conversion of ABEIscAMP to ABEIsAMP under conditions likely to exist in the cell may present problems in an intracellular assay. A derivative, substituted in the C_6 or C_8 of the purine ring may inhibit this conversion. Initial experiments with pholasin suggest that it can be used as a sensitive bioluminescent indicator of oxygen radical production from single cells, and as a label in our energy transfer assay capable of producing light under conditions compatible with the chemistry of living cells.

ACKNOWLEDGEMENTS

We would like to acknowledge the financial support of the MRC and ARC. We thank Jan Knight for *Pholas dactylus*.

REFERENCES

1) Campbell, A.K. & Hallet, M.B. (1983) Measurement of intracellular Ca^{2+} and oxygen radicals in PMN-erythrocyte 'ghost' hybrids. *J.Physiol.(London)* 338, 537-550.
2) Campbell, A.K. (1983) 'Intracellular Ca^{2+}, its Universal Role as Regulator', *John Wiley & Sons, Chichester*.
3) Campbell, A.K. & Patel, A. (1983) A homogeneous immunoassay for cyclic nucleotides based on chemiluminescence energy transfer. *Biochem. J.* 216, 185-194.
4) Hallett, M.B. & Campbell, A.K. (1983) Two distinct mechanisms for stimulation of oxygen-radical production by PMN. *Biochem. J.* 216, 459-465.
5) Siddle, K., Kane-Maguire, B. & Campbell, A.K. (1973) The effects of glucagon and insulin on adenosine 3':5' cyclic monophosphate concentrations in an organ culture of mature rat liver. *Biochem. J.* 132, 765-773.

EVALUATION OF LUMINESCENT IMMUNOASSAY METHODS FOR URINARY STEROIDS

R. Salerno*, G. Moneti**, A. Magini*, A. Tommasi* and M. Pazzagli*

**Endocrinology Unit, University of Florence, Italy*
***Mass Spectrometer Center, University of Florence, Italy*

INTRODUCTION

Determination of urinary steroid hormones is an important tool in the investigation of some physiopathological conditions, eg, Cushing's Syndrome. In our laboratory we have developed homogeneous non-separation methods and separation methods using different separation systems of the bound from the free fraction.

MATERIALS AND METHODS

All reagents have been described previously (1). Conjugation of CL tracers to steroid derivatives was performed according to Kohen et al (2). The antisera raised against cortisol, testosterone-17-glucuronide, estriol-16-glucuronide and "Total" estrogens were obtained using a protein-carrier steroid conjugate as immunogen; the same molecule was then conjugated to an isoluminol derivative (ABEI). The pregnanediol-glucuronide and estrone-glucuronide LIA methods were developed using monoclonal antibodies (3) (Fig. 1). A Berthold Biolumat LB 9500 Luminometer on-line with the computer was used to monitor the kinetics of the chemiluminescent reactions. In all methods reported in Table 1, diluted urine samples were used; in fact it has been demonstrated that in biological samples non specific interferences for urine dilutions of more than 1:100 are negligible.

ISBN 0-12-426290-2

TABLE 1

Measurement of urinary steroid metabolites by luminescent immunoassay (LIA) methods

Hormone Assayed	Steroid Tracer	Sensitivity (pg/tube)	Separation System
Urinary free cortisol	Cortisol-21-HS--ABEI	8-500	Charcoal
Estriol-16-glucuronide	Estriol-16-gluc-ABEI	10-500	Homogeneous
Testosterone-17-glucuronide	Testosterone-17-gluc-ABEI	8-1000	Coated tubes
Total estrogens	Estradiol-17-HS-ABEI	3.5-500	Homogeneous
Pregnanediol-3-glucuronide	Pregnanediol-3 gluc-ABEI	39-10000	Charcoal
Estrone-3-glucuronide	Estrone-3-gluc-ABEI	3.9-500	Charcoal

RESULTS

The specificity of LIA methods has been demonstrated previously (1,2). The specificity of the monoclonal antibody to pregnanediol-glucuronide has been studied by measuring diluted aliquots of early morning urinary samples of a normal menstrual cycle with the LIA method and a Mass-spectrometric-FAB method; both were then compared with a conventional GLC method (Fig. 3). The precision of the hetero-geneous method (solid phase-immunobeads and dextran coated charcoal) has been evaluated from precision profiles of the dose-response curves (Fig. 2).

CONCLUSIONS

The homogeneous LIA methods show many practical advantages,

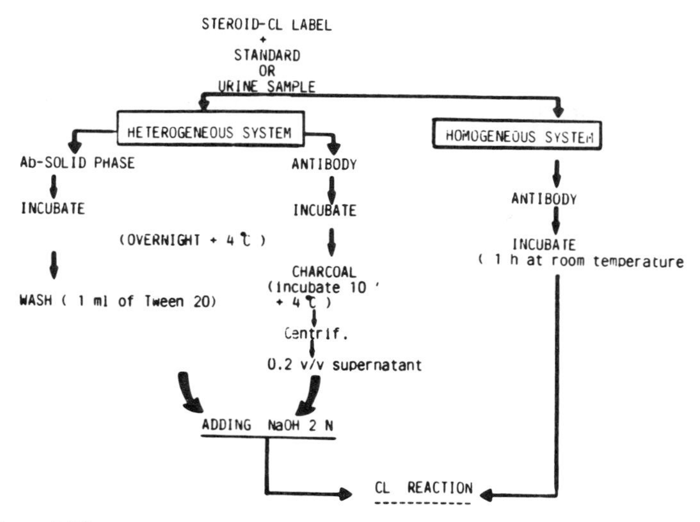

Fig. 1 LIA procedures in homogeneous and heterogeneous systems.

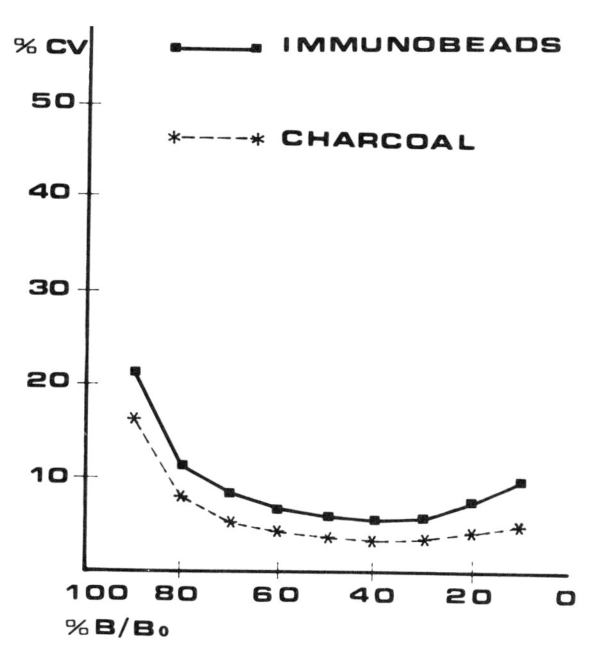

Fig. 2 Precision profiles for charcoal and solid phase LIA. Values (for duplicate determinations) were calculated from the mean of ten standard curves.

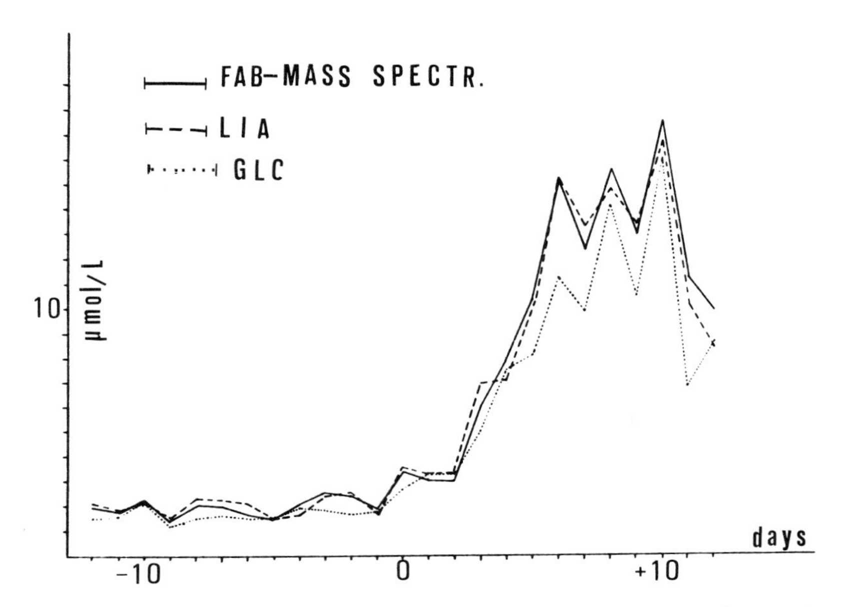

Fig. 3 Measurement of Pregnanediol-3-glucuronide (Pd-Gl) in early morning urine samples of a normal menstrual cycle using different assay procedures.

eg, they can be automated (2) (4). However, they have some disadvantages, eg, susceptibility to non specific interferences with amounts of urine >1 μl, and only a few CL tracers show a significant enhancement in the presence of antibody.

The heterogeneous LIA methods introduce a separation step in the procedure (charcoal or solid phase) which removes non specific interferences for up to 10 μl of urine.

Precision profiles of dose-response curves demonstrated that better results are obtained using the dextran-coated charcoal and it was more convenient in terms of both cost and precision characteristics.

ACKNOWLEDGEMENTS

Dr Salerno is the recipient of a fellowship from Lega italiana per la lotta contro i tumori-sez di Firenze. We thank the University of Florence for a research grant, and Bouty SpA, Milan, Italy for technical assistance.

REFERENCES

1) Pazzagli, M. et al (1981). Luminescent immunoassay of cortisol-2: development and validation of the immunoassay monitored by chemiluminescence. J. steroid biochem. 14, 1181-1187.

2) Kohen, F. et al (1979). An assay procedure for plasma progesterone based on antibody-enhanced chemiluminescence. FEBS Letter 104, 201-205.

3) Eshhar, Z. et al (1981) Use of monoclonal antibodies to pregnanediol-3-glucuronide for the development of a solid phase chemiluminescence immunoassay. Steroids 38, 89-109.

4) Pazzagli, M. et al (1984). On-line computer analysis of the kinetics of chemiluminescent reactions: application to luminescent immunoassay. In "Computers in Endocrinology" (Eds. D. Rodbard and G. Forti) 162-169, Raven Press, New York.

CHEMILUMINESCENT ACRIDINIUM ESTERS AS LABELS IN IMMUNOASSAY

I. Weeks and J.S. Woodhead

Department of Medical Biochemistry
Welsh National School of Medecine
Heath Park, Cardiff CF4 4XN, Wales, U.K.

INTRODUCTION

Since their introduction, immunoassays have relied mainly upon the use of radioactive isotopes as labels.The disadvantages of such labels are well recognised and are tolerated because no universally applicable, non-isotopic alternative has been developed, however certain chemiluminescent molecules have sensitivities of detection superior to that of 125-I.

During the last four years we have been investigating the use of acridinium esters as immunoassay probes in order to develop high-sensitivity immunoassays for polypeptides.

CHEMILUMINESCENT REACTION OF ACRIDINIUM ESTERS

In contrast to luminol derivatives, aryl acridinium esters undergo a chemiluminescent reaction in the presence of dilute alkaline hydrogen peroxide without the need for a catalyst. The reaction consists of a concerted multiple bond clevage mechanism via a dioxetanone intermediate to yield a vibronically excited molecule of N-methylacridone.The excited reaction product is dissociated from the rest of the original molecule prior to photonic emission and as a consequence the chemiluminescence quantum yield of the reaction is relatively independent of structural changes to the molecule at R (fig.1).

PROTEIN LABELLING WITH ACRIDINIUM ESTERS

The labelling of protein molecules is facilitated by the chemical modification of the molecule at R such that conjugation can be achieved without the destruction of the immunoreactivity of the protein.Recently we designed and synthesised such a label (1) which can be reproducibly

ISBN 0-12-426290-2

incorporated into proteins within 15 minutes at pH 8 to yield stable immunoreactive derivatives of high specific activity This property is due to the presence of an N-succinimidyl moiety which reacts rapidly with lysine residues under mild aqueous conditions.

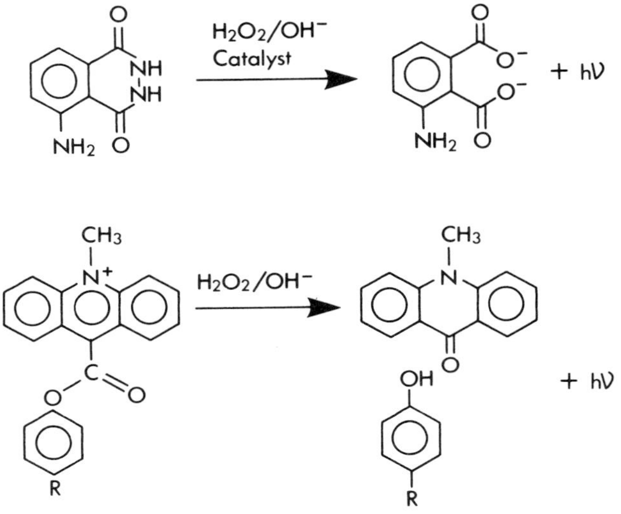

Fig.1 Chemiluminescence of luminol and an acridinium ester.

IMMUNOASSAYS

The majority of immunoassays which we have developed using this technique have consisted of two-site immunochemiluminometric assays (ICMA).The superiority of labelled antibody systems over labelled antigen systems is well recognised and the use of this assay mode enables full advantage to be taken of the high specific activities attainable using acridinium esters.Further, incubation times are short in such assays and complement the rapid measurement of chemiluminescence emission ($<$10 s) to yield very fast assays. Reagent stability is in excess of a year at 4°C.

Two-site ICMAs using both polyclonal and monoclonal antibodies have been developed for the measurement in serum of alphafetoprotein (AFP), ferritin, thyrotropin (TSH),choriogonadotropin (hCG) and prolactin.Examples of performance data from inter-assay precision profiles are shown in table 1.

TABLE 1

Performance data of some two-site ICMAs

Analyte/Units	Sensitivity	Working range/(%CV)		Incubation (hours)
AFP (kiu/l)	1.3	20- >246	(10)	1
Ferritin(ug/l)	0.8	6-467	(15)	1
TSH (mU/l)	0.004	0.06-6	(10)	3

The TSH-ICMA is particularly important in view of the consequences of its extremely high sensitivity (fig.2).

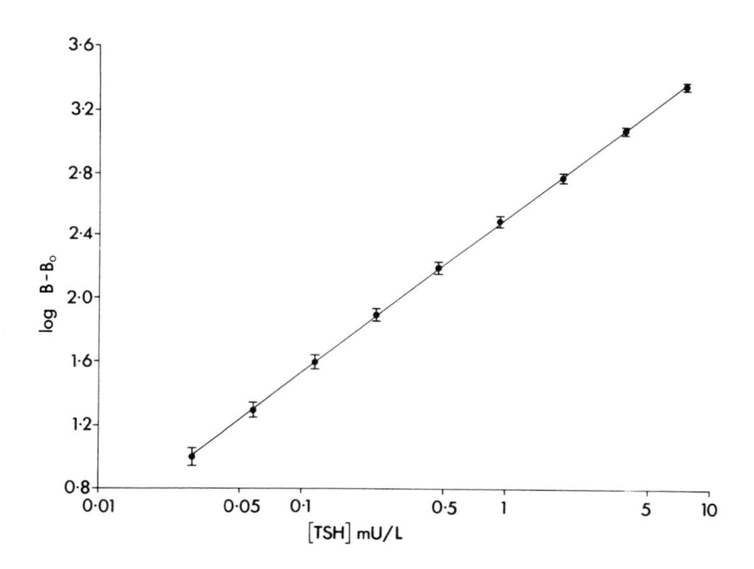

Fig.2 Dose-response relationship of the TSH-ICMA.

Initial studies have shown that this assay is capable of discriminating hyperthyroid patients from euthyroid subjects which is impossible using conventional 125-I based immunoassays.This makes the assay invaluable as a first-line test of thyroid function in addition to being an important tool in the study of the pathophysiology of the hypothalamic-pituitary-thyroid axis.

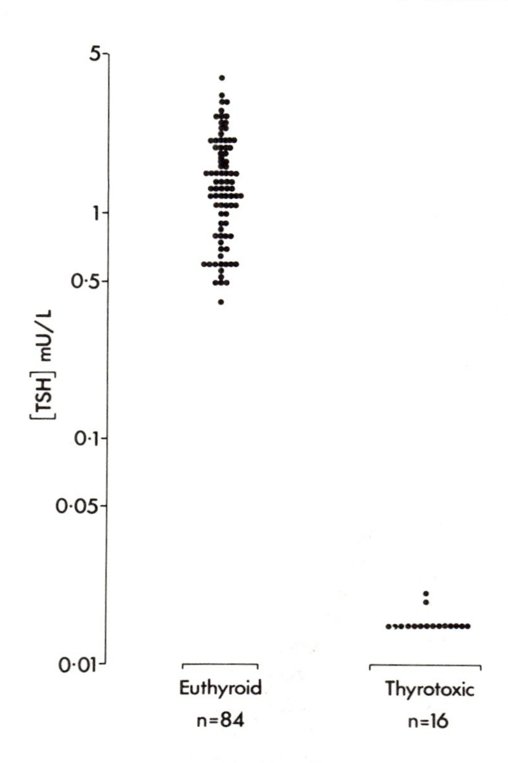

Fig.3 TSH levels in euthyroid and thyrotoxic subjects.

CONCLUSION

Chemiluminescent acridinium esters are capable of producing stable, high specific activity labelled proteins for the development of sensitive immunoassays.The universality of this labelling technique is desirable for the routine acceptance of this new technology.

REFERENCES

1.Weeks,I.,Beheshti,I.,McCapra,F.,Campbell,A.K. and Woodhead, J.S.(1983).Acridinium esters as high specific activity labels in immunoassay, Clin.Chem.29,1474-1479.
2.Weeks,I.,Campbell,A.K. and Woodhead,J.S.(1983).Two-site immunochemiluminometric assay for human α_1-fetoprotein, Clin.Chem.29,1480-1483.
3.Weeks,I.,Sturgess,M.,Siddle,K.,Jones,M.K. and Woodhead,J.S. A high sensitivity immunochemiluminometric assay for human thyrotropin.(1984).Clin.Endocrinol.20,489-495.

ROUTINE LUMINESCENCE IMMUNOASSAYS FOR HAPTENS AND PROTEINS

W.G. Wood

Klinik für Innere Medizin, Medizinische Hochschule Lübeck, D-2400 Lübeck 1, FRG.

INTRODUCTION

The aims of this study were to introduce and update luminescence immunoassays for routine use, eventually replacing existing radioimmunoassays. The article is divided up into individual assay components and is summarised using data from compound precision profiles for the assays described. Full details of the assays used as examples are described in detail in this book.

MATERIALS AND METHODS

The labels used in this study were pyruvate kinase (E.C. 2.7.1.40) for bioluminescence as previously published (1) and diazoluminol and N-(4-aminobutyl-N-ethyl)-isoluminol hemisuccinamide (ABEI-H), the latter being commercially available (LKB, Turku, SF), for chemiluminescent assays. The luminometers used were the LKB-1250 for bioluminescence and either the LKB-1251 or Berthold LB-950 (Berthold, Wildbad, D) for chemiluminescence measurements. Data processing was performed off-line using a CBM 8096 desk top computer. Solid phases used were either microcrystalline cellulose (MCC) as 20 µm particles or 6.4 mm diameter polystyrene balls (Spherotech Kugeln, Fulda, D or Precision Plastic Balls, Chicago, USA). All water was double distilled. The chemiluminescent oxidation system used alkaline peroxide and microperoxidase MP-11 (Sigma, Munich, D.) at pH 13. The bioluminescence measurements were made with the LKB ATP-monitoring kit, using a kinetic measurement over 60 s. The pyruvate kinase was used to generate ATP from ADP and phosphoenol pyruvate, i.e. the reaction was run in the reverse direction with excess ADP and phosphoenol pyruvate.

ISBN 0-12-426290-2

RESULTS

Transferrin Assay - Optimisation as a function of time.

The first luminescence assay was the enzyme enhanced luminescence immunoassay (EELIA) (1) which used transferrin coupled to pyruvate kinase as label and a second antibody coupled to $NaIO_4$ activated MCC. Samples were diluted 1:800 before assay and centrifugation and wash steps were both needed which did not make the assay easy to handle. The time taken for a 100-sample assay was around 9 h, all measurements being performed manually on the LKB-1250. Although this assay was in routine use for several months, the following weak points were noted: certain sera despite the dilution step inhibited the pyruvate kinase which resulted in falsely elevated values. All reagents had to be made up fresh as neither azide nor merthiolate could be used as antimicrobial agents. Attempts at using either penicillin or aminoglycosides as alternatives did not give the expected results as resistent bacteria were present!

The EELIA was replaced by a chemiluminescent assay (CELIA) in which the transferrin was labelled with diazoluminol and the ATP-monitoring system replaced by an alkaline peroxide/peroxidase system. The reagent stability problem was solved as azide could be added to the MCC and tracer solutions. The time needed for a 100 tube assay was reduced to 6 h, an integration time of 30 s being used for the light output. The incidence of falsely elevated results was reduced.

The next change was to a SPALT (solid phase antigen luminescence technique) (2) assay in which the label used was diazoluminol coupled to the second antibody, the solid phase antigen being transferrin covalently coupled to a polystyrene ball (3). The assay time per 100 samples was reduced to 5 h concurrent with the introduction of the LKB-1251 25-sample luminometer. The final change was the replacement of the diazoluminol by ABEI-H and the introduction of 20 or 60-well reaction trays together with a horizontal rotator and semi-automatic wash system (Abbott trays + Pentawash II and Heidolph rotator - Abbott diagnostics). This resulted in a 3 h assay, the LKB-1251 being replaced by the Berthold LB-950 as routine assays of 200-300 tubes were no longer practical on the LKB-1251. The current assay has a mean cv under 6% in the range 1-8 g/1 compared with a cv of 11.8% for the EELIA in the same range. These data are taken from compound precision profiles derived from 1580 duplicate values for the SPALT and from 247 data pairs for the EELIA.

Ferritin - The effect of changing the label

The ferritin ILSA (immunoluminometric labelled second antibody assay) was developed using a goat anti-human ferritin coated polystyrene ball, a rabbit anti-human ferritin as sandwich antibody and a donkey anti-rabbit antibody labelled with diazoluminol. This was chosen for two reasons, namely the use of a universal label and because attempts at labelling the rabbit anti-ferritin with diazoluminol resulted in loss of immunoreactivity. The assay consisted of an overnight incubation followed by six hours incubation (4+2 h) on the second day, and had a lower detection limit of 12 µg/l when using a 100 µl sample.This assay was unacceptable for routine use. By labelling the rabbit anti-human ferritin with ABEI-H via an active ester, the assay time could be reduced to 6 h, the sample size to 20 µl and the lower detection limit to 2 µg/l. This assay was suitable for routine use and the mean cv between 10 and 1000 µg/l was 4.8% from 2240 data pairs. Corresponding data from the ferritin ILSA was 8% for 227 duplicate values. The ferritin ILMA (immunoluminometric assay) using the ABEI-H label has performed successfully in the Wellcome BC 08 external quality assessment scheme and gave more precise results than the radioimmunoassay it replaced.

Thyroglobulin - The effect of changing the assay type

As human thyroglobulin was available in sufficient amounts, the first assay developed was a SPALT using thyroglobulin coupled to polystyrene balls as solid phase antigen. The assay took 24 h, used a 100 µl sample and had a lower detection limit of 10 µg/l using an ABEI-H labelled anti-rabbit IgG as label. After discovering that the thyroglobulin antibody bound to both sides of the sandwich in an ILMA it was labelled with ABEI-H and the corresponding assay was developed. Using the same sample size, an incubation time of 5 h was sufficient to reduce the detection limit to 3 µg /l.The mean cv for the range 15-1000 µg/l was 5.1% for 667 duplicate values. This assay is in routine use and is 43 hours shorter than the immunoradiometric assay it replaced.

CONCLUSIONS

This article describes the "evolution" of 3 assays over a period of up to two and a half years. All three perform acceptably both in routine use and in external quality assessment schemes.

The basic lessons learnt during this time include: a) separation of serum and label - i.e. no CELIA, except after extraction, (b) solid phase techniques using reaction partners covalently coupled to polystyrene balls and (c) labelling with ABEI-H via an N-hydroxysuccinimide active ester as this gives rise to a hydrophilic product which retains its immunoreactivity, even at high substitution rates.

The choice of assay remains either ILMA for molecules with more than one antigenic determinant, or SPALT for haptens or for components with high concentrations in serum, and where the antigen for immobilisation is available in sufficient quantities.

Finally, it has been shown that robust luminescence immunoassays have their place in the routine laboratory, and do not need "trained" personnel outside the normal technician to perform them.

REFERENCES.

1. Fricke, H., Strasburger, C.J. and Wood, W.G. (1982) Enzyme enhanced luminescence immunoassay (EELIA) for serum transferrin, J.Clin.Chem.Clin.Biochem. 20:91-94
2. Wood, W.G. et al. (1982). Solid phase antigen luminescence techniques (SPALT) for the determination of insulin insulin antibodies and gentamicin levels in human serum. J.Clin.Chem.Clin.Biochem. 20:825-831
3. Wood, W.G. and Gadow, A. (1983). Immobilisation of antibodies and antigens on macro solid phases - Part 1 A comparison between adsorptive and covalent binding. J.Clin.Chem.Clin.Biochem. 21:789-797

BLOOD SPOT 17-HYDROXYPROGESTERONE LUMINESCENCE IMMUNOASSAY

W. Klingler, O. Haupt, and R. Knuppen

Institut für Biochemische Endokrinologie
Med. Hochschule, 2400 Lübeck
Federal Republic of Germany

INTRODUCTION

Congenital adrenal hyperplasia (CAH) is caused by a partial or total deficiency of one of the enzymes involved in the biogenesis of mineralo- and glucocorticoids. The most frequent form is the deficiency of the 21-hydroxylase leading to an excessive increase of androgen production in the adrenal gland and to a reduced formation of aldosterone and cortisol (1,2). An important indicator of this disease is 17-hydroxyprogesterone (17-OHP). As in newborns a considerable increase of this steroid in the blood proves the 21-hydroxylase deficiency, we developed a luminescence immunoassay for measuring 17-OHP in heel prick blood that makes possible wide-scale screening for congenital adrenal hyperplasia.

MATERIAL AND METHODS

Synthesis of 17-hydroxyprogesterone-3-(O-carboxymethyl)oxime-aminobutylethyl-isoluminol (17-OHP-ABEI)

6[N-(4-aminobutyl)-N-ethyl]amino-2,3-dihydrophthalazine-1,4-dione was synthesized according to the method of Schroeder *et al.* (3). 17-hydroxyprogesterone-3-(O-carboxymethyl)oxime (75 mg) was dissolved in 2.0 ml of dry dioxane and 0.044 ml of tri-n-butylamine was added. After cooling to

ISBN 0-12-426290-2

10°C isobutyl chloroformate (25.4 mg) was added and the solution was stirred at 10°C for 6o h. This reaction mixture was added dropwise to a solution of aminobutylethyl-isoluminol (ABEI, 51.3 mg) dissolved in 12 ml dioxane/water (1:1). The pH was kept between 8.3 and 8.5 during this procedure. The solution was stirred for 3.5 h. Then 5o ml of water was added and the solid material was filtered off and washed with water. The product (m.p. 147-155°C) was identified by IR and elemental analysis. The proposed structure of the synthesized molecule is shown in Fig.1.

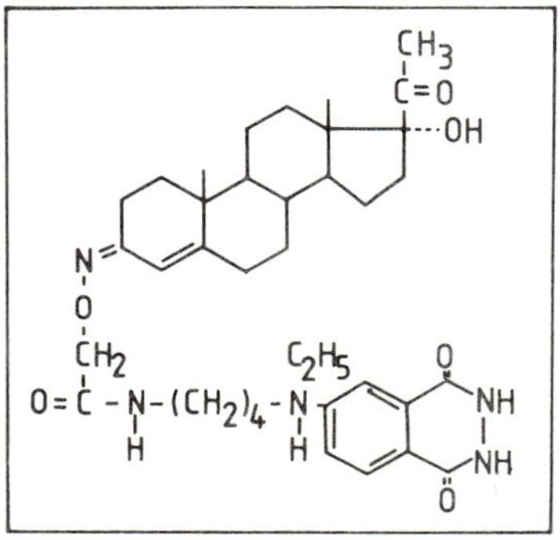

Fig.1 Proposed structure for 17-hydroxyprogesterone-3-carboxymethyloxime-aminobutylethyl-isoluminol

Blood collection

Blood is collected by heel prick in newborns at the 5th day of life. One drop of o.o3 - o.o4 ml dried on filter paper is sufficient for the measurement.

Assay procedure

Standard solutions are o, o.75, 1.5, 3, 6, 12, and 24 ng 17-hydroxyprogesterone/ml ethanol. Two filter paper disks of 3 mm diameter plus o.o25 ml standard solution or two blood spots of 3 mm diameter are incubated with 1 ml of phosphate buffer (o.o1 M, pH 7.4). Then the buffer is extracted twice with 2 ml of ethylacetate/n-hexane (2:8) and after decantation the organic phase is evaporated to dryness. Phosphate buffer (o.1 ml, o.1 M, pH 7.o), o.1 ml of 17-OHP-ABEI dissolved in the same phosphate buffer (5o pg/tube), and o.1 ml of diluted antiserum raised in rabbits against 17-hydroxyprogesterone-3-carboxy-methyloxime-BSA are added to the extracts. After incubation over-night at 4°C the bound/free separation is per-formed by adding second antibody suspension (o.2 ml). The luminescence measurement in the pre-cipitate is carried out as described earlier (4).

RESULTS AND DISCUSSION

The sensitivity of the assay is o.5 $\pm$ o.2 (SD) ng/ml calculated from four dose-response curves. The method is specific. Only progesterone and 11-deoxycortisol show a cross-reactivity of about 5% to the antibody.

17-hydroxyprogesterone in different concen-trations (o.5-6 ng/ml) was added to the blood of a healthy volunteer. The regression line for the expected and found values is y=o.85x-o.31 with the correlation coefficient of r=o.964 (Fig.2).

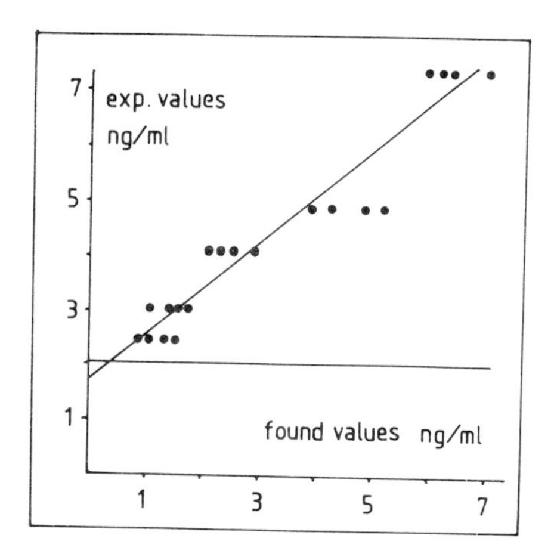

Fig.2 Measurement of different concentrations of 17-OHP added to the blood of a healthy volunteer

Blood of a normal volunteer was mixed with different quantities of the blood from a young adrenal hyperplasia patient. The regression line for the expected and found values is y=1.2x+o.9o with the correlation coefficient of r=o.926 (Fig.3)

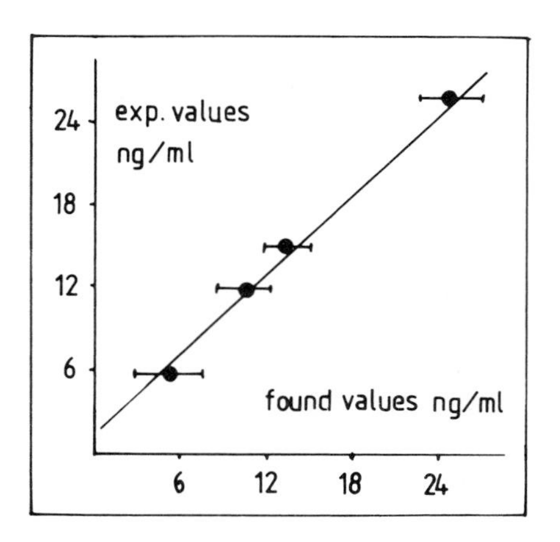

Fig.3 Measurement of 17-OHP in blood of a healthy volunteer mixed with different quantities of blood from a adrenal hyperplasia patient

We examined 225 blood samples from newborns. The distribution of the values compared to a radioimmunoassay method developed in our laboratory is shown in table 1.

TABLE 1

Distribution of 17-OHP values in newborns determined by radioimmunoassay (n=5019) and luminescence immunoassay (n=225)

Values (ng/ml)	RIA %	LIA %
< 3.5	79.o	72.o
3.5-6.5	15.2	18.3
6.5-12.5	4.8	8.o
12.5-25	o.9	1.7
25-5o	o.o6	o.o
> 5o	o.o4	o.o

Although we have not yet found any homocygotic CAH cases the method seems to be applicable for routine measurement. In blood spots of children and adults with proved adrenal hyperplasia always high 17-OHP levels are found.

REFERENCES

1. Pang, S.,Murphey, W.,Levine, L.S.,Spence, D.A., Leon, A.,LaFranchi, S.,Surve, A.S., and New, M.I. (1982). A pilot newborn screening for congenital adrenal hyperplasia in Alaska, J Clin Endocrinol Metab 55, 413-42o.
2. Cacciari, E.,Balsamo, A.,Cassio, A., Piazzi, S.,Bernardi, F.,Salardi, S.,Cicognani, A., Pirazzoli, O.,Zappulla, F.,Capelli, M., and Paolini, M. (1982). Neonatal screening for congenital adrenal hyperplasia using a microfilter paper method for 17-hydroxyprogesterone radioimmunoassay, Horm Res 16, 4-9.
3. Schroeder, H.R.,Boguslaski, R.C.,Carrico, R.J., Buckler, R.T. (1978). Monitoring specific protein-binding reactions with chemiluminescence, Methods Enzymol 57, 424-445.
4. Klingler, W.,Haupt, O.,von Postel, G., Knuppen, R. (1983). Immunoassay of unconjugated estriol in serum of pregnant women monitored by chemiluminescence, Steroids 42, 123-136.

CHEMILUMINISCENT IMMUNOASSAY OF SALIVARY TESTOSTERONE IN NORMAL SUBJECTS AND HIRSUTE WOMEN

R.Galard,R.Catalán,A.Lucas,S.Schwartz & J.Castellanos

Department of Biochemistry.Hormone Laboratory. Vall Hebrón General Hospital.Barcelone.Spain.

INTRODUCTION

Recent reports suggest that the determination of steroid hormone concentration in saliva (1-4) could well become the method of choice for assesing endocrine function.
These methods can provide a useful index of unbound steroids in plasma.Also the collection procedure is non-invasive,allowing frequent sampling over long time periods.
Salivary testosterone levels have been described for normal subjects,hirsute women,and men with testicular disfunction (2,5,6,7) using Radioimmunoassay(RIA). These assays involved ether extraction and were validated in terms of specificity, sensitivity and precision.
In recent years chemiluminescent markers(8-9) have been used to develop and alternative non-isotopic immunoassay (CIA) and on this basis some authors have reported steroid concentrations in human urinary and plasma samples(10-13)
In this paper we describe salivary testosterone levels in normal subjects and hirsute women using chemiluminiscence as the end point.

MATERIAL AND METHODS

3-CMO-ABEI-Testosterone was obtained through the courtesy of Dr.W.P.Collins of the King's College Hospital,London. Testosterone and microperoxidase were purchased from Sigma St.Louis,MO. Antiserum against Testosterone-19-Carboximethyl-ether-BSA, from RSL, Lab. CA, USA. The assay buffer was 0.05 M sodium phosphate pH 7.5 containing gelatin (1 g/l) and sodium azide (1 g/l). Light measurements were made with a

ISBN 0-12-426290-2

luminometer LKB-Wallac model 1250 connected to an integra - tion unit. The hydrogen peroxide was injected with a Hamilton diluter.

Subjects. Controls: 17 normal men and 15 normal women aged 20-40 years and not receiving any form of hormonal contraception. Patients: 15 matched idiopatic hirsute women.

Sample collection. Saliva was collected spitting in a glass vial between 8-10 a.m. The saliva was centrifuged and stored at -20ºC until assayed. *Sample extraction* . 1 ml of saliva was extracted three times with free peroxide diethyl ether. The dry residue was dissolved in 500 µl of buffer and aliquots of 100 µl were used for the CIA.

Chemiluminiscent immunoassay procedure. 100 µl of standards (3.5-200 pg) or sample extracts were transferred to glass tubes and incubated with specific antiserum (100 µl at 1: 60,000 dilution) and labelled hormone (100 µl ,70 mV) for 30 min at room temperature,and 30 min at 4ºC. NSB and blank tubes were set up. Charcoal was added (0.5 mg/tube, 200 µl) to all tubes except total. After centrifugation the supernatant was removed and transferred to a polystyrene tubes. Sodium hydroxide (100 µl, 5N) was added and the tubes were placed in a water bath at 60ºC for 1 hour. After cooling microperoxidase (100 µl, 2 mg/ml) was added. The oxidant (100 µl of 30% H_2O_2 dil. 1/100) was injected and the ligth measured for 10 sec. Luminometer readings were expressed as the percentage of B/B_0 and plotted vs. log-dose concentra - tion on logit-log graph paper. (Fig.1)

Statistical analysis of data was performed according to the Student's "t" test.

RESULTS

a) Validation of the method. The *Sensitivity* defined as the least amount of steroid detectable at the 95% confidence level was 4 pg/tube. *Precision;* The inter and intraassay coefficient variation were CV= 7.9% (n=12) and 11.8% (n=10) respectively. *Specificity:* The cross reactivity with similar steroids is shown in table I. Fig 2 depicts the coefficient relation (r=0.86) between CIA and RIA. Specificity of the RIA was tested with the Abraham's method. *Recovery* on ex - traction exceeded 95% in all samples and it was therefore deemed unnecessary to monitor for recovery in routine practice.

b) Clinical results for normal subjects and hirsute women are shown in Fig.3.

TABLE I
Cross reactivity of related steroids

	RIA	CIA
5 α-Dihydrotestosterone	18 %	21 %
5 α-Androstan-3 α,17 β-diol	3 %	7 %
Androstenedione	0.5 %	1 %
5 α-Androstan-3, 17.dione	0.18 %	0.44 %
Androsterone	0.09 %	0.05 %
Oestradiol-17 β	0.08 %	0.2 %
Progesterone	0.06 %	0.15 %
Dehydroepiandrosterone	0.04 %	0.15 %

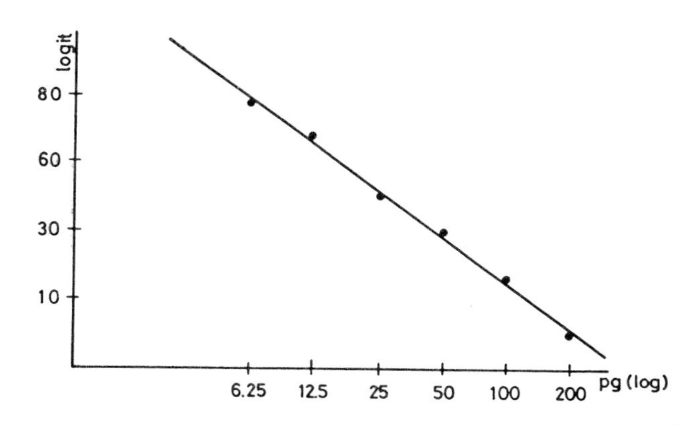

Fig.1 Representative dose response curve for Testosterone.

DISCUSSION

Our validity criteria obtained in the measurement of salivary testosterone with the CIA procedure were similar to those described by some authors using this method to measure other steroids in urine and plasma (10-12-13).
The sensitivity achieved (4 pg) approached that obtained by RIA (3.5 pg). The same occurred with respect to intraassay

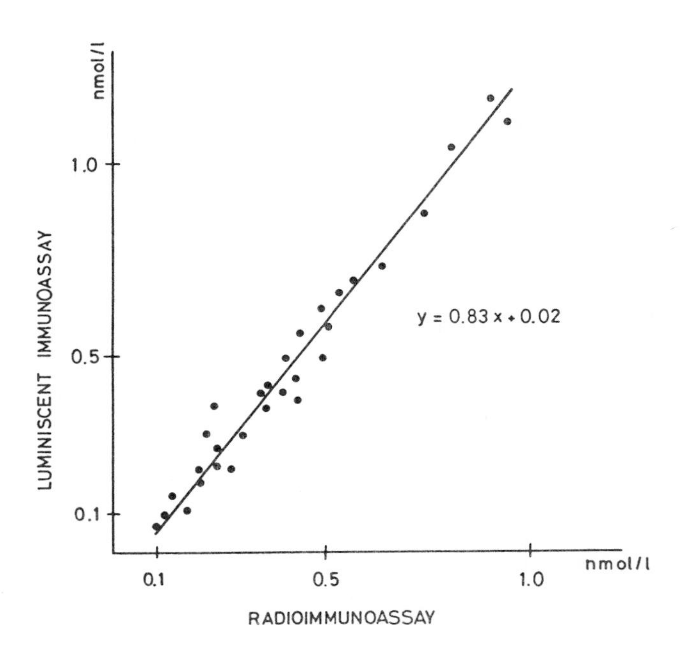

Fig2. Linear regression between CIA and RIA salivary testosterone levels

(RIA:5.8% ; CIA :7.9%) and interassay (RIA : 8.7% ; CIA : 11.8%) variations.
Using a polyclonal antibody and a heterologous immunoassay we cannot find any important difference in cross reactivity between both procedures.Finally, we have demonstrated a good correlation between RIA and CIA results obtained from normal s and from hirsute women.
Our clinical findings are in agreement with those published by other authors using RIA (5-6). We could find a significant difference in salivary testosterone levels between sexes and between normal and hirsute women. It is interesting to point out that the salivary testosterone levels found in hirsute women were even higher than those found in normal men.
Clinically the measurement of testosterone in saliva using CIA offers the opportunity of having an index of free testosterone levels in plasma through a non-isotopic method .
In conclusion we think that CIA can be a feasible alternative to the salivary testosterone RIA analysis in the study of hiperandrogenisms.

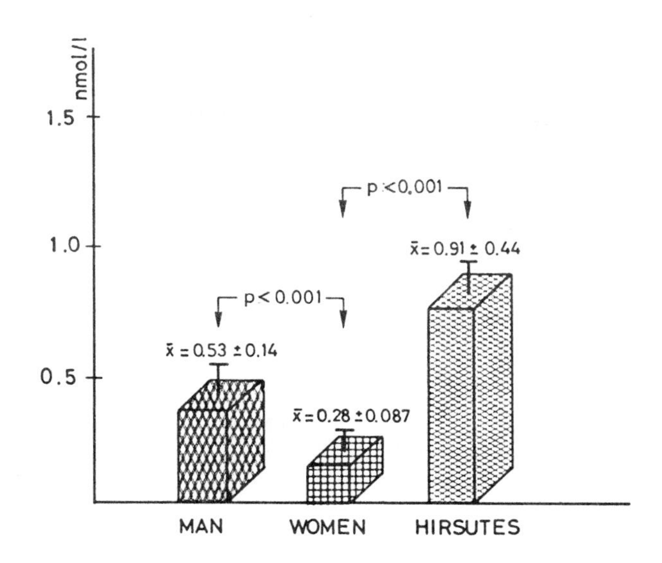

Fig.3 Testosterone levels (mean and s.d) in saliva of normals and hirsute women. Statistical significance between groups is shown.

ACKNOWLEDGEMENTS

We are grateful to Miss M$^{\underline{a}}$.J. Fernandez for her excellent technical assistance.

REFERENCES

1. Wang,C.,Plymate,S.,Nieschlag,E. and Alvin Paulsen,C. (1980). Salivary Testosterone in men:Further evidence of a direct correlation with free serum Testosterone, *J.Clin.Endocrinol.Metab.* 53 , lo21-1024.
2. Baxendale,P.M.,Reed,M.J. and James.V.H.T(1980).Testosterone in saliva of normal men and its relationship with unbound and total Testosterone levels in plasma, *J.Endocrinol* 87, 46 p-47 p.

3. Luisi,M.,Franchi,F.,Kicovic,P.M.,Silvestri,D.,Cossu,G., Catarsi,A.L.,Barletta,D. and Gasperi,M. (1981). Radio - immunoassay for Progesterone in human saliva during the menstrual cycle, *J.Steroid Biochem.*14 ,1069-1073.
4. Vining,R.F.,McGinley.R.and Rice,B.V.(1983).Saliva Estriol measurements: An alternative to the assay of serum unconjugated Estriol in assesing feto-placental fuction. *J.Clin.Endocrinol.Metab.* 56 ,454-460.
5. Luisi,M.,Bernini,G.P.,Del Genovese,A.,Virindelli,R., Barletta,D.,Gasperi,M. and Franchi,F.(1980).Radioimmunoassay for "free" Testosterone in human saliva, *J.Steroid Biochem.* 12 ,513-516.
6. Walker,R.F.,Wilson,D.W.,Read,G.F. and Riad-Fahmy,D.(1980) Assessment of testicular function by the radioimmuno - assay of Testosterone in saliva, *Int.J.Androl.* 3 ,105-120.
7. Baxendale,P.M.,Jacobs,H.S. and James,V.H.T.(1982). Salivary Testosterone: Relationship to unbound plasma Testosterone in normal and hyperandrogenic women *Clin.Endocrinol.*16 ,595-603.
8. Pazzagli,M.,Kim,J.B.,Messeri,G.,Martinazzo,G.,Kohen,F., Franceschetti,F.,Monetti,G.,Salerno,R.,Tommasi,A. and Serio,M.(1981).Evaluation of different progesterone - isoluminol conjugates for chemiluminescence immunoassay, *Clin.Chim.Acta* 115, 287-296.
9. Pazzagli,M.,Messeri,G.,Caldini,A.L.,Monetti,G.,Matinazzo G.and Serio,M.(1983).Preparation and evaluation of Steroid Chemiluminiscence tracers, *J.Steroid Biochem.*19 , pB,407-412.
10. Kohen,F.,Pazzagli,M.,Kim,J.B. and Lidner,H.R.(1980) . An immunoassay for plasma cortisol based on Chemiluminescence *Steroids* 36 ,421-437.
11. Kohen,F.,Kim,J.B.,Barnard,G. and Lidner,H.R.(1980) . An assay for urinary Estriol-16 α-Glucuronide based on antibody-enhanced Chemiluminescence,*Steroids* 36 405 - 419.
12. Pazzagli,M.,Kim,J.B.,Messeri,G.,Marinazzo,G.,Kohen,F., Francheschetti,F.,Tommasi,A.,Salerno,R.and Serio,M(1981) Luminescent immunoassay (LIA) for Progesterone in a heterogeneous system,*Clin.Chim.Acta* 115 ,287-296.
13. Barnard,G.,Collins,W.P.,Kohen,F. and Lindner,H.R.(1981) The measurement of urinary Estriol-16 α-Glucuronide by a solid-phase quemiluminescence immunoassay,*J.Steroid Biochem.*14 ,941-948.

OPTIMISATION OF MACRO-SOLID-PHASE LUMINESCENCE IMMUNOASSAYS USING COVALENTLY IMMOBILISED ANTIGENS AND ANTIBODIES

A. Gadow, C.J. Strasburger and W.G. Wood

Klinische Laboratorien der Klinik für Innere Medizin
Medizinische Hochschule Lübeck
D-2400 Lübeck 1, F.R.G.

INTRODUCTION

In spite of the numerous developments in the field of luminescence immunoassays (LIA) most of the assays remain model systems.

LIAs in routine use are as yet limited to specialised laboratories; No widespread and accesible system being available at the present time.

This presentation is aimed to show the development of robust chemiluminescence immunoassays which are easy to handle and suitable for routine laboratory use. Both the demands upon these assays and the associated problem complexes are summarised.

REQUIREMENTS FOR ROUTINE LIAS

The radioimmunoassay (RIA) has "fixed" the rules for good routine immunological methods during the past 25 years.

A LIA suitable for routine use should fulfil the following requirements:

1) Easy to handle and high reliability.
2) Relatively short incubation and measuring times.
3) Similar precision and lower detection limits to RIA.
4) Low reagent and component costs.

Replacement of the radioligand by a nonisotopic one gives no sufficient advantages when the points mentioned above are not fulfilled at the same time.

ISBN 0-12-426290-2

Many LIAs described in the literature are somewhat cumbersome. Workers used to working with RIA have up to now not accepted the LIA as an RIA-alternative.

BIO- OR CHEMILUMINESCENCE-IMMUNOASSAYS AS ROUTINE METHOD?

Bioluminescence immunoassays based on coupled enzymatic reactions are able to measure extremely low concentration (attomol range) (1,2). Main disadvantages are the cost of the reagents, the long measuring times (kinetic measurements), instability of reagents and the complex handling. Furthermore these systems show all disadvantages well known to enzyme-immunoassays.

Chemiluminescence-labels show a much better stability than enzyme-labels. The chemiluminescent reaction velocity can be controlled by altering the reagent concentrations. The main disadvantages are the low efficiency and quenching effects.

In practice, chemiluminescence-systems show evident advantages over bioluminescence-systems. The detection limits for chemiluminescence markers in the femtomol range are usually sufficient.

SIMPLIFYING OF LIA HANDLING WITH MACRO-SOLID-PHASES

By introduction of a macro-solid-phase (e.g. polystyrene ball) pipetting steps can be reduced, centrifugation can be avoided and the possibility of "sandwich-type" assays (cf. IRMA) is given.

Many LIAs described in the literature use conventional liquid-phase techniques for bound-free separation (3) or microparticles as solid-phases (4,5). These systems can be regarded as a retrograde step.

In this study a macro-solid-phase in form of balls (6-7 mm diameter) for immobilisation of antibodies and antigens was chosen. Coated tube technology was not considered because of limitation in the choice of measuring equipment (different cuvettes for different luminometers). On the other hand more coating solution is needed and there are problems in storage.

In extensive studies glass-, nylon- and polystyrene balls were tested to see if they were suitable for antibody and antigen immobilisation. A comparison between adsorptive and covalent techniques was made (6). The following system has been found to be best: polystyrene balls were coated with a poly-phe-lys copolymer. The phenylalanine residues interact with the ball-surface and the outward-pointing lysine-ε-amino groups are suitable for glutaraldehyde activation followed by covalent attachment of antigens and antibodies.

This immobilisation technique gives a lot of advantages especially due to the stability to detergent washing (e.g. tween 20) to reduce unspecific binding effects. It should be mentioned here that the authors cannot confirm the often heard fact that polymer balls are unsuitable due to their property of "catching" light inside the polymer-structure thus giving problems with the luminescence measurements. Using polystyrene balls no such effect was observed, with nylon balls a small effect (5% of the reagent blank) can occur.

USE OF MULTI-WELL REACTION VESSELS AND A MULTI-WASH SYSTEM

The assays developed in this laboratory are of a heterogeneous type, therefore it is necessary to wash during the assay procedure.

To shorten the most time consuming single tube washing it was decided to use the combination of polystyrene balls with a multi-well reaction vessel together with a horizontal rotator (170-200 rpm) and a multi-wash system (cf. Abbott diagnostic products hepatitis kits with Pentawash II system). With such a system washing of 60 samples is possible in under a minute! The resulting luminescence immunoassays are easy to handle even for long assay runs. The transfer of the polystyrene balls into the measuring cuvettes is easy when using a special cuvette holder.

FINAL PREPARATIONS BEFORE MEASUREMENT OF LARGE NUMBERS OF SAMPLES

When carrying out measurement of an assay run with over 50 tubes a shift phenomena was often observed. The precision was better at the start of the run than at the end. The light yield also varied.

The coefficient of variation obtained using dry

balls was found to be more than double that obtained when the balls were kept covered with physiological saline (dry balls: c.v.=9%, moist balls: c.v.=7%, balls in physiological saline: c.v.=4%).

Covering all balls with physiological saline before measurement enhanced precision and avoided shift effects.

CONCLUSIONS

This article shows that by optimising the assay conditions - these including the choice of solid-phase, coating technique and method of incubation - luminescence immunoassays for routine use can become reality as shown by assays for over 30 parameters in routine use in the authors' laboratory.

REFERENCES

1. Wannlund, J., Deluca, M. (1982). A Sensitive Bioluminescent Immunoassay for Dinitrophenol and Trinitrotoluene. *Anal.Biochem.* 122, 385.
2. Strasburger, C.J. et al. (1983). Lumineszenz Immunoassays-Alternativen zum Radioimmunoassay. *Ärztl. Lab.* 29, 75-82.
3. Pazzagli, M. et al. (1981). Luminescent Immunoassay of Cortisol-2. Development and Validation of the Immunoassay Monitored by Chemiluminescence. *J. Steroid Biochem.* 14, 1181-1185.
4. Weeks, I., Campbell, A.K., Woodhead, J.S. (1983). Two-Site Immunochemiluminometric Assay for Human Alpha-1-Fetoprotein. *Clin. Chem.* 29, 1480-1483.
5. Wood, W.G. et al. (1982). Solid Phase Antigen Luminescent Immunoassays (SPALT) for the Determination of Insulin, Insulin Antibodies and Gentamicin Levels in Human Serum. *J.Clin.Chem. Clin.Biochem.* 20, 825-831.
6. Wood, W.G., Gadow, A. (1983). Immobilisation of Antibodies and Antigens on Macro Solid Phases - A Comparison Between Adsorptive and Covalent Binding. *J.Clin.Chem.Clin.Biochem.* 21, 789-797.

SOLID-PHASE LUMINESCENCE IMMUNOASSAYS FOR THYROID PARAMETERS

A. Gadow and W.G. Wood

Klinische Laboratorien der Klinik für Innere Medizin
Medizinische Hochschule Lübeck
D-2400 Lübeck 1, F.R.G.

INTRODUCTION

Nonisotopic immunoassays have replaced radioimmunoassays to differing extents, mainly according to the analyte concentration.

This contribution is aimed to demonstrate that routine chemiluminescence immunoassays have been developed for the main thyroid parameters. With the help of isoluminol-derivative labelled antibodies and antigens in combination with different immunological methods and macro-solid phase technology, assays for thyroxine (T_4), triiodothyronine (T_3), thyroxine binding globulin (TBG), thyrotropin (TSH), thyroglobulin (Tg) and antibodies to thyroglobulin (anti-Tg) were developed.

METHODS

a.) Preparation of chemiluminescence-labelled antibodies and antigens

In all cases ABEI-H (N-(4-aminobutyl-N-ethyl)-isoluminol hemisuccinamide (LKB, Munich) was coupled via an active ester (prepared with dicyclohexylcarbodiimide and N-hydroxysuccinimide) (1) to amino-groups of antibodies and antigens. Purification was by Sephadex G-25 gel-chromatography for antibody conjugates or thinlayer-chromatography for hapten-conjugates (e.g. T_4-ABEI-H).

ISBN 0-12-426290-2

b.) Immobilisation of antibodies and antigens to polystyrene balls.

Adsorptive immobilisation of antibodies was performed as already published (2). For covalent immobilisation of antibodies, proteins and protein-hapten conjugates to polystyrene balls, the poly-phe-lys glutardialdehyde technique was used (3).

c.) Initiation and measurement of chemiluminescence reaction

The luminol-chemiluminescence reaction was initiated under alkaline conditions (pH 13) using H_2O_2 (0.3%) and microperoxidase-MP11 (5 μmol/l).

Luminometers used were the LKB 1251 and the Berthold LB 950. Injection of reagents and integration of the light output signal over 20 sec. was performed automatically after previous programming of injection- and measurement-parameters.

d.) Immunological methods

CELIA - The CELIA (chemiluminescence immunoassay) (4) is similar to the common competetive RIA. Chemiluminescence labelled and unlabelled antigen compete for a limited number of free antibody binding sites (e.g. T_3- and T_4-CELIA).
ILMA - The ILMA (immunoluminometric assay) (4) was used for the determination of TSH, TBG, and Tg and is analogous to the immunoradiometric assay (IRMA).
SPALT - The SPALT (solid phase antigen luminescence technique) was used for the determination of T_4, T_3, Tg and anti-Tg. In a first step sample or standard antigen react with the antigenspecific first antibody. In the second step unreacted antibodies are bound to an excess of solid phase coupled antigen. After a wash step a chemiluminescence labelled species-specific antibody is allowed to bind to the solid phase bound first antibody. After a final wash step the chemiluminescence reaction is initiated in the luminometer.

The main component of the SPALT is the solid phase antigen. In the case of Tg- and anti-Tg-SPALT thyroglobulin was directly coupled to the solid phase. For T_4- and T_3-SPALT protein-hapten conjugates have to be used (e.g. transferrin-T_4 and transferrin-T_3). Important is that the protein-hapten conjugate for hapten SPALT-assays must be different from the immunogen used for antibody-production (A. Gadow, unpublished results), otherwise crossreactions with protein and bridge elements occur and there is insufficient displacement by high sample antigen concentrations.

RESULTS

a.) Assay features

Parameter	Method	Total time of incubation	Detection limit	Correlation coefficient with RIA
T_4	SPALT	2 h	<5 μg/l	r=0.914 (n=40)
T_4	CELIA	2 h	10 μg/l	r=0.845 (n=58)
T_3	SPALT	3 h	<0.2 μg/l	r=0.96 (n=30)
T_3	CELIA	3 h	0.2 μg/l	r=0.975 (n=30)
TSH	ILMA	5 h	0.7 mU/l	r=0.98 (n=112)
TBG	ILMA	1 h	<1 mg/l	r=0.884 (n=85)
Tg	ILMA	5 h	<5 μg/l	r=0.978 (n=60)
Tg	SPALT	24 h	5 μg/l	r=0.98 (n=30)
a-Tg	SPALT	4 h	/	/ /

b.) Mean coefficient of variation from compound precision profile results

Assay	Range		Assays	Patients	Mean c.v.
TSH-ILMA	0.8-25	mU/l	n=12	n=212	6.5%
TBG-ILMA	16-32	mg/l	n=16	n=162	6.7%
Tg -ILMA	2-1000	μg/l	n=10	n=153	6.5%
T_4 -CELIA	41-132	μg/l	n=18	n=234	10.2%
T_4 -SPALT	5-200	μg/l	n= 8	n=132	7.5%
T_3 -SPALT	0.3-6.9	μg/l	n= 6	n= 68	8.2%

c.) Established reference ranges for thyroid parameter determination with LIA

Parameter	Method	Patients*	Mean	Reference range
T_4	SPALT	n=80	71µg/l	45 - 110µg/l
T_3	SPALT	n=52	1.4µg/l	0.6- 2.2µg/l
TBG	ILMA	n=68	23.6mg/l	15 - 30 mg/l
Tg	ILMA	n=91	26.9µg/l	0 - 70 µg/l
TSH-basal	ILMA	n=36	1.2mU/l	<0.7- 4 mU/l
TSH-after TRH stim.	ILMA	n=16	12.4mU/l	5 - 15 mU/l
T_4/TBG	-	n=45	2.6	1.6- 4
TBG	ILMA	n=50 (pregnancy)	67.7mg/l	/
T_4/TBG	-	n=20 (pregnancy)	2.3	/

*euthyroid blood donors

CONCLUSIONS

The main thyroid function parameters can be determined using the chemiluminescence immunoassays described above. The assays are easy to handle and automate. Clinical evaluations, examination of precision and correlation with established RIAs gave ample proof of the reliability of all assays.

Comparison between T_4-CELIA and T_4-SPALT show that the SPALT-method is the better technique for T_4-measurement, especially where only low affinity antibodies are available. Here many problems arise with immobilisation of antibodies to the macro-solid phase.

The ILMA-technique is the method of choice for measuring molecules with more than one antibody binding site. At extremely low analyte concentrations the need for good matched pair antibodies is most important (e.g. TSH-ILMA where an optimal combination giving a sensitivity under 0.1mU/l has not yet been found).

The SPALT-method for protein determination is ideal where the protein is available both in pure form and at higher serum concentrations. A SPALT assay for Tg was set up, but a long preincubation time was necessary to give the required sensitivity.

Our good results show the need for robust commercial kits, so that a widespread application of these techniques can be realised.

ACKNOWLEDGEMENTS

The excellent techniqual assistance of Mrs. G. Donovang-Müller and Mrs. J. Jäger is acknowledged.

REFERENCES

1. Gadow, A. et al. (1984). Synthesis and Evaluation of Luminescent Tracers and Hapten-Protein Conjugates for Use in Luminescence Immunoassays with Immobilised Antibodies and Antigens. *J.Clin.Chem.Clin.Biochem.* 22, in print.
2. Von Klitzing, L. et al. (1982). Comparison between Adsorption and Covalent Coupling of Proteins to Solid Phases Using Different Polymers as Support. *Fres. Z. Anal.Chem.* 311, 356-357.
3. Wood, W.G., Gadow, A. (1983). Immobilisation of Antibodies and Antigens on Macro Solid Phases - A Comparison Between Adsorptive and Covalent Binding. *J.Clin.Chem.Clin.Biochem.* 21, 789-797.
4. Wood, W.G. et al. (1984). An Evaluation of Four Different Luminescence Immunoassay Systems: CELIA (chemiluminescent immunoassay), SPALT (solid-phase antigen luminescence technique), ILMA (immunoluminometric assay) and ILSA (immunoluminometric labelled second antibody assay). *J.Clin.Chem.Clin.Biochem.* 22, in print.

CHEMILUMINESCENCE IMMUNOASSAYS FOR ALPHA FOETOPROTEIN FOR USE IN EARLY PREGNANCY AND TUMOUR MONITORING

J. Haritz and W.G. Wood

Klinik für Innere Medizin, Medizinische Hochschule Lübeck, D-2400 Lübeck, FRG.

INTRODUCTION

Serum alpha foetoprotein (AFP) measurements have become a routine part of pregnancy monitoring between the 10th and 20th weeks of pregnancy to check for neural tube defects which may result in the decision to interrupt the pregnancy. AFP determinations have also proved useful in following up patients with primary hepatoma and other tumours with yolk-sac elements such as teratomas. This article describes the setting up of a routine luminescence immunoassay for AFP.

MATERIALS AND METHODS

Sheep anti AFP was purchased from Seward, London, GB and rabbit anti-AFP from Dakopatts, Glostrup, DK. ABEI-H for labelling was obtained from LKB, Munich, D and polystyrene balls (6.4 mm diameter) from Spherotech Kugeln, Fulda, D. The sheep antibody was immobilised on the polystyrene balls and the rabbit antibody labelled with ABEI-H via an active ester (1). Table 1 shows the assay flow diagrams for the immunoluminometric assay (ILMA) and immunoluminometric labelled second antibody assay (ILSA).

The ILSA was set up as labelling the rabbit antibody with diazoluminol led to a product without immunoreactivity. In this case, donkey anti-rabbit IgG (Wellcome RD 17, Wellcome Diagnostica, Burgwedel, D) was labelled with diazoluminol and used as tracer, (1).

RESULTS

Figure 1 shows the AFP values in women between the 12th and 40th weeks of pregnancy. The values up to the 20th week are shown as solid squares as this is the "period of decision".

ISBN 0-12-426290-2

TABLE 1

Incubation schemes for the alpha foetoprotein ILMA and ILSA

Parameter	ILMA	ILSA
Serum / sample µl	100	100
Assay buffer µl	100	100
Antibody coated ball	1	1
Incubation step / temp (h/°C)	18/RT[+]	3/RT
Wash steps	2 x 5 ml 0.15 mol/1 NaCl with 150 µl/1 Tween 20	
Rabbit anti-AFP µl	200 (ABEI-H)	200
Incubation step / temp (h/°C)	4/37	2/RT
Wash steps	As above	
Donkey anti rabbit IgG µl	---	200
Incubation step / temp (h/°C)	---	2/37
Wash step	---	as above

Transfer balls to cuvettes, cover with 300 µl 0.15 mol/1 NaCl, transfer to luminometer and measure for 20 s.

Key: [+]RT = room temperature (20 - 25°C)

In Figure 1 the two specially treated patients were cases of interest. The one with 321 kU/1 AFP aborted 2 days after measurement after foetal death had occurred. The case with 265 kU/1 gave birth to twins in the 40th week of pregnancy.

The ILMA was compared with an enzyme immunoassay for AFP (Abbott Diagnostics), the correlation on 220 patients and volunteers being good. Higher values were measured lower in the ILMA. The correlation data for the regression line y = a + bx (EIA = x, ILMA = y) were: r = 0.962, a = 3.26 and b = 0.698. Reference ranges established on 132 blood donors with normal transaminases and blood smear were: EIA - up to 7 kU/1, ILMA up to 8 kU/1 AFP. The mean value for the ILMA for the blood donors was 3.14 kU/1, the median value being 3,08 kU/1. Comprehensive reference ranges for the ILSA were not established.

The main diagnostic use for the AFP assay is in the diagnosis and follow up of patients with hepatomas or yolk-sac component tumours. Table 2 shows a few values from selected patients with diverse tumours. All values were pre-operative. One interesting set of values came from a female patient who showed decreasing AFP values over an eight month period. These fell from ver 2000 kU/1 to under 100 kU/1 at death. On autopsy, a large necrotic tumour was found in the liver, which explained the AFP results.

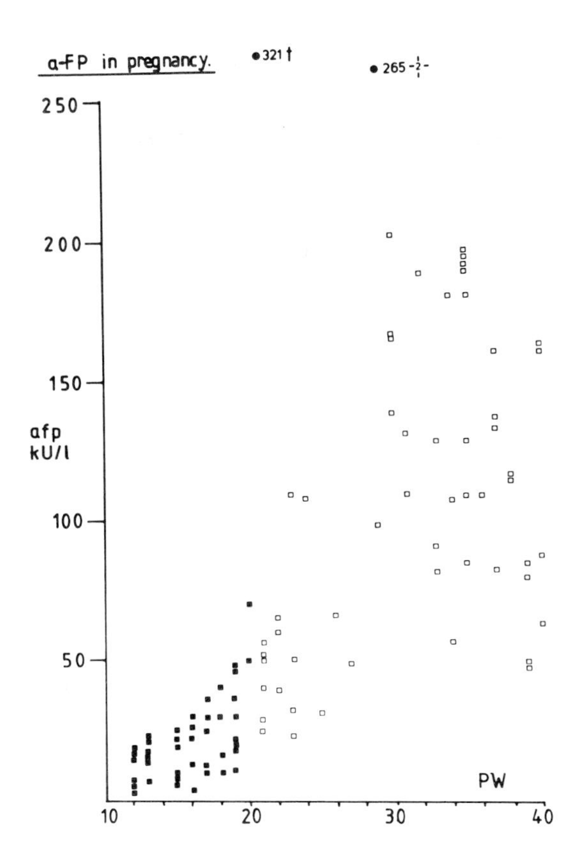

Fig 1. Distribution of AFP values in the 12th - 40th weeks of pregnancy. AFP values are on the ordinate, pregnancy week (PW) on the abscissa. The two numerical values are explained in the text.

TABLE 2

AFP levels in serum of patients with various tumour types.

Patient No.	Tumour type	AFP kU/1
D-5081	Hepatoma	211500
D-5104	Hepatoma	224
D-5105	Necrosing Hepatoma	115
D-5111	Hepatoma with secondaries	10800
D-5115	Liver metastases	3.5
D-5114	Pancreas	n.d.+

TABLE 2 cont.

Patient No.	Tumour type	AFP kU/1
D-5125	Colon with liver metastases	2.5
D-5121	Germinal cell tumour - testis	3.2
D-5107	Ovary	2.1
D-5128	Phaeochromocytoma	1.5
D-5132	Prostate Gland	5.1
D-5116	Melanoma	3.1
D-5119	Hypernephroma	3.6
D-5130	Fibrosarcoma	4.2

Key: $^{+}$n.d. not detectable

CONCLUSIONS

The ILMA developed was suitable for routine use in following up tumour patients and surveying pregnancy between the 10th and 20th weeks. In this study, AFP levels in pregnancy were followed over the second and third trimester to check assay performance.

The ILSA was a compromise and showed in the length of incubation times how diazotisation reduces the immunoreactivity of antibodies. As in the case of the ferritin assay, it was only possible to label the donkey anti-rabbit IgG with diazoluminol and retain some immunoreactivity.

REFERENCES

1. Gadow, A. et al. (1984). Synthesis and evaluation of luminescent tracers and hapten-protein conjugates for use in luminescence immunoassays with immobilised antibodies and antigens. Part II of a critical study of macro solid phases for use in immunoassay systems. J.Clin.Chem.Clin.Biochem. 22: In Print

SOLID PHASE LUMINESCENCE IMMUNOASSAYS FOR ACUTE-PHASE PROTEINS SUITABLE FOR ROUTINE PAEDIATRIC APPLICATION

Heidi-Susanne Krausz and W.G. Wood

Klinik für Innere Medizin, Medizinische Hochschule Lübeck, D-2400 Lübeck 1, FRG.

INTRODUCTION

The acute-phase proteins orosomucoid (AGP), transthyretin (TBPA) and c-reactive protein (CRP) were chosen as parameters to monitor acute phases of an illness, either as an isolated episode, or superimposed upon a chronic illness. Accent has been given here to the time dependance of the incubation steps as well as to the investigation of shift phenomena during measurement.

MATERIALS AND METHODS

Antisera were purchased from Dakopatts, Glostrup, DK, Behringwerke, Marburg a.d. Lahn, D and Seward Laboratories, London, GB, for all three parameters. Donkey anti-rabbit IgG was purchased from Wellcome Diagnostica, Burgwedel, D. All labelling was performed with ABEI-H (LKB, Munich, D).

The assay types were: AGP and CRP - ILMA (1) and TBPA - SPALT (2). All assays were measured in an LB 950 250/300 sample luminometer (Berthold, Wildbad, D), using an alkaline peroxide / pseudoperoxidase oxidation system, integrating the signal over 20 s. Table 1 shows the assay flow sheets.

TABLE 1

Assay flow charts for AGP, CRP and TBPA

Parameter	AGP	CRP	TBPA
Sample dilution 1:	1000	300	1000
Sample volume µl	100	200	100
Assay buffer µl	100	-	100
Preincubation min	-	-	20
Incubation with solid phase antibody or antigen min	60	60	40

ISBN 0-12-426290-2

TABLE 1 cont.

Parameter	AGP	CRP	TBPA
Wash step	2 x 5 ml physiological saline with 150 µl/1 Tween 20		
Second incubation min	60	60	60
Wash step	as above in all assays.		

The polystyrene balls were transferred to measuring cuvettes 300 µl 0.15 mol/1 NaCl added and loaded into the LB 950.

In the case of the AGP and CRP assays the DAKO antibodies were labelled, Seward and Behringwerke antibodies being on the balls respectively. In the case of TBPA, the antigen on the ball was human TBPA, the first antibody being from Behringwerke and the labelled second antibody from Wellcome.

RESULTS

Table 2 shows the main features of the assays in terms of established reference ranges and quality control parameters.

TABLE 2.

Reference ranges and relevant quality control parameters

Parameter	AGP	CRP	TBPA
Units	g/1	mg/1	g/1
Reference range[+]	0.4-1.7	0-5	0.1-0.7
Lower detection limit	0.2	1	0.03
Quality control			
Working range (cv under 10%)	0.3-5.0	2-160	0.1-2.5
Mean cv in working range (%)	6.8	4.2	6.4
No of samples used for cv	117	99	89

[+]The reference range was established on blood donors with normal aminotransferases and normal blood smear. For AGP, 85 donors were tested, for CRP and TBPA 80 and 92 donors.

As the TBPA SPALT had the possibility of a pre-incubation step before addition of the TBPA-coated ball, the effects of such a step upon assay sensitivity was checked. The effect of time on binding of antigen to the antibody-coated balls was checked using the CRP ILMA, keeping the labelled antibody incubation at 60 min. Variation of the second incubation step between 30 and 120 min did not affect the assay dynamic range. Assay shift was checked on the AGP ILMA by varying the time the balls were kept in NaCl before being transferred to the luminometer. Results are in Table 3.

TABLE 3

Effect of preincubation, time of first incubation and delay time before measurement on the TBPA, CRP and AGP assays

a. Effect of preincubation on assay sensitivity - TBPA SPALT

Standard concentration	Preincubation time (min)		
	30	20	10
Unspecific binding	232+	310	339
Zero standard	8071	7500	9200
0.063 g/l standard	4500	5400	7290
0.125	3500	4200	5450
0.250	3030	3580	4100
0.500	2830	3100	3400
1.00	2660	2540	2940
2.00	2490	2350	2570

+Counts are 20 s integral on the Berthold LB 950 divided by 1000. They are the mean of duplicate determinations.

b. Effect of the first incubation on the dynamic range - CRP

Standard concentration	First incubation time (min)		
	90	60	30
Zero standard	378+	470	500
4.75 mg/l standard	1323	1280	1125
9.50	1584	1562	1382
19.0	1700	1869	1737
39.0	1860	2117	1930
78.0	2007	2394	2296
156	2131	2504	2582
Values of control sera			
K-34 (normal)	0.83	0.85	0.67
D-6055 (elevated)	26.6	29.7	24.4

c. Effect of balls standing in NaCl before measurement - AGP

Standard concentration	Delay time in NaCl (min)		
	45	120	210
Zero standard	1270+	1200	1401
0.16 g/l standard	2250	2100	2070
0.31	2500	2500	2650
0.62	3200	3250	3210
1.25	4400	4500	4500
2.50	5700	5400	5730
5.00	6260	6400	6700
Value of control serum			
BW 1 (normal)	0.59	0.64	0.66

+The counts in b. and c. are as in a. above.

CONCLUSIONS

Rugged and stable assays for three acute-phase proteins have been developed and tested under routine conditions. All 3 assays can be performed in under three hours and are also suitable for paediatric use, needing only the serum from a haematocrit tube for all 3 parameters after the PCV has been measured. The incubation times chosen in Table 1 lay in the middle of a "stable" area, i.e. small changes in assay conditions did not affect the results.

Although the CRP assay gave identical results with a nephelometric assay (Behringwerke), the AGP assay delivered results which were consistantly 10-15% higher than the nephelometric assay from the same firm. This was probably due to the standards used, the CRP standard being from Behringwerke and the AGP standard from Sigma.

REFERENCES

1. Wood, W.G. et al. (1984). An evaluation of four different luminescence immunoassay systems: CELIA (chemiluminescent immunoassay), SPALT (solid-phase antigen luminescence technique), ILMA (immunoluminometric assay) and ILSA (immunoluminometric labelled second antibody) - Part III of a critical study of macro solid phases for use in immunoassay systems.
J.Clin.Chem,Clin.Chem. 22: In Print
2. Wood, W.G. et al. (1982). Solid phase antigen luminescent technique (SPALT) for the determination of insulin, insulin antibodies and gentamicin levels in serum.
J.Clin.Chem.Clin.Biochem. 20:825-831

A TWO HOUR CHEMILUMINESCENT ASSAY FOR AUSTRALIA ANTIGEN

S. Peitzner and W.G. Wood

Klinik für Innere Medizin, Medizinische Hochschule Lübeck, D-2400 Lübeck 1, FRG.

INTRODUCTION

The aim of this study was to develop a rapid, reproducible and sensitive chemiluminescence immunoassay to test for Hepatitis-B surface antigen in platelet-rich plasma and similar blood products with a short shelf-life, and which normally have to be infused without being screened. The test which was developed was checked against two commercial enzyme immunoassays from Behringwerke and Organon for discrepancies, either false positive or false negative results.

MATERIALS AND METHODS

Anti-HBsAg was either purchased from Dakopatts, Glostrup, DK or from the Dutch Blood Donor Centre, Rotterdam, NL. Anti-HBsAg was kindly donated by Dr. A. Voller, Nuffield Inst. Regent's Park, London, GB and by Dr. P.D. Weston, Wellcome Research Laboratories, Beckenham, GB. ABEI-H for labelling was obtained from LKB Munich, D and polystyrene balls from Spherotech Kugeln, Fulda, D.

The final assay scheme chosen is shown in Table 1.

TABLE 1
Flow diagram for the HBsAg 2 hour assay

100 µl sample or control serum
100 µl assay buffer
1 anti-HBsAg coated ball

incubate 45 min on rotator

wash with 5 ml Tween 20 / saline

incubate as above, wash, transfer ball to cuvette and measure in luminometer.

ISBN 0-12-426290-2

The limits of the assay were set as follows: In each assay 7 negative control sera from single donors were used to set the discrimination limits. These were set at +3 standard deviations (s) for borderline and +5s for positive values. In each assay two positive control serum pools were used to check the ratio positive/negative controls. In this way, the precision of each assay determined the limits.

RESULTS

Figure 1 shows the comparison between the HBsAg ILMA and the Organon enzyme immunoassay for Australia Antigen, (n = 295)

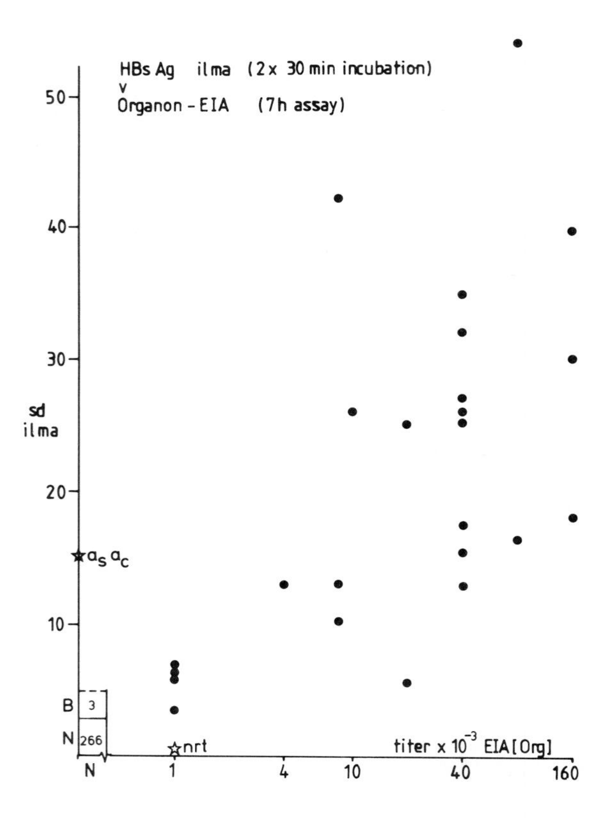

Fig. 1. Correlation between ILMA (y) and EIA (x). Key: N - negative, B - borderline, nrt - negative on re-test, $a_s a_c$ - anti-HBsAg and anti HBcAg positive in EIA.

In testing different combinations of antibody on the solid phase and antibody labelled with ABEI-H, only one combination was suitable for the rapid assay, namely the antibody from Dr. Voller on the solid phase with the Wellcome antibody as label. Other combinations were either totally unsuitable, or were only suitable for assays taking several hours.

In a further study in which 325 sera were tested in the immunoluminometric assay (ILMA) and in the Behringwerke EIA, 7 sera were rated positive only in the EIA and 2 in the ILMA alone. On retesting, 5 of the EIA sera were negative, 1 remained positive and one became borderline. The 2 sera in the ILMA remained positive (5.8 s and 7.2 s) on retesting.

CONCLUSIONS

From Fig. 1 it can be seen that the standard deviation value in the ILMA (ordinate) correlate roughly with the titer on the abscissa given by the Organon kit. In only two cases of 295 were discrepancies to be found, one of which disappeared on retesting, the other perhaps being explainable in that the weak anti-HBsAg and anti-HBcAg positive serum may have contained low amounts of HBsAg which were not seen in the EIA.

Although the HBsAg molecule is large, the combination of antibodies used played an important part in the development of rapid assays, a point shared by many ILMA-type set-ups.

ACKNOWLEDGEMENTS.

The authors would like to thank Dr. A. Voller and Dr. P.D. Weston for supplying the antibodies used in this assay. Special thanks go to Dr. A. Vosberg, Lübeck, without whose help the comparison of methods would not have been possible.

DEVELOPMENT OF A SENSITIVE DIRECT SOLID-PHASE COMPETITIVE ENZYME LUMINESCENT ASSAY FOR SERUM AND SALIVARY CORTISOL

A. Roda, S. Girotti, S. Lodi, S. Preti and A. Piacentini

Istituto di Scienze Chimiche, Facoltà di Farmacia,
Università degli Studi di Bologna
Bologna, Italy

INTRODUCTION

It is well known that the plasma cortisol levels are an important test for monitoring the glucoticoid production. Normal adrenal activity displays a marked circadian rhythm and usually more than one plasma sample is required. Previous studies showed that the salivary cortisol is well correlated with plasma cortisol.

The methods until now available such as RIA are sensitive but are expensive and require special equipment. Alternative methods to RIA have recently been proposed, including enzymeimmunoassay (1), ampliphied enzymatic methods (2) and fluorescent (FIA) or luminescent (LIA) immunoassay (3). Until now few of them have fulfilled the requisite of sensitivity and accuracy previously shown by RIA. In the present paper we report the development of highly sensitive enzyme luminescent immunoassay based on a competitive principle able to measure cortisol both in plasma and salivary samples (4).

MATERIALS AND METHODS

Cortisol-carboxymethyloxime derivate (CMO) was synthetized in our laboratory using conventional previously published methods (5). Antisera were raised in rabbits against cortisol-3-CMO-BTG; this derivative was prepared using previously described methods (mixed anhydride) (6). The immunization protocol was similar to that previously reported (7), and antisera with adequate titre, affinity

ISBN 0-12-426290-2

and specificity were obtained 3-5 months after the first immunization. The IgG rich fraction was isolated by salting (Na_2SO_4) precipitation following a slight modification of the method described by Axen *et al.* (8).

The purified antibody was immobilized on polystyrene plastic balls (6.4 mm i.d.). An appropriate dilution of IgG was incubated with the beads (1 mg/100 beads) in 0.1 M phosphate buffer, pH 8 for 4 h at 4°C. The solution was removed and BSA solution containing 0.01% of thiomersal was added and incubated for 2 hrs. The solution was discarded and the balls were washed with bidistilled water, dried under vacuum and stored at 4°C until use.

The mixed anhydride method was selected to conjugate cortisol-CMO with HRP (9). The conjugate was purified by gel exclusion-chromatography on Sephadex G-100. The molar ratio steroid/HRP was evaluated by adding ^3H-labeled cortisol-CMO to the reaction mixture.

The method developed is based on a competitive principle, i.e., the sample or cortisol standard and a fixed amount of the "label", 20-40 pg/tube, was incubated with the specific antibody. At the end of incubation (2 h at 4°C) the solid beads (bound fraction) were washed twice with 4 ml of water. The enzyme activity of the bound fraction was measured adding 2 ml of "substrate solution" (0.1 M phosphate buffer, pH 6.0 containing 2 mg of OPD and 1 µl of 30% H_2O_2) to each tube. After incubation for 30 min in the dark at 25°C the reaction was stopped with 200 µl of 4 M H_2SO_4 and adsorbance recorded at 492 nm. The luminescent detection was based on the use of luminol/H_2O_2. The washed plastic balls were added to a tube containing the "luminescent solution (1 ml of 0.01 M phosphate buffer pH 7.0; 0.1 ml of 0.01 M luminol; 0.5 ml of 0.01 M H_2O_2). The light emission occurs in a few seconds and its variation vs time is inversely proportional to the mass of cortisol in the sample.

RESULTS AND DISCUSSION

The antibody used is extremely specific for cortisol and the cross reactivity of other steroids is less than 10%. The stability of the immobilized antibody was followed by weekly assays of the same standard both at 4 and 25°C. The beads are stable at 4°C for more than six months while at 25°C only for 2-3 months. The reproducibility of the immobilized procedure was found to be excellent; the adopted procedure allows adsorption to the bead of antibody able to complex ∿50% of the "label".

With the procedure used approximatively 1 mole of HRP bond 2-3 moles of cortisol. In this way, at the final dilution of this label in the assay tube, not more than 10 pg of cortisol is present. When luminescent detection is used it is possible to further reduce the mass of the label to 2 pg/tube. This high specific activity of the "label" justifies the elevated sensitivity of the present method. The stability of these "labels" is high when stored in concentrated forms (1 mg/ml), at 4°C and in presence of 0.1 M gelatine. Once diluted for the assay (1 μg/ml) the stability is reduced to few days.

A linear response was obtained over a wide concentration range when the percent of bound antigen was plotted against the concentration on a logit-log plot. Representative standard curves are reported in Fig. 1, using colorimetric , luminescent detection, EIA , and RIA.

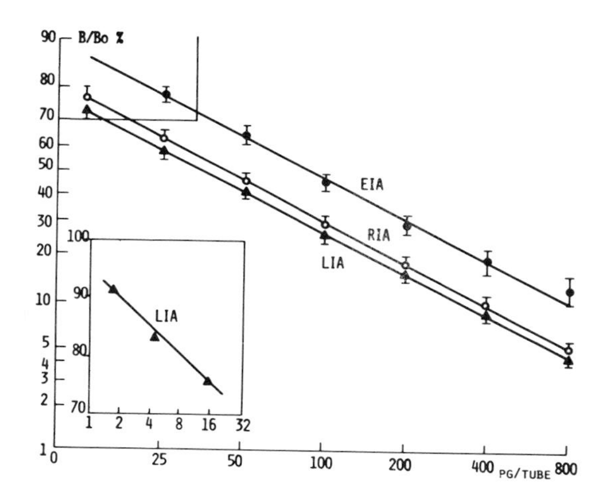

Fig. 1 Standard curve for the cortisol assay using different immunological methods.

The detection limits defined as the last detectable concentration resulting in a response measuring two standard deviations away from the zero dose response are reported in Table 1 for both detection systems used. In all assays the luminescent method also proved to be more sensitive when compared with RIA and EIA.

The method here proposed satisfies the normal criteria of precision and accuracy and a comparison with RIA showed excellent agreement (r>0.95). Salivary levels are also well correlated with those in serum and are ∿10 times lower.

TABLE 1

Comparison of chemiluminescent, colorimetric and radioimmunological methods for cortisol

Method	Sensitivity pg/tube	Reproducibility CV% (n=5)	Added pg/ml	Recovery (%)
chemiluminescent	1	10	50 400	98
colorimetric	10	9	500 400	101
RIA	5	8	50	107

The luminescent method shows an adequate sensitivity for measuring salivary cortisol on a small size sample (100μl). The assay here described represents an improvement in steroid analysis thanks to the combination of high sensitivity-rapidity. The use of both colorimetric and luminescent detection allows accurate measurement of cortisol in serum, whilst the luminescent method is more appropriate for saliva in which cortisol is present at lower concentration. In addition the proposed method is rapid, no separation steps are required and the luminescence detection is faster than the colorimetric one and requires inexpensive equipment (luminometer). Cortisol levels in saliva therefore may provide a good indication of adrenal activity cf. those in plasma and evidence suggests that this test will play a role in screening for adrenal disfunction and in differential diagnosis of defects in the hypothalamo-tuitary-adrenal axis.

1. Scharpe, S.L., Cooreman, W.M., Blomme, W.J. and Laekeman, G.M. (1976). Quantitative enzyme immunoassay: current status, *Clinical Chemistry* 22, 733-743.
2. Nicolas, J.C., Boussioux, A.M., Descamps, B. and Crastes, De Paulet, A. (1979). Enzymatic determination of estradiol and estrone in plasma and urine, *Clinica Chimica Acta* 92, 1-9.

3. Whitehead, T.P., Kricka, L.J., Carter, T.J.N. and Thorpe, G.H.G. (1979). Analytical luminescence: its potential in the clinical laboratory, *Clinical Chemistry* 25, 1531-1546.
4. Riad-Fahmy, D., Read, G.F., Walker, R.F. and Griffiths, K. (1982). Steroids in saliva for assessing endocrine function, *Endocrine Reviews* 3, 367-396.
5. Roda, A., Roda, E., Festi, D., Aldini, R., Mazzella, G., Sama, C. and Barbara, L. (1978). A radioimmunoassay for lithocholic acid conjugates in human serum and liver tissue, *Steroids* 32, 13-24.
6. Erlanger, B.F., Borek, F., Beiser, S.M. and Liebermann, S. (1957). Steroid-protein conjugates. Preparation and characterization of conjugates of bovine serum albumin with testosterone and with cortisone, *Journal of Biological Chemistry* 228, 713-721.
7. Roda, A. and Bolelli, G.F. (1980). Production of a high-titre antibody to bile acids, *Journal of Steroid Biochemistry* 13, 449-454.
8. Axén, R., Porath, J. and Ernback, S. (1967). Chemical coupling of peptides and proteins to polysaccharides by means of cyanogen biromide, *Nature* 214, 1302-1304.
9. Rajikowski, K.M., Cittanova, N., Desfosses, B. and Jayle, M.F. (1977). The conjugation of testosterone with horseradish peroxidase and a sensitive enzyme assay for the conjugate, *Steroids* 29, 701-713.

PREPARATION OF MICROPEROXIDASE AND ITS USE AS A CATALYTIC LABEL IN LUMINESCENT IMMUNOASSAYS.

R.A. Stott, G.H.G. Thorpe, *L.J. Kricka and T.P. Whitehead

*Department of Clinical Chemistry, Wolfson Research Laboratories and *Department of Clinical Chemistry, University of Birmingham, Edgbaston, Birmingham, UK*

INTRODUCTION

Catalytic labels such as microperoxidase (MP) offer potentially greater sensitivity than conventional luminescent labels in immunoassay. Commercially available MPs are highly active catalysts for the oxidation of luminol, but these preparations contain either non- haem peptides or ammonium salts and are unsuitable for use in coupling reactions. We report a method of preparing highly purified haem octapeptide (8MP) and hexapeptide (6MP) and the use of 8MP and 6MP in a catalyst labelled luminescence immunoassay.

MATERIALS AND METHODS

Pepsin, Trypsin, Thermolysin, horse heart cytochrome C, glutaraldehyde, microperoxidase (8MP and 11MP) were obtained from Sigma Chemical Company Ltd. Rabbit anti AFP, peroxidase conjugated anti AFP and AFP standard serum were obtained from Mercia Brocades. Luminol (Aldrich Chemical Company Ltd) was converted to its sodium salt prior to use (1). ABTS was obtained from Boehringer Corporation Ltd. Polyacrylic tubes and polystyrene cuvettes were from Sarstedt. Polystyrene beads were from Northumbria Biologicals Ltd. Precast IEF plates were from L.K.B. Instruments Ltd. The luminometer used in this work was built in the laboratory. (2).

Luminescent Microperoxidase Determination Stock reagent: sodium luminol (2mg/100 ml) and hydrogen peroxide (11 μl of

ISBN 0-12-426290-2

30% solution/100 ml) in boric acid - Na OH buffer (0.1 mol/1 PH 8.6).

1 ML of a freshly prepared 1/10 dilution of stock reagent (in boric acid - Na OH 0.1 mol/1 pH 8.6) was injected into the cuvette containing the sample (10 μl or 1 bead).

Preparation of Impure Microperoxidases 9MP was made via trypsin digestion of cytochrome C according to Plattner *et al* (3). 11MP was made via pepsin digestion, and 8MP via digestion with pepsin and trypsin, according to Kraenenbuhl *et al.* (4).

6MP was made via thermolysin digestion in a method according to Matsubara *et al* (5). 8MP (5 mg) was dissolved in ammonium bicarbonate (1 ml, 0.1 mmol/1) containing calcium chloride (0.2 mmol/1). A further 1 ml of ammonium bicarbonate was added, followed by thermolysin solution (300 μl, 0.1 mg/ml in ammonium bicarbonate - calcium chloride). The digestion mixture was lyophilised and stored at 4°C.

Purification of Microperoxidases The digestion mixtures were subjected to gel filtration at 4°C on a Sephadex G50 column eluted with 0.1 mol/1 sodium bicarbonate solution. Separation was monitored at 280 nm and 405 nm and appropriate fractions pooled and lyophilised.

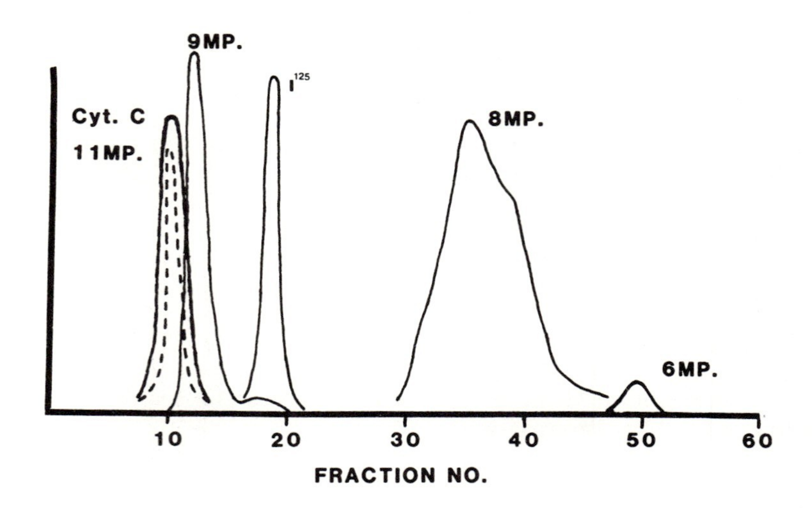

Fig 1 Elution of Microperoxidases from Sephadex G25 at 4°C

The method gave little resolution between cytochrome C and haem peptides larger than 11MP, while 9MP was not separated from non haem peptides. The small peptides, 8MP and 6MP, were retarded via an ionic interaction and eluted at between 2 and 3 times the elution volume for radioiodine (Fig 1). Each pure product focussed as a single red band when subjected to isoelectric focusing on precast slab gels. All the peptides were efficiently desalted and therefore the 8MP and 6MP were suitable for use in labelling antibodies.

CATALYST LABELLED IMMUNOASSAY

Microperoxidase immunoglobulin conjugate: This was prepared using a modified one step glutaraldehyde reaction. A 2.5% glutaraldehyde solution was made and stored at room temperature for several hours prior to use. 500 µl of antibody was added to 1 ml of phosphate buffer. This was mixed and 400 µl of microperoxidase solution (8MP, 5 mg/ml in water) added. 100 µl of glutaraldehyde solution was added and the solutions mixed for 5 hours. The product was purified by gel filtration on Sephacryl S200 eluted using phosphate buffered saline (0.015 mol/l pH 7.2 containing 0.1 mol/l NaCl). (Figure 2). The immunoglobulin peak was pooled.

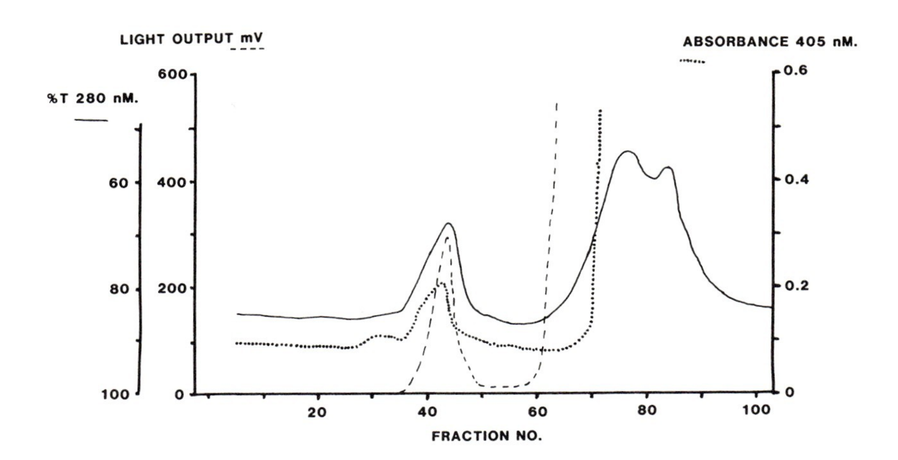

Fig 2 Elution of conjugate from Sephacryl S200

Immunoassay using polystyrene beads The catalyst labelled immunoassay was modified from a sandwich EIA for AFP (6). Polystyrene beads were coated with anti AFP and free binding sites blocked with albumin and Tween 20. Standards were prepared by dilution of the commercial standard in phosphate buffered saline (0.015 mol/l pH 7.2 containing 0.1 mol/l saline) containing bovine serum albumin (1%) and Tween 20 (0.05%). Beads were placed in the springs of a Kone immunoassay module. Standard or sample (50 μl) and PBS-BSA-Tween diluent (950 μl) were pipetted into tubes and pre heated to 37°C. The beads were agitated in sample for one hour and then washed in PBS-Tween. Peroxidase conjugate was diluted 1/1000 in PBS-BSA-Tween; microperoxidase conjugate was diluted 1/10 in PBS-Tween and prewarmed to 37°C. The washed beads were agitated in conjugate solution for 1 hour at 37°C. Beads treated with peroxidase conjugate were washed twice in PBS-Tween and then incubated at 37°C for 30 minutes in ABTS solution (1.1 g/l ABTS, 2.5 mmol/l peroxide in acetate phosphate buffer (0.1 mol/l acetate, 0.05 mol/l phosphate pH 4.2, 0.01% Tween 20). Beads treat; ed with microperoxidase conjugates were washed twice in deionised distilled water and transferred to individual cuvettes. 1 ml of luminescent reagent was added and light output measured.

Immunoassays using polyacrylic tubes An assay was developed on tubes using a similar format to the bead method (7). Polyacrylic tubes were coated with anti AFP and blocked with albumin. The tubes were emptied and washed once with PBS-Tween using a semi automated tube washing device. Samples and standards were diluted 1/20 and incubated in the tubes at 37°C for 1 hour. Each tube was washed 5 times in PBS-Tween and diluted conjugate added. After 1 hour incubation at 37°C the tubes were washed 5 times in PBS-Tween and colorimetric reagent or luminol reagent added.

RESULTS AND DISCUSSION

Coated tubes provided a more sensitive assay than could be achieved on beads: precision was markedly better and the slope of the standard curve greater. Figure 3 shows a comparison of the results obtained for the four assay methods.

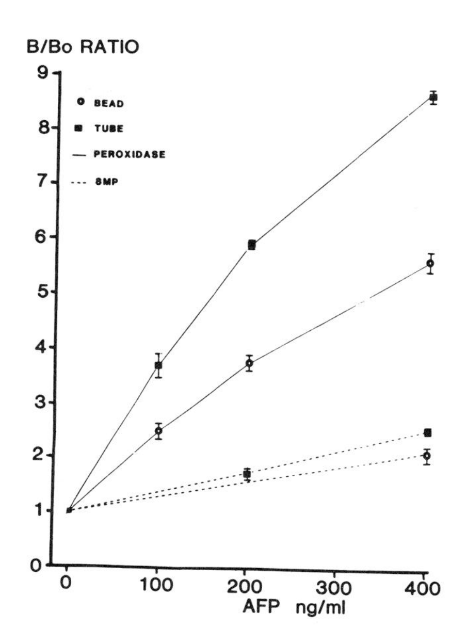

Fig 3 Comaprison of catalytic labelled luminescence immunoassay with enzyme immunoassays

The catalyst labelled immunoassay using polstyrene beads did not give adequate precision. Integration of the light output gave better precision than peak height measurements, however the light emission was low and the slope of the standard curve poor. Manual addition of luminescent reagents into the cuvette prevented formation of air bubbles and movement of the beads, however, the light output was low and difficult to measure.

Luminescent immunoassays on polyacrylic tubes resulted in higher light output. Due to the absence of entrapped air bubbles, automatic reagent addition could be used and therefore the precision was much better than with beads.

The conjugate contained only 1 microperoxidase for 6 immunoglobulin molecules, therefore the sensitivity of the assay was poor due to competition between labelled and unlabelled antibodies. Similarly the catalytic activity of haem in the conjugate was reduced 200 fold as a result of coupling to proteins.

Haem can form both hydrophobic and electrostatic bonds with suitable molecules, therefore the non specific binding of haem labelled assays can be a significant proportion of total conjugate binding. The microperoxidase conjugates also resulted in high non-immunological binding which was not prevented via the inclusion of detergents and carrier proteins in assay reagents.

Microperoxidase labels are not likely to provide sensitive sandwich immunoassays because of the high non specific binding of conjugates and the low catalytic activity when coupled to proteins, although the reduction in activity may result in homogeneous assays for small analytes.

ACKNOWLEDGEMENTS

The financial support of the DHSS is gratefully acknowledged.

REFERENCES

1. Ham, G., Belcher, R., Kricka, L.J., and Carter, T.J.N., (1979). Stability of trace iodine solutions. Anal. Lett. 12, 535-541.
2. Bunce, R.A., Carter, T.J.N., Kennedy, J.H., Kricka, L.J. and Whitehead, T.P., (1978). Apparatus and method for luminescent determination of concentration of analyte in a sample. British Patent 790094.
3. Plattner, H., Wachter, E., and Grobner, P., (1977). A hemenonapeptide tracer for electron microscopy. Preparation, characterisation and comparison with other heme-tracers, Histochem., 53, 223-242.
4. Kraehenbuhl, J.P., Galardy, R.E., and Jamieson, J.D., (1974). Preparation and characterisation of an immunoelectron microscope tracer consisting of a haem-octapeptide coupled to Fab., J. Expl. Biol, 139, 208.
5. Matsubara, H., (1966). Observation on the specificity of thermolysin with synthetic peptide. Biochem. Biophys. Res. Comm., 24, 427.
6. Hill, P.G., O'Toole, A., Kricka, L.J., Carter, T.J.N., and Whitehead, T.P., (1981). Optimisation of an automated enzyme immunoassay for α-fetoprotein using polystyrene beads as solid phase. J. Clin. Chem. Clin. Biochem., 19, 545.
7. Stott, R.A.W., (1984). Microperoxidase as a catalytic label in chemiluminescence immunoassay. Thesis submitted for the degree of Master of Science at the University of Birmingham.

SOLID PHASE ANTIGEN LUMINESCENCE TECHNIQUE (SPALT) ASSAYS FOR CORTISOL AND GENTAMICIN DETERMINATION IN SERUM

C.J. Strasburger, A. Gadow and W.G. Wood

Klinik für Innere Medizin, Medizinische Hochschule, D-2400 Lübeck 1, FRG

INTRODUCTION

The SPALT principle developed in this laboratory offers a simple, sensitive and precise method for the determination of haptens and proteins (1). This article deals with the development and optimisation of SPALT assays for cortisol as a steroid hormone, and gentamicin as an example for drug monitoring using chemiluminescence immunoassays.

MATERIALS AND METHODS

Donkey anti rabbit-IgG serum (Wellcome RD 17) was purified by PEG precipitation and subsequent DEAE chromatography. The purified γ-globulin was labelled either with diazoluminol or preferably with an active ester of ABEI-H (LKB). This labelled second antibody was used in both assays.

Anti gentamicin serum was raised in rabbits, rabbit anti cortisol serum was kindly donated by Dr M. Pazzagli, University of Florence, Italy. Both antisera were used without further purification.

6.4 mm diameter polystyrene beads (Spherotech, Fulda, FRG) were used as solid phase and activated using the poly-phenylalanine-lysine- pentan-1,5-dial method (2). Incubations were carried out in plates of 20 or 60 wells on a horizontal rotator, washing being performed using the Pentawash II system (Abbott).

Chemiluminescence was measured in an alkaline micro-peroxidase- H2O2 system by integration of the light signal for 20 seconds. LKB 1251 (25 samples, LKB-Wallac, Turku, Finland) and LB 950 (300 samples, Berthold, Wildbad, FRG) luminometers were used.

ISBN 0-12-426290-2

RESULTS

For the *immobilisation of haptens* to polystyrene beads it was found to be necessary to use a protein- hapten complex, as otherwise the binding of antibodies to the hapten was sterically impossible. As polyclonal antisera do not only contain antibodies against the hapten but also against the "bridge" and protein part of the immunogen, the solid phase antigen complexes have to be composed from components different of the immunogen.

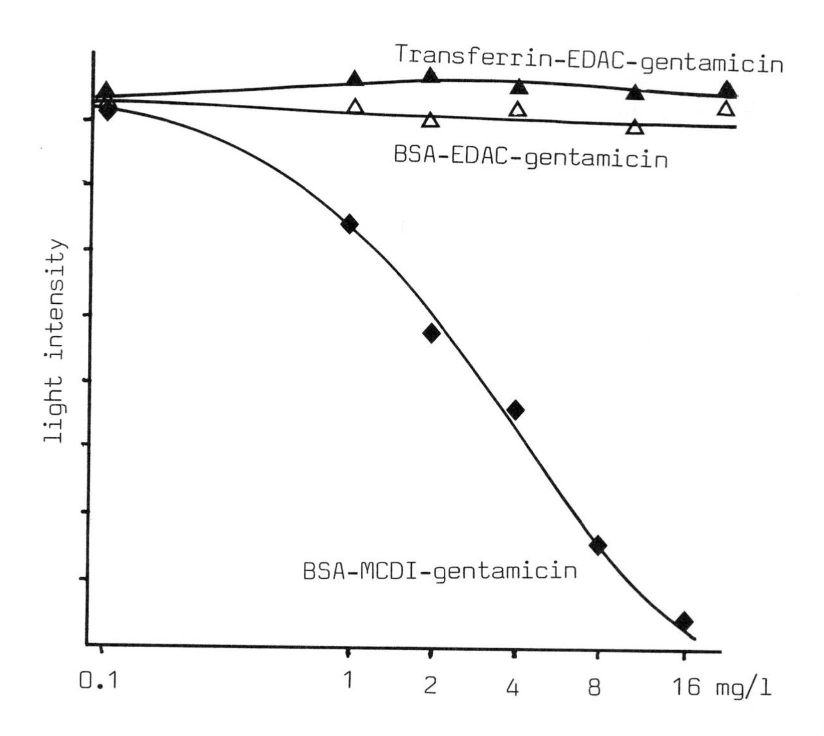

Fig. 1 Gentamicin SPALT: Effect of the conjugate chosen as solid phase antigen. The conjugate used for immunisation of the rabbit was transferrin-EDAC-gentamicin.

Incubation times for the assays were varied and optimised at ambient temperature without agitation as well as using a horizontal rotator. The results showed that a reduction of 50% for the first incubation between solid phase antigen and first antiserum, and a reduction by approximately 30% for the incubation with the labelled antiserum is possible.

Optimised assay procedures for both assays are shown in table 1.

Table 1

FINAL ASSAY PROCEDURES

	Cortisol SPALT	Gentamicin SPALT
Amount of sample	100µl (1:40 diluted in assay buffer)	50µl (1:450 diluted in assay buffer)
Assay buffer	0.1 M phosphate buffer pH8 cont. 1.5% Na salicylate	0.05 M Tris-HCl pH7.5
First antiserum	100µl rabbit anti cortisol (1:3000 dilution)	150µl rabbit anti gentamicin (1:375 dilution)
1. incubation*	40 min	30 min
1. wash step	2x5 ml 0.15 molar NaCl cont. 150µl/1 Tween 20	
Second antiserum	200µl labelled donkey anti rabbit-IgG (1:2000)	
2. incubation*	40 min	
2. wash step	as above	
Measurement	transfer bead to cuvette and measure for 20 s in luminometer	

* Incubations are carried out on a horizontal rotator at ambient temperature.

Theoretical sensitivity was calculated as B_0- 2s from a 10 fold determination of the 0-standard. The lower detection limit was 20.3 nmol/l for cortisol and 0.27 mg/l for gentamicin respectively.

Precision is expressed as mean C.V. from a compound precision profile. The cortisol SPALT showed a mean C.V. of 6.4% (range 30-1360 nmol/l, n=58), the gentamicin SPALT performed even better with 4.8% (0.2-8.1 mg/l, n=96).

Correlation with reference methods showed a regression coefficient of r=0.908 between gentamicin SPALT and RIA. As correlation between SPALT and DuPont ACA was good (r=0.969) and RIA and DuPont ACA showed a poor regression coefficient of r=0.918 it was concluded that the laboratorys own RIA delivered inaccurate results. This was later confirmed by external quality control. Correlation between cortisol SPALT and RIA is shown in figure 2.

External quality control was checked by participation in Wellcome BC 08 survey for cortisol (mean deviation from "all method mean" =0.7s, 12 samples) and INSTAND quality control programm for gentamicin (mean deviation from "true" defined values =12.6%).

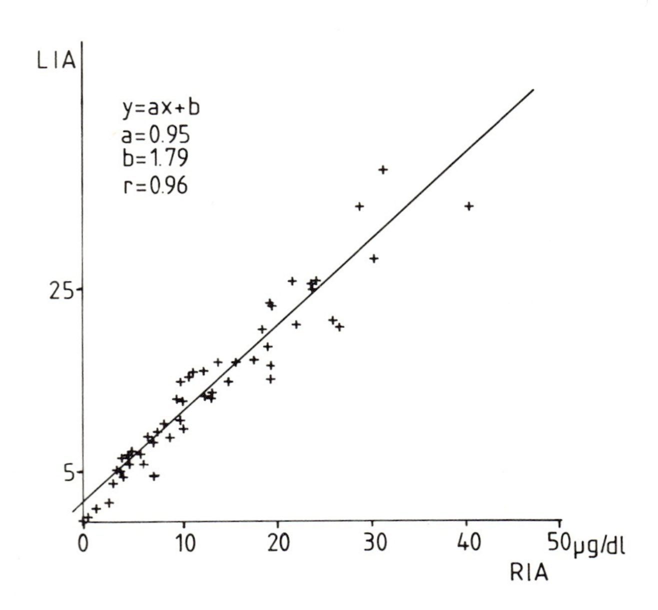

Fig. 2 Correlation between cortisol SPALT and RIA (n=58)

CONCLUSIONS

The SPALT principle avoids contact of sample and label, thus explaining the fact that serum additives such as EDTA, heparin and citrate as well as haemolysis or lipaemia do not significantly affect the assay results.

Both assays are equivalent to RIA in terms of precision, sensitivity, accuracy and ease of handling. The cortisol SPALT is sensitive enough to measure "free" cortisol levels in saliva.

All reagents used in the assays can be stored without a significant loss of assay stability for at least a year.

External quality control data show these methods to be suitable for routine use.

REFERENCES

1. Wood W.G. et al. (1982). Solid Phase Antigen Luminescent Immunoassays (SPALT) for the Determination of Insulin, Insulin Antibodies and Gentamicin Levels in Human Serum, J Clin Chem Clin Biochem 20:825-831
2. Wood W.G. and A. Gadow (1983) Immobilisation of Antibodies and Antigens on Macro Solid Phases- A Comparsion Between Adsorptive and Covalent Binding, J Clin Chem Clin Biochem 21:789-797

AN IMMUNOASSAY FOR SERUM THYROXINE EMPLOYING ENHANCED LUMINESCENT QUANTITATION OF HORSERADISH PEROXIDASE CONJUGATES

Gary H G Thorpe*, Eileen Gillespie*, Robert Haggart*, Larry J Kricka† and Thomas P Whitehead*

*Department of Clinical Chemistry, Wolfson Research Laboratories, Queen Elizabeth Medical Centre, and †Department of Clinical Chemistry, University of Birmingham, Birmingham, B15 2TH UK

INTRODUCTION

Light emission from the horseradish peroxidase (HRP) catalysed oxidation of cyclic diacylhydrazides, such as luminol (1), isoluminol (2) or 7-dimethylaminonaphthalene-1, 2-dicarbonic acid hydrazide (3) is enhanced by synthetic firefly D-luciferin (4,5-dihydro-2-(6-hydroxy-2-benzothiazolyl)-4-thiazole-carboxylic acid) (4).

Conjugates of HRP with antigens and antibodies also catalyse these enhanced luminescent oxidations and this has been exploited analytically in immunometric assays using HRP as the label (5). The light emission from an enhanced luminescent reaction is relatively high decays slowly and provides a rapid and sensitive assay for peroxidase conjugates either free in solution or bound to solid supports (6).

We present here a method for directly incorporating an enhanced luminescent end-point into a competitive heterogeneous enzyme immunoassay for thyroxine based on thyroxine-HRP conjugates, and details of its validation using serum samples.

MATERIALS AND METHODS

Synthetic firefly D-luciferin, luminol, Tris-HCl buffer, KCl and hydrogen peroxide (30% w/v) were purchased from Sigma Chemical Company. Luminol was purified by conversion to its sodium salt (6). Light measurements were performed in a luminometer based on a side-window photomultiplier

ISBN 0-12-426290-2

tube (EMI type 9731A, 94 mA/lumen) using photocurrent measurement. Serum specimens with a range of thyroxine concentrations determined using radioimmunoassay were supplied by the Clinical Chemistry Department, Queen Elizabeth Medical Centre.

Immunoassay Procedure An Enzymun-Test T4 kit (BCL Ltd, East Sussex, UK) was used according to the manufacturers instructions. The assay involved incubation of anti-thyroxine coated plastic tubes with sample (20 µl), HRP-labelled thyroxine, 8-aniline-1-naphthalenesulphonic acid and buffer for two hours at room temperature. After aspiration and washing, conjugate bound to tubes was quantitated using colorimetry or enhanced luminescence.

Colorimetry 2-2'-Azino-di-(3-ethylbenzthiazoline-6-sulphonate)/perborate substrate was added to each tube and incubated for 1 hour in the dark at room temperature. Tube contents were transferred to cuvettes and the absorbance measured at 420 nm using a Shimadzu UV-240 recording spectrophotometer.

Enhanced luminescent end-point Tubes were aspirated to dryness and 0.6 ml Tris buffer (0.01 M, pH 8.0) KCl (0.15M) added to each tube. The luminescent reaction was initiated by adding 10 µl firefly D-luciferin (3.6 mM) in the Tris buffer and 0.4 ml of a mixture of luminol (3.1 mM) and hydrogen peroxide (7.3 mM) in the Tris buffer. The intensity of light emission after 30 seconds was then measured.

RESULTS

Calibration Curves Dose-response curves for thyroxine using an enhanced luminescent and a colorimetric end-point are illustrated in Figure 1. The enhanced luminescent end-point produced the more sensitive assay despite the fact that the light detector only monitored 50% of the active area of the assay tube.

Precision An estimate of the within-assay variation for the enhanced luminescent immunoassay was obtained by analysing replicate samples of serum specimens within a single analytical run. Good precision was obtained at thyroxine concentrations at the upper limit (mean value = 154 nmol/l, n = 12, coefficient of variation (CV) = 8.8%) and lower regions (mean value = 80 nmol/l, n = 13, CV =

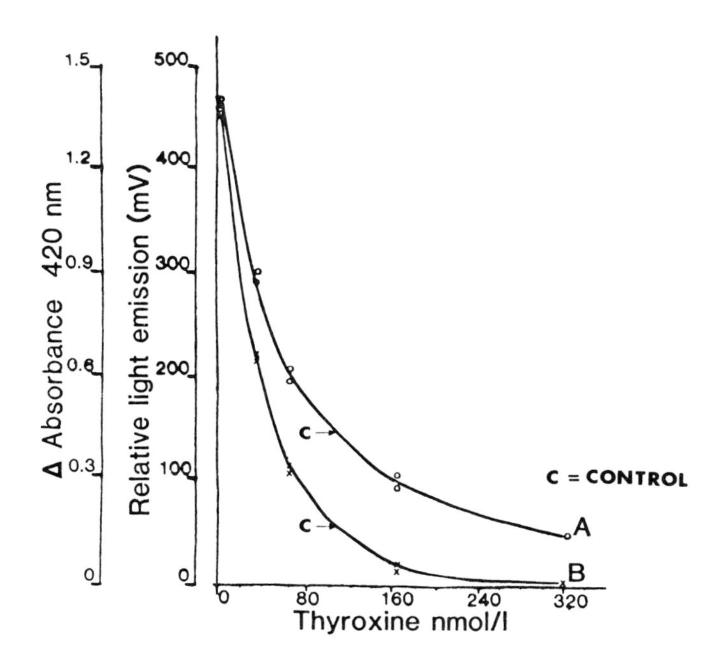

Fig. 1 Dose - response curves for serum thyroxine using A) a colorimetric or B) an enhanced luminescent immunoassay

8.8%) of the clinical range.

Kinetics Of Light Emission Light emission with the enhanced reaction is prolonged and relatively constant over several minutes (Figure 2). For convenience light intensity after 30 seconds was recorded, however more rapid or alternative measurements could be employed. The kinetics of the light emitted also eliminate any requirement for rapid reproducible initiation of luminescent reactions on solid supports in front of the photodetector.

Comparison of serum thyroxine levels determined by enhanced luminescent immunoassay, enzyme immunoassay or radioimmunoassay The concentrations of thyroxine as determined by enhanced luminescent immunoassay showed a good correlation with results obtained by enzyme immunoassay (Figure 3) (r = 0.97; regression line y = 1.018x - 6.33) and by radioimmunoassay (r = 0.95; y = 1.027x - 4.6) The thyroxine results for a control serum supplied with the kit also showed good agreement with the assigned value.

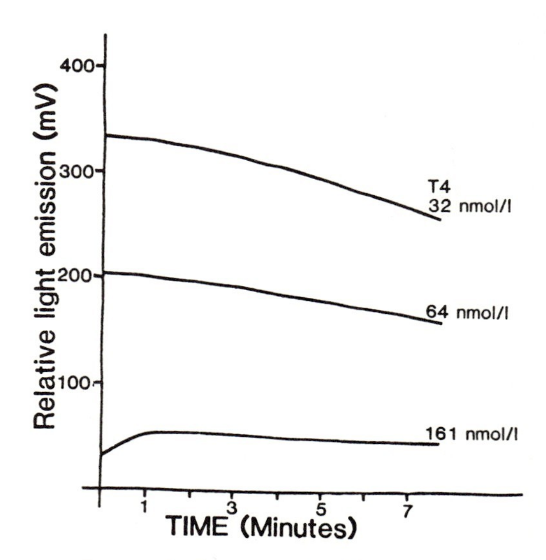

Fig. 2 Kinetics of light emission for thyroxine standards assayed using an enhanced luminescent immunoassay

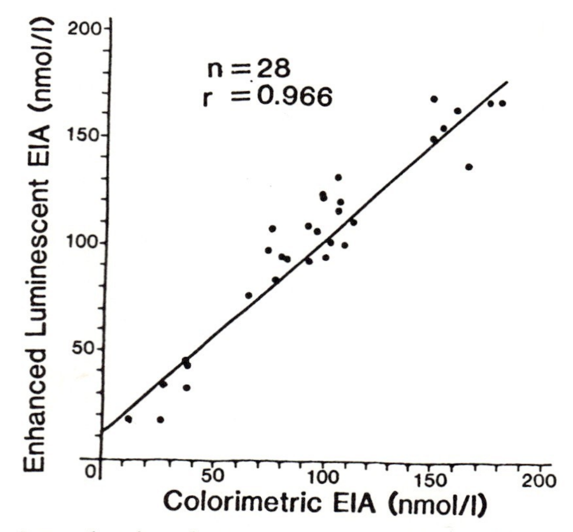

Fig. 3 Correlation between enhanced luminescent immunoassay and enzyme immunoassay results for thyroxine

DISCUSSION

The enhanced luminescent immunoassay for thyroxine is rapid, sensitive, precise and the results compare favourably with those obtained by other immunoassays. Re-optimisation of the enhanced assay in conjunction with sensitive instrumentation, efficient light collection and purpose designed solid supports should further improve the assay.

A variety of luminescent reactions can be used to monitor immunoassays (7,8,9). The main advantages of the enhanced luminescent immunoassay based on HRP labels are (i) the enzyme label introduces an amplification step, (ii) reagents are readily available, (iii) the light emission decays slowly and thus it is not necessary to initiate the reaction in front of the photodetector as in the case for luminescent immunoassays which produce flashes of light, and (iv) the enhanced luminescent assay procedure is simple, sensitive and rapid and does not require rapid reproducible mixing or lengthy preincubations used in some other luminescent immunoassays (10).

Finally, the enhanced assay has general applicability to enzyme immunoassays based on HRP labels. In the case of thyroid-function testing immunoassays based on HRP labels are known for tri-iodothyronine, thyroid binding globulin (11) thyroid stimulating hormone (1) and free thyroxine (12), thus it would be possible to develop a full range of luminescent immunoassays for thyroid-function testing analogous to that described here for thyroxine.

ACKNOWLEDGEMENTS

The financial support of the Department of Health and Social Security is gratefully acknowledged.

REFERENCES

1. Tsuji, A., Maeda, M., Arakaura, H., Matsuoka, K., Kato, N., Naruse, H., and Irie, M. (1978). Enzyme immunoassay of hormones and drugs using fluorescence and chemiluminescence reaction. *In* "Enzyme Labelled Immunoassay of Hormones and Drugs" (Ed S.B. Pal) pp 326-339. Walter de Gruyter, Berlin.
2. Pronovost, A.D., and Baumgarten, A. (1982). Comparison of chemiluminescence and absorptiometry in enzyme immunoassays for protein quantification, *Experientia* 38 304-306.

3. Gundermann, K-D., Wulff, K., Linke, R., and Stahler, F. (1982). Enzyme immunoassays using horseradish peroxidase as a label assayed by chemiluminescence with luminol derivatives as substrates. *In* "Luminescent Assays: Perspectives in Endocrinology and Clinical Chemistry" (Eds M. Serio and M. Pazzagli) pp 157-161. Raven Press, New York.
4. Carter, T.J.N, Groucutt, C.J., Stott, R.A.W., Thorpe, G.H.G.H., and Whitehead, T.P., (1982). *UK Pat. Applic.* 8206263.
5. Thorpe, G.H.G., Haggart, R., Kricka, L.J., and Whitehead, T.P. (1984). Enhanced luminescent enzyme immunoassays for rubella antibody, IgE and digoxin 'in preparation'.
6. Whitehead, T.P., Thorpe, G.H.G., Carter, T.J.N., Groucutt, C., and Kricka L.J. (1983). Enhanced luminescence procedure for sensitive determination of peroxidase-labelled conjugates in immunoassay, *Nature* 305, 158-159.
7. Kricka, L.J., and Thorpe, G.H.G. (1981). Luminescent immunoassay, *Ligand Review* 3, 17-24.
8. Collins, W.P., Barnard, G.J., Kim, J.B., Weerasekera, D.A., Kohen, F., Eshhar, Z., and Lindner, H.R., (1983). Chemiluminescence immunoassays for plasma steroids and urinary steroid metabolites. *In* "Immunoassays for Clinical Chemistry" (Eds W.M. Hunter and J.E.T. Corrie) pp 373-397. Churchill Livingston, Edinburgh.
9. Weeks, I., Beheshti, I., McCapra, F., Campbell, A.K., and Woodhead, J.S. (1983). Acridinium esters as high-specific-activity labels in immunoassay, *Clin. Chem.* 29, 1474-1479.
10. Weerasekera, D.A., Kim, J.B., Barnard, G.J. and Collins, W.P. (1983). Measurement of serum thyroxine by solid-phase chemiluminescence immunoassay, *Ann. Clin. Biochem.* 20, 100-104.
11. Bernutz, C., Rohler, M., and Horn, K. (1982). A new solid phase enzyme-immunoassay for total thyroxine (T_4), triiodothyronine (T_3) and thyroxine-binding globulin (TBG), *Fresenius Z. Anal. Chem.* 311, 353-354.
12. Weetall, H.H., Hertl, W., Ward, F.B., and Hersh, L.S. (1982). Enzyme immunoassay for free thyroxin, *Clin. Chem.* 28, 666-671.

A NEW IMMUNOASSAY FOR ECDYSTEROIDS BASED ON CHEMILUMINESCENCE

Lutz Reum*$, Wolfgang Klingler** and Jan Koolman*

* *Physiol.-Chem. Inst., Deutschhausstr. 1-2, D-3550 Marburg*
** *Inst. Biochem. Endocrinologie, Ratzeburger Allee 160 D-2400 Lübeck 1*
$ present address: Mallinckrodt Diagnostica, von-Hevesy-Str. 1-3, D-6057 Dietzenbach 2

INTRODUCTION

Ecdysteroids, first detected as moulting hormones of insects (1) receive increasing medical interest since it became apparent that they occur and probably function as hormones not only in arthropodes but also in other representatives of lower animals (2). Some of them such as helminth species *Schistosoma mansoni, Brugia malayi and Ochocerca volvulus* are of prime medical importance as parasites of man.

Ecdysteroids are commonly detected and quantified in biological fluids by Radioimmunoassay (RIA), which was first established by Borst and O'Connor (3) in 1972. Following this method all subsequent investigators have used radiolabelled compounds as tracer molecules. However the RIA has several disadvantages, the work with radioisotopes requires special precautions and is restricted by governmental regulations. These disadvantages prompted us to develop a new immunoassay which avoids any radioactivity.

METHODS AND RESULTS

Ecdysone was modified to its carboxymethoxime derivative and covalently attached to amino butylethyl-isoluminol (ABEI) according to Kohen et al. (4) to form a chemiluminescent tracer. The ecdysone-ABEI conjugate was purified by High Performance Liquid Chromatography (HPLC) using

ISBN 0-12-426290-2

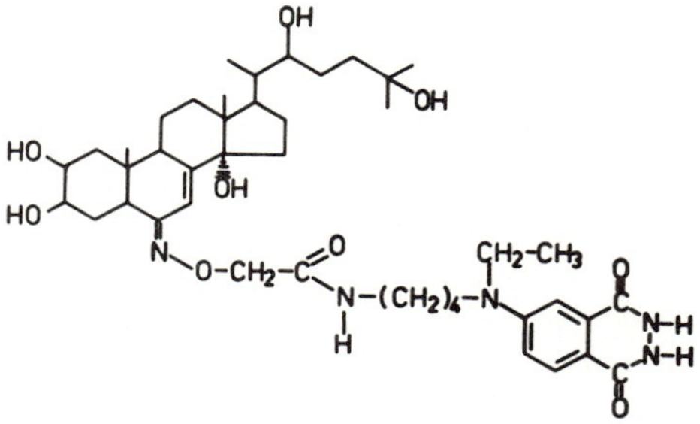

Fig. 1 Ecdysone-6-carboxymethoxime-ABEI

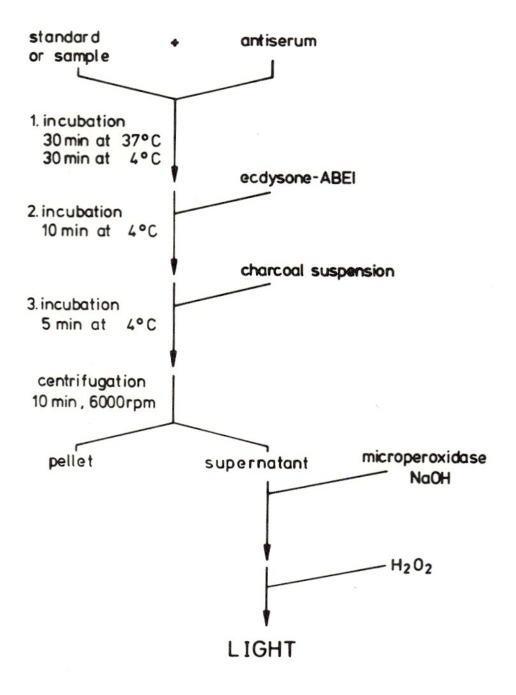

Fig. 2 Flowdiagram of the chemiluminescence immuno assay

reversed phase material.
The purification was monitored by UV-light and luminescence measurement. On TLC-plate the ecdysone-ABEI molecule showed a single spot with a Rf-value of 0.64, while ecdysone had a Rf of 0.69, ecdysone-6-carboxymethoxime 0.20 and ABEI 0.00 in the solvent system chloroform/methanol 60/40 v/v.
The UV spectrum of the tracer revealed a maximum at 284 nm (ABEI) and at 254 nm (ecdysone-6-carboxymethoxime).
The proposed structure of the tracer molecule is shown in figure 1. The tracer was recognized by antibodies when incubated with the anti-ecdysteroid antisera DUL-1 and DBL-1 (5) and could be displaced by an increasing amount of unlabelled ecdysone.
The flowdiagram of the CIA is illustrated in figure 2. Microperoxidase and sodiumhydroxide were added to the supernatant and the luminescence measured after the injection of hydrogenperoxide in a photon counter over 30 sec in the integration mode.

DISCUSSION

To develop a new nonisotopic immunoassay for ecdysteroids we coupled aminobutylethyl-isoluminol to ecdysone-6-carboxymethoxime to get the luminescent tracer. The CIA described here principally functions like a RIA with slight differences shown in fig. 2.
A preincubation of antibodies with standard or sample is necessary because the ecdysone-ABEI derivative has a higher affinity towards the anti-ecdysteroid antibodies than ecdysone itself due to the additional oxime bridge in the molecule (fig. 1). This oxime bridge was also a part of the immunogen used for antibody production. It therefore will be recognized by the antibodies. A short tracer incubation is used to avoid any exchange of antibody bound sample or standard against the tracer molecule, which would cause a decrease of assay sensitivity. In addition a rapid separation method is necessary to eliminate the free antigen. Dextran-coated charcoal treatment was found to be a suitable procedure. More than 98 % of the free ecdysteroids are adsorbed during the incubation time.
After centrifugation the chemiluminescence is measured in the supernatant. More than 90 % of the resulting light is emitted during the first 30 seconds. A shorter light measurement led to an increase of the standard deviation between replicates while a longer counting period has no effect on the result.
Competition curves obtained with anti-ecdysteroid antisera

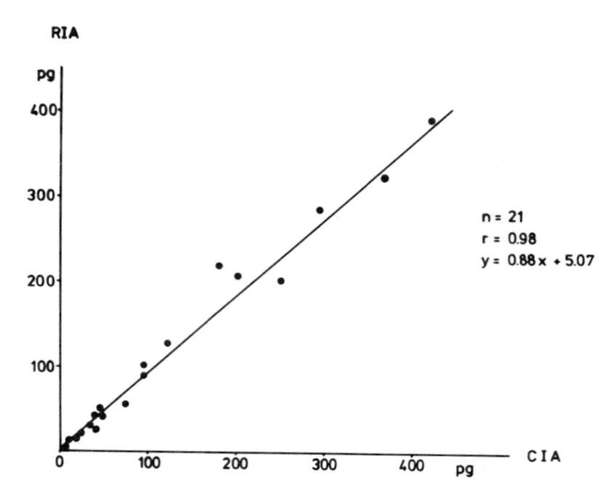

Fig. 3 Comparison of ecdysteroid levels determined by RIA and CIA

were quite similar in the RIA and CIA and the crossreaction of different ecdysteroids was nearly identical.
Compared to the RIA the CIA showed a broader range of detection (ca. 20 pg - 11 ng).
Ecdysteroid levels of the hemolymph of blowfly larvae were determined by RIA and CIA after methanolic extraction of the biological tissue. Figure 3 showed a good agreement between the two assays.
In general we observed that the CIA is faster than the RIA (80 samples can be processed within 2 hours)and more precise than the RIA.

REFERENCES

1. Butenandt,A. und Karlson,P. (1954). Über die Isolierung eines Metamorphosehormons der Insekten in kristallisierter Form, *Z. Naturforsch.* 9b, 389-391.
2. Karlson,P. (1983). Why are so many hormones steroids ? *Hoppe-Seyler's Z. Physiol. Chem.* 364, 1067-1087.
3. Borst,W. and O'Connor,J.D. (1972). Arthropod moulting hormone: Radioimmuo assay. *Science* 178, 418-419.
4. Kohen,F. et al. (1981). Development of a solid-phase chemiluminescence immunoassay for plasma progesterone, *Steroids* 38/1, 73-88
5. Reum,L., Haustein,D. and Koolman,J. (1981) Immunoadsorption as a means for the purification of low molecular weight compounds: Isolation of ecdysteroids from insects. *Z. Naturforsch.* 36c, 790-797.

ENZYME IMMUNOASSAY MONITORED BY CHEMILUMINESCENCE REACTION USING BIS(2,4,6-TRICHLOROPHENYL)OXALATE-FLUORESCENT DYE

A. Tsuji, M. Maeda and H. Arakawa

School of Pharmaceutical Sciences, Showa University
Hatanodai, Shinagawa-ku, Tokyo, Japan 142

INTRODUCTION

In recent years, enzyme immunoassay (EIA) using appropriate enzyme as label instead of radioisotopes has become a clinically useful technique for the measurement of physiological concentrations of many clinically important compounds hormones and drugs. Very often the final determination of enzymatic activity in EIA is performed by colorimetric or fluorimetric method. In order to increase the sensitivity of EIA, the use of chemiluminescence (CL) or bioluminescence reactions is currently being investigated in several laboratories. We have reported the CL EIAs of cortisol (1) and dehydroepiandrosterone (2) using peroxidase as label and luminol-H_2O_2 as substrate for the assay of enzymatic activity. In this report new CL EIAs of 17α-hydroxyprogesterone (17-OHP) and thyroxine (T_4) using glucose oxidase (GOD) as the label enzyme and the peroxyoxalate CL reaction for the assay of enzymatic activity are presented.

MATERIALS AND METHODS

Materials 17-OHP, T_4, bis(2,4,6-trichlorophenyl)oxalate

ISBN 0-12-426290-2

(TCPO), 8-anilinonaphthalene-1-sulfonic acid (ANS), and other chmicals were of reagent grade. Anti-17-OHP serum, four kinds of anti-T_4 sera, and goat anti-rabbit IgG serum were used in this study. 17-OHP-GOD conjugate and four T_4-GOD conjugates were prepared by the mixed anhydride or glutaraldehyde method according to the literature. Double antibody solid phase (DASP) beads were prepared by coating polyacetal beads with purified anti-rabbit IgG antibody.

EIA Procedure Anti-serum and enzyme labeled conjugate were diluted with 0.05 M phosphate buffer containing 0.1% BSA and 0.9% sodium chloride (pH 7.0). The outline of EIA for T_4 is as follows :

Step 1: Dried blood sample disc (3 mm, i.d.) (or standard disc) — 1
Anti-T_4-hemiglutarate-BSA serum (1 : 10^5) — 200 μl
T_4-GOD conjugate (1 : 2000) — 100 μl
DASP bead — 1

↓ incubate for 3 hr at 37°C
wash with saline (2 ml x 3)

Step 2: 0.5 M Glucose solution (0.01 M acetate buffer, pH 5.1) — 300 μl

↓ incubate overnight at 4°C

Step 3: The resultant reaction solution — 100 μl
5 mM TCPO-ethyl acetate solution — 200 μl
0.02% ANS solution — 100 μl
Measurement of chemiluminescence intensity

In the case of 17-OHP, at the 1st step the sample disc was incubated with anti-17-OHP serum and DASP bead, and the 2nd step 17-OHP-GOD conjugate was added and incubated for 2 hrs at room temperature. After washing the beads with 0.1% Tween 20 solution, the CL assay was carried out as described above.

RESULTS AND DISCUSSION

Chemiluminescence EIA of T_4 In this EIA system, GOD was used as label enzyme. The DASP bead was used to separate the bound and free fractions after immune reaction. The CL determination of GOD is as follows:

$$\text{D-Glucose} + H_2O + O_2 \xrightarrow{\text{GOD}} \text{D-Gluconic acid} + H_2O_2$$

$$H_2O_2 + \text{TCPO} \rightarrow \text{Dioxetandione} + 2\ 2,4,6\text{-Trichlorophenol}$$

$$\text{Dioxetandione} + \text{ANS} \rightarrow \text{ANS* (excited state)} + 2\ CO_2$$

$$\text{ANS*} \rightarrow \text{ANS} + \text{light}$$

The CL intensity was measured with an Aloka Luminescence Reader (a photon counter) (waiting time: 15 sec, integrating time: 6 sec).

In order to test the sensitivity homologous and heterologous systems using four kinds of anti-T_4 sera and T_4-GOD conjugates were examined. The bridge heterologous system using anti-T_4-serum against T_4-hemiglutarate-BSA and T_4-GOD conjugate prepared by glutaraldehyde method showed the highest sensitivity. A good standard curve was obtained in the range from 0.25 to 20 ug/dl. The CV% ranged from 2.1 to 4.8%. The concentration of T_4 in one disc of dried blood sample could be determined by this EIA method. The correlation coefficient between RIA and EIA was 0.91 (n = 50). In the preliminary mass-screening test for congenital hypothyroidism in neonates, 492 samples of newborn babies were assayed by this EIA and RIA. Though the values obtained by this EIA were lower than those obtained by RIA, the distribution pattern was similar.

Chemiluminescence EIA of 17-OHP GOD was also used as label enzyme. Anti-17-OHP serum was prepared by using 17-OHP-3-O-carboxymethyl oxime-BSA conjugate and 17-OHP-GOD conjugate was also prepared by using 17-OHP-3-O-carboxymethyl oxime and GOD. The standard curve was constructed in the range from

0.25 to 100 pg/assay tube. The CV% ranged from 1.7 to 8.8% (n = 5). The detection limit of this EIA was 0.25 pg/assay tube corresponding to 5 pg/ml (16 fmol/ml). This value is lower than the detection limits of RIAs reported in the literature . The correlation coefficient between this EIA and RIA was 0.97.

In newborns, measurement of 17-OHP is used in the initial diagnosis of congenital adrenal hyperplasia (CAH), due to steroid 21-hydroxylase deficiency. To facilitate mass-screening for CAH, we have examined to apply this CL EIA of 17-OHP to mass-screening test for CAH. Good results were obtained in the pilot study.

Conclusion The CL reaction of TCPO-H_2O_2-ANS system has been used for the assay of GOD activity in EIAs of 17-OHP and T_4. These CL EIAs using GOD offer considerable advantages for clinical diagnosis. Because of the high sensitivity of the CL method, only a few microliters of samples are needed for each determination: this makes them suitable for screening tests of neonatal dried blood samples. From the results of pilot study, it may be concluded that the EIAs of T_4 and 17-OHP are useful tools for the congenital adrenal hyperplasia and hypothyroidism screening.

The CL EIA system reported here may be applicable to various EIAs using H_2O_2-generating enzyme system as the label enzyme.

1. Arakawa, H., Maeda, M. and Tsuji, A. (1979). Chemiluminescence enzyme immunoassay of cortisol using peroxidase as label, Anal. Biochem., 97, 248-254.

2. Arakawa, H., Maeda, M. and Tsuji, A. (1981). Chemiluminescence enzyme immunoassay of dehydroepiandrosterone and its sulfate using peroxidase as label, Steroids, 38, 453 - 464.

THE ROUTINE DETERMINATION OF METALLOPROTEINS IN BIOLOGICAL FLUIDS USING CHEMILUMINESCENCE IMMUNOASSAYS

Bettina Tode and W.G. Wood

Klinik für Innere Medizin, Medizinische Hochschule Lübeck, D-2400 Lübeck 1, FRG.

INTRODUCTION

The three metalloproteins studied in this article are: ferritin (F), transferrin (TF) and caeruloplasmin (CP). All three are closely linked with iron metabolism, storage and transport, although caeruloplasmin is better known for its copper content and oxidase activity.

The ferritin assay was developed as an ILMA (1), the transferrin and caeruloplasmin assays as SPALT (2). All three assays were suitable for paediatric use, duplicate determinations of all 3 parameters needing less than 100 µl serum.

The accent has been laid upon the study of blood donors, patients with Crohn's disease and with ulcerative colitis, as well as in tumour patients, in which one common link is the disturbance of iron metabolism.

MATERIALS AND METHODS.

Antibodies to transferrin and caeruloplasmin were raised in rabbits against the purified human protein. Ferritin antisera were purchased from Dakopatts, Glostrup, DK., (rabbit anti-ferritin) and from Atlanta (pelfreez), Heidelberg, D., (goat anti-human liver ferritin). ABEI-H for labelling was obtained from LKB Munich and polystyrene balls for immobilisation of reaction partners from Spherotech Kugeln, Fulda, D. Donkey anti-rabbit IgG for the SPALT assays was obtained from Wellcome Diagnostica, Burgwedel, D.

Labelling of antiserum gamma-globulin fractions was done via an active ester of ABEI-H and N-hydroxysuccinimide (3). Tracer purification was performed over Sephadex G-25 or Ultrogel AcA 202 columns. Table 1 shows final assay details.

ISBN 0-12-426290-2

TABLE 1

Flow schemes of the routine assays for ferritin, transferrin and caeruloplasmin

Parameter	Ferritin	Transferrin	Caerulo-plasmin
Sample dilution 1:	1	1000	25
sample volume µl	20	50	40
Preincubation min	--	10	15
Incubation with solid phase min/temp °C	180/RT^{+}	50/RT	45/RT
Wash step	2 x 5 ml 0.15 mol/1 NaCl with 150 µl/1 Tween 20		
Tracer volume µl	200	200	200
Tracer incubation (min / temp °C)	150	120	60
Wash steps	As above		

Transfer balls to cuvettes, cover with 300 µl 0.15 mol/1 NaCl and load luminometer. Initiate light reaction and integrate over 20 s.

Key: $^{+}$RT = room temperature (20 - 25°C)

RESULTS

Table 2 shows the established reference ranges and quality control parameters and Table 3 the results from the groups of patients studied together with first-time blood donors.

TABLE 2

Reference ranges and quality control parameters

Parameter	F	TF	CP
Units	µg/1	g/1	g/1
Reference range	30-300	2-4	0.15-0.65
Working range (cv 10%)	5-1000	0.5-6.0	0.15-1.5
Mean cv in working range	4.5	6.0	5.8
No of samples used for cv	570	564	105
Lower detection limit	2	0.5	0.08

The coefficients of variation (cv) were derived from precision profiles of several assays using the mean of duplicates.

The inter-assay cv for the ferritin assay derived from control sera run in 50 consecutive assays were: 7.24% at 16.3 µg/1, 6.82% at 45.7 µg/1 and 6.95% at 133 µg/1.

TABLE 3

Levels of ferritin and caeruloplasmin in different groups

Experimental Group	Ferritin	Caeruloplasmin
Blood donors -first time		
Range of values+	5-240	0.06-1.01
Median value	58	0.26
No. of elevated values	0	3
No. of reduced values	16	4
No. in group	84	45
Crohn /colitis patients		
Range of values	2-911	0.14-0.53
Median value	18.8	0.30
No.of elevated values	4	0
No. of reduced values	39	1
No. in group	65	12
Tumour bearers		
Range of values	3-2440	0.13-2.34
Median value	152	0.44
No. of elevated values	14	12
No. of reduced values	4	1
No. in group	42	35

Key: +The units of concentration are as in Table 2. The median value was used as the distribution of values was not normal.

In the case of the tumour patients an interesting fact was seen as far as the caeruloplasmin values were concerned. The method used for comparing the caeruloplasmin SPALT was an assay in which the oxidase activity was measured, i.e. the biological activity of the molecule. The substrate used was p-phenylene diamine. The values from this assay were normal or reduced in the case of 20 tumour patients tested, whereas in the SPALT no reduced values were found, the majority of values being elevated or in the upper reference range. This divergence of biological and immunological reactivity may be of importance in assessing the status of tumour patients.

The serum ferritin levels of the Crohn / colitis group were mainly under the lower reference limit for persons with normal iron status. The transferrin levels in serum, which are not listed here, were correspondingly elevated. These results are in accord with impaired iron resorption and increased intestinal haemorrhage in these patients.

CONCLUSIONS

The assays which have been described here are all in routine use, transferrin for over 2 years (4), ferritin and caeruloplasmin for 9 months. All are set up using the reaction trays, horizontal rotator and wash system from Abbott Diagnostics (Pentawash II and Heidolph T-190 rotator).

The role of metalloproteins in disease states warrants their routine determination in certain patient groups, not as a screening or blanket determination. A combination of metalloprotein and acute-phase protein determinations seems to give clinically useful data, together with tumour markers (carcinoembryonal antigen and tissue polypeptide antigen) in the case of tumour patients. An ongoing study in this laboratory has the goal of finding optimal combinations of tests for different illnesses, at the same time keeping costs to an acceptable minimum.

REFERENCES.

1. Wood, W.G. et al. (1984). An evaluation of four different luminescence immunoassay sytems: CELIA (chemiluminescence immunoassay), SPALT (solid phase antigen luminescence technique), ILMA (immunoluminometric assay) and ILSA (immunoluminometric labelled second antibody). Part III of a critical study of macro solid phases for use in immunoassay systems. J.Clin.Chem.Clin.Biochem. 22: In Print
2. Wood, W.G. et al. (1982). Solid phase antigen luminescence techniques (SPALT) for the determination of insulin insulin antibodies and gentamicin levels in human serum. J.Clin.Chem.Clin.Biochem. 20: 825-831
3. Gadow, A. et al. (1984). Synthesis and evaluation of luminescent tracers and hapten-protein conjugates for use in luminescence immunoassays with immobilised antibodies and antigens. Part II of a critical study of macro solid phases for use in immunoassay systems. J.Clin.Chem.Clin.Biochem. 22: In Print
4. Fricke, H., Strasburger, C.J. and Wood, W.G. (1982). Enzyme enhanced luminescence immunoassay (EELIA) for serum transferrin, J.Clin.Chem.Clin.Biochem. 20: 91-94

TRANSFERRIN AND CERULOPLASMIN MEASUREMENT IN HUMAN SEMINAL PLASMA BY A CHEMILUMINESCENT METHOD

C.Orlando*, A.L.Caldini*, T.Barni*, W.G.Wood**, B.Tode**, G.Fiorelli*, G.Forti* and M.Serio*

*Endocrinology Unit,University of Florence,Italy
**Klinik fuer Innere Medizin,Medizinische Hochschule, Luebeck,FRG

INTRODUCTION

It has been demonstrated that Sertoli cells in monolayer culture (1) produce transferrin (TF) and ceruloplasmin (C). Moreover seminal TF was found to be reduced in infertile subjects in comparison to normal controls (2).

In attempting to investigate if seminal concentrations of these proteins could be used as an "in vivo" index of Sertoli cell function, we measured TF and C by a chemiluminescent method in seminal plasma of patients affected by infertility of different origin.

MATERIALS AND METHODS

The subjects studied were divided in the following groups:

A) Azoospermic subjects affected by congenital absence of deferent ducts and seminal vesicles
B) Vasectomized subjects after at least 1 year from vasectomy
C) Subjects affected by azoospermia due to seminiferous tubule damage without obstruction
D) Unselected oligozoospermic subjects,sperm density $<20 \times 10^6$ /mL
E) Fertile normospermic subjects, sperm density $>20 \times 10^6$ /mL

ISBN 0-12-426290-2

TF and C were measured by SPALT (Solid Phase Antigen Luminescent Technique) methods (3), in which the antigen (TF or C) was covalently linked to polystirene balls (0.5 cm diameter). The antisera used were rabbit anti-human TF and rabbit anti human C. The tracer used was ABEI-H (LKB,Bromma) which was converted into an hydroxysuccinimide ester and then coupled to a second antibody(donkey anti-rabbit IgG) by the Bolton - Hunter method. The assay buffer was a Tris/PBS buffer, 0.025 mol/L, pH 7.4, containing 0.25 g/L BSA and 0.05% Tween20. The standard curves ranged from 0.125 ug/mL to 8 ug/mL for TF and from 0.08 ug/mL to 0.5 ug/mL for C. The assay procedures for TF and C are summarized in Table 1.

TABLE 1

Assay procedures for TF and C

		TF	C
Standard	ul	50	50
or seminal plasma	uL/dilution	50/1:20	50/undiluted
Buffer	uL	50	50
1° Antibody	uL/dilution	100/1:500	100/1:2000
PREINCUBATION	min/°C	10/RT	10/RT
Solid phase	ball	TF	C
INCUBATION	h/°C	1/RT	1/RT
Wash*		1mLx2	1mLx2
Tracer	uL/dilution	200/1:200	200/1:100
INCUBATION	h/°C	2/RT	2/RT
Wash*		1mLx2	1mLx2

*Wash : NaCl 0.9% containing 0.015% Tween20

Measurement procedure : The ball was transferred to a measuring cuvette and 0.2 mL of NaOH 0.8 N and 0.1 mL of microperoxidase 5 umol/L were added. The light emission was measured by a Biolumat LB9500 (Berthold) using the integration mode over a 10 sec interval after the automatic injection of 0.1 mL of hydrogen peroxide 0.15 %.

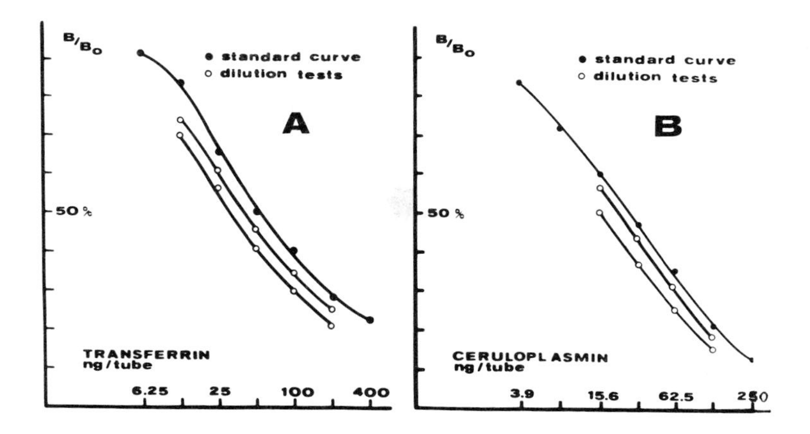

Fig. 1 A) Transferrin standard curve and dilution tests (10, 5, 2.5, 1.25 and 0.625 uL/tube of seminal plasma). B) Ceruloplasmin standard curve and dilution tests (100, 50, 25 and 12.5 uL/tube of seminal plasma).

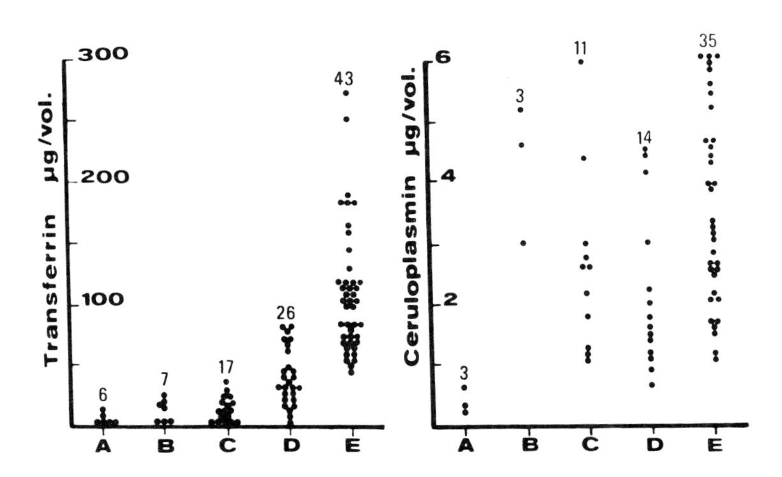

Fig. 2 Transferrin and ceruloplasmin seminal levels in normal and infertile patients. Results are expressed in ug/ total volume of ejaculate.

RESULTS

In Fig.1 A and B are shown typical standard curves and dilution tests respectively for TF and C. Recovery experiments (n=10), performed by adding different concentrations of TF and C to seminal plasma, resulted (m ± S.D.) 103 % ± 11 for TF and 102% ± 10 for C. The within (n=30) and between (n=15) assay variation were respectively (CV) 9.9% and 14.9% for TF and 10.1% and 15.2% for C. The sensitivity of the methods was 6 ng/tube for TF and 4 ng/tube for C. TF and C seminal levels in normal and infertile patients are reported in Fig.2.

CONCLUSIONS

The SPALT methods for TF and C measurement in seminal plasma showed good characteristics of reproducibility, accuracy and sensitivity.

No difference was found in C seminal levels among the groups of patients studied, with the only exception of group A, in relation to the low volume of ejaculate of these subjects. On the other hand seminal TF was always found below the lowest limit of normal range in the groups A,B,C.

These data seem to suggest that seminal TF could be a direct and useful index of Sertoli cell function.

ACKNOWLEDGEMENTS

This paper has been supported by a research grant from the University of Florence.

REFERENCES

1. Mather, J.P. et al (1983). The hormonal and cellular control of Sertoli cell secretion, J. steroid Biochem 19, 41-51.
2. Holmes, D.H. et al (1982). Transferrin and gonadal disfunction in man, Fertility and Sterility 38,600-604.
3. Strasburger,C.J. et al (1982).Luminescence immunoassays for proteins and haptens. in "Radioimmunoassay and Related Procedures in Medicine". pp.757-777, IAEA,Vienna.

CHEMILUMINESCENCE IMMUNOASSAY FOR THE DEMONSTRATION OF HBs AG USING A FLUORESCEIN ISOTHIOCYANAT LABELLED ANTI HBs.

H. Berthold

Institut für Virologie
Universität Freiburg
Hermann Herder Str. 11
West Germany

Introduction: The aim of this study was to investigate, if Fluorescein isothiocyanate (FITC) can be used as label in a chemiluminescence immunoassay. A system in which FITC was activated to light emission was described 1974 in the Journal of Physical Chemistry (1974, 78, 1681 - 1683). In this paper H_2O_2 an NaOCl were used for chemiluminescence activation. However, the smallest amount of of FITC, which was required in this system was 2 x 10^{-4} mol, a concentration which is a priori to high for the use in an luminescence immunoassay. Therefore at first the activation of FITC by these substances had to be activated. It was found, that NaOCl alone in a concentration of 0,3 % (volume/weight) is sufficient, to induce the light-immission of FITC down to a concentration of 10^{-10}. In the following the suitibility of an FITC label was investigated in a solidphase immunoassay for HBs Ag.

Material and methods: Solid phase consisted of anti HBs coated polystyrol beads (ABBOTT Lab.) or polystyrol tubes, which had been coated with anti HBs from the goat (BEHRING Werke, Marburg). The tracer we used, was different hepatitis hyperimmunglobulin preparations (HBIG ABBOTT, BIOTEST, KABI), which were labelled with FITC (SIGMA) according to the classical methods. Finally monoclonal anti HBs (Hybridtech) and affinitychromatographically purified anti HBs from the goat (ELECTRONUCLEONICS) were labelled with FITC.
Sera: HBs Ag pos. reference sera with known subtype were used. The concentration of HBs Ag in these sera was between 5000 to 50 000 neg/ml. Titration of HBs Ag pos. and neg. control sera was

ISBN 0-12-426290-2

performed in PBS and normal human sera.

Assay: 0,2 ml of the HBs Ag pos. sera or the correspondent HBs Ag neg. control were incubated with the solid phase over night at roomtemperature, then 5 times rinsed with PBS and reincubated with 0,2 ml tracer, which was the FITC labelled HB - IG in dilutions from 1 : 50 up to 1 : 300. After a second incubation of 2 hours at 37 C° the solid phase was once again intensively rinsed. FITC dependent luminescence was induced by injecting of 0,2 to 0,3 ml of 0,3 % NaOCl to the solid phase. Luminescence was measured with a BIOLUMAT 9500 T (LAB. Prof. BERTHOLD), lightimmission was measured and integrated over 10 s. after the activation.

Results: Table 1 demonstrates the detection of HBs Ag titrated in normal human serum down to a concentration of 1,0 ng/ml. Table 2 demonstrates the detection of HBs Ag titrated in PBS down to 0,5 ng/ml. In this experiment the pos./neg. ratios between the measured counts of the pos. serum against the counts of the negative serum, which was titrated likewise, is indicated.

After these encouraging results, some problems of this test should be mentioned. These difficulties seems not to be specific for the here presented luminescence immunoassay using the FITC label. The problem of the nonspecific binding of the tracer to the solid phase - which is indeed a problem of this test - is a general problem of luminescence immunoassays. Free FITC - like other luminescence tracers - has an extreme affinity to polystyrol, this is demonstrated on table 3: if FITC solutions are incubated with coated or uncoated polystyrol beads, an extreme binding of free FITC, especially to the uncoated beads is observed. This nonspecific binding of the FITC, - i.e. the FITC labelled tracer - how can it be reduced? The theoretical answer is: By augmentation of the specific anti HBs antibody within the total fo FITC labelled immunoglobulin. The HBIG, we have used, contains only about 3 % of anti HBs antibody, the rest is in regard to this system nonspedific. In order to solve this problem of nonspedific binding, we can refer to recent experiments with monoclonal and also with anti HBs purified by affinitychromatography labelled with FITC. These tracers could be diluted up to 1 : 300. They gave excellent results. The background, which was in the first mentioned experiments, using the TITC labelled HBIG, 800 - 1000 cts/10 s, was reduced to 40 - 100 cts/10 s, meanwhile the counts of the pos. controls were not influenced. With this decreased background the pos./neg. ratio increased significantly by what the sensitivity of course increased likewise: HBs Ag was detected down to concentrations of 0,1 ng/ml.

Summary: A solidphase luminescence immunoassay for the detection of HBs Ag is described. The tracer is FITC -labelled anti HBs. FITC is activated by NaOCl. With this system HBs Ag can be detected down to a concentration of 0,1 ng/ml. The sensitivity of the test was hampered by the relatively high background. The use of monoclonal or anti HBs purified by affinitychromatography as FITC labelled tracer - instead of normal hepatitis hyperimmunglobulin (HBIG) increased significantly the sensitivity of this test.

TABLES

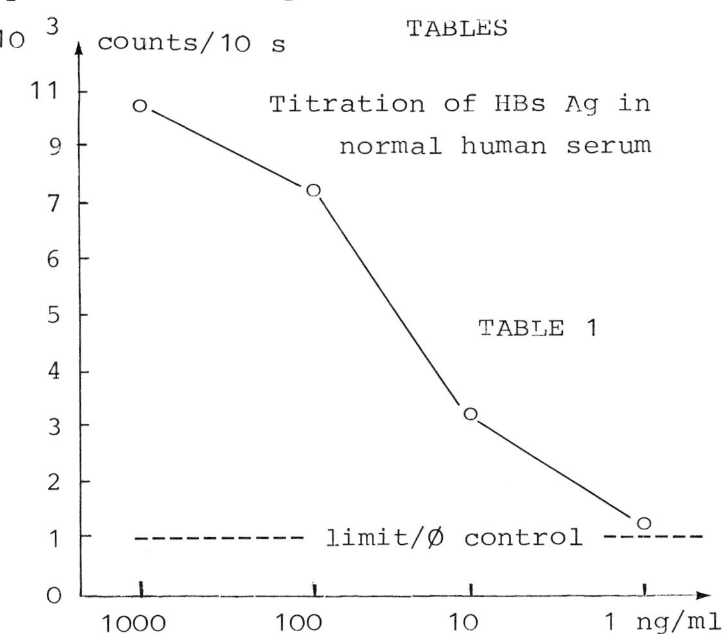

TABLE 1

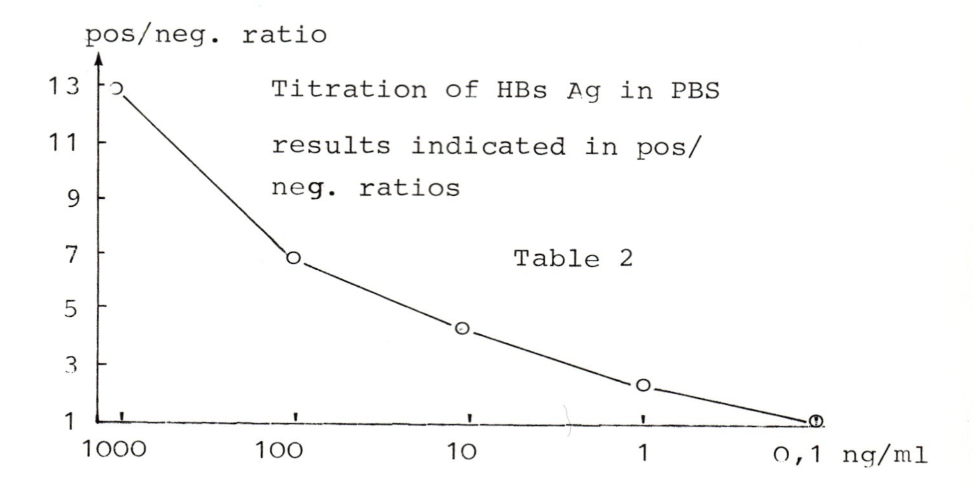

Nonspecific binding of free FITC to polystyrol beads

FITC mol	uncoated	coated/BSA
10^{-2}	130 000	8 000 cts/10 s
10^{-3}	14 000	4 300
10^{-4}	7 800	1 400
10^{-5}	1 700	1 300
10^{-6}	1 600	470

Beads had been incubated over night with the FITC dilutions. Luminescence activation by 0,3 % NaOCl. Table 3

SYNTHESIS OF A NOVEL BIOLUMINESCENT CONJUGATE OF PROGESTERONE FOR IMMUNOASSAY

J. Hughes, Frances Short and V.H.T. James

Department of Chemical Pathology, St. Mary's Hospital Medical School, Paddington, London, W2 1PG

INTRODUCTION

Currently, many steroids are measured by immunoassay, using radioistopes as labels. Methods using non-isotopic labels, e.g. particle, enzymic, fluorescent and chemiluminescent, have been developed and evaluated, but no single serious competitor to radioimmunoassay has yet emerged. Immunoassays using a bioluminescent label may offer advantages such as homogeneity, increased sensitivity of label, and shorter assay time. Using progesterone as a model, the use of such labels is being investigated, using a novel progesterone-NAD conjugate synthesised in our laboratory.

MATERIALS AND METHOD

Preparation of Conjugate

Progesterone-3-0-carboxymethyloxime (Steraloids:Croydon, U.K.) was linked to N^6-[(6-aminohexyl)-carbamoyl methyl]-NAD (Sigma:Poole, U.K.) using the mixed carbonic- carboxylic acid anhydride method of Allen and Redshaw (1). The resultant conjugate was isolated by preparative layer chromatography on silica gel, using 0.1M Ammonium acetate pH 6.5 : Ethanol (1:4) as solvent; eluted overnight in water at 4^oC, and centrifuged (1500g, 4^oC, 1 hours). The supernatant was freeze-dried, redissolved in water, and stored at -20^oC. The yield of conjugate was 20%.

ISBN 0-12-426290-2

Bioluminescent Assay

The conjugate was diluted in 0.1M phosphate buffer, pH 7.0. Either 60ul or 100ul of sample was reduced with 20ul Glucose-6-Phosphate Dehydrogenase (2.5 units per ml, containing 14mM Glucose 6-phosphate). After 10 minutes, 20ul (for 60ul sample) or 30ul (for 100ul sample) of NADH monitoring reagent (LKB : Croydon, U.K.) was added, and bioluminescence measured for 60 seconds, after a 30 second delay.

Antisera Dilution Curves

These were prepared in the presence or absence of 320 p.mol progesterone per tube, using 4p.mol of conjugate per tube. After appropriate incubations, bioluminescence was measured as above. Solid-phase antisera or IgG-fractions were prepared by coupling to immunobeads (BioRad : Watford, U.K.). Dilution curves were prepared as before; the immunobeads were isolated by centrifugation and aspiration, and bioluminescence of the antibody-bound conjugate was measured.

Homogeneous Immunoassay

Ethanolic progesterone standards were dried, whilst conjugate plus antiserum were incubating for 30 minutes. 60ul of the incubation mixture was added to each standard, and mixed. Bioluminescence was measured after 30 minutes.

Heterogeneous Immunoassay

Ethanolic progesterone standards were dried, and 100ul of immunobead suspension was added per tube, followed immediately by 100ul conjugate per tube, then mixed. After an overnight incubation at $4^{o}C$, 1ml of buffer was added per tube, followed by centrifugation (1500g, $4^{o}C$, 30 minutes). The supernatant was aspirated, and the immunobeads resuspended in 100ul buffer, prior to measurement of bioluminescence.

RESULTS

Characterisation of Conjugate

Figure 1 shows the proposed structure of the conjugate. The u.v. spectrum agreed with the theoretical spectrum. The conjugate displaced tritiated progesterone from specific

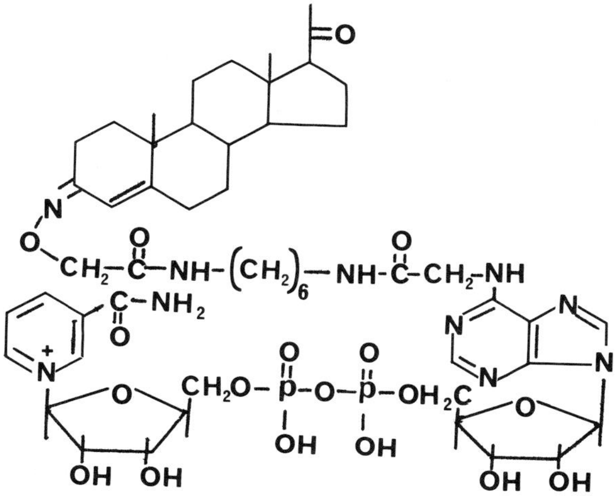

Fig. 1 Structure of Conjugate

antiserum, retaining from 2.5 to 4.7% of the immunoreactivity of progesterone. The NAD derivative used for conjugation was chosen because it has the same coenzymic activity with dehydrogenases as NAD, and is linked to progesterone at a position away from the active site. However, the conjugate only retained 8% of the bioluminescent activity of NAD. Peak light emission occured 30 seconds after addition of monitoring reagent, remaining stable for 5 minutes. A linear standard curve of bioluminescence was obtained from 0.08-800 p.mol of conjugate. Further structural evidence was obtained from n.m.r. and infra-red spectroscopy. It has remained stable for 12 months.

Homogeneous Immunoassay

The maximum amount of progesterone which did not interfere with bioluminescence was found to be 320p.mol per tube, and so was chosen when setting up the antisera dilution curves. 2 antisera inhibited conjugate bioluminescence, presumably by sterically hindering coenzymic activity, but progesterone did not reverse this inhibition. One antiserum, raised in rabbits against progesterone-1-hemisuccinyl-BSA, enhanced bioluminescence twofold. The degree of enhancement was affected by progesterone, so giving the basis for a homogeneous assay. Maximum enhancement was achieved by pre-incubating conjugate with antibody. At least 4p.mol of conjugate was needed per tube, with antiserum at a final dilution of 1:16. A standard curve, taking 2 hours to produce, is shown in Figure 2.

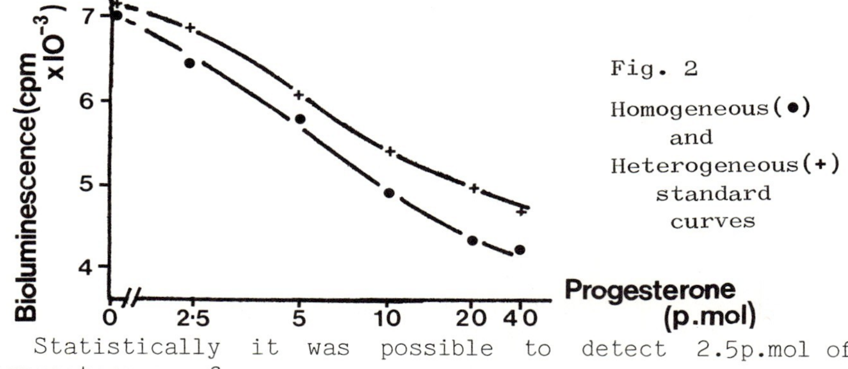

Fig. 2 Homogeneous (•) and Heterogeneous (+) standard curves

Statistically it was possible to detect 2.5p.mol of progesterone from zero.

Heterogeneous Immunoassay

This was developed to try to reduce the amounts of antisera used and improve precision. Progesterone only affected conjugate binding to two of the coupled antisera, out of nine tried. Coupled immunobeads were diluted with unreacted beads, to maintain 100ug of beads per tube, and to give lower background interference than when the manufacturers diluent beads were used. An overnight incubation lowered non-specific binding. The standard curve obtained using a coupled goat antiserum is shown in Figure 2. These assays required immunobead concentrations of 1:4, and precision and sensitivity were worse than for the homogeneous curves.

CONCLUSIONS

This novel conjugate can be incorporated into both homogeneous and heterogeneous assay systems, but the need to use at least 4p.mol of conjugate per assay tube makes these methods less sensitive than comparable immunoassays. Synthesis of more active conjugates may produce assays of use in the clinical laboratory.

REFERENCE

1. Allen, R.M. and Redshaw, M.R.(1978). The use of homologous and heterologous ^{125}I-radioligands in the radioimmunoassay of progesterone, **Steroids** 32 467-485

CALCIUM-SENSITIVE PHOTOPROTEINS AS BIOLUMINESCENT LABELS IN IMMUNOASSAY

Ashok Patel and Milton J. Cormier

Bioluminescence Laboratory, Department of Biochemistry, University of Georgia, Athens, GA 30602, USA.

INTRODUCTION

Homogeneous, i.e., non-separation, immunoassays based on chemiluminescence energy transfer from ABEI-labelled antigens to fluorescent labelled antibodies have been recently reported that are generally applicable to both small and large antigens (1,2). Even with the low chemiluminescence quantum yields of the iso-luminol derivatives (0.1-5%) used and the low energy transfer efficiency of the immunoassay system, the sensitivity of these homogeneous immunoassays was as good or better than conventional radioimmunoassays (3).

Calcium-sensitive photoproteins isolated from marine coelenterates have both a high luminescence quantum yield (∿13%) and an efficient *in vivo* energy transfer system for shifting the wavelength of the emitted light from the blue to green region (4). Aequorin, the photoprotein isolated from the hydromedusan *Aequorea forskalea*, is the best-known and most readily available photoprotein. It is a single polypeptide chain (MW 20,000), produces blue light (λ_{max} ∿ 470 nm) and has been a very useful tool in experimental biology (5). Here we report the use of these photoproteins, especially aequorin, as a sensitive non-isotopic label for immunoassays.

ISBN 0-12-426290-2

Methods

Synthesis of progesterone-aequorin conjugate: Progesterone was coupled to aequorin (purified from the jellyfish *Aequorea forskalea* by the method of Blinks et. al. (6)) using the N-hydroxysuccinimide ester ("activated") of progesterone-11α-hemisuccinate. The progesterone-aequorin conjugate was purified by gel filtration using Sephadex G-25 and stored at -20°C.

Luminescence measurements: Aequorin was assayed by injecting 100 μl of photoprotein assay buffer (100 mM $CaCl_2$ in 10 mM Tris, pH 8) into 50-250 μl sample of aequorin and the luminescence produced in the first 10 S measured using a commercial luminometer (Biolumat LB 9500T; Berthold Labs., Wildbad, W. Germany).

Luminescence immunoassay: For progesterone was carried out using the progesterone-aequorin conjugate as tracer and solid-phase (diazocellulose) antibodies raised to progesterone-11α-hemisuccinate-BSA conjugate in rabbits. The assay buffer was 50 mM Tris, 3 mM EGTA, 2 mM EDTA, 0.1% BSA, pH 7.4. Separation of unbound label was achieved by centrifugation and an aliquot of the supernatant assayed for luminescence of unbound label.

Results

Under optimized conditions, the light emitted by aequorin increased linearly with the amount of aequorin (Fig. 1). 5×10^{-19} mol of aequorin could be detected with ease (70 ± 10 counts/10 sec). The luminescence activity was calculated to be 1.4×10^{20} luminescence counts/mol with a buffer blank of 100 ± 6 counts/10 sec.

The light yield of the progesterone-aequorin conjugate was decreased by about 40% during the conjugation reaction. This loss could have been due to calcium contamination of the DMF which was dried over molecular sieve.

Fig. 2 shows a bioluminescence immunoassay standard curve for progesterone which has a range from 0.01 to 10 pmol.

No deterioration of either immunological or luminescence activity of the progesterone-aequorin conjugate stored at -20°C was observed over a six month period. Neither was loss of activity observed when the conjugate was stored at 4°C for 2 weeks or room temperature for 48 hours.

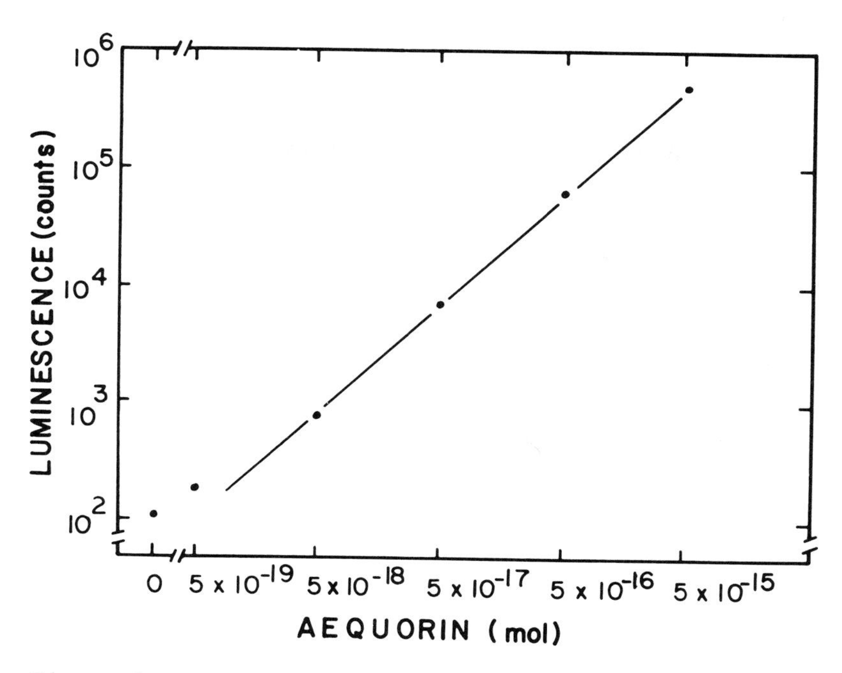

Figure 1. Detection of aequorin using luminescence.

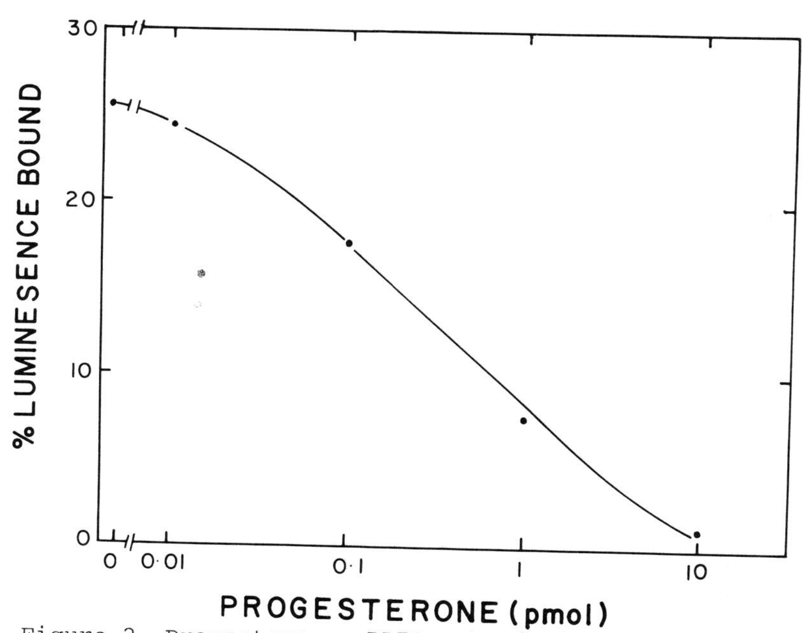

Figure 2. Progesterone BLIA. standard curve.

Discussion

The results presented here demonstrate that despite aequorin's high reactivity, it can be used as a sensitive and stable label for immunoassays. Advantages of using aequorin as a label are not only it's high luminescence quantum yield but that the luminescence can be initiated by the addition of Ca^{2+}. Therefore, there are less problems of noise from the chemical reaction which could otherwise be a problem with chemiluminescent systems such as isoluminol derivatives and acridinium esters (3).

Clearly the usefulness of aequorin, and other calcium-sensitive photoprotein-labelled antigens, will be in the development of sensitive homogeneous assays using their efficient energy transfer mechanisms and applications of these assays in measurement of chemical changes in intact cells (3).

ACKNOWLEDGEMENTS: We are grateful to Richard O. McCann for the purification of aequorin.

REFERENCES

1. Patel, A., et. al. (1983). Chemiluminescence energy transfer; a new technique applicable to the study of ligand-ligand interactions in intact cells, Anal. Biochem., *129*, 162-169.
2. Patel, A. and Campbell, A. K. (1983). Homogeneous immunoassays based on chemiluminescence energy transfer, Clin. Chem., *29*, 1604-1609.
3. Campbell, A. K. and Patel, A. (1983). A homogeneous immunoassay for cyclic nucleotides based on chemiluminescence energy transfer, Biochem. J., *216*, 185-194.
4. Cormier, M. J. (1981). *Renilla* and *Aequorea* bioluminescence. In "Bioluminescence and Chemiluminescence" 225-233. (DeLuca, M. and McElroy, W. D., Eds.), Academic Press, N. Y.
5. Ashley, C. C. and Campbell, A. K. (1979). "Detection and Measurement of Free Ca^{2+} in Cells." Elsevier, North Holland.
6. Blinks, J. R., et. al. (1978). Practical aspects of the use of aequorin as a calcium indicator. Assay, preparation, microinjection and interpretation of signals, Methods in Enzymology, *57*, 292-330.

V PHAGOCYTOSIS

CHEMILUMINESCENCE AS AN APPROACH TO THE STUDY OF PHAGOCYTE BIOCHEMISTRY AND HUMORAL IMMUNE MECHANISMS

Robert C. Allen[1]

Departments of Pathology and Clinical Investigation
Brooke Army Medical Center, and the
U.S. Army Institute of Surgical Research,
Fort Sam Houston, Texas 78234 U.S.A.

PHAGOCYTE MICROBICIDAL ACTION AND NATIVE CHEMILUMINESCENCE

Stimulation of granulocytic leukocytes, monocytes, and macrophages by particle phagocytosis, lectins, or chemical stimuli results in the activation of O_2-redox metabolism. This "respiratory burst" activity, characterized by increased glucose dehydrogenation via the hexose monophosphate shunt and increased non-mitochondrial O_2 consumption, is necessary for effective microbicidal action (1,2). In essence, stimulation results in the activation of an oxidase (or oxidases) that catalyzes the reduction of O_2 to yield $\cdot O_2^-$, H^+, and H_2O_2 (3-5). These products of O_2 reduction can participate in radical reactions ultimately resulting in oxygenation of biological substrates (6,7). H_2O_2 is also a substrate for myeloperoxidase (MPO), a potent microbicidal haloperoxidase of polymorphonuclear leukocytes (PMNL) and monocytes capable of broad spectrum oxygenation activity (8).

The oxygenation of biological substrates by activated phagocytes can yield electronically excited products capable of relaxing to ground state by photon emission. This

[1] *The opinions or assertions contained herein are the private views of the author and are not to be construed as reflecting the views of the Department of the Army or the Department of Defense.*

ISBN 0-12-426290-2

activity is responsible for the native luminescence associated with phagocyte microbicidal function (9,10). As such, phagocyte chemiluminescence (CL) is O_2-dependent and correlates with hexose monophosphate activity. This relationship is illustrated clinically in chronic granulomatous disease, a syndrome of severe and recurrent bacterial and fungal infections (11). The PMNL of these patients are capable of phagocytosis, but do not show normal respiratory burst metabolism. Thus, microbicidal action and CL are greatly diminished (12,13).

Native CL has been successfully applied as an approach to analysis of phagocyte function. However, as an analytic technique it lacks high sensitivity. Native substrate oxygenations are typically low quantum yield reactions; that is, the yield of photons per given number of substrate oxygenations is relatively low. As such, relatively large numbers of phagocytes, typically 10^6 to 10^7 are required for testing. Additionally, there is the problem of substrate variability; the molecular composition of microbes varies. Consequently, the substrates oxygenated and their related quantum yields vary relative to the microbe tested. Thus, interpretation and comparison of native CL data should be restricted to a given microbe.

CHEMILUMINIGENIC PROBING

The limitations associated with the native CL approach to analysis of phagocyte oxygenation activity are obviated by chemiluminigenic probing. A chemiluminigenic probe (CLP) is an organic substrate whose oxygenation results in a relatively high yield of excited product (14-16). As such, the sensitivity for detecting oxygenation activity is increased in proportion to the increase in quantum yield over native substrates. For example, sensitivity can be increased by greater than 10^3 using either cyclic hydrazides or acridinium salts as CLPs. The CLP should also be physically and chemically compatible with the type of biological oxygenation activity measured, and should be nontoxic to the biological system at the concentration employed.

Substrate uniformity is also imposed through use of CLP's. The increase in sensitivity allows CL measurements using concentrations of phagocytes that would not yield a measureable native CL. Furthermore, the CL emission spectrum is that of the excited CLP oxygenation product (Biggley, Allen, and Seliger, unpublished observation). Consequently, the variable of native substrate quantum yield is effectively eliminated allowing intercomparison of CL data using different stimuli.

Chemiluminigenic probing also provides an ultrasensitive approach to differential analysis of biological oxygenation activities. By selecting CLP's with different chemical reactivities and physical characteristics it is possible to analyze the nature and location of the oxygenating agent or agents involved. This point is illustrated through use of luminol (5-amino-2,3-dihydro-1,4-phthalazinedione) and lucigenin (10,10'-dimethyl-9,9'-biacridinium dinitrate; DBA) for detection of the oxygenation activities of stimulated PMNL. Figures 1 and 2 depict the CL responses of 40,000 density-gradient purified PMNL (17) using luminol and lucigenin as the CLP's respectively. In figure 1, the upper curve

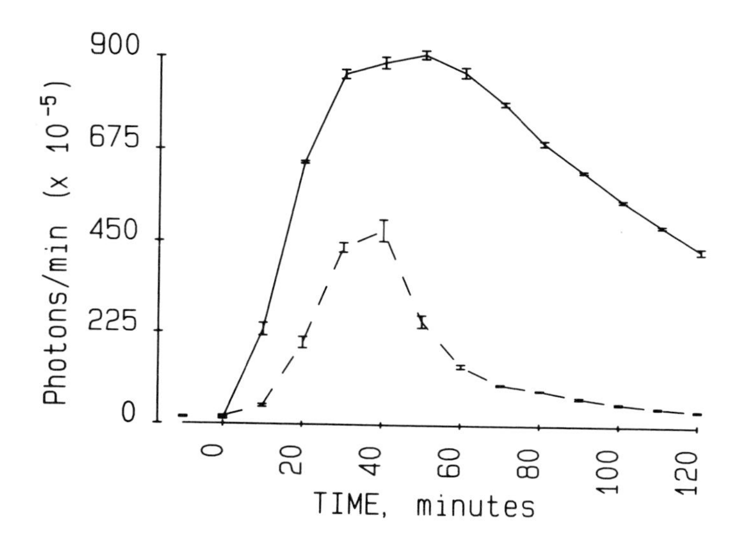

Fig. 1. The luminol-dependent CL responses of 40,000 PMNL in 2.0 ml complete veronal buffer (17) containing 50 μM luminol. Runs are in triplicate with mean ± standard error shown. The solid line depicts the response to COZ (25 μg/ml final) and the dashed line depicts the response of PMA (50 nM final). The stimuli were added at time zero.

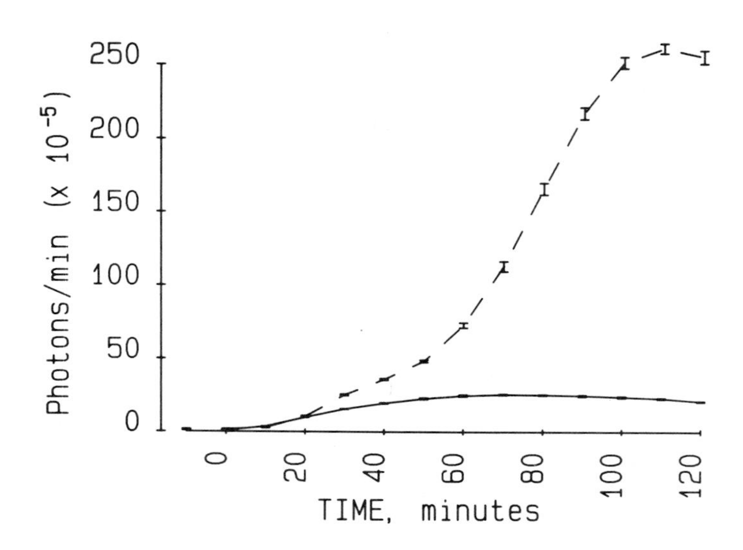

Fig. 2. The lucigenin-dependent CL responses of 40,000 PMNL. The conditions and symbols are as described for figure 1 except that 50 μM lucigenin was employed as the CLP.

(solid line) depicts the luminol-dependent response using complement-opsonified zymosan (COZ) as the stimulus (18). COZ is phagocytosed by the PMNL with phagolysosome formation, acidification, and participation of the MPO system. The lower curve (dashed line) depicts the luminol-dependent response using phorbol-12-myristate-13-acetate (PMA) as stimulus. The chemical stimulus PMA causes specific degranulation and activation of respiratory burst metabolism; however, there is no phagocytosis or phagolysosomal formation (19-21).

As depicted in figure 2, when lucigenin is employed as the CLP, the magnitude of the CL responses using COZ (solid line) and PMA (dashed line) are inversely related to those obtained using luminol. Furthermore, the temporal pattern of CL velocity is specific for the combination of CLP and stimulus employed (16). As will be described in future reports, these activities can be further analyzed and differentiated through use of inhibitors.

MEASUREMENT OF PHAGOCYTE OXYGENATION IN UNSEPARATED WHOLE BLOOD

A complete description of the methodology for measurement of phagocyte oxygenation activity using unseparated biological fluids cannot be presented in this short overview. However, it should be pointed out that the high sensitivity gained by chemiluminigenic probing allows measurement of the oxygenation activities of specimens containing less than 10^3 total phagocytes. As such, phagocyte oxygenation activity can be measured with good precision using 0.5 microliter equivalent of whole blood.

A number of recent reports have described the direct measurement of phagocyte oxygenation from whole blood using luminol (22-25) and lucigenin (23) as the CLP. In considering these methods, it is important to appreciate that hemoglobin absorption of the emitted CL can present a major difficulty with regard to quantifying the specific oxygenation activity per phagocyte. This difficulty can be essentially obviated by specimen dilution. Furthermore, high dilution also serves to diminish the concentration of plasma factors that might influence phagocyte function. We have described the use of 10 μl, i.e. a 1 to 200 dilution, and 1 μl, i.e. a 1 to 2000 dilution, equivalent specimens for testing (23,17). DeSole *et al*. have also suggested use of 5 μl whole blood with adjustment of the CL measurement based on a calculated hemoglobin absorption factor (26). We presently use 0.5 μl whole blood, i.e., a 1 to 4000 dilution, for routine clinical testing.

KINETIC ANALYSIS OF MICROBE OPSONIFICATION BASED ON PHAGOCYTE STIMULATION

Acute host protection against microbial infection requires the interaction of both humoral and phagocyte components of the immune system. In essence, this humoral-phagocyte axis of defense operates as an information-effector mechanism. Humoral recognition of the microbe results in opsonification and generation of chemotactic agents. The phagocyte responds to this information by chemotaxis to the site of infection, recognition of the opsonified microbe, phagocytosis, and activation of metabolism as required for microbicidal action.

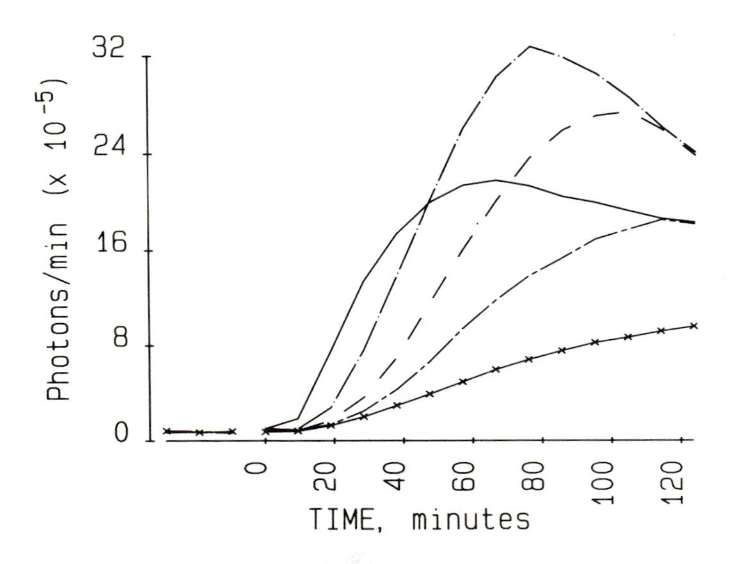

Fig. 3. Opsonin-dependence of the PMNL oxygenation response to P. aeruginosa. Each curve depicts the CL response of 25,000 PMNL in 2.0 ml complete veronal buffer containing 0.5 μM luminol using 5 μl (solid line), 1.25 μl (dot-dashed line), 0.313 μl (dashed line), and 0.078 μl (long-short dashed line) of antiserum. P. aeruginosa, 1.6 x 10^6 CFU of Habs type #13, were added at time zero.

This information-effector relationship linking opsonification to phagocyte stimulation also provides an in vitro approach to quantifying the microbe-specific opsonic capacity of a serum by the resulting stimulation of phagocyte CL. Earlier studies based on the native CL response of PMNL have demonstrated the validity of this approach (27-28).

The previously described CLP approach is ultrasensitive and allows continuous monitoring of the velocity of phagocyte oxygenation activity. As such, this approach is ideally suited to functional analysis of the kinetics of microbe opsonification. If opsonification is by a microbe-specific IgG mechanism, the rate of activated PMNL oxygenation of CLP to yield CL will be a function of the concentration of opsonin and the number of microbes and PMNL tested. Therefore, when the concentrations of microbes, PMNL, and CLP are not rate limiting, the concentration of opsonic IgG will influence the rate of PMNL stimulation

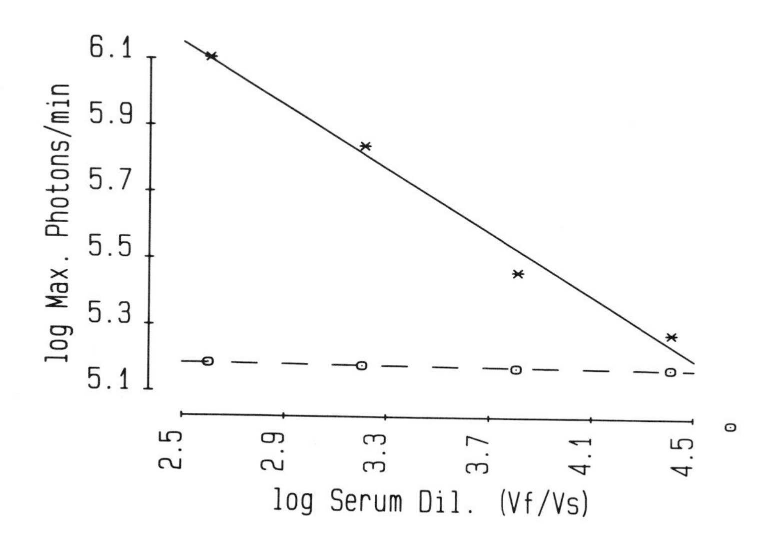

*Fig. 4. Double log plot of the peak CL velocity versus antiserum dilution (final volume/initial serum volume). The *'s depict the antiserum data derived from figure 3. The o's depict the results using different dilutions of preimmune serum.*

yielding CL as described by the equation:

$$L' = k' \, [IgG]^{i}$$

where L' is the peak CL velocity (photons/min), k' is the proportionality constant, and [IgG] is the concentration of antigen-specific IgG, and i is the order of the reaction with respect to opsonin (29).

The effect of microbe-specific antiserum on the rate of PMNL oxygenation of luminol following stimulation with *Pseudomonas aeruginosa* is depicted in figure 3. Four different dilutions of antiserum were tested. The bottom curve marked by x's depicts the lectin-like stimulation of PMNL by *P. aeruginosa* in the absence of serum. Although not shown, the response to equivalent dilutions of pre-immune serum were below that obtained using 0.078 µl of immune serum.

The microbe-specific IgG opsonins of antiserum are polyclonal, and are directed against several microbial antigens. As such, the concentration of opsonin can be expressed relative to serum dilution, i.e. $L' = k'D^{-i}$ or by log conversion $\log L' = \log k' - i \log D$, where D is the dilution (final volume/initial serum volume). Since D is the reciprocal expression of [IgG], the sign of i is changed to negative. Figure 4 is the double log plot of peak CL velocity versus serum dilusion. The antiserum data, depicted as *'s, were derived from figure 3. The results using preimmune serum are depicted as o's, and the o to the far right not falling on the grid depicts the response in the absence of serum. The non-rate limiting condition with respect to PMNL and microbes is well maintained only during the initial 28 minute interval of testing, and therefore, this interval was used for the CL velocity measurements as shown in figure 4.

REFERENCES

1. Sbarra, A.J. and Karnovsky, M.L. (1960). The biochemical basis of phagocytosis. I. Metabolic changes during ingestion of particles by polymorphonuclear leukocytes. *J. Biol. Chem.* 234, 1355-1362.
2. Rossi, F., Romeo, D. and Patriarca, P. (1972). Mechanism of phagocytosis-associated oxidative metabolism in polymorphonuclear leukocytes and macrophages. *J. Reticuloendothel. Soc.* 12, 127-149.
3. Iyer, G.Y.N., Islam, D.M.F., and Quastel, J.H. (1961). Biochemical aspects of phagocytosis. *Nature* 192, 535-541.
4. Babior, B.M. (1980). The Role of Oxygen Radicals in Microbial Killing by Phagocytes. *In* "The Reticuloendothelial System. Vol. 2. Biochemistry and Metabolism".
5. Allen, R.C. (1980). Free-radical production by reticuloendothelial cells. *In* "The Reticuloendothelial System. Vol. 2. Biochemistry and Metabolism". (Eds A.J. Sbarra and R.R. Straus). pp. 309-338. Plenum Publ. Co., New York.
6. Klebanoff, S.J. and Clark, R.A. (1978). The Neutrophil: Function and Clinical Disorders. North-Holland Publ. Co., Amsterdam.
7. Allen, R.C. (1979). Reduced, radical, and excited state oxygen in leukocyte microbicidal activity. *Frontiers in Biology.* 48, 197-233.
8. Klebanoff, S.J. (1968). Myeloperoxidase-halide-hydrogen peroxide antibacterial system. *J. Bacteriol.* 95, 2131-2138.

9. Allen, R.C., Stjernholm, R.L. and Steele, R.H. (1972). Evidence for the generation of (an) electronic excitation state(s) in human polymorphonuclear leukocytes and its participation in bactericidal activity. *Biochem. Biophys. Res. Commun.* 47, 679-684.
10. Allen, R.C. (1979). Chemiluminescence from eukaryotic and prokaryotic cells: Reducing potential and oxygen requirements. *Photochem. Photobiol.* 30, 157-163.
11. Mills, E.L. and Quie, P.G. (1980). Congenital disorders of the function of polymorphonuclear neutrophils. *Rev. Infect. Dis.* 2, 505-517.
12. Allen, R.C., Stjernholm, R.L., Reed, M.A., Harper, T.B., III, Gupta, S. and Steele, R.H. (1977). Correlation of metabolic and chemiluminescent responses of granulocytes from three female siblings with chronic granulomatous disease. *J. Infect. Dis.* 136, 510-518.
13. Allen, R.C., Mills, E.L., McNitt, T.R., and Quie, P.G. (1981). Role of myeloperoxidase and bacterial metabolism in chemiluminescence of granulocytes from patients with chronic granulomatous disease. *J. Infect. Dis.* 144, 344-348.
14. Allen, R.C. and Loose, L.D. (1976). Phagocyte activation of a luminol-dependent chemiluminescence in rabbit alveolar and peritoneal macrophages. *Biochem. Biophys. Res. Commun.* 68, 245-252.
15. Allen, R.C. (1981). Lucigenin chemiluminescence: A new approach to the study of polymorphonuclear leukocyte redox activity. *In* "Bioluminescence and Chemiluminescence: Basic Chemistry and Analytical Applications". (Eds M.A. DeLuca and W.C. McElroy). pp. 63-73, Academic Press, New York.
16. Allen, R.C. (1982). Biochemiexcitation: Chemiluminescence and the study of biological oxygenation reactions. *In* "Chemical and Biological Generation of Excited States". (Eds W. Adam and G. Cilento). pp. 309-344, Academic Press, New York.
17. Allen, R.C. (1982). Direct quantification of phagocyte activity in whole blood: A chemiluminigenic probe approach. *In* "XI International Congress of Clinical Chemistry". (Eds E. Kaiser, F. Gabl, M.M. Müller, and P.M. Bayer). pp. 1043-1058, Walter deGruyter and Co., Berlin, New York.
18. Muller-Eberhard, H.J. (1975). Complement. *Ann. Rev. Biochem.* 44, *697-724.*
19. White, J.G. and Esensen, R.D. (1974). Selective labilization of specific granules in polymorphonuclear leukocytes by phorbol myristate acetate. *Am. J. Pathol.* 75, 45-54.

20. Repine, J.E., White, J.G., Clawson, C.C. and Holmes, B.M. (1974). The influence of phorbol myristate acetate on oxygen consumption by polymorphonuclear leukocytes. *J. Lab. Clin. Med.* 83, 911-920.
21. DeChatelet, L.R., Shirley, P.S., and Johnston, R.B., Jr. (1976). Effect of phorbol myristate acetate on oxidative metabolism of human polymorphonuclear leukocytes. *Blood* 47, 545-554.
22. DeChatelet, L.R. and Shirley, P.S. (1981). Evaluation of chronic granulomatous disease by a chemiluminescence assay of microliter quantities of whole blood. *Clin. Chem.* 27, 1739-1743.
23. Allen, R.C. and Pruitt, B.A., Jr. (1981). Humoral-phagocyte axis of immune defense in burn patients: Chemoluminigenic probing. *Arch. Surg.* 117, 133-140.
24. Fischer, H., Ernst, M., Maly, F.E., Kato, T., Wokalek, H., Heberer, M., Maas, D., Peskar, B., Rietschel, E.T. and Standinger, H. (1982). Chemiluminescence assays in the diagnosis of immune and hematological disease. *In* "Luminescent Assays: Perspectives in Endocrinology and Clinical Chemistry'. (Eds M. Serio and M. Pazzagli). pp. 229-241, Raven Press, New. York.
25. Selvaraj, R.J., Sbarra, A. J., Thomas, G.B., Cetrulo, C.L. and Mitchell, G.W., Jr. (1982). A microtechnique for studying chemiluminescence response of phagocytes using whole blood and its application to evaluation of phagocytes in pregnancy. *J. Reticuloendothel. Soc.* 31, 3-16.
26. DeSole, P., Lippa, S. and Littarru, G.P. (1983). Whole blood chemiluminescence: A new technical approach to assess oxygen-dependent microbicidal activity of granulocytes. *J. Clin. Lab. Automation.* 3, 391-400.
27. Hemming, V.G., Hall, R.T., Rhodes, P.G., Shigeoka, A.O. and Hill, H.R. (1976). Assessment of group B streptococcal opsonins in human and rabbit serum by neutrophil chemiluminescence. *J. Clin. Invest.* 58, 1379-1387.
28. Allen, R.C. (1977). Evaluation of serum opsonic capacity by quantitating the initial chemiluminescent response from phagocytizing polymorphonuclear leukocytes. *Infect. Immun.* 15, 828-833.
29. Allen, R.C. and Lieberman, M.L. (1984). Kinetic analysis of microbe opsonification based on stimulated polymorphonuclear leukocyte oxygenation active. *Infect. Immun.* (in press).

LUMINESCENCE METER FOR THE KINETIC MEASUREMENT OF PHAGOCYTIC ACTIVITY AND OTHER LUMINESCENT REACTIONS

Eric Schram, Henri Roosens and Patrick De Baetselier

Instituut voor Moleculaire Biologie
Vrije Universiteit Brussel
Brussels, Belgium

INTRODUCTION

One of the difficulties that have been encountered in the past in the development of dedicated instrumentation for the assay of luminescent reactions has been the constant evolution of the methodology in this rather new field. The recent development in the applications of phagocyte luminescence is an example of such an evolution. A consensus seems however to have been reached as to the basic prerequisites with which modern luminometers should conform, i.e.

- temperature control
- provision for injection of reagents
- mixing, either with an independant device, or by the convection created by injection
- monitoring of the slope, peak and plateau, as well as integration

In the last years the applications of luminometry have been characterized by a growing interest in phagocyte luminescence. Since this phenomenon was described by Allen *et al.* in 1972 (1) it has been shown that phagocytes respond to a wide variety of substances of immunological or pharmacological importance. It has also become clear that the kinetic pattern of the luminescence response may differ as a function not only of the activating substance, but also of the luminescent substrate used as an amplifier (luminol, lucigenin). For this reason it is essential that the reaction be followed over a sufficient length of time. However, when the measurements have to be performed sequentially, the slowness of the reaction precludes the assay of a high number of samples. Moreover, the reactivity of phagocytes is likely to change within the duration of the measurements and parallel assays offer therefore a definite advantage.

ISBN 0-12-426290-2

In the accompanying paper by De Baetselier *et al.* (this volume) the accent is laid on the fact that luminescence is not only a method of choice to check the function of phagocytes but also a very adequate way of assaying substances acting on phagocytes and not always easy to determine by other means. This of course implies the availability of a source of phagocytes with known characteristics. On the other hand, when applying the method to large numbers of samples it occurred to us that the counting capacity of the instrumentation rapidly becomes the limiting factor. This led us to the development of a simple instrument, based on a novel concept and originally designed for the present purpose but which can also be used for all other luminescence assays.

BASIC CONCEPTS

Continuous monitoring

Continuous monitoring of the reaction has now become customary in most luminescent assays and is even essential for many applications. Luminometers must therefore be conceived in order to satisfy this prerequisite, taking into account that several types of kinetic patterns are likely to be encountered:

a) rapid reactions: e.g. immunoassays in which luminol or acridine derivatives are used as a label, assay of H_2O_2 produced in oxidase catalyzed reactions, etc. In situ injection of the reagents is compelling in these cases and the luminescence must be integrated over a short time interval (a few seconds).

b) steady state luminescence: e.g. assay of ATP with firefly luciferase. Checking of the steadiness of the plateau is advisable in order to detect possible interfering reactions.

c) slow reactions: slowly increasing or decreasing luminescence is first of all encountered in coupled reactions, e.g. in the use of firefly luciferase for the assay of creatine kinase, ATPase, hydrolases, etc. and in the coupling of bacterial luciferase to some dehydrogenases. The curves obtained with activated phagocytes present a more complex behaviour and several parameters must be taken into account:

- lag period
- ascending slope
- height at maximum
- time at maximum
- total emitted light

Although for slow reactions the luminescence needs to be measured over a longer period, this does not necessarily means that it must be recorded *continuously*. Indeed, discrete measurements at regular time intervals make it possible to monitor several simultaneous reactions with a single detector as discussed in the next paragraph.

Simultaneous and parallel monitoring of multiple luminescent assays

Number of samples: If a single detector is used the time devoted to each sample will necessarily decrease when the number of samples increases. This sets an upper limit to this number, more especially when the level of luminescence is low. Six appears to be an optimal figure in this respect allowing for the assay of samples in duplicate together with a blank and/or reference sample(s). It should also be born in mind that time lost in switching samples should be reduced as much as possible

Sample-changers: Two types have been used so far for the assay of phagocyte luminescence

a) turn-tables (Lumac, LKB, Packard): for repeated counting sequences it is important that the turntable can be programmed for series involving varying numbers of sample vials.

b) conveyor (Berthold): in the system developed by Berthold the length of the conveyor can be adjusted to the number of samples to be counted.

All the above systems can be combined with automatic injection. Beside these dedicated instruments scintillation counters in the out of coincidence mode are still used by many experimentators. Such counters are highly sensitive but their main drawback is the lack of injection device and of thermostatization at the adequate temperature.

The critical factors encountered with sample-changers have to do

a) with the delay between successive countings; this holds more particularly for scintillation counters.

b) with the fact that shaking of the tubes, especially in the case of phagocyte suspensions, is not always advisable.

Parallel detectors: The use of several detectors in parallel which amounts to the use of several single detector instruments, makes it possible to bias some of the disadvantages of sample-changers: continuous monitoring of several samples without interruption, absence of shaking, possibility of starting countings individually, etc. This mode of operation has been adopted in the Biolumat LB 9505 counter manufactured by Berthold.

Rotating detector: The use of parallel detectors obviously increases the investment per assay, which for phagocyte luminescence is already high due to the duration of the measurements. We therefore chose to use a single detector but instead of moving the samples, we decided to switch the detector from sample to sample. This solution offers several advantages: reduced delay between successive measurements (less than one second), no shaking of the tubes, easy temperature control (see below), lower investment.

Temperature control

Temperature control has not always been given the necessary attention during the early years of luminescence analysis, probably because the optimal temperature for the classical firefly and bacterial systems happens to be around room temperature. For the firefly system the recommended temperature is usually 23°C. In our experimental conditions we found it to be closer to 19°C (2), the quantum efficiency being 6 % lower at 23°C. This means that a multipurpose luminometer should be adjustable for temperatures ranging from below room temperature up to ca. 40°C. When using a sample changer the whole sample compartment must of course be kept at the desired temperature, which is less easy to achieve than in the case of a multiple sample cell holder as used in our set-up. Even in the absence of any external regulation, thermal conduction in such a holder is sufficient in order that the temperature remain uniform for all samples counted in parallel, and their results are thus fully comparable.

TECHNICAL DESCRIPTION

Initial prototype

In order to check the feasibility of our method a very simple system was built, consisting of a four-place sample holder and a rotating disk masking all sample vials except one (see figure 1). Rotation occurs discontinuously with a speed of 1 rev/s, this means that it takes only a fraction of a second to switch from one sample to the next one. The vials are of the standard type, i.e. 12 x 47 mm. The rotating mask can be set manually in any selected position in order to perform single assays.

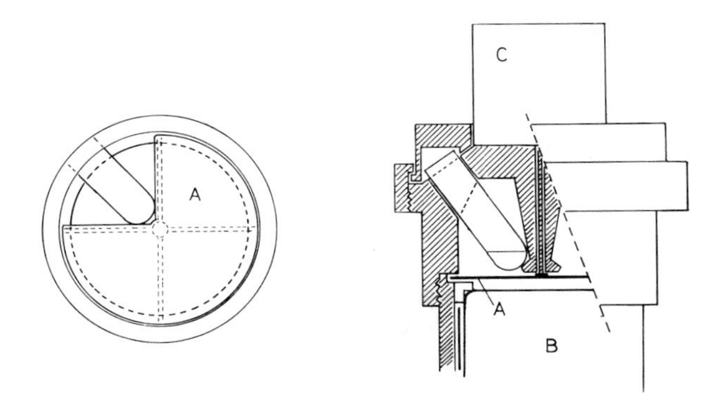

Fig. 1 Cross-section of cell-holder showing rotating mask (A), photomultiplier (B) and motor with encoder (C)

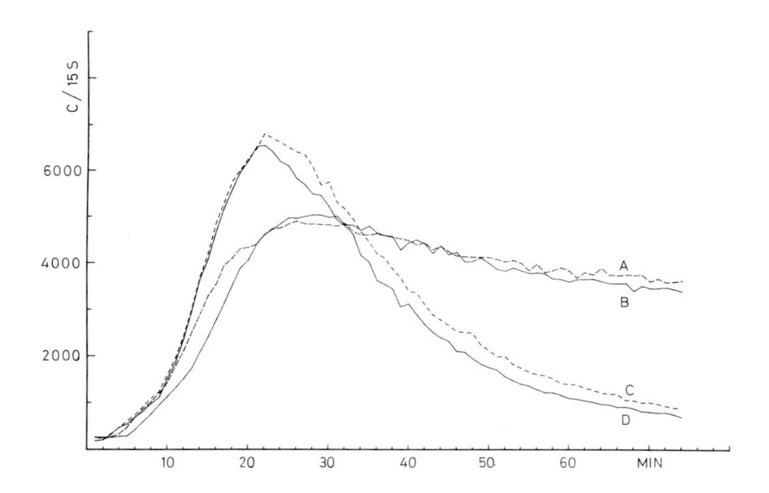

Fig. 2 Parallel plottings of whole blood (2 microliters) luminescence after addition of phorbol myristate acetate (PMA) in the presence of lucigenin (A and B) and luminol (C and D)

Luminescence is detected by means of a 2" unselected EMI photomultiplier tube type 9635 B. The counting time can be set from 1 to 99 s but for most measurements the preferred interval will usually range between 3 and 15 s. With this simplified instrument the results are listed in succession without processing.

In order to check the efficiency of the system a photon source was used consisting of a CCl_4 quenched scintillating solution containing a tritium labelled molecule, as described earlier (3). Two different sources were used, one made up in a regular scintillating solution, the other made up in a solution prepared by Nuclear Enterprises and emitting around 540 nm, i.e. close to the emission maximum of the firefly reaction. The efficiency could not be established in an absolute way but the signal to noise ratio was found to be higher than that of the Biolumat LB 9505 counter used in our laboratory. The same photon source was used to check two other characteristics of our counter:

a) cross-talk: although the activity of the source amounted to 500 times the background level no significant spill-over could be observed between neighbouring sample holes

b) variability of efficiency among sample holes: the observed variation between sample-holes amounts to $\pm$ 10 % and must of course be taken into account in the calculations.

Updated instrument

The positive results obtained with the simplified system described above have prompted us to complete our instrument with some essential features, in order to make it suitable for routine use.

a) first of all the shape of the cell holder was modified in order to accomodate *six* vials in vertical position around two concentric rotating cylinders; the inner cylinder acts as a *reflector* and when its slit coincides with that of the outer one the light originating from the individual cells will be reflected at right angle towards the window of the photomultiplier; before opening the cover the slits are brought in opposite position.

b) the more compact shape of the holder facilitates its *thermostatization* which is achieved by means of Pelletier cells; these cells will automatically act as heater or cooler as a function of the surrounding temperature. The temperature oscillations recorded with this system amount to only a few hundredths of a degree between 16° and 40°C

c) *injection of reagents:* adapters for automatic or

manual injection can easily be fixed on top of the cover above the individual vials.

d) *modes of operation:* the instrument may be operated either in the single mode or in the group mode; in the first case the instrument can be used in any position as a single sample detector; in the latter case any number of tubes up to six may be chosen (empty holders will then not be monitored)

e) *processing of results:* in order to make the results easily comparable all countings are expressed in counts/min. Optional summation of the actual counting figures allows for easy integration of any part of the luminescence curve by a a simple substraction.

f) *electronics:* the electronic sub-assemblies are under control of a Zilog microprocessor (Z-8671). Besides its other functions this microprocessor also allows for the direct coupling of the instrument to a plotter without the use of an external microcomputer. The counter is equipped with a serial output RS 232 and a parallel output Centronix compatible.

The several features mentioned here are at present part of a laboratory set-up and will be incorporated in the instrument that is now being assembled for routine applications (patent applied for).

ACKNOWLEDGEMENTS

This work was supported by the National Fund for Scientific Research (Belgium), contract nr 3.0096.83

REFERENCES

1. Allen, R.C., Stjernholm, R.L. and Steele, R.H. (1972). Evidence for the generation of (an) electronic excitation state(s) in human polymorphonuclear leukocytes and its participation in bactericidal activity, *Biochem. Biophys. Res. Commun.* 47, 679-684.
2. Ahmad, M. and Schram, E. (1981). Analytical aspects of the firefly luciferase reaction kinetics. *In* "Bioluminescence and Chemiluminescence" (Eds M.A. DeLuca and W.D. McElroy), pp. 435-441. Academic Press, New York.
3. Schram, E., Demuylder, F., De Rycker, J. and Roosens, H. (1976). On the use of scintillation spectrometers for chemiluminescence assay in biochemistry. *In* "Liquid Scintillation, Science and Technology" (Eds A.A.Noujaim, C. Ediss and L.I. Wiebe), pp.243-254. Academic Press, New York.

GENERATION OF MACROPHAGE-HYBRIDOMAS FOR THE STUDY OF MACROPHAGE CHEMILUMINESCENCE

P. De Baetselier, L. Brys, E. Vercauteren, L. Mussche, R. Hamers, and E. Schram

Algemene Biologie, Instituut voor Moleculaire Biologie, Vrije Universiteit Brussel, St-Genesius-Rode, Belgium.

INTRODUCTION

Cells of the monocyte/macrophage lineage (MΦ cells) play a cardinal role in various immunological processes, among these mainly antigen catabolism and antigen presentation. Furthermore, their function is controlled by immunoregulatory molecules secreted by activated lymphocytes (collectively designated as lymphokines). It is thus of importance to probe accurately the functional activity of MΦ cells as well as their regulation. An extreme sensitive method to monitor MΦ activity is provided by chemiluminescent measurements. Indeed the activation of MΦ cells through chemical or immunological agents is associated with the production and release of highly reactive oxygen metabolites such as superoxide anions (O_2^-) singlet oxygen (O_2^*) and hydrogenperoxide (H_2O_2). These active oxygen components are unstable and their degeneration results in light emission i.e. chemiluminescence (CL), which can be amplified through addition of exogeneous luminescent substrates such as luminol and lucigenine (chemiluminescent probes or CLP) (1). Cl measurements allow as such to measure the biological activity of phagocytes in different individuals for the diagnosis of immunological deficiencies. Moreover, this assay system allows to probe the effect of different agents on MΦ cell activation and function, implying that CL measurements could be adopted to screen analytically various immunological and pharmacological agents. In order to relate quantitatively MΦ cell CL emission to the presence of either triggering or modulating agents the quantity and

ISBN 0-12-426290-2

functional capacity of responding MΦ cells should be held constant. This obligatory condition is difficult to satisfy, primarily because of the extreme heterogeneity of the cell types analysed, the various stages of differentation of primary macrophages and the inability of primary macrophages to grow in culture. Hence the standardization of MΦ dependent chemiluminometric signals is hampered by the intrinsic heterogenicity of the MΦ cells analysed. The availability of uniform cloned populations of MΦ cell lines might simplify to a great extent to study of MΦ-mediated CL emission and the modulation of this emission. Hence we have generated different continuous MΦ-like cell lines, which manifest different MΦ specific characteristics including generation of active oxygen components and we have screened these lines for MΦ-mediated CL measurements.

IMMORTALIZATION OF MΦ FUNCTIONS BY SOMATIC HYBRIDIZATION

We have adopted the somatic hybridization technology (2) to generate continuous cell lines expressing macrophage phenotypes and macrophage functions. This technology is based on cell fusion between a normal somatic partner and a malignant cell partner, resulting in a hybrid which will manifest different properties derived from the two parents. A successful hybrid will inherit the malignant properties of the tumoral partner (namely continuous growth in vivo and in vitro) and functional properties of the normal somatic partner.

The generation of stable functional hybrids through somatic cell hybridization requires a stable, fusogenic tumor cell line, which resembles phenotypically the normal somatic cell partner. We have isolated such a fusogenic cell line, derived from the murine lymphosarcoma J774.2. This 8-azaguanine resistant cell line (designated J774-C2E2-HAT) is a very efficient fusogenic tumor cell line for PEG somatic hybridization with different murine macrophages (i.e. splenic macrophages, peritoneal macrophages). These macrophage hybridomas thus generated exhibit in contrast to the parental tumor fusion partner, a variety of macrophage functions such as phagocytosis, lymphokine secretion and production of reactive oxygen metabolites.

MΦ CELL LINES AS A CELLULAR REAGENT FOR CL PROBING

The production of reactive oxygen metabolites by different MΦ cell lines after stimulation with various triggering agents can be easily assessed by chemiluminogenic probing using luminol as a CLP. Chemiluminescent responses can be induced through a variety of agents such as zymosan, yeast, microbes, parasites, phorbol myristate acetate and tumor cells. Certain lines, however, cultured in normal medium, will produce a weak or no CL in response to the different triggering agents tested. In contrast, when culturing these cells in a medium containing soluble factors from activated lymphocytes (lymphokines), a strong luminescence can be produced following stimulation with triggering agents. Hence these different continuous macrophage-like cell lines may provide an appropriate material to probe macrophage specific triggering substances and macrophage specific modulating products.

a) Methodology

For determinations of chemiluminescence, MΦ cells were transferred from tissue culture petri dishes to lumacuvettes (PST cuvettes, Lumac 4960) at a concentration of 1 to 2 x 10^4 cells in 1 ml of culture medium in the presence or absence of modulators. The lumacuvettes were subsequently incubated for two or three days. Alternatively MΦ cells could be transferred to microtiterplates (1 to 2 x 10^3 cells in 200 μl of culture medium) and cultivated during 2 or 3 days with or without modulators. The cups containing the responding cells were then cut and transferred to lumacuvettes. The medium in the lumacuvettes or microtiter wells was removed and replaced with veronal-buffered solution. CL emission after addition of a CLP (Luminol) and a triggering substance was recorded in a 6-channel Biolumat apparatus from the Berthold company. The kinetics of the light emission were recorded in 6 parallel experiments over a total of half an hour.

b) Examples

Probing opsonine activity Opsonization is the humoral immune mechanism for identifying and labelling foreign antigens in a manner that ensures its recognition by MΦ cells. Antigen-specific immunoglobulins serve as the heat-stable or immune opsonine (Fc-mediated opsonisation), whereas the activities of the classical and alternative pathways of complement are required for heat-labile opsonisation (C3-mediated opsonisation). Contact recognition of opsonized antigens via Fc and C3b receptors on the MΦ membrane results in phagocytosis, activation of O_2 redox metabolism and finally CL emission. Measurement of CL allows as such the evaluation of the opsonic capacity of humoral serum components i.e. antigen specific antibodies and complement.

The assessment of either Fc-mediated opsonisation or C3-mediated opsonisation through CL emission using macrophage cell lines is demonstrated in Table 1.

TABLE 1

Detection of opsonic activity.

Cell line	antigen	opsonising agent	Maximum cpm$\times 10^{-3}$
2C11-12	Zymosan	-	10
	Zymosan	complement (1 unit)	110
2C11-D3	Micrococcus	-	12
	Micrococcus	antisera (1 µl)	40
2C11-12	Plasmodium	-	20
	Plasmodium	antisera (1 µl)	130
LA5-9	Trypanosomes	-	15
	Trypanosomes	antisera (1 µl)	110

These results clearly indicate that the presence of complement components or antibodies directed against microbes or parasites significantly enhances MΦ mediated CL emission.

Probing immunomodulators We have selected a number of MΦ cell lines which exhibit a low or no responsiveness towards different stimuli such as zymosan, micrococci, PMA and tumor cells. We found, however, that when these cells were cultured in medium containing soluble factors released by activated lymphocytes, these cells acquired the ability to produce CL in response to different stimuli. Thus this method might be useful for the in vitro assay of biological liquids from in vitro or in vivo origin containing macrophage activating substances (MAF). The detection of such substances in the supernates of activated lymphocytes or in the serum of BCG-PPD sensitized animals (as a source of interferon-γ) is presented in Table 2. In this experiment the 2C11-12 cells or LA5-9 cells were preincubated with either 5% lymphokine supernate (LK-SN) or 1% serum containing interferon-γ (IFNγ, 10 units). After two days of incubation, these cells were tested for CL responsiveness towards either zymosan or PMA.

TABLE 2

Detection of macrophage activating activity

Cell line	treatment	triggering agent	Maximum cpmx10^{-3}	stimulation index
2C11-12	-	Zymosan	40	-
	LK-SN	Zymosan	500	12.5
	IFN-γ	Zymosan	475	11.8
2C11-12	-	PMA	100	-
	LK-SN	PMA	900	9
	IFN-γ	PMA	1200	12
LA5-9	-	Zymosan	20	-
	LK-SN	Zymosan	300	45
	IFN-γ	Zymosan	40	2
LA5-9	-	PMA	55	-
	LK-SN	PMA	400	7.2
	IFN-γ	PMA	110	2

The data outlined in Table 2 demonstrate that during incubation with biological fluids, the MΦ cells manifestly acquired CL responsiveness, indicating that the tested fluids contained factors which modulate MΦ cell differentiation and activity. Besides identifying and quantifying MAF activity in biological fluids, these MΦ cell lines can also discriminate between different MAF activities (i.e. MAF lymphokines). Indeed, as shown in Table 2, the 2C11-12 cell line responds to activating factors present in LK-SN or IFN-γ serum, while LAS-9 responds mainly to lymphokines present in LK-SN. Thus the discrimination between different lymphokines is possible using selective responsive MΦ cell lines.

In addition to products secreted by activated lymphocytes, other factors modulate MΦ-cell activity such as bacterial substances (LPS or lipopolysaccharide). Using the 2C11-12 and LA5-9 cell lines, we have analysed the combined effects of LPS and LK-SN on the induction of CL responsiveness (Table 3).

TABLE 3

Analysis of Lymphokine/LPS-induced Activation

Line	Treatment	Triggering agent	Maximum cpmx10^{-3}
2C11-12	LK-SN	Micrococcus	100
	LPS	Micrococcus	220
	LK-SN+LPS	Micrococcus	750
LA5-9	LK-SN	Micrococcus	250
	LPS	Micrococcus	110
	LK-SN+LPS	Micrococcus	0

These results show that two different MΦ modulators i.e. lymphokines and LPS, may act together to activate synergistically MΦ cells. This interaction between lymphokines and LPS has also been reported for the tumoricidal activity of MΦ cells (3). Yet this synergistic effect appears not to be a general phenomenon, since with a different MΦ-cell line i.e. LA5-9, the interaction between LPS and LK-SN results in a complete inhibition of CL emission. These results again indicate that different lines respond differentially to certain modulators.

Probing cell surface antigens It has been reported that antibody-induced CL could be adopted for the detection of cell surface antigens (4). We have analysed the effect of rabbit anti-MΦ (anti-LA5-9, anti-2C11-12) antisera and monoclonal anti-MΦ reagents (Mab anti-MΦ) on the CL emission of the cell line 2C11-12. Hereby the intrinsic capacity of the antibodies to induce CL was tested as well as their capacity to modulate CL after addition of a triggering agent (i.e. micrococcus). The results outlined in Table 3 indicate that most antibody preparations drastically inhibit the micrococcus-induced CL emission. Only one monoclonal antibody manifested the capacity to trigger CL emission on the 2C11-12 cells.

TABLE 4

Detection of membrane antigens

Line	Antibody	Max cpmx10^{-3}	Inhibition of CL emission(% suppression)
2C11-12	rabbit IgG	0	5
	anti-LA5-9	0	35
	anti-2C11-12	0	78
2C11-12	control Mab	0	7
	Mab anti-MΦ1	0	56
	Mab anti-MΦ2	20	61
	Mab anti-MΦ3	0	45
	Mab anti-MΦ4	0	45

These results demonstrate that CL measurement is a sensitive and simple method for the detection of cell surface antigens and for the screening of antibodies reacting with these antigens.

Probing pharmacological activity Conceivably this assay system could be adopted to analyse the effect of different drugs which manifest antagonistic or synergistic activities on MΦ-mediated CL emission. This possibility is exemplified in Table 5, where we have analysed the effect of an anaesthetic drug i.e. lidocaine on the CL emission of 2C11-12 cells.

TABLE 5

Effect of lidocaine on CL emission

Line	Treatment	Triggering agent	Maximum cpmx10^{-3}
2C11-12+LK-SN	-	PMA	300
	lidocaine(30µm, 10')	PMA	0
	-	zymosan	200
	lidocaine(30µm, 10')	zymosan	0
2C11-12+IFN-γ	-	PMA	350
	lidocaine(30µm, 10')	PMA	0
	-	zymosan	220
	lidocaine(30µm, 10')	zymosan	0

These results illustrate the potential of this assay system to screen for pharmacological agents that interfere with MΦ function and activation.

DISCUSSION

The assay system described above uses continuous macrophage cell lines as a cellular material to probe analytically various biological and biochemical substances. Such monoclonal cell lines offer many advantages over heterogeneous MΦ cell populations on account of following characteristics :
- uniformity, conventional MΦ populations are notoriously difficult to standardize and no two cell populations, even when obtained from the same individual or animal, can be considered to be identical. Monoclonal MΦ cell lines are uniform, so that the way is open to obtain cellular reagents of definable reactivity, a matter of great importance for diagnostic work.
- availability, MΦ cell lines can be maintained indefinitely in culture without loss of functional activity. Thus virtually unlimited quantities of homogeneous, functionally defined MΦ cells can be produced.
- specificity, MΦ cell lines can be monitored for their selective reactivity towards well defined triggering or modulating agents. As such they can be used to detect selectively certain agents present in complex biological mixtures such as culture supernates or serum. Such selective identification and quantification is impossible to perform with heterogeneous populations since different MΦ subpopulations might interact with different stimulators or modulators.

Finally, besides the analytical screening assay applications, these different continuous macrophage-like cells may provide a useful model for the study of the enzymes, membrane receptors, immune and non-immune agents involved in the generation of a respiratory burst by monocytic cells, as well as for the study of the role of lymphokines in this process.

ACKNOWLEDGEMENTS

This work was supported by a grant of the ASLK Kankerfonds and an FKFO grant nr. 3.0096.83 (to E. Schram).
P. De Baetselier is a fellow of the NFWO and L. Mussche a fellow of the IWONL.

REFERENCES

1. Sbarra, A.J. and Karnovsky, M.L. (1960). The biochemical basis of phagocytosis. I. Metabolic changes during ingestion of particles by polymorphonuclear leukocytes, *J.Biol.Chem.* 234, 1355-1362.
2. Köhler, G., Howe, S.C., and Milstein, C. (1976). Fusion between immunoglobulin-secreting and non-secreting myeloma cell lines. *Eur.J.Immunol.* 6, 292-295.
3. Pace, J.L. and Russell, S.W. (1981). Activation of mouse macrophages for tumor cell killing, *J.Immunol.* 126, 1863-1867.
4. Peterhans, E., Albrecht, H. and Wyler, R. (1981). Detection of H-2 and Sendai virus antigens by chemiluminescence, *Journal of Immunological Methods* 47, 255-302.

SELECTIVE UNRESPONSIVENESS OF ALVEOLAR MACROPHAGES (AM) IN PULMONARY ALVEOLAR PROTEINOSIS (PAP)

Joachim Müller-Quernheim[+], Rudolf E. Schopf[++], Peter Benes[++], Roman Rubin[+] and Volker Schulz[+]

[+]*Department of Pneumology and* [++]*Department of Dermatology University of Mainz, Langenbeckstr., 65 Mainz, FRG*

Introduction

PAP is a rare disease of unknown etiology characterized by the deposition of proteinaceous material in the distal air spaces. This material is produced by pneumocytes II and resembles the surfactant. A hypothesis concerning the immunopathology of this disease is based on the observation of decreased phagocytosis by alveolar macrophages. It is supposed that this results in a hampered clearance of surfactant which leads to the accumulation of protein-rich material in the lung. Other investigators claimed a decreased production of oxidizing agents by AM to be involved in the high frequency of opportunistic infections seen in this patients (1). There is some evidence that AM acquire those defects in the lung and it is suggested that abundant proteins found in bronchoalveolar lavage fluid and serum of patients with PAP are involved in the immunopathology of PAP (2).

In a case of PAP we obtained bronchoalveolar lavage fluid, serum and AM during the symptomatic period of the disease. We addressed the question of whether macrophage defects are inherited in AM or are acquired in the distal air spaces. For this purpose the responses of AM and peripheral blood monocytes (PM) to challenges with several agents such as phorbol myristate acetate (PMA) and zymo-

ISBN 0-12-426290-2

san (Z) were measured by chemiluminescence (CL). Our results demonstrate that AM of PAP phagocytize and that there is a normal oxidative burst during phagocytosis. Challenges by membrane stimuli, however, are not responded with an oxidative burst. This unresponsivness can be transferred by certain fractions of lavage fluid and serum of PAP to normal PM. This indicates that AM defects observed in PAP are caused by abundant proteins.

MATERIALS AND METHODS

Bronchoalveolar lavage and CL were performed as previously described (3). To elicit CL the following concentrations of the stimulating agents have been used: zymosan (250 ug/ml), PMA (600 nM), Con A (10 ug/ml), and aggregated immunoglobulin (Ig) (80 ug/ml). All other methods employed were performed as described in standard text books.

RESULTS

AM obtained during the symptomatic period of PAP phagocytize erythrocytes in the same order of magnitude as AM of a healthy control (table 1).

TABLE 1

Phagocytosis of ^{51}Cr-labelled sheep red blood cells (SRBC) by AM of PAP and of control[+]

AM/SRBC	PAP AM	Control AM
1/400	4,773	4,156
1/200	3,240	2,967
1/100	1,111	1,396

[+] counts per minute per 50,000 AM

The oxidative burst in AM of PAP depends upon the stimulant used. Z which is normally phagocytized in PAP activates oxidative metabolism of AM in PAP. The turnover equals that obtained in control AM. If, however, stimulants are used which act via membrane receptors such as PMA, Con A, and Ig it is not possible to activate AM of PAP. The control AM can easily be activated by those agents (table 2). In contrast to the control AM, the AM of PAP can be rendered refractory to the challenge by Z by an 2 hours preincubation in the reaction vial (data not shown).

TABLE 2

CL⁺ of AM of PAP and of control upon challenge with different stimulants

stimulant	PAP AM	Control AM
Z	23,214	28,892
PMA	8,303	61,160
Con A	3,517	34,821
Ig	5,089	23,035

+ counts per 10 seconds per 200,000 AM at peak time

The reaction pattern seen in AM of PAP (high response after stimulation with Z, and no response after stimulation with PMA) can be transferred to PM of healthy donors by lavage fluid or by serum of PAP. Controls do not exhibit these properties. By ion exchange chromatography a lavage fraction eluting at 3 to 10 % 0.15 M NaCl and a serum fraction eluting at 24 to 38 % 0.15 M NaCl can be found containing the suppressive factors. Corresponding fractions of healthy donors do not alter CL of PM in any respect (data not shown).

DISCUSSION

The experiments shown in the tables give clear-cut evidence, that - at least in the patient tested by us - AM phagocytize normally and produce normal amounts of oxidizing species during this process. Thus, hampered phagocytosis and lacking capacity to produce oxidizing species cannot be a constant cause of frequent opportunistic infections.

CL is a well-suited tool to investigate disturbances of macrophage metabolism as in the case of PAP. Stimulants as Z and PMA are competing for the same postreceptoral enzyme pool. We can demonstrate that it is possible to activate this enzyme pool in AM of PAP by the phagocytosis of Z. Alterations in the membranes of those AM dislink receptors and enzyme pool. As a consequence challenges via those receptors cannot be responded with an oxidative burst. Extreme membrane alterations as elicited by attachment to plastic surface render AM of PAP refractory to Z. This indicates an abnormal susceptibility to those challenges.

The *in-vivo* membrane alterations of PAP AM are caused by serum factors and factors found in the lavage fluid. The hampered response upon membrane signals as e.g. PMA, Ig, and complement may contribute to a decreased production of oxidizing agents by these cells after *in-vivo* activation followed by fequent infections with opportunistic species. By means of CL it was possible to propose a mechanism leading to increased susceptibility to these infections.

Our results support the clinical procedure of therapeutical bronchoalveolar lavage, by which those suppressive factors can be washed out of the lungs by large volumes of 0.15 M NaCl.

REFERENCES

1. Harris, J.O. (1979). Pulmonary Alveolar Proteinosis Abnormal in vitro Functions of Alvellar Macrophages, *Chest* 76, 156-159.
2. Bell, D.Y. and Hook G.E.R. (1979). Pulmonary Alveolar Proteinosis: Analysis of Airway and Alveolar Proteins, *Am Rev Respir Dis* 119, 979-990.
3. Müller-Quernheim, J.,Rubin, R.,Leimer, L. and Ferlinz, R. (1983). Studien zur Funktion von Makrophagen in der Sarkoidose I. Supprimierter oxidativer Metabolismus von Makrophagen bei der Sarkoidose in Korrelation zur Entzündung, *Prax Klin Pneumol* 37, 1130-1133.

HUMAN POLYMORPHONUCLEAR LEUKOCYTE LUMINOL CHEMILUMINESCENCE IDENTIFIES ALBUMIN AS AN OXYGEN RADICAL SCAVENGER

Mary E Holt and A K Campbell*

*Departments of Rheumatology and Medical Biochemistry**
Welsh National School of Medicine, Cardiff CF4 4XW, Wales.

Introduction

Polymorphonuclear leukocyte (PMN) luminol chemiluminescence is modified in vitro by many medium constituents (1). We have investigated the mechanism by which albumin the major extracellular protein, modifies luminol chemiluminescence of resting and stimulated human PMN. Our results demonstrate a direct interaction of albumin with the oxygen radical generating pathway.

Materials and Methods

PMN were isolated from normal venous blood and suspended in HEPES buffered Krebs medium pH 7.4 (2). Chemiluminescence was measured using a specially constructed research grade luminometer (2). 5×10^5 PMN were pre incubated at 37°C for 5 min in 500 ul medium pH 7.4 containing 2×10^{-5} M luminol and 0.2% (v/v) DMSO (and albumin or Na azide as indicated). Resting (unstimulated) PMN chemiluminescence was measured before cells were stimulated by addition of opsonised Zymosan (Sigma) (2) or the chemotactic peptide N formyl met leu phe (Sigma) in 500ul medium pH 7.4 also preincubated at 37°C for 5 min. Results of chemiluminescence assays performed in the presence of human serum albumin (Boehringer) or sodium azide were expressed as a percentage of the value expected in the absence of either.

ISBN 0-12-426290-2

RESULTS AND DISCUSSION. Stimulation of PMN by chemotactic peptide in the presence of luminol resulted in a rapid increase in light emission after a lag of 20 sec. The chemiluminescence was clearly biphasic, the early phase producing a shoulder at 60 sec on the larger second phase which reached a maximum at 180 - 240 sec. For opsonised Zymosan (1mg/ml) following a lag of 30 sec a gradual rise to a single peak at 420 sec was observed (2).

Albumin in the range 0.01 - 4% (w/v) inhibited resting chemiluminescence, and maximum chemiluminescence following stimulation by chemotactic peptide (Fig 2) or opsonised Zymosan. This inhibition did not result from alteration of pH, impairment of light transmission, altered cell viability or contaminants in the albumin preparation but was reduced by 30% when the albumin induced fall in extracellular free calcium was corrected by dialysis of albumin against HEPES buffered Krebs medium pH 7.4.

The possibility remained, however, that albumin might reduce luminol chemiluminescence either by quenching of the excited state or by direct binding of luminol. Albumin would then be expected to inhibit luminol chemiluminescence under all conditions. Analysis of the chemiluminescence traces revealed that while the larger second phase was always inhibited by albumin the smaller early phase for both stimuli was infact enhanced by low concentrations of albumin (0.05 - 0.5% w/v). Albumin also enhanced chemiluminescence resulting from the reaction of luminol with O_2^- and H_2O_2 produced by the oxidation of xanthine catalysed by xanthine oxidase (2). Similarly the comparable inhibitory effect of albumin on resting and second phase chemiluminescence taken together with the enhancement of the early phase by low concentrations of albumin indicated that the overall decrease in PMN luminol dependent chemiluminescence was not attributable to impaired activation of the cells by these stimuli.

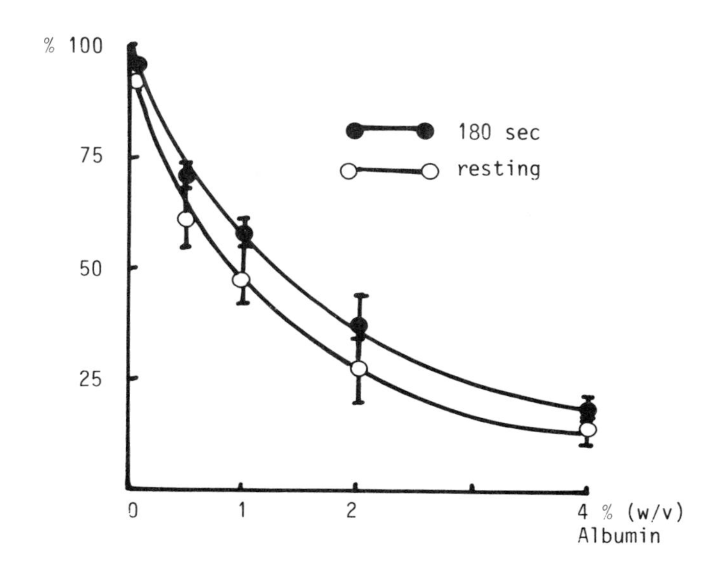

Fig 1.
Effect of albumin or resting chemiluminescence (○) or maximum chemiluminescence (●) following stimulation of 5×10^5 PMN by 0.25 uM chemotactic peptide. (mean ± S.E.M., n=3).

The most likely explanation for the inhibition is therefore modification of the oxygen radical generating pathway by albumin, oxygen radical scavenging. This hitherto unrecognised property in the major extracellular protein requires further investigation if oxygen radicals are to be established as mediators of inflammation (2). Furthermore the differential effect of albumin on the two phases of stimulated PMN chemiluminescence suggests that a different radical predominates in each. Na azide 1-50 uM was found to similarly preferentially decrease the second phase suggesting that radicals in this phase are dependent on the presence of myeloperoxidase.

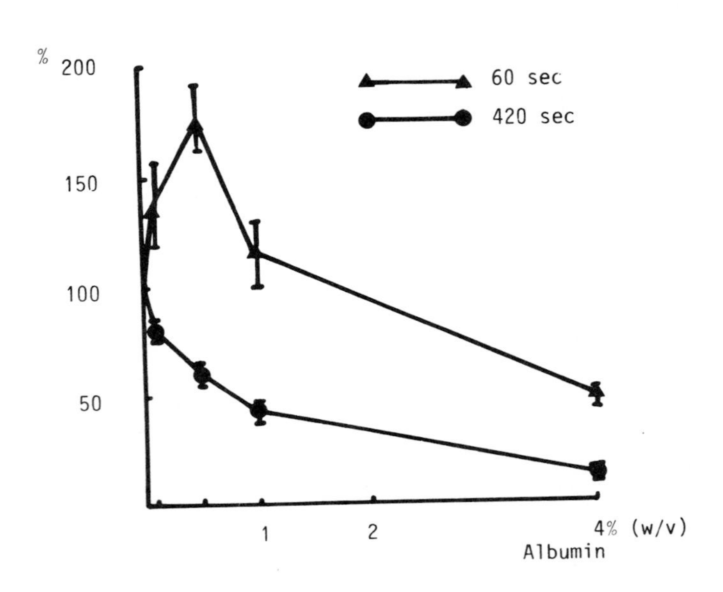

Fig 2.
Effect of albumin on early (▲) and second (●) phases of chemiluminescence of 5 x 10^5 PMN stimulated by 1mg/ml opsonised Zymosan.

Acknowledgments and References

We gratefully acknowledge the financial support of the Arthritis and Rheumatism Council for Research.

1. Campbell A.K. Holt M.E. Patel A (1984). Chemiluminescence in Medical Biochemistry, in Recent Advances in Clinical Chemistry VolIII Ed K.G.M.M. Alberti and C.P. Price, Edinburgh, Churchill Livingstone.
2. Holt M.E., Ryall M.E.T., Campbell A.K. (1984). Albumin inhibits human polymorphonuclear leucocyte chemiluminescence: Evidence for Oxygen Radical Scavenging. B.J. Exp Path 65: 2, 231-241.

MONOCYTES ARE THE EFFECTOR CELLS IN TUMOR TARGET CELL INDUCED CHEMILUMINESCENCE: NK CELLS DO NOT GENERATE ACTIVATED OXYGEN SPECIES

M. Ernst, A. Lange+, A. Havel, J. Ennen,
A.J. Ulmer, and H.-D. Flad

Borstel Research Institute, D-2061 Borstel, FRG

+Ludwik Hirszfeld Institute of Immunology
and Experimental Therapy, Wroclaw, Poland)

INTRODUCTION

In our previous work we were able to show that cytotoxicity of lymphokine-activated mouse macrophages against tumor target cells is accompanied by the generation of activated oxygen species as measured by luminol dependent chemiluminescence (CL) (1). As others we found that hydrogen peroxide participates causally in the killing of many target cell lines (2). Since it is still a matter of discussion whether natural killer (NK) cells originate from the monocyte/macrophage lineage or from the T lymphocyte lineage we investigated the relation between the generation of luminol dependent CL and the NK activity of human peripheral blood mononuclear cells (MNC) and subpopulations thereof. Recently it has been reported that human NK cells form activated oxygen species when killing tumor target cells and that NK-activity can be measured by both isotope release from target cells and by luminol dependent CL (3).

In our present work we demonstrate that isolated active NK cells do not show any CL but that the target cell-induced CL in human cell populations, exerting lytic activity against K562 tumor target cells requires the presence of monocytes.

MATERIAL AND METHODS

Effector cells

MNC were isolated from human heparinized blood by discontinuous density gradient centrifugation on Percoll (4).

Phagocyte-depleted MNC were obtained by treating MNC suspensions with iron carbonyl (4). These primarily non-phago-

ISBN 0-12-426290-2

cytic MNC were then incubated for 60 min at 37°C in plastic tissue culture flasks. The non-adherent cells obtained by this procedure will be termed phagocyte-depleted MNC.

T lymphocytes were isolated from MNC by a rosetting method as described recently by Ulmer et al. (4).

Percoll subfractionation of MNC and phagocyte-depleted MNC by continuous density gradient centrifugation was performed according to Ulmer et al.(4). 8 fractions with densities from 1.030 g/ml to 1.096 g/ml were obtained.

Following a 1 h incubation at 37°C of MNC suspensions in plasma coated plastic flasks adherent MNC were harvested by means of a rubber policeman.

Cell sorter-derived MNC subpopulations were obtained using a cytofluorograf system 50H (Ortho) to which non-labeled MNC and anti-Leu11a-FITC labeled MNC suspensions were applied. Monocytes could be gained by scatter-gated sorting of cells with high intensity 90° scatter signals. Leu11a(+) cells were sorted according to the fluorescence signals of cells within the lymphocyte window defined by the appropriate forward scatter and 90° scatter signals.

NK assay

A 14 h NK assay using ^{75}Se (Seleno-methionine) labeled K562 cells was performed (5). The number of effector cells required to achieve 40% specific isotope release was defined as 1 lytic unit (LU).

CL measurement

All cells used in CL measurements were suspended in CL medium (Dulbecco's modified Eagle's medium modified for CL) obtained from Boehringer, Mannheim, FRG. Normally 4 x 10^5 effector cells were suspended in 300 µl CL medium and incubated for at least 90 min at 37°C. Ten min before CL measurement, 10 µl of a luminol solution (2mg/ml) were added to the effector cells. CL was detected in a 6 channel Biolumat (model LB 9505, Berthold, FRG) measuring CL at 37°C. As trigger signals for the CL measurements we used K562 cells in a volume of 100 µl (typically containing 2 x 10^6 target cells) and as phagocytic stimulus we used 10 µl of a suspension of latex beads (d = 1.09 µm, Sigma), at a stock concentration of 5%.

RESULTS AND CONCLUSION

When we investigated in parallel CL activities (induced by latex or K562 target cells) and the NK activity of various

TABLE 1

CL and NK activity of cell sorter derived MNC subpopulations

Donor	Effector cells	Monocytes	Chemiluminescence		NK activity	
		ANAE$^+$ cells %	Counts $*10^{-3}$ /10^5 effector cells/10min	Ratio to non sorted control cells	LU/10^7 cells	Ratio to non sorted control cells
H.B.	MNC	10	1607	1.0	55	1.0
	Leu11a(+) cells	20	1438	0.89	215	3.9
	Leu11a(-) cells	1	46	0.03	7	0.13
H.B.	phagocyte-depleted MNC	1.5	13	1.0	29	1.0
	Leu11a(+) cells	1.5	30	2.3	1309	45
	Leu11a(-) cells	0.5	5	0.38	14	0.48
J.S.	MNC	23	1198	1.0	89	1.0
	Leu11a(+) cells	8	156	0.13	2079	23.3
	Leu11a(-) cells	9	371	0.31	40	0.45
G.D.	phagocyte-depleted MNC	4	213	1.0	64	1.0
	Leu11a(+) cells	1	44	0.21	1268	19.8
	Leu11a(-) cells	5	114	0.54	78	1.22

MNC subpopulations we obtained the following results:

1. T cells had 60% of MNC NK-activity but only 2.5% of target cell or latex inducible CL-activity of MNC.

2. Phagocyte depleted MNC showed a little increase of NK-activity but both latex and target cell induced CL-activity were only 2-5% of MNC CL-activities.

3. Reconstitution of phagocyte depleted MNC with increasing amounts of adherent cells resulted in superadditive increases of CL responses but in no changes of NK-activity.

4. Density gradient centrifugation on Percoll of phagocyte depleted MNC yielded fractions with up to 13 fold enriched NK-activity whereas both target cell induced and latex induced CL-activities were below 5% of MNC.

5. Cell sorter derived MNC subpopulations enriched for anti-Leu11a(+)-cells were highly enriched in NK-activity (up to 45 fold) whereas latex and target cell induced CL was below 5% of MNC CL (see table 1).

6. Reconstitution of cell sorter derived Leu11a(+) cells with small amounts of cell sorter derived monocytes yielded again superadditively enhanced K562 induced CL responses.

From these results we conclude that human NK cells do not generate activated oxygen species. The target cell induced CL in NK cell populations requires the presence of phagocytes as CL-effectors: It might be the result of a positive cooperation of target cell recognizing NK cells and CL mediating monocytes.

REFERENCES

1. Ernst, M., Lang, H., Fischer, H., Lohmann-Matthes, M.-L., Staudinger, H. (1981). Chemiluminescence of cytotoxic macrophages. In Bioluminescence and Chemiluminescence (Eds M. DeLuca and W. McElroy) pp. 609-616. Academic Press, New York
2. Nathan, C.F., Silverstein, S.C., Brukner, L.H., and Cohn, Z.A. (1979). Extracellular cytolysis by activated macrophages and granulocytes, J.Exp.Med. 149, 100-113.
3. Helfand, S.L., Werkmeister, J., and Roder, J.C. (1982). Chemiluminescence response of human natural killer cells, J.Exp.Med. 156, 492-505
4. Ulmer, A.J., Scholz, W., Ernst, M., Brandt, E., and Flad, H.-D. (1984). Isolation and subfractionation of human peripheral blood mononuclear cells (PBMC) by density gradient centrifugation on Percoll. Immunobiology, in press.
5. Leibold, W. and Bridge, S. (1979). ^{75}Se-release: a short and long term assay system for cellular cytotoxicity, Z.Immun.-Forsch. 155, 287-311.

THE USE OF CHEMILUMINESCENCE TO INVESTIGATE MECHANISMS OF ACTIVATION OF POLYMORPHONUCLEAR LEUCOCYTES BY MONOSODIUM URATE CRYSTALS

P N Platt and I Bird

University Department of Rheumatology, Royal Victoria Infirmary, Newcastle Upon Tyne

INTRODUCTION

The crystal induced diseases have emerged as a group of diseases linked by common pathogenetic mechanisms. We have used the technique of luminol dependent chemiluminescence (LDCL) to study the interaction of monosodium urate crystals, the crystals found in the synovial fluid of patients with acute gout, with polymorphonuclear leucocytes (PMNLs). The crystals described by McCarty and Hollander (1) were predominantly contained within PMNLs and this lead to the suggestion that PMNLs were of central importance in the production of imflammation and tissue damage in these conditions.

The technique of LDCL by measurement of oxygen related radicals allows detection of activation of PMNLs at an early stage. We have used LDCL to investigate mechanisms of activation of PMNLs by monosodium urate crystals in particular the role of divalent cations, calcium and magnesium and agents that modify concentrations of cyclic AMP.

MATERIALS AND METHODS

A leucocyte suspension was prepared by dextran sedimentation techniques (2) from heparinised normal human peripheral blood. Contaminating red cells were lysed by TRIS-ammonium chloride. The cells were counted and suspended at a concentration of 1.5 x 10^6 per ml in medium. The reaction mixture consisted of 1 ml of leucocyte suspension, 100 ul of a standard luminol

ISBN 0-12-426290-2

solution (20mg/100ml), 100 ul of a solution of the test substance to which 100 ul of a 1% weight/volume suspension of monosodium urate crystals were added. Monosodium urate crystals were produced by the method described by Seegmiller et al (3). The chemiluminescence (CL) response was monitored by an LKB 1250 luminometer connected to a chart recorder. All experiments were performed with the reaction mixture at 37° C. The CL results were measured in terms of peak values, which were found to correlate closely with the area under the curve (corr. coef. = 0.99, $p < 0.001$). LDCL results for 12 replicate samples assayed using a single batch of PMNLs showed a coefficient of variation of 2.2%. All experiments were completed using single batches of PMNLs.

RESULTS

The addition of 100 ul of a 1% w/v suspension of monosodium urate crystals leads to a increase in LDCL after a lag phase of approximately 25 seconds. This reaches a peak at 3-5 minutes, declining to a plateau phase which gradually returns to the baseline. (FIG 1).

FIG 1. TYPICAL LDCL RESPONSE CURVE TO MONOSODIUM URATE CRYSTAL STIMULUS

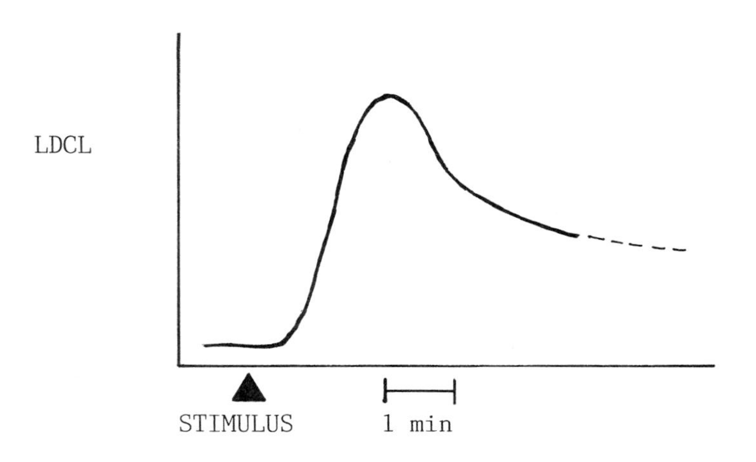

A dose dependent increase in LDCL was seen in response to increasing concentrations of monosodium urate crystals.

Studies using divalent cation free medium indicate that LDCL occured in the absence of divalent cations and in the presence of 10 M EDTA and EGTA, however the response is enhanced (2-3 fold increase) by the presence of magnesium at physiological conditions (0.5-1mmol) (FIG 2). An enhanced response in the presence of calcium ions was only seen at supra-physiological concentrations.

Dose response studies using theophylline (FIG 3), prostaglandin E1 and B-2 agonists show step-wise reductions in response with increasing concentration. In the case of theophylline and prostaglandin E1 a 50% reduction was seen at concentrations between 10^{-4}M and 10^{-5} M. Dose response studies using carbachol to reduce cyclic AMP levels show an enhanced response at concentrations of 10^{-5} M and 10^{-6} M.

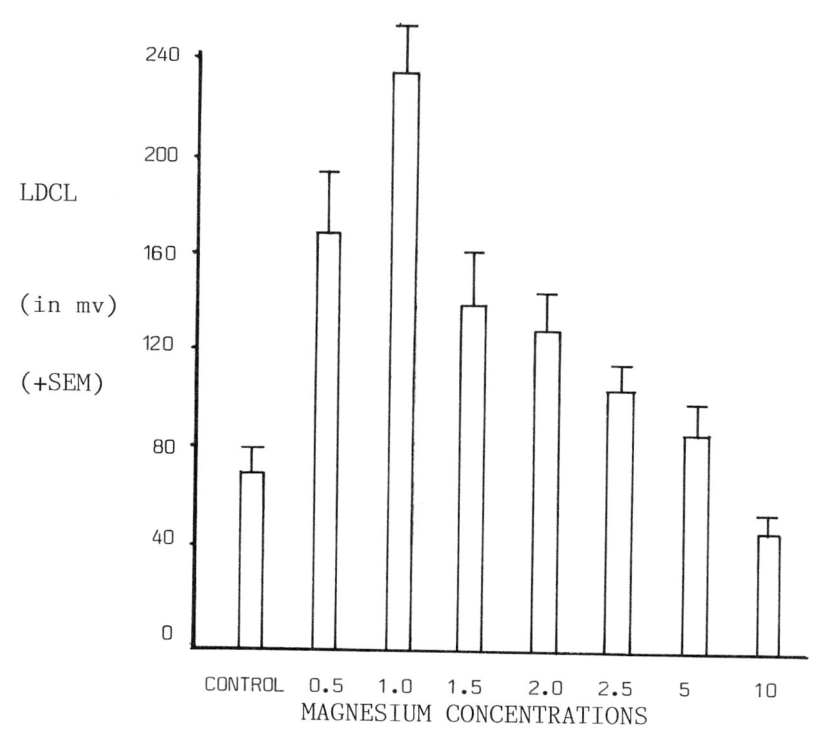

FIG 2. THE EFFECTS OF MAGNESIUM CONCENTRATION ON LDCL RESPONSE TO MONOSODIUM URATE CRYSTALS

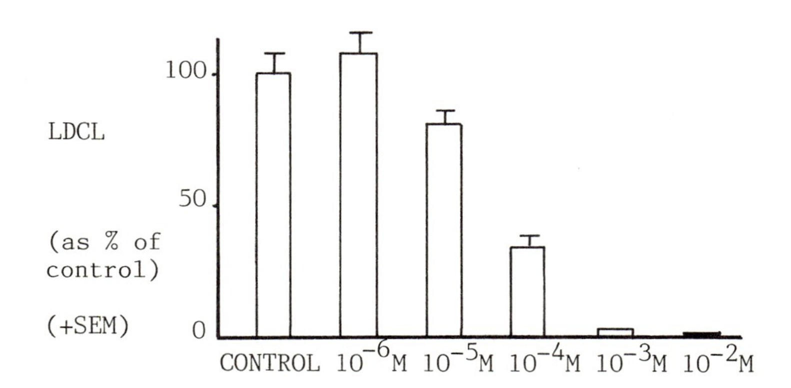

FIG 3. THE EFFECTS OF THEOPHYLLINE ON THE LDCL RESPONSE TO MONOSODIUM URATE CRYSTALS

CONCLUSIONS

LDCL appears to be a useful technique for the investigation of the interaction of monosodium urate crystals and PMNLs. The LDCL response to monosodium urate crystals occurred in the absence of divalent cations but was enhanced by magnesium in physiological conditions. Agents that increase cellular levels of cyclic AMP reduce the LDCL response to monosodium urate crystals. Carbachol, an agent capable of reducing cellular levels of cyclic AMP, enhanced the LDCL to monosodium urate crystals.

REFERENCES

1. McCarty D J, Hollander J L. (1961) Identification of urate crystals in gouty synovial fluid. Ann Int Med 54, 452-60
2. Boyum A. (1968) Isolation of mononuclear cells and granulocytes from human blood. Scand J Clin Lab Invest 97, 77-89
3. Seegmiller J E, Howell R R, Malawista S E. (1962) The inflammatory reaction to sodium urate. JAMA 180,469-75

CELLULAR CHEMILUMINESCENCE AS A TOOL IN THE STUDY OF SURFACE ANTIGENS, ANTIBODY SPECIFICITIES AND FC RECEPTORS

E. Peterhans, T. Arnold, M. Grob and G. Bertoni

Institute of Virology, University of Zurich, CH-8057 Zurich, Switzerland

INTRODUCTION

Antibodies are unique within the wide array of agents that are capable of eliciting in phagocytic cells CL. In vivo, the induction of reactive oxygen generation by the vast majority of agents is controlled by preventing contact between stimulus and phagocytic cell. By contrast, antibodies and phagocytes are constantly in contact with each other; therefore, antibody-induced reactive oxygen generation must be strictly controlled. This control is exerted by the immunological specificity of antibody, i.e. only antibody bound to its target antigen triggers the cells to produce reactive oxygen. Binding to antigen changes the antibody molecule in a way that enables the antibody Fc portion to tightly bind to Fc receptors (FcR) (1). These receptors then transmit in phagocytic cells a signal(s) that results in the production of reactive oxygen species. In the case of bovine polymorphonuclear leucocytes (PMNL), it has been shown that CL can be activated by FcR for IgG_1 and IgG_2 (2). The present work demonstrates that CL measurement is a simple tool in the study of both Fab-epitope and Fc-FcR interactions.

ISBN 0-12-426290-2

MATERIALS AND METHODS

Indicator cells: Bovine polymorphonuclear leucocytes (PMNL) were prepared according the method of Carlson and Kaneko (3) and suspended at 4 x 10^5/ml in phenol red-free Hanks' balanced salt solution buffered with 25 mM HEPES and supplemented with 7.5 µl of 4% bovine serum albumin saturated with luminol. Target cells: Cultures of embryonic bovine lung cells were infected with *bovine herpes virus type I (BHV-I)* at a multiplicity of 1. At 14-16 hours post-infection, the cells were detached from the culture dishes and coated with antibodies as described (4). To stimulate CL, target cells were added to PMNL indicator cells and CL was measured as described (4).

RESULTS AND DISCUSSION

Two recognition events must occur before antibody is able to elicit CL. Fab must bind to its target epitope and Fc to FcR on the surface of CL-positive cells. Figure 1 illustrates schematically two ways of making use of this situation for analytical purposes. When antibody bound to antigen is added to CL-positive cells, a burst of light emission will occur only when Fc of antigen-bound

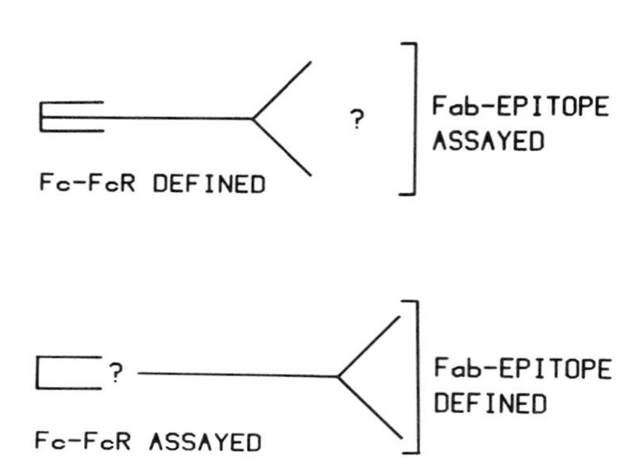

Fig. 1 The two recognition events in the antibody molecule analyzed by CL measurement.

antibody is able to activate the cell by binding to FcR. On the other hand, using antibody compatible with FcR of CL positive cells, the Fab-epitope interaction may be studied. That kind of approach is exemplified in the experiment shown in Fig. 2. To bovine PMNL were added BHV-I-infected target cells coated with bovine or rabbit-anti-BHV-I antibody of similar anti-viral activity as estimated in a neutralization assay. Target cells coated with bovine (A), but not rabbit (B) anti-viral antibodies were able to induce CL, suggesting a species-specific failure of rabbit antibody to activate FcR-dependent CL in bovine PMNL. This

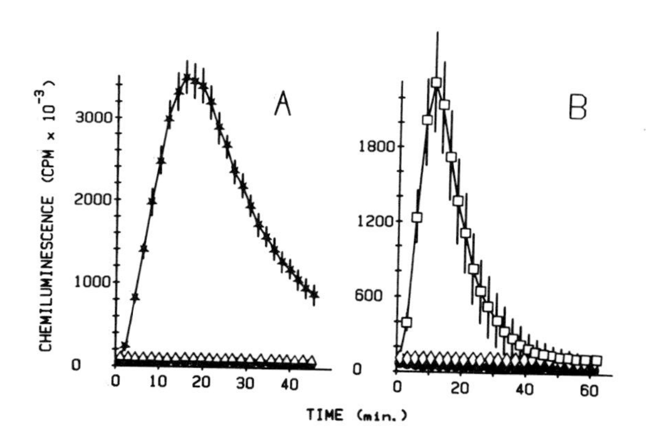

Fig. 2 The induction of CL in bovine PMNL by antibody-coated BHV-I infected target cells. To stimulate CL, 20'000 target cells were added to PMNL. Three samples were run of antibody-coated cells and one sample mock-stimulated with PBS.

	first antibody	second antibody
★ BHV-I targets	bovine-anti BHV-I	PBS
△ BHV-I targets	bovine-anti BHV-I	rabbit-anti-bovine Ig
◊ BHV-I targets	rabbit-anti BHV-I	PBS
□ BHV-I targets	rabbit-anti BHV-I	bovine-anti-rabbit Ig

● *cells mock-stimulated with PBS*

interpretation was further supported by the effects on CL of BHV-I-infected target cells coated, in sequence, with bovine and rabbit (A) and rabbit and bovine antibodies (B), respectively. Coating cells having bovine antibody on their surfaces with rabbit antibody led to the neutralization of CL (A). In contrast, when the sequence of antibodies was reversed, the target cells were able to induce CL. Earlier work had shown that the effects on CL of the antibodies was specific, requiring the presence on the target cells of the respective antigens (4).
The failure of rabbit antibody to induce CL in bovine PMNL correlates with its failure to mediate cytotoxicity (5), indicating the functional significance of CL measurement as an FcR assay. Moreover, the present CL assay is highly sensitive in the detection of surface antigens without requiring radioactive label (4).

ACKNOWLEDGEMENTS

This work was supported by the Swiss National Science Foundation Grant No. 3.429.0.83. We thank Drs. R. Wyler and R. Keller for discussions and C. Gerber for secretarial assistance.

REFERENCES

1. Dickler, H.B. (1982). *Mol.Immunol.* 19, 1301-1306.
2. Mossmann, H. *et al.*(1981). *Adv.Exptl.Med.Biol.* 137, 279-291.
3. Carlson, G.P., and Kaneko, J.J. (1973). *Proc. Soc.Exptl.Biol.Med.* 142, 853-856.
4. Weber, L., and Peterhans, E. (1983). *Immunobiol.* 164, 333-342.
5. Rouse, B.T. *et al.* (1976). *Infect.Immun.* 13, 1433-1441.

CHEMILUMINESCENCE STUDIES ON THE ADHESIN-RECEPTOR INTERACTION OF DIFFERENT STRAINS OF E. COLI AND MOUSE MACROPHAGES

K.-H. Büscher*, V. Klimetzek+ and W. Opferkuch*

*Department of Medical Microbiology and Immunology, Ruhr-Universität, Bochum, FRG
+Department of Immunology and Oncology, Pharma - Research - Centre, Bayer AG, Wuppertal, FRG

INTRODUCTION

Gram-negative bacteria possess different types of adhesins. Beside their role as colonization factors, these adhesins can also be involved in the elimination of bacteria by phagocytes as was suggested for mannose-sensitive (MS) and mannose-resistant (MR) adhesins of *E.coli* (1). Strains with MS-adhesins adhere to macrophages (MØ) and induce chemiluminescence (LdCl), whereas strains with MR-adhesins lack such activity. It was therefore thought that LdCl could be used for the discrimination of these two types of adhesins. During reinvestigation of the LdCl induction by *E.coli* carrying MS- and/or MR-adhesins evidence was found that even MR-adhesins could be responsible for the induction of LdCl.

MATERIAL AND METHODS

Animals. 2-4 month old *B6D2F1* mice were used.

Bacteria. *E.coli* (strain D 509, V 7724, D 539 and B 880) were grown for 18 h in Müller-Hinton-Broth (BBL, Cockeysville, Maryland, USA). After washing with PBS the bacteria were adjusted to a density of 1 x 10^9 bacteria/ml and stored on ice until use.

Macrophages. Peritoneal cells were harvested from normal mice or mice which were injected i.p. with *C.parvum* (0.35 mg/mouse, Wellcome, Beckenham, UK), or with 1 ml 10% proteose peptone (Difco, Michigan, USA). Purification of MØ by teflon adherence was performed as described earlier (2). The MØ were suspended in Dulbecco's-PBS containing 10 mg

ISBN 0-12-426290-2

glucose and 10 mg albumin/ml.

Chemiluminescence assay. Luminol-dependent chemiluminescence (LdCl) using 1 x 10^6 MØ in suspension was done as described earlier (2).

Enzyme determinations. Superoxide dismutase and catalase activity was determined according to Schwartz *et al.* (3).

RESULTS AND DISCUSSION

E.coli carrying MS-adhesins induced a LdCl of about 20 min. duration in MØ of *B6D2F1* mice. The magnitude of the responses was dependent on the ratio of bacteria to MØ, as demonstrated for *E.coli* D 509 in figure 1. At a constant ratio of *E.coli* to MØ the responses increased with the activation state of MØ (table 1).

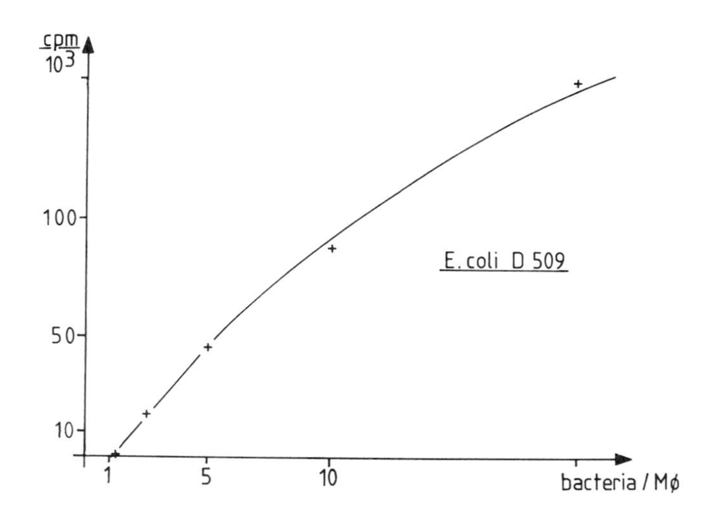

Fig. 1 Relationship between peak of LdCl and the ratio of bacteria to proteose peptone activated MØ. Results are shown for E.coli D 509.

Bacteria expressing MS- and MR-adhesins simultanously behaved similarily in regard to these variations, as shown for *E.coli* D 539.

E.coli bearing MR-adhesins, of which V7724 is an example, were also capable of inducing LdCl of MØ. However, the MR-strain B 880 seemed to lack this activity (table 1). Further investigations showed that this was an encapsulated strain. Partial removal of capsular polysaccharides by

TABLE 1

LdCl of Activated MØ Induced by E.coli

strain	adhesins	LdCl (counts x 10^{-6}/20 min) proteose/peptone MØ	*C.parvum* MØ
D 509	MS	1.83	70.0
D 539	MR/MS	0.71	17.3
V 7724	MR	0.40	5.6
B 880	MR	0.08	0.0

heating lead to a significant LdCl. Therefore encapsulation might protect bacteria against recognition by phagocytes.

The differences in the magnitude of LdCl induced by the various strains cannot be explained by the amounts of oxygen radical scavenging enzymes associated with the bacteria, as the activities of superoxide dismutase (SOD) and catalase found did not correlate with the magnitudes of LdCl (table 2). Furthermore, the activities of the two enzymes are too low to affect LdCl, because much higher amounts of pure enzymes are needed to suppress LdCl. In addition, the presence of 2 x 10^7 *E.coli* B 880 did not alter the LdCl pattern induced by strain D 509. Therefore, differences in the activity of certain receptors on MØ or differences in the density of stimulating epitopes (4) on the bacteria are responsible for the different reactions.

TABLE 2

SOD and Catalase Activity of E.Coli

strain	adhesins	SOD (mU/10^8	catalase bacteria)
D 509	MS	280	620
D 539	MR/MS	400	870
V 7724	MR	450	1,670
B 880	MR	230	290

The participation of MR-adhesins in the induction of LdCl could also be shown with the MR/MS-strains by pretreatment of bacteria with mannose. This treatment inhibited 80-90 % of the LdCl response of the MS-strain, whereas only 0-30 % inhibition was found when *E.coli* D 539 was tested. This

TABLE 3

Inhibition of LdCl by Pretreatment of Bacteria with 100 mM Mannose

strain	% inhibition resident MØ	proteose/peptone MØ	*C.parvum* MØ
D 509	85.0	90.0	79.1
D 539	0.0	24.0	31.8
V 7724	0.0	0.0	0.0

indicates that strains carrying both types of adhesins can induce LdCl by means of their MR- as well as their MS-adhesins (table 3).

The extent to which they participate in LdCl is dependent on the population of MØ investigated, indicating that LdCl might be useful to study differences in the expression or the functional state of MØ membrane receptors.

REFERENCES

1. Blumenstock, E.,Jann, K. (1982). Adhesion of piliated *E.coli* strains to phagocytes: Differences between bacteria with mannose-resistant pili and those with mannose-sensitive pili, *Infect. Immun.* 35, 264-269
2. Klimetzek, V.,Schlumberger, H.D. (1984). Age-dependent modulation of superoxide and hydrogen peroxide secretion and chemiluminescence by in vitro cultivated macrophages. *In* "Oxygen Radicals in Chemistry and Biology" (Eds. W. Bors, M. Saran and D. Tait). pp. 887-891. de Gruyter, New York.
3. Schwartz, C.E.,Krall, J.,Norton, L.,McKay, K.,Kay, D., Lynch, R.E. (1983). Catalase and superoxide dismutase in *Escherichia coli*, *J. Biol. Chem.* 258, 6277-6281
4. Büscher, K.-H.,Klimetzek, V.,Opferkuch, W. (1984). Effect of different opsonins on the production of oxygen radicals and chemiluminescence of macrophages. *In* "Oxygen Radicals in Chemistry and Biology", see ref. (2). pp. 893-896.

GRANULOCYTE CHEMILUMINESCENCE INDUCER (GCI) : A CYTOKINE PRODUCED BY MITOGEN-STIMULATED MONONUCLEAR CELLS (MNC)

F.-E.Maly[§] and A.Kapp[+]

[§] *Max-Planck-Institut für Immunbiologie und*
[+] *Dermatologische Klinik, Freiburg, West Germany*

INTRODUCTION

Granulocytes (PMN) are capable of the production of $^{\cdot}O_2$, H_2O_2 and $^{\cdot}OH$ by an ectoenzyme (NAD(P)H - Oxidase). These are potent oxidizing agents and are involved in their defence function, but may also lead to host tissue damage(1). Other functions of granulocytes like migration and antibody-dependent cytotoxicity have been shown to be affected by interleukins such as leukocyte inhibitory factor(LIF), interleukin 1(IL-1) and colony-stimulating factor(CSF)(2,3,4). Here we show that mitogen-activated mononuclear cells produce an interleukin-like peptide(s) which elicits intense chemiluminescence (CL) from granulocytes. This activity was consequently termed granulocyte chemiluminescence inducer (GCI).

MATERIALS AND METHODS

Human Granulocytes were prepared for CL measurement as described earlier(5). *Mononuclear Cells* were isolated via a ficoll gradient. Enrichment/depletion of adherent cells was achieved by adherence to plastic dishes.

Cell Cultures. Cells were cultured for 24 hrs in serum-free Eagle's medium at 37 °C and 8% CO_2. The mitogen Phythaemagglutinin(PHA) was added at a concentration of 5 ug/ml.

Chemiluminescence measurements were performed in a 6-channel luminometer(Berthold) at 37°C.

ISBN 0-12-426290-2

RESULTS

Supernatants of PHA-stimulated MNC(PHA-MNC-SN), when added to isolated PMN, evoked a slow but long-lasting CL in the presence of Lucigenin(or Luminol, not shown). Superoxide dismutase(100 μg/ml) inhibited more than 90% of this response. Controls (supernatant of unstimulated MNC(MNC-SN), medium + PHA(M+PHA) or medium alone(M) elicited only background CL. With high concentrations of PHA-MNC-SN (25-50%), the 2 hr-integral of CL was comparable to the one induced by phorbol myristate acetate(PMA) at a high concentration (1 μg/ml)(Fig.1).

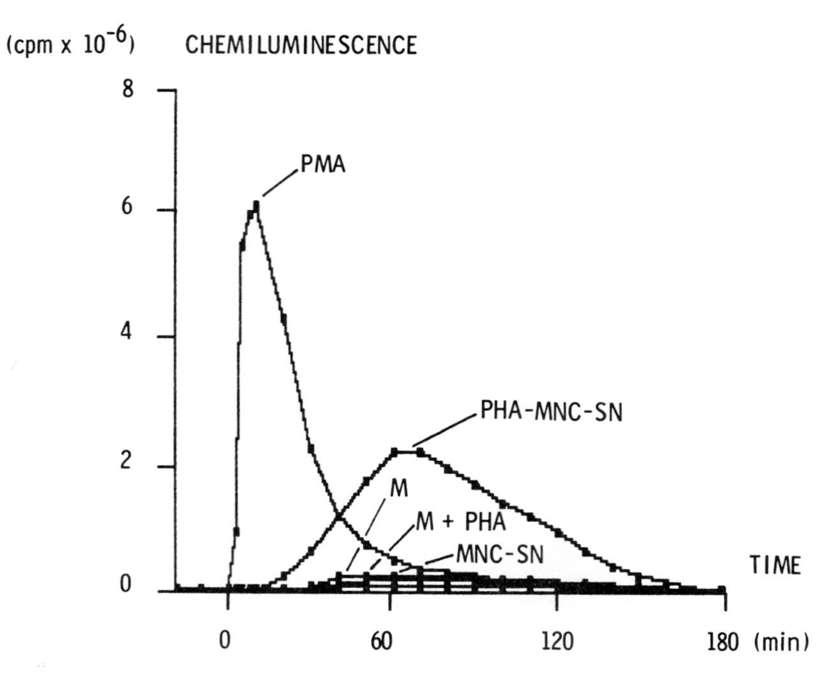

Fig.1 CL of PMN induced by supernatants. $1x10^6$ PMN/ vial were used. Supernatants were added at time 0 at a final concentration of 50%.

The CL caused by PHA-MNC-SN was marginal at 5% final concentration, increased then and reached a plateau over 25%.

The chemilumigenic activity in PHA-MNC-SN appeared unchanged after freezing and thawing and was stable to heating to 56°C for 1hr. After dialysis with PBS, activity was partially retained (30%). Treatment with trypsin resulted in diminished activity (50%). Upon gel filtration on Sephadex G-200 using PBS as elution buffer, the chemilumigenic activity was eluted in 3 distinct peaks corresponding to MW's of 50±5, 20±5 and below 10 kDa (Fig.2).

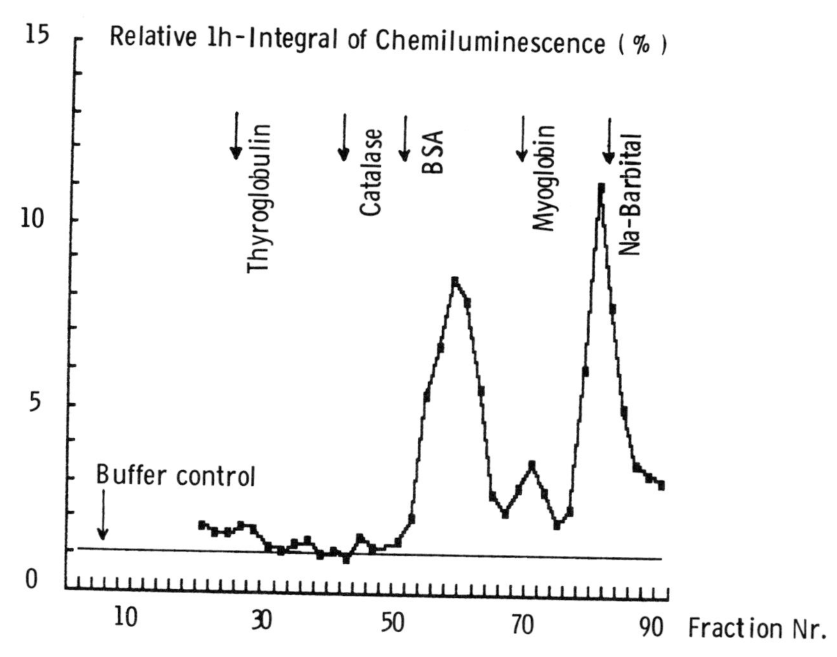

Fig.2 Distribution of GCI-activity after gel filtration of PHA-MNC-SN on Sephadex G-200. Marker proteins a-e indicated by arrows.

After removal of adherent cells from the MNC, the non-adherent cells showed a drastically reduced capacity to produce GCI upon stimulation with PHA. The remaining adherent cell population was capable of substantial production of GCI, but did not account for the total amount of GCI produced by unseparated MNC. The remainder of GCI-activity was therefore produced upon interaction of adherent and non-adherent cells (data not shown).

DISCUSSION

The presented data show that one or several products of PHA-avtivated MNC possess the capacity to activate granulocyte chemiluminescence,i.e. to stimulate the granulocyte's production of reactive oxygen species and its oxygenation potential. The molecule(s) responsible for this activity (GCI) appears to be derived directly from adherent cells but also from an interaction between adherent and non-adherent cells.
Mitogen-activated MNC are known to produce a number of cytokines,some of which do affect granulocytes(LIF,IL-1,CSF).From their MW's,LIF and GCI can be concluded to be different.IL-1 and CSF, however,have MW's similar to GCI.The question, whether GCI-activity is an additional property of IL-1 or CSF or of an interleukin not known to act on PMN,or is caused by a previously undescribed molecule, is presently being investigated.
Granulocyte oxygenation activity is important in the defence against microorganisms,parasites and tumor cells,but may also lead to host tissue damage,if triggered inappropriately.The existence of a cytokine controlling granulocyte oxygenation activity demonstrates a new way of granulocyte activation.This pathway may be operative in the physiological defence function of these cells,but may also constitute a new mechanism of pathological granulocyte-mediated damage.

REFERENCES

1. Allen,R.C. (1982).*In* "Chemical and biological generation of excited states"(Eds W.Adam and G.Cilento),Academic Press,pp. 310-341.
2. Luger,T.A. et al. (1983). *J.Immunol.* 131(2), pp. 816 - 820.
3. Rocklin,R.E. (1975). *J.Immunol.*114,1161 pp.
4. Vadas,M.A. et al. (1983). *J.Immunol.* 130(2), pp.795 - 799.
5. Maly,F.-E. et al. (1983). *Immunobiol.*164, pp. 90 - 97.

MEASUREMENT OF LUMINOL-DEPENDENT LEUKOCYTE CHEMILUMINESCENCE ORIGINATED FROM INTRA- AND EXTRACELLULAR EVENTS.

C. Dahlgren, G. Briheim[1], and O. Stendahl

Department of Medical Microbiology and
[1]Department of Infectious Diseases,
University of Linköping,
S-581 85 LINKÖPING, SWEDEN

INTRODUCTION

The generation of chemically reactive molecules as a result of oxidative activation in polymorphonuclear leukocytes (PMNL) can be measured as light emission or chemiluminescence (1). Addition of luminol to PMNL can amplify the chemiluminescence response (2), but as a result of luminol addition the mechanism for light emission changes from involving both O_2^- and myeloperoxidase (MPO)(3), to be totally dependent on the MPO-H_2O_2-system (4,5). Luminol acts as a bystander substrate for the oxidative species generated during activation of the PMNL. It is, however, not clear whether the oxidation takes place extracellularly or if luminol enters the PMNL and becomes oxidized intracellularly. This investigation was designed to provide insight into the role of extra- and intracellular events in luminol-dependent chemiluminescence.

METHODOLOGY

Chemiluminescence of human polymorphonuclear leukocytes (PMNL), isolated from healthy laboratory personnel and from an MPO-deficient patient, as a response to formylmethionyl-leucyl-phenylalanine (fMLP) or phorbolmyristate acetate (PMA) was measured either in a liquid scintillation counter or in a modified luminometer. Samples for chemiluminescence were obtained by adding 0.6 ml Krebs-Ringer phosphate buffer (KRG) containing luminol and PMNL to disposable polypropylene tubes. The tubes were allowed to stand until a

ISBN 0-12-426290-2

stable background of chemiluminescence was obtained. To activate the system, a stimulus was added and the light emission was recorded. Chemiluminescene in cellfree systems was obtained by addition of MPO, purified from human leukocytes, to KRG containing luminol and H_2O_2. In some experiments KRG contained human serum albumin (1% w/v).

RESULTS AND DISCUSSION

PMNL chemiluminescence response. Addition of fMLP to PMNL, resulted in a bimodal stimulation of chemiluminescence (Fig. 1). A number of factors influence the time course of the response, and the magnitude of the two peaks (4,6,7). After adding PMA to the PMNL there was a lag phase before any measurable signals were recorded. The peak activity was reached after 20-25 min (Fig. 1).

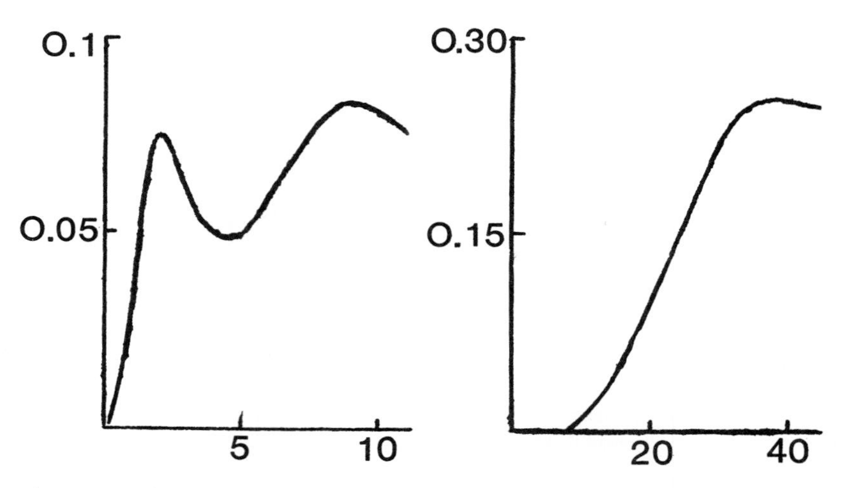

Fig. 1. Time trace of chemiluminescence emitted from normal PMNL when exposed to 10^{-7}M fMLP (left) or 10^{-7}M PMA (right). Abscissa: Time of study (min), ordinate: Chemiluminescence (arbitrary units).

Chemiluminescence from MPO-deficient PMNL. MPO-deficient PMNL required addition of MPO in the extracellular fluid to produce chemiluminescence despite a pronounced production of O_2^-, demonstrating the requirement of MPO for the chemiluminescence response. With MPO-deficient PMNL in the presence of MPO, only the initial peak of chemiluminescence as a response to fMLP was observed (Fig. 2). In the response to PMA, the peak activity was reached much earlier in MPO-deficient cells than in normal PMNL (Fig. 2).

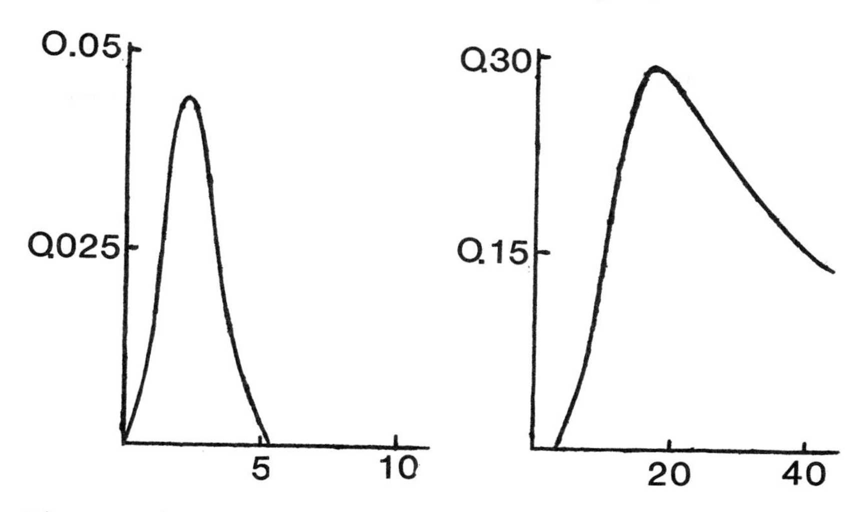

Fig. 2. Time trace of chemiluminescence emitted from MPO-deficient PMNL, in the presence of purified MPO, when exposed to fMLP (left) or PMA (right). Abscissa: Time of study (min), ordinate: Chemiluminescence (arbitrary units).

Since MPO added to MPO-deficient PMNL probably will remain extracellularly, the results indicated that the initial peak in the fMLP induced response, and the early part of the PMA induced response in normal PMNL was due to extracellular events involving H_2O_2 and MPO.

Effects of human serum albumin (HSA). To further analyse the chemiluminescence response, HSA (1% w/v) was added to the measuring systems. In cellfree MPO-H_2O_2-systems, we found that HSA totally inhibited light emission. HSA is a large molecular protein, that does not readily gain access to intracellular sites of PMNL, indicating that HSA could possibly be used to determine the importance of extra- and intracellular events in luminol-dependent chemiluminescence in PMNL. The presence of HSA resulted in a pronounced decrease in the light emitted from MPO-deficient PMNL with added MPO. From the peak values of the PMA induced response it could be calculated that the extracellularly generated chemiluminescence was reduced by more than 90% in the presence of HSA. The presence of HSA had a pronounced effect also on the response of normal PMNL, in that the initial peak in the fMLP response was reduced while the second peak was unaffected. In response to PMA, the peak activity was unaffected by the presence of HSA, but the initial rate was decreased.

CONCLUSIONS

The results indicate that the first peak in an fMLP induced PMNL response, is a result of extracellular and the second peak a result of intracellular reactions of the MPO-H_2O_2-system. Most of the PMA induced response in normal PMNL was due to intracellular events. Furthermore, since luminol seems to diffuse into phagocytic cells, this diffusion may create a problem, in quantitative measurements, in that the reaction may be limited not by generation of oxidative metabolites, but by the diffusion of luminol into the cells.

REFERENCES

1. Allen, R.C., Stjernholm R.L., and Steele R.H. (1972). Evidence for the generation of an electronic exitation state(s) in human polymorphonuclear leukocytes and its participation in bactericidal activity, *Biochem. Biophys. Res. Commun.* 47, 679-684.
2. Allen, R.C. and Loose, L.D. (1976). Phagocytic activation of a luminol-dependent chemiluminescence in rabbit alveolar and peritoneal macrophages. *Biochim. Biophys. Res. Commun.* 69, 245-252.
3. Rosen, H. and Klebanoff, S.J. (1976). Chemiluminescence and superoxide production by myeloperoxidase deficient leukocytes. *J. Clin. Invest.* 58, 50-60.
4. Dahlgren, C. and Stendahl, O. (1983). Role of myeloperoxidase in luminol-dependent chemiluminescence of polymorphonuclear leukocytes. *Infect. Immun.* 39, 736-741.
5. De Chatelet, L.R., Long G.D., Shirley, P.S., Bass, D.A. Thomas, M.J., Henderson, F.W. and Cohen, M.S. (1982). Mechanism of the luminol-dependent chemiluminescence of human neutrophils. *J. Immunol.* 129, 1589-1593.
6. Dahlgren, C. and Stendahl, O. (1982). Effect of in vitro preincubation of polymorphonuclear leucocytes on formylmethionyl-leucyl-phenylalanine induced chemiluminescence. *Infect. Immun.* 37, 34-39.
7. Bender, J.G. and van Epps, D.E. (1983). Analysis of the bimodal chemiluminescence pattern stimulated in human neutrophils by chemotactic factors. *Infect. Immun.* 41, 1062-1070.

SUPERNATANT FROM THROMBIN-ACTIVATED PLATELETS INHIBITS THE NEUTROPHIL CHEMILUMINESCENCE

D.Del Principe[*], A.Menichelli[*], W.De Matteis[*], R.De Santis[*], A.M.Pentassuglio[*] and A.Finazzi-Agrò[+]

[*]*Department of Pediatrics and* [+]*Institute of Biological Chemistry, University of Rome, Italy*

INTRODUCTION

The measurement of chemiluminescence(CL)has been used to assess the phagocytic function of polymorphonuclear cells(PMNs).Many investigators have employed luminol in order to amplify the CL response(1).

Human blood platelets(PLTs)play an important role in non specific immune mechanisms by generating chemotactic factors and reactive oxygen species, which affect PMNs(2).

The present investigation was undertaken to measure the interaction between human activated PLTs and PMNs by means of luminol-enhanced CL.

MATERIALS AND METHODS

Suspensions of PLTs were prepared by centrifugation of platelet-rich plasma.PMNs were isolated as reported(3).Zymosan particles were opsonized by incubation with autologous plasma for 30 min at 37°C and then washed 3 times in saline.CL was measured with a luminometer(Picolite,Packard).

The reaction mixture consisted of suspensions containing 5×10^6 PMNs,opsonized-Zymosan(opZ),1mg/10^6

ISBN 0-12-426290-2

PMNs and 200 µl supernatant from resting or thrombin(1 U/ml)activated PLTs(10^9),giving a final volume of 1 ml.

Samples(100 µl)were read after addition of luminol(20 µl,250 µM)and were incubated at 37°C between readings;samples were counted for 1 min.

O_2 consumption was measured with an O_2 electrode(Yellow Springs Instruments)(3).

O_2^- generation was measured by the SOD-inhibitable reduction of cyt.C(3).

Spectrophotometric assay of MPO was determined in Triton X100 lysed cells and in intact cells(3).

RESULTS

PMNs phagocytizing opZ emit a burst of light,which is inhibited in the presence of supernatant from stimulated PLTs(Fig.1).

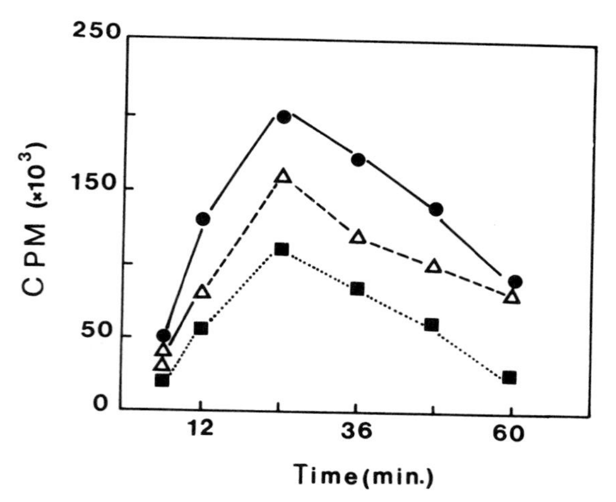

Fig. 1 Effect of platelet supernatant on chemiluminescence by opZ stimulated PMNs.

●——● PMNs + opZ

△---△ PMNs + opZ + resting supernatant

■···■ PMNs + opZ + stimulated supernatant

The inhibition was dose-dependent. The supernatant from $5x10^9$ PLTs caused a 90% inhibition. In the PLT supernatants no catalase or SOD activity was found. The cytoplasmatic enzyme leakage was negligible (LDH discharge<2%). The inhibitory activity was lost when the supernatants were boiled. The supernatants from PLTs did not affect the respiratory burst of phagocytizing PMNs, as determined by O_2 consumption and O_2^- production.

MPO activity, elicited by ConA or PMA, was greatly reduced when PMNs were incubated with supernatant from activated PLTs. Unstimulated PLT supernatants were almost inactive. The difference was less evident when Triton X100 was used to release MPO (Fig.2)

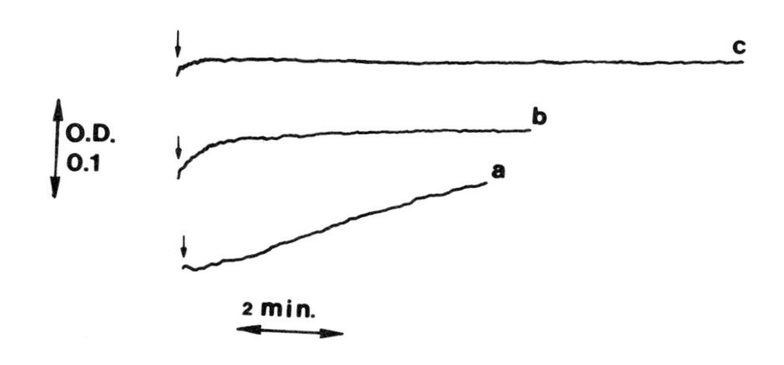

Fig. 2 Myeloperoxidase activity of PMNs. PMNs were stimulated with PMA in the absence (a) and presence of supernatant from resting (b) or stimulated (c) platelets. The oxidation of o-dianisidine was followed at 450 nm.

DISCUSSION

The absolute requirement of the luminol system for MPO, H_2O_2 and Cl^-, suggested that the initial reaction might involve formation of hypochlorous acid, which in turn would serve as the immediate oxidant for the luminol(1,4). A decrease of PMN CL by PLT supernatant could result from an inhibition of the production of O_2^- and H_2O_2, from the presence of scavengers, from an inhibition of MPO release or activity, or from an interaction with luminol. Under these experimental conditions neither the respiratory burst of stimulated PMNs, nor the production of O_2^- were affected by the addition of PLT supernatant.

Instead the lower MPO activity in the presence of PLT supernatant after stimulation with ConA or PMA could indicate an effect on MPO release. It is worth recalling that PLTs release components stimulating the motility of PMNs, which in turn can induce changes in PLT activity. It is possible that the activity observed by us represent a negative feed back mechanism.

REFERENCES

1. De Chatelet, L.R., Long, G.D., Shirley, P.S., et al. (1982). Mechanism of the luminol-dependent chemiluminescence of human neutrophils, *J. Immunol.* 129, 1589-1593.
2. Becker, E.L. (1980). Chemotaxis, *J. Allergy Clin. Immunol.* 66, 97-105.
3. Babior, B.M. and Cohen, H.J. (1981). Measurement of neutrophil function: phagocytosis, degranulation, the respiratory burst and bacterial killing. *In* "Leukocyte Function" (Ed M.J. Cline). pp 1-38. Churchill Livingstone, New York.
4. Weiss, S.J., Klein, R., Slivka, A., et al. (1982). Chlorination of taurine by human neutrophils, *J. Clin. Invest.* 70, 598-607.

RELIABILITY OF THE DETECTION OF HUMAN MONOCYTE CHEMILUMINESCENCE IN MONONUCLEAR CELL SUSPENSION (*)

A. Müller, P. Schuff-Werner, M. Wurl and G.A. Nagel

Dept. Internal Medicine, Div. Hematology/ Oncology, University Clinic Göttingen, Robert-Koch-Str. 40, D-3400 Göttingen (FRG)

(*) supported by BMFT grant no. 0384 100

The majority of previous studies on monocyte chemiluminescence (CL) have been conducted using mononuclear cell (MNC) suspensions (1;2). In order to find out whether lymphocytes interfere with production or detection of human monocyte CL the CL-response of MNC, adjusted to 1 x 10^5 monocytes per CL-assay by non-specific esterase stain, was compared with that of 1 x 10^5 isolated monocytes purified either by adherence or by Percoll gradient centrifugation.

MATERIALS AND METHODS

Isolation of Mononuclear Cells (MNC). MNC were isolated from leucocyte-rich buffy coat by Ficoll-Isopaque gradient centrifugation. For further monocyte enrichment a discontinuous gradient centrifugation method using hyperosmotic (325-340 mosm) Percoll solutions (Pharmacia, Uppsala, Sweden) was employed (3). Thus 35 $\pm$ 14% of the Esterase-stained MNC prior to isolation were recovered giving a monocyte suspension containing 73.7 $\pm$ 3.7% Esterase-positive cells.

Simultaneous to the Percoll isolation procedure another method for monocyte isolation based upon the reversible Mg^{2+}-dependent adherence of human monocytes to plasma fibronectin (4) was performed. The monocyte yield obtained by this method was

ISBN 0-12-426290-2

46.4 ± 17% related to the number of Esterase-positive MNC before isolation, the purity was 93.1 ± 2.4% ad judged by Esterase stain.

Chemiluminescence Measurement. CL was detected using a six-channel luminescence analyzer (Biolumat, Model LB 9505, Berthold, Wildbad, Germany), connected on-line with a microcomputer (IMCA, Rosdorf, Germany). Data analysis included the calculation of the maximum CL-response (Peak-CL), the total number of counts registered in certain time limits (Integrals) and the time of Peak-CL (t_{max}). Furthermore the kinetics of the CL-intensities were plotted.

A typical CL-assay was performed as follows: all reaction components were suspended or dissolved in Hepes-buffered HBSS without HCO_3^- (CL-Medium) Monocyte suspensions were generally adjusted to 5 x 10^5/ml using non-specific Esterase stain. 200 µl cell-suspension and 100 µl of a 4 x 10^{-5}M solution of either Luminol or Lucigenin were preincubated for 10 min at 30°C and background CL was recorded. The reaction was started by the addition of 100 µl particle suspension (5 mg/ml) or 100 µl of a 0.4 µg/ml Phorbol-myristate-acetate solution (PMA).

RESULTS

Luminol-enhanced Measurements. The time course of the CL-response of 1 x 10^5 Esterase-positive cells in mononuclear cell suspension in the presence of 10^{-5}M Luminol is shown in Fig. 1. The CL kinetics of isolated monocytes did not show substantial differences. When the CL of Esterase-stained mononuclear cells was compared with that of adherence as well as Percoll-isolated monocytes in 12 experiments no significant differences were found with any of the stimuli employed as far as the parameters Peak-CL and Integral between 0 and 30 min (Integral 0-30) were concerned. Only upon stimulation of adherence purified monocytes with opsonized Staph.aureus did the maximum CL-response occure clearly earlier (Table 1). Upon comparison of the CL-response in each experiment by relating the results of adherence- and Percoll-purified monocytes to those obtained with MNC (values ex-

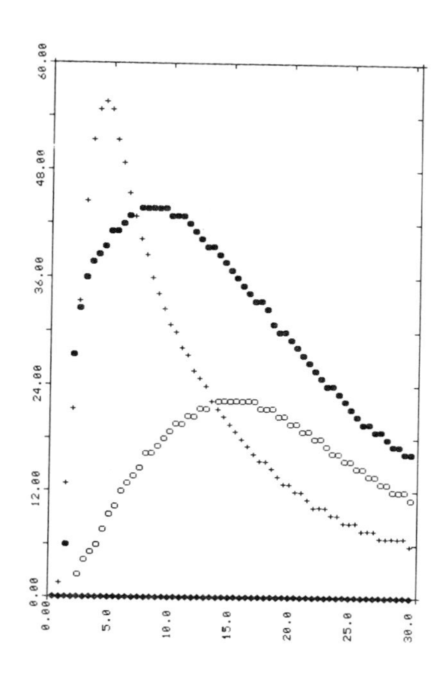

Fig. 1 Time course of the Luminol-dependent CL of 10^5x Esterase-stained cells in MNC-suspension using different stimuli.

- *o opsonized zymosan (0.5 mg per tube)*
- *● opsonized Staph.aureus (0.5 ml per tube)*
- *+ phorbol-myristate-acetate (0.1 µg/ml)*

COMPARISON OF LUMINOL ENHANCED CL OF MNC AND PURIFIED MONOCYTES

	E^+ MNC*	E^+ PIM*	E^+ APM*	E^+ MNC	E^+ PIM	E^+ APM	E^+ MNC	E^+ PIM	E^+ APM
PEAK - CL (CPM x 10^3)	2315 ± 417	100 ± 500	2149 ± 352	4348 ± 1378	3618 ± 1643	3342 ± 1635	5392 ± 1330	4894 ± 1300	4668 ± 846
% OF MNC	100	99,5 ± 16,9	94,9 ± 19,8	100	80,3 ± 12,9	74,2 ± 15,0	100	97,7 ± 29,4	93,4 ± 25,4
$\int_0^{30}$ (COUNTS x 10^6)	51,5 ± 9,4	48,0 ± 11,1	48,9 ± 7,6	95,0 ± 26,8	74,2 ± 24,5	57,9 ± 19,6	68,9 ± 12,9	51,5 ± 13,9	57,4 ± 14,7
% OF MNC	100	100,4 ± 15,3	97,9 ± 17,7	100	78,1 ± 12,4	60,4 ± 12,7	100	94,5 ± 23,3	86,5 ± 16,2
T MAX (MIN.)	15,4 ± 1,8	15,7 ± 2,8	16,2 ± 3,3	9,7 ± 2,6	9,3 ± 2,7	6,1 ± 0,7	5,2 ± 1,8	5,2 ± 2,3	3,9 ± 0,5
STIMULATION	+ 0,5 MG OPS. ZYMOSAN			+ 0,5 MG OPS. STAPH.AUREUS			+ 0.1 µL/ML PHORBOL-MYRISTATE ACETATE		

* MNC: MONONUCLEAR CELLS PIM: PERCOLL ISOLATED MONOCYTES APM: ADHERENCE PURIFIED MONOCYTES

Table 1

pressed as % of MNC in Table 1) only the Staph. aureus induced CL of isolated monocytes appeared to be about 20% (Percoll-monocytes(and about 25% (adherence-monocytes) less than that of MNC.

Lucigenin-enhanced Measurements. Similar to the results in the luminol-amplified system, there were no significant differences between the Lucigenin dependent CL-response of the three monocyte populations as measured in seven experiments. Analysing the results in detail, an accelerated kinetics of the PMA stimulated purified monocytes was obvious, as is demonstrated by a shorter t_{max} interval in comparison to MNC. The Staph. aureus induced CL-response of isolated monocytes was again diminished but upon direct comparison of the results of every single experiment only the Peak-CL of adherence-treated monocytes was continually lower than the Peak-CL of MNC.

DISCUSSION

The fact that the CL-response of mononuclear cells, adjusted to a definite monocyte number by Esterase stain, was in the same range as the CL-values of purified monocytes indicates that quenching or other inhibitory effects through autologous lymphocytes, do not occur. Nevertheless, the altered time course of isolated-monocyte-CL in some experiments and quantitative differences in Staph.aureus induced CL might point to a more specific role of lymphocytes in regulating human monocyte oxidative microbicidal metabolism.

REFERENCES

1. Nelson, R.D., Mills, E.L., Simmons, R.L. et al. (1976). Chemiluminescence response of phagocytizing human monocytes, *Infect. Immun.* *14 (1), 129-134*
2. Braun, D.P., Harris, J.E., Maximovich, S. et al. (1981). Chemiluminescence in peripheral blood mononuclear cells of solid tumor cancer patients, *Cancer Immunol. Immunother.* *12, 31-37*
3. Fluks, A.J. (1981). Three-step isolation of human blood monocytes using discontinuous density gradients of Percoll, *J. Immunol. Methods* *41, 225-233*
4. Bevilacqua, M.P., Amrani, D., Mosesson, M.W. et al. (1981). Receptors forcold-insoluble globulin (plasma fibronectin) on human monocytes. *J. Exp. Med.* *153, 42-60*

DEPENDENCE OF POLYMORPHONUCLEAR LEUCOCYTE LUMINOL-DEPENDENT CHEMILUMINESCENCE ON OXYGEN

MB Hallett*, SW Edwards, SCL Barrow and AK Campbell+

Departments of Surgery and Medical Biochemistry,+ Welsh National School of Medicine, Heath Park, Cardiff, CF4 4XN, UK.*

INTRODUCTION

We have shown that oxygen radical production by polymorphonuclear leucocytes (PMN) can be activated by two distinct mechanisms, one dependent upon a rise in intracellular Ca^{2+}, producing external radicals, and the other independent of a rise in intracellular Ca^{2+} producing intralysosomal radicals (1,2). A crucial feature of our studies has been the use of chemiluminescence to measure both intracellular Ca^{2+}, using a calcium-activated photoprotein, obelin, and to oxygen radical production, using luminol (3,4). Whereas the validity of the Ca^{2+}-activated photoprotein is established, verification of luminol-chemiluminescence as a monitor of oxygen radical production has not yet been performed. The two most widely reported methods for assessing oxygen radical produced have been luminol-dependent chemiluminescence and cytochrome c reduction. However the dependence of these methods on the primary substrate of the oxidase, namely oxygen, has not been established. Furthermore, there are several discrepancies between the time course and extent of oxygen radical production as monitored by the two methods. In particular, the consumption of oxygen by stimulated PMN was unaffected by the presence of cytochrome c which became reduced (5), despite the expectation that oxygen would be generated during the reduction of cytochrome c by O_2^-. The reduction of cytochrome c therefore does not appear to be a satisfactory indicator of oxygen radical production.

ISBN 0-12-426290-2

The aim of the work reported in this paper was to establish the validity of luminol-dependent chemiluminescence as an indicator of oxygen radical production, by correlating the chemiluminescence response observed from resting and activated PMN with oxygen concentration and with oxygen consumption rate. A further aim was to compare the affinities of oxygen for resting and stimulated chemiluminescence in an attempt to identify intracellular components of the radical-generating chain.

METHODS

Polymorphonuclear leucocytes were isolated from rat peritoneal exudate as previously described (3). Simultaneous measurement of oxygen concentration and chemiluminescence was made in the specially-built apparatus (6) shown in figure 1.

RESULTS

Dependence of luminol-dependent chemiluminescence from polymorphonuclear leucocytes on oxygen concentration.

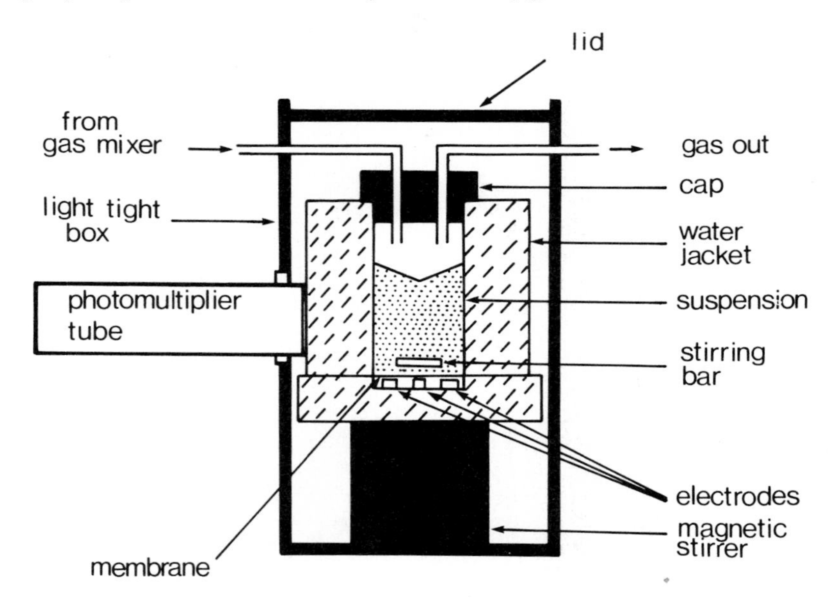

Fig. 1 Apparatus for simultaneous measurement of oxygen and chemiluminescence.

In order to verify that luminol-dependent chemiluminescence is a true indicator of the production of an oxygen metabolite by PMN, it is necessary to demonstrate the dependence of the chemiluminescence for oxygen. The rate of chemiluminescence from resting cells was reduced by 90% in low O_2 medium. Restoring the air completely restored the chemiluminescence rate. Luminol-dependent chemiluminescence from cells stimulated either with unopsonised latex beads ($4x10^9 ml^{-1}$) or a chemotactic peptide (N-formyl-methionyl-leucyl-phenylalanine, 1μM) plus cytochalasin B ($5μg ml^{-1}$) similarly was inhibited by low oxygen and was restored on re-aeration. The effects on PMN chemiluminescence were not attributable to reduced cell viability because (i) no reduction in cellular ATP occurred at low O_2 concentration (up to 200s) (ii) no increase in LDH release on trypan blue uptake occurred and (iii) the chemiluminescence responses were completely restored by re-aeration. Resting and luminol-dependent chemiluminescence from PMN therefore reflect the production of an oxygen metabolite.

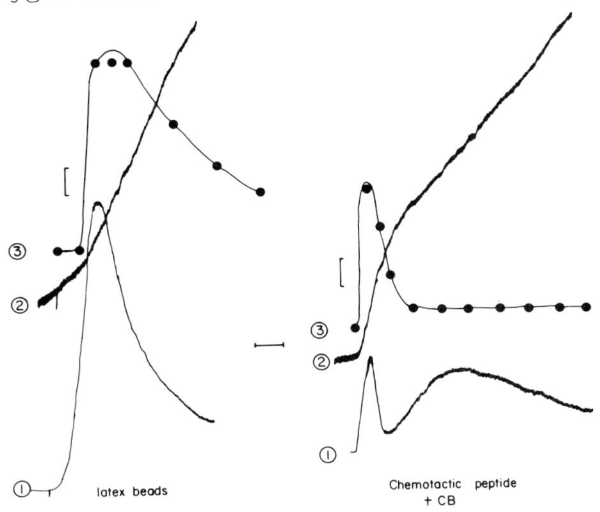

Fig. 2 Simultaneous measurement of O_2 consumption and luminol CL stimulated by a. latex beads ($4x10^5$/ml), b. chemotactic peptide + cytochalasin B. 1. O_2 production rate 2. O_2 consumption 3. O_2 consumption rate. Scale ►60s; ▲ LX; 1, $4.2x10^4$ cps; 2, 1.4nmol O_2; 3, 0.4nmol/m; CP+CB; 1, $4x10^4$ cps; 2, 1.4nmol O_2; 3, 2nmol/m

Unlike cytochrome c, whose molar reduction is of a similar order to PMN oxygen consumption, it can be estimated from the quantum yield and the integrated chemiluminescence response, that during a typical experiment less than 10^{-11} mole of luminol react whereas $\sim 10^{-8}$ moles of oxygen are consumed. Luminol would not significantly change in concentration nor would it affect the event being monitored. Thus luminol would act as a true indicator. The relationship between oxygen consumption and luminol-dependent chemiluminescence measured simultaneously therefore reflects the true relationship between oxygen concentration and production of the oxygen metabolite that reacts with luminol. Unopsonized latex beads caused an increase in both the rate of cxygen consumption and the rate of luminol-dependent chemiluminescence after a lag phase (Fig 2a).

DISCUSSION

These results show that luminol-dependent chemiluminescence fulfils two essential criteria for an indicator of PMN oxygen radical production, namely 1 dependence upon oxygen and 2 co-incidence between oxygen consumption and luminol signal. Further they demonstrate the potential significance of local oxygen concentration *in vivo* in regulating radical production.

REFERENCES

1. Campbell, A.K. and Hallet, M.B. (1983). J. Physiol 338, 537-550.
2. Hallett, M.B. and Campbell, A.K. (1983). Biochem J. 216, 459-465.
3. Hallett, M.B., Luzio, J.P. and Campbell, A.K. (1981). Immunology 44, 565-576.
4. Hallett, M.B. and Campbell, A.K. (1982). Nature 295, 155-158.
5. Segal, A.W. and Meshulam, T. (1974). FEBS Lett 100, 27-32.
6. Edwards, S.W., Hallett, M.B. and Campbell, A.K. (1983). Biochem J. 217, 851-854.
7. Hallett, M.B. and Campbell A.K. (1984). Biochem Soc Trans (in press).
8. Hallett, M.B., Edwards, S.W. and Campbell, A.K. (1984). In "Cellular Chemiluminescence" (Ed K. Von Dyke) CRC Press (in press).

PERIOPERATIVE ZYMOSAN INDUCED CHEMILUMINESCENCE OF POLYMORPHONUCLEAR LEUKOCYTES IN SURGICAL CANCER PATIENTS

Hans Konrad Schackert, Michael Betzler, Georg F. Zimmermann, Hans-Peter Geisen, Lutz Edler[+], G. Harald Geelhaar, Christian Herfarth

Department of Surgery, University of Heidelberg D-6900 Heidelberg, W.Germany

[+]*Dep. of Biostatistic, German Cancer Research Centre D-6900 Heidelberg, W.Germany*

INTRODUCTION

Increased activity of phagocytic cells of patients with solid tumors has been reported (2). Several factors which may be responsible for polymorphonuclear leukocytes' (PMNL) activation are discussed. Lymphokines, extracts from the cell wall of bacteria, immune complexes and tumor cells can activate phagocytic cells to a bactericidal or tumoricidal state (1). There is a respiratory burst in activated PMNL which leads to the generation of electronically excited oxygen radicals. Relaxation and photon emission generates chemiluminescence which can be measured after amplification with luminol (3,4).

AIMS OF THE STUDY

The aims of this study were to evaluate zymosan induced and luminol amplified chemiluminescence (CL) in different surgical patient groups. To investigate the influence of surgical trauma and tumor stage CL-activity (CLA) was measured pre- and postoperative in control patients and patients with different stages of malignant tumors.

MATERIAL

PMNL-activity was evaluated in 221 hospitalized surgical patients. Patients with benign non inflammatory diseases

ISBN 0-12-426290-2

served as controls. There were three cancer patient groups (colorectal, esophagus/stomach, cancer of different origin), one inflammatory diseases group and miscellaneous. Except of the miscellaneous' group which had an average age of 43.8 years all the groups had an average age ranging from 50.6 (controls) to 64.9 (colorectal cancer) years.

METHODS

Venous blood was drawn preoperatively at the day of operation and 10 days postoperatively between 7 a.m. and 9 a.m. 1 ml of blood was diluted with 4 ml of DMEM (Dulbecco's modification of Eagle's medium). 1o U heparin/ml whole blood was used as an anticoagulant. White cells were counted and differentiated. Polymorphonuclear leukocytes and monocytes were considered as CL producing cells. 10 µl luminol were added to 500 µl diluted blood and incubated for 10 minutes at 37°C. Background activity was measured for 1 minute and the CL reaction was initiated adding .5 mg non opsonized zymosan (suspended in 10 µl DMEM). Dynamic CLA was measured in intervals of 30 seconds for 4o minutes using a BIOLUMAT LB 9505 (Fa. Berthold, Wildbad, W.Germany). Data out of simultaneous six channel measurement were transferred to an apple II computer, stored on floppy discs and processed. The shape of the CLA curve is characterized by 4 basic parameters: total 40 min integral (CLA), peak level which is the maximum 30 sec interval activity, peak time and the integral to peak time as a percentage of the CLA. 4 additional parameters were created. In order to determine the parameters which contribute at most to the discrimination between the groups we formulated this as a multivariate analysis of variance (MANOVA) problem (5). A stepwise elimination procedure created total integral (CLA) and peak level as the parameters with the best effect of discrimination. Therefore all further evaluations were based on total integral. Preoperative CLA was not correlated with or limited by the number of CL producing cells. Therefore CLA was related to 1000 CL producing cells.

RESULTS

Neither age nor sex specific differences could be found in the control and in the infectious disease group. In comparison to the preoperative level there was a significant increase (Wilcoxon signed rank test) of postoperative CLA in the controls (n=34/CLA-pre 66.0/CLA-post 73.8/p .02). Similar results were obtained in the infectious disease group

(36/74.2/80.7/p .10). Cancer patients had significantly (Wilcoxon rank sum test) elevated preoperative CLA levels (66/79.5/p .044) compared to the controls. The group of all cancer patients showed a slight postoperative decrease of CLA (79.5/78.1). There was a significant difference between CLA levels of curatively operated cancer patients (curative) and palliatively operated cancer patients (palliative)(Fig.1).

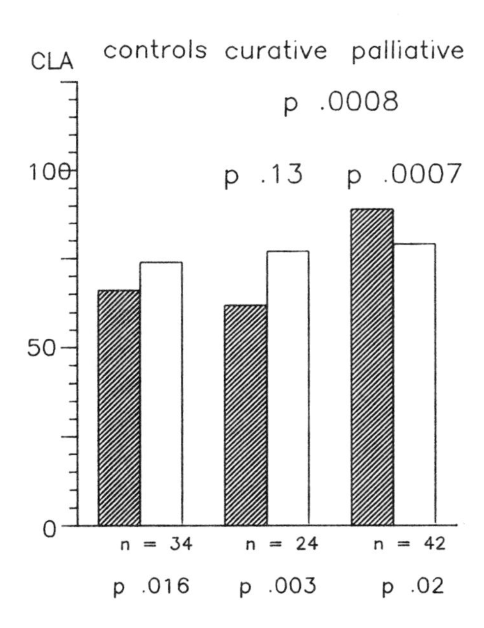

Fig. 1 Chemiluminescence-activity (CLA) in controls, curatively and palliatively operated cancer patients.

Preoperative CLA of the palliative group was significantly elevated (42/89.2) compared to the controls (p .0007) and to the curative group (p .0008). The CLA of the curative group (24/62.5/76.8/p .003) increased postoperative in the same manner as the CLA in the controls. The palliative group was the only one with a postoperative decrease of CLA (89.2/78.9/p .02). All cancer subgroups (colorectal, esophagus/stomach, different origin) showed the same tendency.

To study the influence of progressive tumor growth onto CLA three groups were created:

1. Curatively operated cancer patients without metastases (cur-MO), n=23.
2. Palliatively operated cancer patients without metastases (pall-MO), n=16.

3. Palliative operated cancer patients with metastases (pall-M1) n=23.

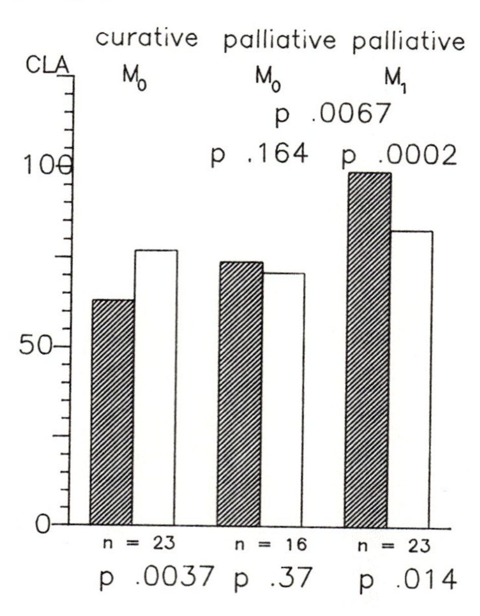

Fig. 2 Chemiluminescence-activity (CLA) in progressive tumor growth. (Hatching: CLA-preoperative)

Pall-M1 patients showed a significant higher CLA level (99.0) in comparison to pall-MO patients (73.9/p .007) and to cur-MO patients (66.3/p .0002). There was a significant postoperative increase (66.3/77.4/p .004) in the cur-MO group, a slight decrease (73.9/71.2) in the pall-MO group and a significant postoperative decrease in the pall-M1 group (99.0/82.8/p .01)(Fig. 2).

With an increasing preoperative CLA there was a tendency towards a decrease of postoperative CLA in the control and infectious disease group (correlation coefficient r= -.53), palliative cancer group (r= -.67) and all patients (r=-.59).

CONCLUSIONS

Chemiluminescence as a phenomenon of phagocytic cell activation was investigated in different surgical patient groups. Malignancy does not generally activate CLA. Activation state of CL producing cells is significantly increased in progressive tumor growth (metastases). Generalization of the tumor causes a maximum CLA. Increase of tumor surface and tumor mass in metastatic tumor growth may have a stimulation effect on CLA. Surgical trauma seems to influence CLA in dependence of the preoperative level. There is a postoperative elevation of low preoperative values and a diminuation of high preoperative values.

In the early postsurgical period the effect of tumor mass reduction on CLA may be too low compared to the effect of the surgical trauma.

This data indicate a relationship between CL producing cells and tumor stage. There is the possibility to discriminate between local tumor growth and generalization. How far this method might serve as an useful tumor marker will be the subject of further longterm follow up studies in cancer patients.

ACKNOWLEDGEMENT

This study was supported by the TUMORZENTRUM HEIDELBERG/ MANNHEIM "Colon-Rektum-Projekt"

REFERENCES

1. Fidler, I.J., Poste, G. (1982). Macrophage-mediated destruction of malignant tumor cells and new strategies for the therapy of metastatic disease, *Springer Semin. Immunopathol.* 5, 161-174.
2. Magarey, C.J., Baum, M. (1970). Reticuloendothelial activity in humans with cancer, *Brit.J.Surg.* 57, 748-752.
3. Allen, R.C., Stjernholm, R.L., Steele, R.H. (1972). Evidence for the generation of an electronic excitation state(s) in human polymorphonuclear leukocytes and its participation in bactericidal activity, *Biochem.Biophys. Res.Commun.* 47, 679-684.
4. Hafeman, D.G., Zoltan, J.L. (1979). Polymorphonuclear leukocyte-mediated, antibody-dependent, cellular cytotoxicity against tumor cells: dependence on oxygen and the respiratory burst. *J.Immunol.* 123, 55-62.
5. Wilks, S.S. (1932). Certain generalizations in the analysis of variance, *Biometrika* 24, 471-494.

ASSOCIATION OF CHEMILUMINESCENCE INDUCED BY PNEUMOCOCCAL ANTIGENS IN WHOLE BLOOD WITH SPECIFIC ANTIBODY TITERS

M. Heberer*, M. Durig*, M. Ernst[1], J. Schlenar[+], P. Erb[+], F. Harder*

*Department of Surgery and +Institute for Microbiology and Hygiene, University of Basel, Switzerland
[1]Department of Cell Biology I, Forschungsinstitut Borstel, Federal Republic of Germany

Bacterial sepsis as a late complication of splenectomy is mainly caused by pneumococci. The risk of fatal infection in asplenic patients is estimated to be 30 to 200 times that in the general population (4).

For this reason, surgery of the spleen became conservative. If splenectomy is not avoidable, splenic tissue is replanted into the greater omentum and prophylactic antibiotics and active pneumococcal vaccination are given (2). The efficacy of these measures is, however, unclear.

Polymorphonuclear granulocytes (PMN) play an important role in the defense of bacterial invaders. Particular antigens such as pneumococci can activate PMN mediated by IgG or via C3b-receptors following complement activation. Chemiluminescence (CL) resulting from pneumococcal stimulation of isolated PMN has been shown to correlate with the concentration of the type specific antibodies of the opsonizing serum (1, 3). We have recently reported the induction of CL by pneumococcal antigens in diluted whole blood samples (5). The correlation of this CL-response with the respective

ISBN 0-12-426290-2

antibody concentrations is the question of the current investigation.

METHODS

5 healthy volunteers were vaccinated with a polyvalent pneumococcal antigen preparation (Pneumovax, MSD, Westpoint, USA). Before and after the vaccination both CL in diluted whole blood samples and antibody concentrations were measured.

CL-assay: Whole blood diluted 1:5, luminol (final concentration 40mg/l), zymosan (1000mg/l), latex (2000mg/l) or pneumococcal capsular polysaccharides as CL-signals.

IgM and IgG antipneumococcal antibodies: Dynatech microelisa plates coated with pneumococcal polysaccharides, goat anti-human IgM and IgG peroxidase conjugates.

RESULTS AND DISCUSSION

After the vaccination CL stimulated specifically by pneumococcal antigens increased (peak: 8 weeks after vaccination), whereas non specifically induced CL-activities did not change (zymosan and latex-controls) (cp. fig. 1). This increase in CL induced specifically by pneumococcal antigens suggested a significance of the corresponding specific antibodies. This hypothesis was supported by the determination of the specific antibodies (cp. fig. 2): IgG and IgM antipneumococcal antibodies were elevated after the vaccination and the over-all kinetics was comparable to the pneumococci induced CL.

Based on these results a positive correlation between immunoglobulin - and CL-levels was expected, however not confirmed by regression analysis.
Three different patterns of response were detected:

1) comparable kinetic of CL and immunoglobulin (cf fig. 3)
2) significant increase in specific CL-activity but only minimal antibody response to the vaccination (cp. fig. 4)
3) increase and subsequent decline of CL with concomitant increase of specific antibodies but subsequent plateau between 8 and 14 weeks.

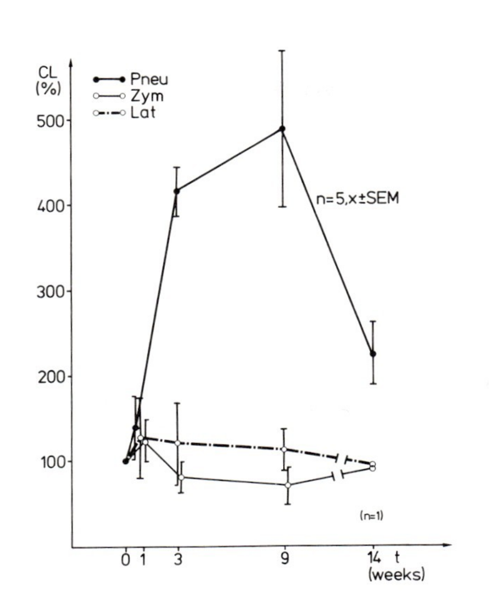

fig. 1

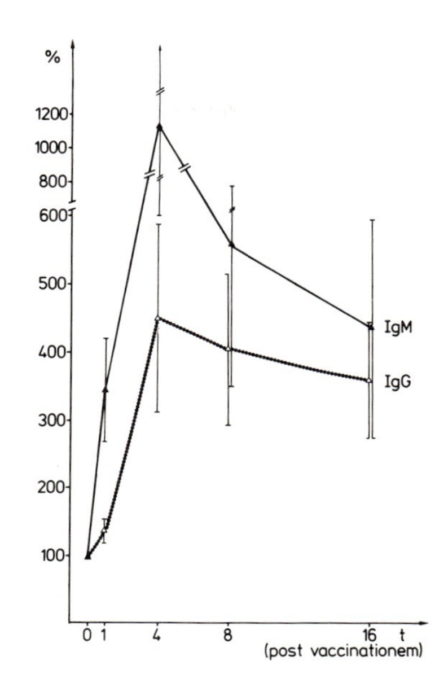

fig. 2

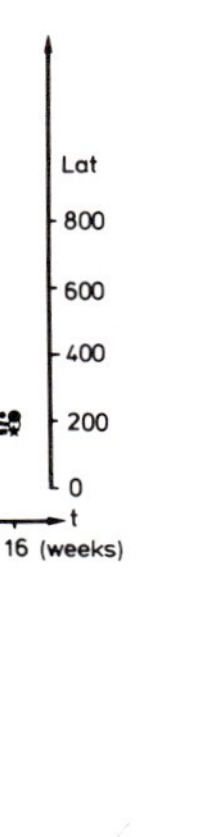

fig. 3

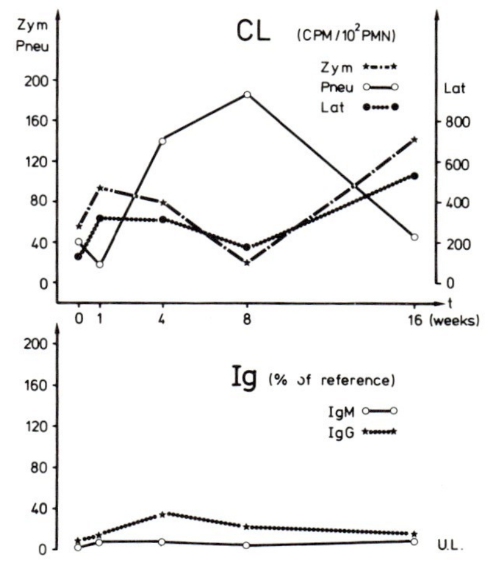

fig. 4

Therefore, the CL-response induced by pneumococcal antigens in diluted whole blood appears to be partly mediated by the respective specific immunoglobulins. It is however no measure of the IgG or IgM-titers, but may provide additional information. Potential further mediators of the CL-response include the factors of the complement system. The complement system can be activated directly by some pneumococcal strains or following interaction of pneumococcal antigens with IgM. From a clinical point these interactions require further evaluation because measurement of granulocyte-activation in whole blood may offer relevant information about an organism's capacity of defense against an invading microorganism.

REFERENCES

1) Bortolussi, R., Marrie, T.J., Cunningham, J., Schiffman, G. (1981) Serum antibody and opsonic responses after immunization with pneumococcal vaccine in kidney transplant recipients and controls. Infect.Immun. 34, 20-25.

2) Conney, D.R., Dearth, J.D., Swanson, S.E., Dawanjee, M.K., Telander, R.L. (1979) Relative merits of partial splenectomy, splenic reimplantation and immunisation in preventing postsplenectomy infection. Surgery 86, 561-569.

3) Gardner, S.E., Anderson, D.C., Webb, B.J., Stitzel, A.E., Edwards, M.S., Spitzer, R.E., Baker, C.J. (1982) Evaluation of sptreptococcus pneumoniae Type XIV opsonins by phagocytosis-associated chemiluminescence and a batericidal assay. Infect. Immun. 35, 800-808.

4) Gelfand, J.A., Grabbe, J.P. (1983) Hyperthermia of acute onset in an asplenic man. N.Engl.J.Med. 308, 1212-1218.

5) Heberer, M., Dürig, M., Wadström, J., Nassenstein, D., Ernst, M., Harder, F. (1984) Chemiluminescence of granulocytes to detect the immune response to pneumococcal vaccination. In: Oxygen radicals in chemistry and biology (Eds: W. Bors, M. Saran, D. Tait) 867-873. Walter de Gruyter, Berlin.

AGENTS EXERTING CONTROL OVER LUMINOL-ENHANCED PHAGOCYTIC CHEMILUMINESCENCE

Rudolf E. Schopf

Department of Dermatology, Johannes Gutenberg University D-6500 Mainz, Federal Republic of Germany

INTRODUCTION

In the past years, chemiluminescence (CL) has become an established tool for the assessment of phagocytic function (1,2) in particular owing to its rapidity and high sensitivity compared to photometric tests (3). Although the molecular mechanisms underlying CL are yet incompletely defined, in contrast to photometric tests, the advantages of CL measuring the total amount of oxygen intermediates generated, outweigh the lack of well-defined specificity of certain oxygen radicals produced. We therefore chose CL as an assay system to study the effects of a number of well-defined agents on the respiratory burst (4).

This report presents data of CL measurements of both human polymorphonuclear leukocytes (PMN) and monocytes/macrophages (MØ) stimulated with zymosan particles under the influence of nonsteroidal anti-inflammatory drugs, protease inhibitors, a calcium channel blocker, transglutaminase inhibitor, β_2 adrenoceptor stimulant, ethanol, acetaldehyde, and for comparison superoxide dismutase. The findings should shed some more light on the mechanisms involved in phagocytic stimulation. We selected zymosan particles to stimulate the cells which has yielded highly reliable results (3). Other stimulants include complement components, immune complexes, chemotactic peptides, calcium ionophores, phorbol esters or bacteria.

ISBN 0-12-426290-2

MATERIALS AND METHODS

Preparation of peripheral blood phagocytes of healthy volunteers and measurements of CL were performed as described (3). Hydrophobic agents were dissolved in dimethyl sulfoxide (DMSO). The final DMSO concentration employed did not interfere with CL. Viability tests were performed either with trypan blue exclusion or determination of lactate dehydrogenase release and found to exceed 98%. Peak CL of zymosan-stimulated phagocytes ($5x10^5$ PMN; 2×10^5 MØ) was recorded after 30 min corresponding to the maximum of response. The cell isolation and short-term culture medium consisted of minimum essential medium (MEM).

RESULTS

The results are summarized in table 1. In each case the dose of the agent suppressing peak CL 50% is given (ID50).

DISCUSSION

Our findings with ASA and NDGA indicate the involvement of arachidonic acid metabolism in phagocytic activation. While ASA inhibits mainly cyclooxygenase, NDGA acts on lipoxygenase pathway(s). The ID50 is about 350 times lower for NDGA compared to ASA stressing the importance of leukotrienes in CL as has been suggested (5).

Chloroquine also suppresses CL, its mechanism, however, is unclear. We included this agent in here because of its potent anti-inflammatory properties. Our findings may be useful to explain part of its action in patients in particular since it is well-known that leukocytes may concentrate the drug up to 2000 times its serum concentration.

The steroids employed also induced suppression of CL. However, no class specificity or rank order was observed possibly suggesting non-specific effects. Furthermore, the short duration of the incubation period was too short to induce changes in cell function via the cytoplasmic steroid receptor and cell nucleus.

The microtubule binding and disrupting agents colchicine, vinblastine and vincristine also in-

TABLE 1

SUPPRESSION OF ZYMOSAN-INDUCED CL BY DIFFERENT AGENTS (PEAK CL OF PMN CORRESPONDING TO 671 ± 144, OF MØ 193 ± 64; CL OF UNSTIMULATED PMN 15.0 ± 10.9, OF MØ 7.21 ± 4.21 x 10^3 COUNTS/10S, MEAN ± SD, N=21. NO. OF EXPERIMENTS IN PARENTHESIS). ID50 (MOLAR CONC.).

Agent	PMN	MØ
ASA* (5)	9.2×10^{-4}	8.0×10^{-4}
NDGA (3)	2.7×10^{-6}	2.3×10^{-6}
Chloroquine (4)	4.2×10^{-6}	3.3×10^{-5}
Hydrocortisone (4)	6.2×10^{-5}	5.3×10^{-5}
DHEA (4)	7.8×10^{-5}	1.0×10^{-4}
Colchicine (4)	1.0×10^{-3}	5.0×10^{-4}
Vinblastine (3)	5.3×10^{-5}	8.7×10^{-6}
Vincristine (3)	8.0×10^{-5}	1.0×10^{-5}
Benzamidine (3)	6.7×10^{-4}	4.3×10^{-4}
DFP (3)	3.3×10^{-4}	2.2×10^{-4}
Verapamil (4)	6.7×10^{-5}	2.6×10^{-4}
Dansylcadaverin(1)	2.4×10^{-5}	4.9×10^{-6}
Fenoterol (3)	1.7×10^{-6}	5.4×10^{-6}
Ethanol (3)	5.3×10^{-2}	7.8×10^{-2}
Acetaldehyde(3)	6.5×10^{-3}	4.9×10^{-3}
SOD (4)	9.3×10^{-7}	2.1×10^{-7}

*ASA= actylsalicylic acid, NDGA=nordihydroguaiaretic acid, DHEA=dehydroepiandrosterone, DFP=diisopropylfluorophosphate, SOD=superoxide dismutase.

hibited CL stressing the importance of the cytoskeleton.

Proteolysis apparently is also of paramount importance as evidenced by the inhibition of CL by benzamidine and DFP, findings in common with other reports (6). The low molecular weight of the inhibitors may suggest events not on the outer cell membrane as soybean trypsin inhibitor and aprotinin caused only marginal suppression (data not shown).

Verapamil demonstrates the importance of calcium channels, a finding that has not been reported with CL before.

Suppression of CL can also be induced by the transglutaminase inhibitor dansylcadaverin similar to inhibition of receptor-mediated endocytosis (7).

The β_2 adrenoceptor stimulating agent fenoterol also suppressed CL indicating the participation of the adenylate cyclase system.

Both ethanol and acetaldehyde suppressed CL in doses associated with inebriation in man, findings of importance in diseases where phagocytes are involved in etiopathogenesis.

Finally, SOD stresses the importance of O_2^- in the assay system. Catalase, by contrast was only marginally inhibitory (data not shown).

It was not possible to compare the agents satisfactorily in cell-free luminogenic systems, as suggested (5).

To summarize, our findings indicate important mechanisms in the generation of CL involving transglutaminases, the cytoskeleton, proteases, arachidonate metabolism, calcium transport as well as modulation by the adenylate cyclase system.

REFERENCES

1. Allen, R.C. et al. (1972). Evidence for the generation of an electronic exitation state(s) in human PMN, BBRC 47, 679-684.
2. Allen, R.C. et al. (1976). Phagocytic activation of a luminol-dependent chemiluminescence in rabbit alveolar Mø, BBRC 69, 245-252.
3. Schopf, R.E. et al. (1984). Measurement of the respiratory burst in human Mø and PMN by NBT reduction and CL, J. Immunol. Meth. 67, 109-117.
4. Babior, B.M. (1978). Oxygen-dependent microbial killing by phagocytes, New Engl.J.Med.298,721-5.
5. Cheung K.et al.(1983). The origin of CL... J. Immunol. 130, 2324-2329.
6. Kitagawa,S. et al. (1980). Evidence that proteases are involved..,J.Clin.Invest. 65, 74-81.
7. Günzler V. et al. (1982). Transglutaminase and polyamine dependence..,FEBS Lett. 150, 390-396.

OPSONO-PHAGOCYTIC VARIATIONS OF SEROTYPES Ia, Ib, II & III OF GROUP B STREPTOCOCCI AS OBSERVED BY NEUTROPHIL CHEMILUMINESCENCE (CL). ROLE OF THE ALTERNATIVE PATHWAY OF COMPLEMENT IN OPSONIZATION OF SEROTYPE III OF GROUP B STREPTOCOCCUS (GBS)

P. Hindocha, Ruth Hill, C.B.S. Wood, Urmila Patel and George Hunt.

The Medical College of St Bartholomew's Hospital & the London Hospital Medical College, Academic Department of Child Health and Department of Microbiology, Queen Elizabeth Hospital for Children, Hackney Road, London, E2 8PS.

INTRODUCTION

In order to better define the role of cell wall components of *GBS* in virulence we have utilised the technique of phagocytic CL to study opsonophagocytosis (an important host defence mechanism) of untreated and enzyme treated NCTC serotypes Ia, Ib, II & III of *GBS*. The role of alternative pathway of complement in defence against *GBS* infection has been evaluated.

MATERIALS AND METHODS

Bacteria

NCTC strains Ia (8190), Ib (8180), II (11079) and III (8184) of human origin (Streptococcal Reference Laboratory, London, UK), after culture in Todd-Hewitt broth (THB) were heat killed, washed and resuspended in Hank's balanced salt solution (HBSS) at 1×10^9 /ml.

Sera. Sera from healthy adults and sick children were used. Classical and alternative complement pathway activity were inhibited by MgEGTA, heating sera to 50°C for 20 mins & 56°C for 30 mins.

Enzymes used. Pancreatic trypsin (Difco Labs UK) and Type V neuraminidase *(Clostridium Perfringens)* Sigma Chemicals.

ISBN 0-12-426290-2

Enzymatic treatment and Opsonization of serotypes of GBS. Serotypes were cultured in THB supplemented with 0.5% trypsin or 0.9u/ml neuraminidase, washed, opsonized and adjusted to 1 x 10^9/ml.

Polymorphonuclear leucocytes (PMNs). Dextran separated PMNs after treating with 0.83% tris buffered ammonium chloride to remove contaminating erythrocytes were washed twice in HBSS and adjusted to 5 x 10^6/ml in HBSS.

Chemiluminescence assay. Luminol dependent phagocytic CL was measured at 37°C in a picolite luminometer (Packard Instruments). Light generated due to phagocytosis of opsonized or unopsonized particles was recorded every 6 mins and a mean value of triplicate results obtained.

RESULTS

Experiments carried out to optimise conditions for maximal CL showed that optimum incubation time was 15 mins. CL increased with increasing serum concentration, however, 14% serum concentration was selected (somewhat arbitrarily) as a convenient concentration that produced significant but not a maximal CL. CL increased as the ratio of bacteria to PMN increased however ratio of 200:1 *(GBS:PMN)* was selected. Visualisation of suspension after CL experiment revealed that positive smears had numerous ingested bacteria which correlated with a high degree of CL whereas negative preparations contained fewer bacteria and diminished CL.

Fig. 1 shows that of pooled serum opsonized serotypes, Ib & II resisted phagocytosis, by contrast serotypes Ia& III were readily phagocytosed.

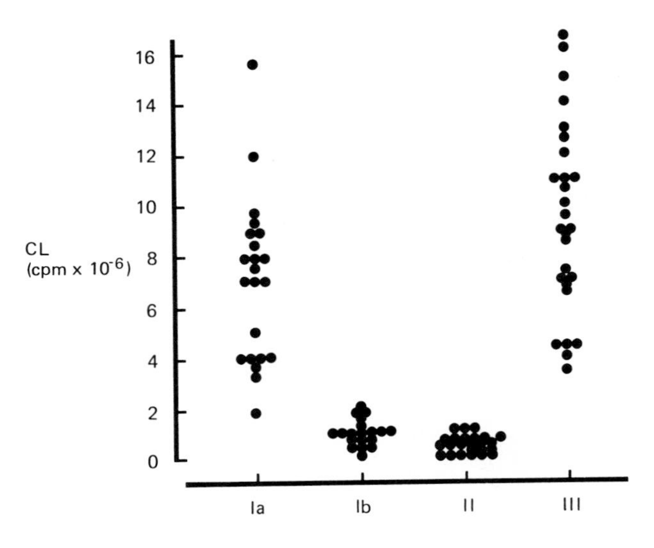

Fig. 1 Variations in the phagocytosis of serotypes by normal adult PMNs.

Fig. 2 shows that serotypes Ib & II resisted opsonization from different sources (different donors) however, Ia & III were uniformly opsonized and phagocytosed with an increased CL.

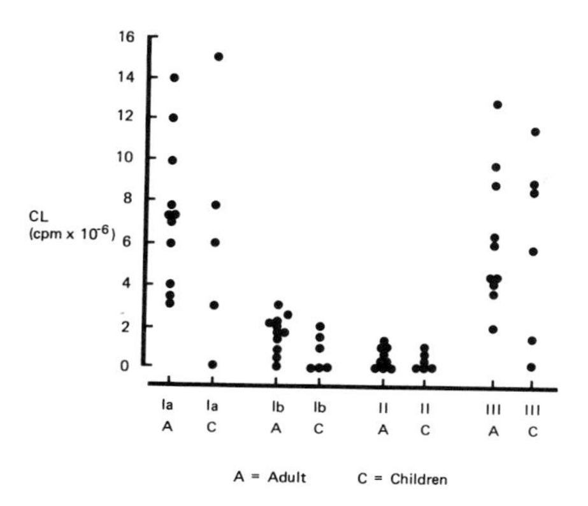

Fig. 2 Serotypes specific variations in opsonization.

Opsonization following neuraminidase treatment had no effect on phagocytosis of all the serotypes tested, however, opsonization following trypsin treatment resulted in increased phagocytosis of all the serotypes though resistant strains were not rendered fully sensitive to opsonization (fig. 3)

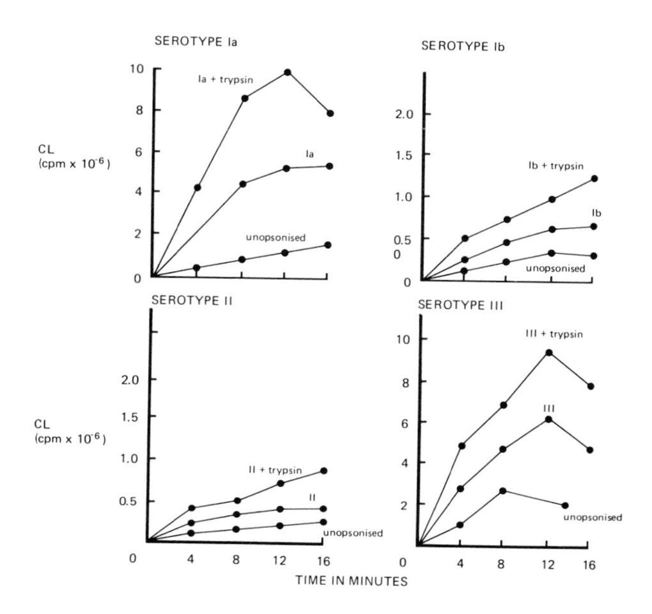

Fig. 3 Effect of opsonization following enzymatic treatment.

Fig. 4 shows that opsonization was little affected by chelation of Ca^{++} using MgEGTA, however, inactivation of factor B by heating serum to 50°C for 20 mins reduced opsonization considerably.

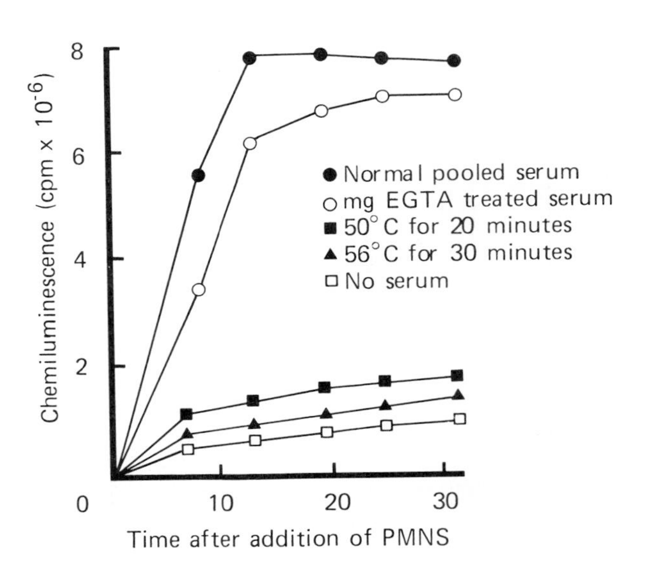

Fig. 4 Effect of MgEGTA chelation of serum, heating serum to 50°C for 20 mins and 56°C for 30 mins on kinetics of opsonization of serotype III.

CONCLUSIONS AND RECOMMENDATIONS

The present studies suggest that protein cell wall components act as anti-phagocytic factors in some strains of *GBS*. Such studies may be useful in determining immunological adjuncts to antibiotic therapy in *GBS* infection. Our studies also show that alternative pathway of complement has an essential role in defence against *GBS* infection.

A COMPARISON OF WHOLE BLOOD AND NEUTROPHIL CHEMILUMINESCENCE IN PATIENTS UNDERGOING CARDIOPULMONARY BYPASS

P. T. Conroy, P. N. Platt, M. J. Elliott

Departments of Cardiothoracic Anaesthesia, Rheumatology and Cardiothoracic Surgery, Freeman Hospital Newcastle Upon Tyne, England

INTRODUCTION

Open heart surgery is becoming safer progressively however, there is still a significant mortality and morbidity to which postoperative infection makes a major contribution. Rates of postoperative bacterial infection are higher in patients who undergo cardiac surgery with cardiopulmonary bypass (CPB) than in patients without CPB (1). Hicks et al (2) demonstrated red blood cell damage during CPB. No clear pattern emerges on the effect of CPB on neutrophil function from the reports in the literature. Reduced chemotaxis, phagocytosis, and killing ability have been reported but findings are not consistent.

Polymorphonuclear neutrophils exhibit an intense luminol dependent chemiluminescence during phagocytosis (3), which has been correlated with neutrophil bacterial killing ability (4). Selvaraj et al (5) reported a micromethod for quantitation of the chemiluminescence response of leucocytes in whole blood. This technique offers a rapid assay applicable to the evaluation of cellular and humoral factors involved in the phagocytic process. We have undertaken a prospective controlled study on the effect of CPB on neutrophil function as measured by whole blood and neutrophil chemiluminescence.

Materials and Methods.

Fifteen adult patients undergoing coronary artery vein graft surgery using a standard anaesthetic and bypass technique were studied . Venous blood samples from a canula in the antecubital fossa or the internal jugular

ISBN 0-12-426290-2

vein were taken at least 24 hours preoperatively, ten minutes after discontinuation of CPB and 24 hours postoperatively. Neutrophil chemiluminescence and whole blood chemiluminescence were measured in each sample. Whole blood chemiluminescence was also measured in venous samples taken before induction of anaesthesia, after induction of anaesthesia, ten minutes after skin incision, five minutes, thirty minutes and sixty minutes after commencement of CPB and at two and seven days postoperatively.

Human Polymorphonuclear Leucocytes (PMNLs) were separated from heparinized whole blood (100 units calcium heparin to 10 ml blood) by dextran sedimentation according to Boyum (6), washed twice in Hanks solution and adjusted to a concentration of $1.5X10^6$ cells per ml Hanks solution.

Neutrophil luminol dependent chemiluminescence was performed by incubation of 1 ml of the above cell suspension with 100 ul of standard luminol solution (20 mg diluted in 1ml of dimethyl sulfoxide , then added to 100ml of phosphate buffered saline) for 5 minutes at 37 C. 100ul of a 1% uniform latex particles suspension were then added as a phagocytic stimulus and the chemiluminescent response measured using an LKB 1250 luminometer.

A whole blood luminol dependent chemiluminescence assay based on a modification of that originally described by Selvaraj et al (6) was performed. Heparinised whole blood (100 units of calcium heparin per 10 ml whole blood) was diluted 1 in 10 in PBS. 100ul of 1% uniform latex particles were then added as a phagocytic stimulus and the response measured . This assay was performed six hours after the blood sample was taken. Previous work in our laboratory on healthy human adult volunteers showed that the maximum whole blood chemiluminescence response was achieved at this time. (fig.1)

Results

Figure 2 shows responses obtained for neutrophil chemiluminescence which was increased significantly postoperatively ($p<0.01$).

To compensate for variations in haematocrit which occurr during CPB, namely marked haemodilution at the onset of and during CPB, whole blood chemiluminescence response was divided by the neutrophil count for that sample .

FIG 1. THE EFFECT OF TIME FROM VENEPUNCTURE TO MEASUREMENT ON WHOLE BLOOD CHEMILUMINESCENCE (20 HEALTHY CONTROL)

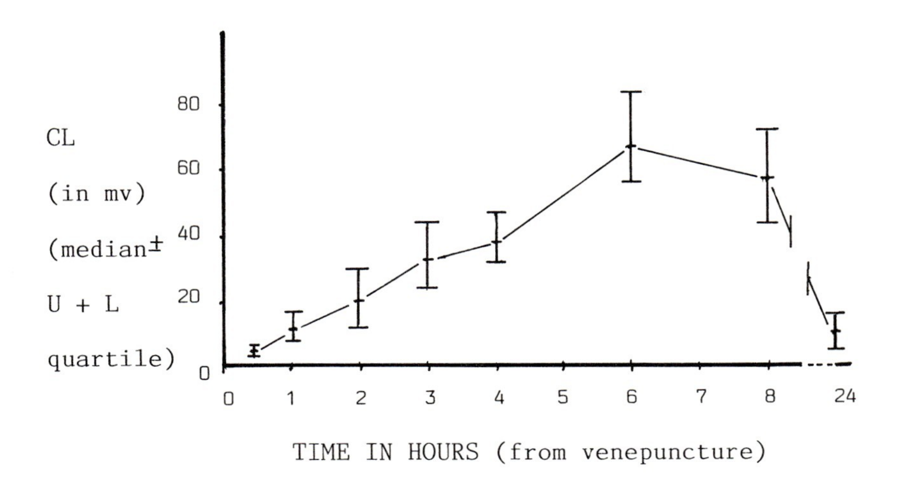

FIG 2. NEUTROPHIL CHEMILUMINESCENCE (CL)

FIG 3. WHOLE BLOOD CHEMILUMINESCENCE CORRECTED FOR NEUTROPHIL COUNT

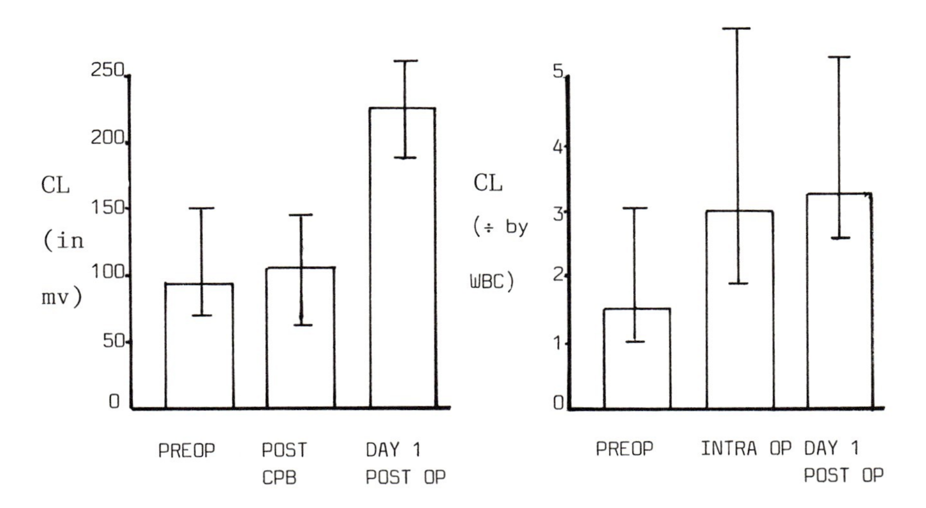

CL results shown as median and upper and lower quartiles

Whole blood chemiluminescence increased intraoperatively ($p<0.01$), and postoperatively ($p<0.01$), as shown in fig. 3.
There was no significant correlation between the results obtained for neutrophil and whole blood chemiluminescence (corr. coeff 0.31 N.S.) or for neutrophil chemiluminescence and whole blood chemiluminescence corrected for neutrophil count (corr. coeff. 0.28 N.S.).

Conclusions.

1. Neutrophil luminol dependent chemiluminescence is enhanced following surgery with CPB.
2. Whole blood luminol dependent chemiluminescence is increased intraoperatively and postoperatively.
3. The poor correlation between neutrophil and whole blood chemiluminescence suggests the techniques measure different parameters.

References.

1. Goodman,J.S. , Schaffner, W. , Collins H.A., et al.(1968). Infection after cardiovascular surgery, The New England Journal Of Medicine vol. 278 no.3, 117-123.
2. Hicks,G.L., Zwart,H.H., Dewall,R.A. (1979) Membrane versus Bubble Oxygenators, The Archives Of Surgery vol. 114 , 1285-1287.
3. Allen,R.C., Loose,L.D., (1976). Phagocytic Activation of a luminol dependent chemiluminescence in rabbit alveolar and peritoneal macrophages. Biochemical and Biophysical Research Communications, 69, 245-52
4. Horan, T.D., English D., Macpherson T.A.,(1982) Association of Neutrophil Chemiluminescence with Microbicidal Activity, Clinical Immunology and Immunopathology 22, 259-269.
5. Selvaraj, R.J., Sbarra, A.J, Thomas, G.B. et al. (1982) A Microtechnique for studying chemiluminescence response of phagocytes using whole blood and its application to the evaluation of phagocytes in pregnancy, The Journal Of The Reticuloendothelial Society 31 3-16.
6. Boyum, A. (1968) Isolation of leucocytes from human blood. Scandinavian Journal Of Clinical and Laboratory Medicine. 97, 77-89

ELUCIDATION OF THE PHAGOCYTOSIS MECHANISM WITH THE AID OF LUMINOUS BACTERIA

Mira Barak,[+] S. Ulitzur[*] and D. Merzbach[+]

[+] *Department of Microbiology, Technion Faculty of Medicine and Rambam Medical Center*

[*] *Department of Food Engineering and Biotechnology, Technion-Israel Institute of Technology, Haifa, Israel*

INTRODUCTION

Phagocytosis of bacteria by leucocytes is accompanied by a burst of non-mitochondrial respiration. Although this oxygen consumption is not required for phagocytosis (1), it is required for the killing of certain bacteria (2,3). In addition, cells from patients with the rare congenital syndrome of chronic granulomatous disease do not show any evidence of the respiratory burst (4) and kill bacteria less efficiently than do normal cells (5).

The efficient microbicidal activity of the polymorphonuclear leucocyte (PMN) depends on the formation of highly reactive products of oxygen metabolism such as superoxide (O_2^-), hydrogen peroxide, singlet oxygen (O_2') and the hydroxyl radical (6,7). The mechanism of activation of the bactericidal system has not been elucidated, and it is not clear if the bacteria are killed extracellularly and then ingested or whether the killing is an intracellular process.

The bacterial luminescence system offers a unique model to study the interaction between the phagocyte and bacteria. The luminescence level reflects both the cellular metabolic activity and the viability of the phagocytised bacteria (8, 9). The bioluminescence test for phagocytosis, being simple rapid and non-destructive to the experimental system, allows a continuous determination of the kinetics of the process. The luminous bacterium *Vibrio cholerae* var. *albensis* resembles the pathogenic *V.cholerae* strains and thus represents a realistic model for phagocytosis. The present communication shows the important role of oxygen in the PMN

ISBN 0-12-426290-2

bactericidal mechanism and presents evidence and estimations for the extent of the extracellular killing.

RESULTS

The effect of PMN on the potential luminescence and viability of *V. cholerae* cells that were incubated under anaerobic conditions was investigated.
Neither the luminescence nor the viability was affected by the PMN under anaerobiosis. The exposure to air after different periods of anaerobic phagocytosis resulted in a prompt burst of luminescence, followed by a rapid decrease in the emitted light. This drop in luminescence occurred only in the presence of PMN; in their absence, the light remained high and steady throughout the experiment.

Microscopic examination revealed that after 1 h most of the bacteria had been ingested during the anaerobic phase. This observation was further strengthened by the fact that most of the bacteria disappeared from the supernatate of both the anaerobic and the aerobic assay systems.
However, while the anaerobically phagocytised bacteria accumulated within the PMN, the aerobically phagocytised bacteria disappeared with time. It should be noted that the method employed for achieving anaerobiosis by addition of dithionite (100 μg/ml) did not interfere with the potential of PMN to act under aerobic conditions.

As we have already shown (9), the fall in the luminescence of *V. cholerae* cells upon the addition of PMN under aerobic conditions began only after a short time lag that was required for the adherence and engulfment process. This pre-ingestion stage did not require the presence of oxygen. Upon shifting from anaerobic to aerobic conditions, the PMN killing apparatus was immediately potentiated, as could be judged from the prompt decrease in luminescence.
Using this data (10) one could estimate the extent of the intra- and extracellular killing of bacteria in the phagocytosis process. There is a difference of 8-10% between the intracellular and the total killing of the bacteria in the phagocytosis assay.

In order to obtain a more direct estimation of the extracellular killing, we performed the following experiments: *V.cholerae* cells were filtered through a 3-μm Millipore filter, and the non-entrapped bacteria were washed out by repeated washings; the entrapped bacteria could not be engulfed because they were trapped in the filter. Approximately 60% decrease in luminescence occurred when such filter-bound bacteria were incubated with PMN for 30 min. This decrease in luminescence could not be accounted for by bacteria that escaped from the filter, since only 0.08% of the bacteria were released from the filter during 1 h of incubation in the absence of PMN (data not shown). Moreover supernate from active PMN were able to reduce the luminescence of opsonised bacteria. It can be seen that the effect of the supernate was proportional to the length of the previous phagocytosis period.

CONCLUSIONS

Phagocytosis of the luminous bacterium *Vibrio cholerae* var. *albensis* caused a similar decrease in both viable count and in the *in vivo* luminescence. The effect of polymorphonuclear neutrophil leucocytes (PMN) on both parameters was an oxygen-dependent process. Exposure of PMN to oxygen caused a prompt decrease in the luminescence of bacteria that had been ingested under anaerobic conditions. As much as 10% of the killing could be attributed to extracellular killing. A decrease in luminescence was caused by cell-free supernates from active PMN suspensions.

Similarly, bacteria entrapped on a membrane filter showed a decrease in luminescence upon addition of active PMN, even though they could not be ingested.

REFERENCES

1. Koch, C. (1978) Bactericidal activity of human neutrophil granulocytes. *Acta Pathologica, Microbiologica, et Immunologica Scandinavica. Section C, Immunology* 226, suppl., 7-62.
2. Inghan, H.R., Sisson, P.R., Middleton, R.L., Narang,H.K., Codd, A.A. and Selkon, J.B. (1981) Phagocytosis and killing of bacteria in aerobic and anaerobic conditions. *Journal of Medical Microbiology* 14, 391-399.
3. Mandell, G.L. (1974) Bactericidal activity of aerobic and anaerobic polymorphonuclear neutrophils. *Infection and Immunity* 9, 337-341.
4. Holmes, B., Page, A.R. and Good, R.A. (1967) Studies of the metabolic activity of leucocytes from patients with a genetic abnormality of phagocytic function. *Journal of Clinical Investigation* 46, 1422-1432.

5. Segal, A.W., Harper, A.M., Garcia, R.C and Merzbach, D. (1982) The action of cells from patients with CGD on *Staphylococcus aureus*. *Journal of Medical Microbiology* 15, 441-449.
6. Babior, B.M. (1978) Oxygen-dependent microbial killing by phagocytes. *New England Journal of Medicine* 298, 659-669. 721-725.
7. Klebanoff, S.J. (1980) Oxygen metabolism and the toxic properties of phagocytes. *Annals of Internal Medicine* 93, 480-489.
8. Barak, M., Ulitzur, S. and Merzbach, D. (1983) Phagocytosis-induced mutagenesis in bacteria. *Mutation Research* 121, 7-13.
9. Barak, M., Ulitzur, S. and Merzbach, D. (1983) The use of luminous bacteria for determination of phagocytosis. *Journal of Immunological Methods* 64, 353-363.
10. Barak, M., Ulitzur, S. and Merzbach, D. (1983) Elucidation of the phagocytosis mechanism with the aid of luminous bacteria. *J. Med. Microbiol* (in press).

OPSONIZATION - CHARACTERISTICS AND MECHANISM WITH THE AID OF LUMINOUS BACTERIA

Mira Barak[+], S. Ulitzur[*] and D. Merzbach[+]

[+]*Department of Microbiology, Technion Faculty of Medicine and Rambam Medical Center*

[*]*Department of Food Engineering and Biotechnology, Technion - Israel Institute of Technology, Haifa, Israel.*

INTRODUCTION

The existing methods for phagocytosis of microorganisms study are based on the direct bactericidal assay, the uptake of radioactive matter or emulsions, the increase in different PMN metabolic parameters or on the specific characteristics of the phagocytized particle. These methods can not provide kinetic information about the short-lasting events occuring during the interaction between phagocytes and bacteria. Recently we have developed a new assay for assessing the different steps of phagocytosis (1). The luminous terrestrial bacterium *Vibrio cholerae* var *albensis* used in this assay resembles the pathogenic *V. cholerae* strains and thus represents a realistic model for phagocytosis. The killing capacity of the phagocytes was determined by the decay of the bacterial *in vivo* luminescence. The luminescence decrease shows a high correlation to both the viable count of the luminous bacteria and to the chemiluminescence produced by the dead luminous bacteria.

The fact that the engulfed luminous bacteria continue to emit light until they are actually killed by the polymorphonuclear phagocytes, together with the technical advantage of the bioluminescence assay being an undestructive test, allow a continuous monitoring of the phagocytosis kinetics.

The present communication uses this bioluminescence (BL) assay to study the kinetics of the phagocytosis process at different PMN/bacteria ratio and to investigate the opsonic requirements and mode of action of this system.

ISBN 0-12-426290-2

RESULTS

The kinetics of the luminescence decrease during the phagocytosis assay shows a typical short delay. After 5-10 min of incubation of the luminous bacteria with PMN the level of the luminescence decreases in a first order type of kinetics for a period of about one hour followed by a slower rate of luminescence decay. It can be revealed that the rate of the decrease in liminescence (K) is almost constant between 10-50 minutes. The value of K, the initial rate of decrease in luminescence (V_0) and the residual luminescence after 60 min of phagocytosis (Lt=60) are dependent on concentrations of both the luminous bacteria and the PMN as well as on the ratio between them. At the same ratio of PMN/bacteria 1:10 or 10:1, increasing the concentration of bacteria & PMN will increase the V_0 and will decrease the L_t=60 .

The concentration of the serum that has been used for opsonization and the period of opsonization affected the kinetics of the phagocytosis. Increasing the concentration of the serum that had been used for opsonization increased the rate of phagocytosis (K) and decreased the ultimate level of luminescence (L_t=60). The duration of the opsonization period had almost no effect on the kinetics of the luminescence although it had a clear effect on the residual luminescence after 60 min of incubation. This discrepancy stems from the fact that the extension of the opsonization period shortened the lag in the active phagocytosis. The effect of the serum concentration and the length of the opsonization period were more pronounced with low ratios of PMN to bacteria than with high ratios of PMN to bacteria.

The efficiency of the opsonization stage depended on the active C3 complement factor in the serum. Heat inactivated serum or EDTA treated serum as well as the removal of immunoglobulins abolished the activity of the PMN on the luminous bacteria. On the other hand the neutralization of the classical complement pathway or the removal of the lysozyme did not affect the PMN activity.

The present study brings evidence that opsonization alters the outer membrane permeability of the gram negative luminous bacteria towards the hydrophobic high molecular weight (MW-1255) (2) antibiotic-actinomycin D.

The effect of the actinomycin D was tested on the *de-novo* synthesis of luciferase in a dark variant of a luminous bacteria. The dark variant *Photobacterium leiognathi* SD-18 has been shown to undergo a prompt induction of the luminescence system upon addition of the DNA-intercalating agent proflavin (3). Fig. 1 shows that actinomycin D has inhibited the proflavin-induced luminescence in opsonized cells, but not in the non opsonized bacteria. The degree of luminescence inhibition was proportional to the actinomycin concentration. Heat inactivated serum (56°C, 30 min) did not increase the susceptibility of *P.leiognathi* SD-18 cells to actinomycin D.

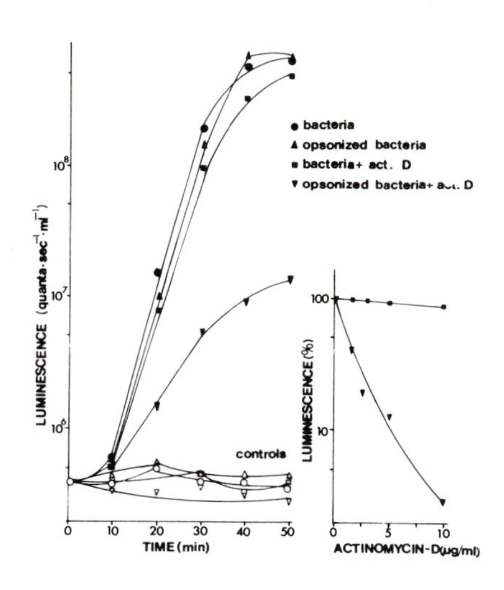

Fig.1 Actinomycin D permeability in opsonized bacteria

P.leiognathi SD-18 cells were grown in ASWRP liquid medium (17.55 g/1 NaCl, 0.75 g/1 KCl, 12.3 g/1 $MgSO_4$ $7H_2O$, 1.45 g/1 $CaCl_2$, 0.075 g/1 $K_2HPO_4 \cdot 3H_2O$, 0.5% peptone, 0.3% yeast extract, 0.02M morpholinopropane sulfonic acid) at 30°C as described by Ulitzur et al.(3). Human serum used was a pool from several healthy donors that was prepared as previously described (1). 10^8 bacteria were opsonized in 5% serum for 30 min at 30°C and washed twice by centrifugation in PBSC buffer (0,8% NaCl,0.12% K_2HPO_4,0.034% KH_2PO_4, 0.15 mmol $CaCl_2$, 0.02 mmol $MgSO_4$, 0.1% bovine serum albumine and 10 mmol glucose) (4). Opsonized and nonopsonized bacteria (10^6 cells/ml) were incubated in 1 ml ASWRP medium containing proflavin (2µg/ml) and different concentrations of actinomycin D. The *in vivo* luminescence was determined with time at 30°C with the aid of a scintillation counter operating without coincidence at ^3H setting. Right side figure gives the percentage of luminescence inhibition by different concentrations of actinomycin D after 60 min of incubation. Controls without proflavin opens symbols, with proflavin solid symbols.

CONCLUSIONS

The bioluminescence test offers an easy and simple method to determine the kinetics of phagocytosis by following the luminescence of the bacteria. Intracellular ingestion of phagocytized luminous bacteria *Vibrio cholerae* var *albensis* caused a decrease in the *in vivo* luminescence, at a rate corresponding to the rate of cell killing. Increased serum concentration enhanced the rate of luminescence decrease while increased opsonization period decreased the lag in the luminescence decrease and increased the maximal decrease in luminescence. At high ratios of bacteria to PMN (100:1), the opsonization period and the concentration of the serum were critical for the luminescence decrease. Immunoglobulins and complement (via both pathways) were necessary for maximal opsonization. Their action for facilitation of phagocytosis is explained by increasing the permeability of the outer membrane of gram negative bacteria, causing a greater susceptibility of the bacteria to the bactericidal effects of different cellular and humoral factors.

REFERENCES

1. Barak, M., Ulitzur, S and D. Merzbach (1983). The use of luminous bacteria for determination of phagocytosis. *J. Immunol. Meth.* 64, 353-363.
2. Leive L. (1965) Actinomycin D sensitivity in *E. coli* produced by EDTA. *Biochem. Biophys. Res. Commun* 18, 13-17.
3. Ulitzur, S., Weiser, I. and S. Yannai (1980) A new, sensitive and simple bioluminescence test for mutagenic compounds. *Mutat. Res.* 74, 113-124.
4. Barak, M., Ulitzur, S. and D. Merzbach (1983) Determination of the serum bactericidal activity with the aid of luminous bacteria. *J. Clin. Microbiol.* 18(2), 248-253.

LECTIN MODIFIED CHEMILUMINESCENCE OF HUMAN MONONUCLEAR CELLS AND MORPHOLOGICAL CORRELATES IN ELECTRON MICROSCOPY

J. Barth[1], U. Welch[2], U. Schumacher[3], and K.G. Ravens[1]

[1]*Department of Internal Medicine, Kiel University, 2300 Kiel*

[2]*Department of Anatomy, Munich University*

[3]*Department of Pathology, University of Kiel, 2300 Kiel*
West Germany

INTRODUCTION

Usually phagocytosis of macrophages is accompanied by a respiratory burst, which can be measured as Chemiluminescence (CL). Earlier investigations (Barth et al (1)) showed that the lectins Concanavalin A (ConA) and wheat germ agglutinin (WGA) induce an early and short burst of metabolic activity in the macrophages. On the subsequent course of the CL-response, however, they acted differently:Zymosan induced CL was inhibited by WGA while ConA increased resulting in a higher and steeper CL-response. In order to investigate whether this phenomenon is due to different incorporation and processing mechanisms of the lectins ferritin-labeled lectins were added to the cells and cells were fixed for electron microscopy and reprocessed.

MATERIAL AND METHODES

Freshly drawn heparinized blood (10 I.E./ml) was diluted with PBS-buffer layered over a Ficoll-Isopaque-gradient and centrifuged for 10 min at 800 g. The mononuclear cell ring was then washed 2 times with Hanks-solution. The number of cells was adjusted to 2000 mononuclear cells/µl. 500 µl of the cell suspension was pipetted into the test tubes of a Berthold Biolumate LB 9505 and 20 µg of luminol, dissolved in 10 µl PBS, was added. Thereupon, 100 µl of ferritin-labeled lectins ConA or WGA-stock solution was added to the samples to give a concentration of 100 µg/ml of ConA and of 10 µg/ml of WGA. For comparison CL during phagocytosis of

ISBN 0-12-426290-2

Zymosan was measured adding of 1 mg Zymosan, suspended in 20 µl of Eagle Hepes medium.
At 3, 15 and 40 min after measurement had started the samples were fixed with 600 µl of a 6%-glutaraldehyde solution in 0,1 m of phosphate buffer for 2 h at 4°C. Then the cell suspensions were centrifuged at 1500 g and embebbed in 6%-Difgo Agar Noble solution. The agar-sediment pieces were repeatedly rinsed in cold phosphate buffer, treated for 2 h with 2% OsO_4, dehydrated in alcohol and embedded in araldite. The thin-sections were contrasted in the usual manner (uranyl and/or lead citrate) and analysed electromicroscopically (Zeiss EM 10).

RESULTS

Both lectins induced an initial respiratory burst which occured some minutes after addition of the lectin to the mononuclear cells (Fig. 1-3), it was considerably higher with ConA compared to WGA. During phagocytosis of Zymosan the picture was different: ConA stimulated the CL-response (Fig. 1), while WGA provoked a pronounced inhibition of the CL-response Fig.(3), while the initial respiratory burst was maintained.
The electron microscopic evaluation presented the following picture. Both lectins demonstrated a labeling of cell membranes after 3 min, with the WGA-labeling being stronger.As early as after 3 min, ferritin-labeled WGA-particles were incorporated into the cell. They could be seen in sub-plasmalemmal vacuoles. At 15 min the vacuoles were strongly labeled by WGA, the cell membranes continued to be labeled (Fig. 4). A similar picture was seen with ConA, only the amount of labeling of cell membranes and vacuoles was considerably weaker. At 40 min the picture resulting was analogous, but a weaker labeling of cell membranes and vacuoles by ConA was seen.
In summary the amount of lectins at the cell membranes was relatively constant during the whole time of measurement. Correlating CL and EM-results one finds that WGA in spite of stronger incorporation into the cell suppresses CL-response whereas incorporation into the cells was weaker with ConA.

CONCLUSION

Although WGA has been used in considerably lower concentration than ConA, WGA resulted in a stronger incorporation into the cells. The reason might be, that the WGA-receptors have a higher density on the cell surface than the ConA-

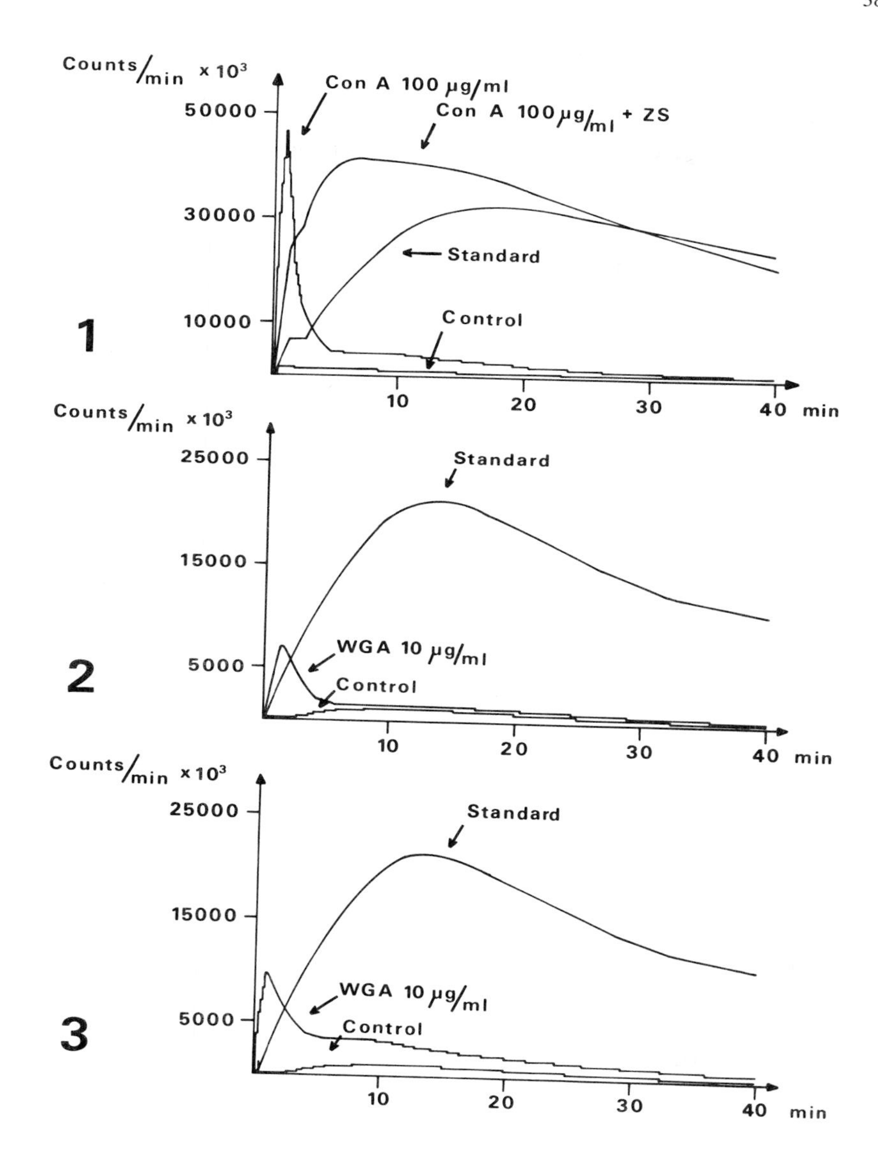

Fig. 1-3: Modulation of CL by ConA and WGA
Figure 1 shows the effects of ConA on CL, figure 2 shows WGA-effects on CL without Zymosan and figure 3 shows the effects of WGA on Zymosan-induced CL.

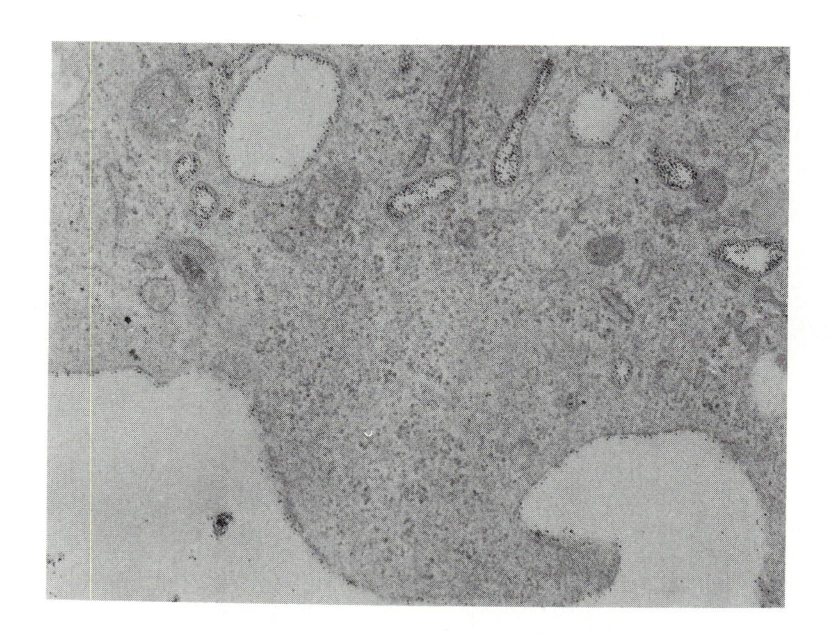

Fig. 4: Electron-micrograph showing ferritin-labeled WGA 15 min after starting measurement of CL. Strong labelling of vacuoles. The outer cell membrane is also labeled.

receptor, another possibility could be a greater affinity of WGA to its receptors in comparison to ConA.The WGA-experiments showed that CL and vacuolar incorporation into the cells were not necessarily correlated with each other, but could also be opposed, as has been demonstrated. A possible cause of this phenomenon might be the influence of lectins on the membraneous Ca^{2+}-metabolism (Weidemann, et. al (2)).

REFERENCES

1. Barth J., Ravens, K.G. and Schumacher, U.(1983):Lectin induced modification of cell metabolic activities as measured by luminol enhanced chemiluminescence in human mononuclear cells.European Journal of Cell Biology, Supp.4,3
2. Weidemann, M.I., Smith, R., Heavy, T. and Alandeen, S. (1980): On the machanism of the generation of chemiluminescence by macrophages. Behring Inst. Mitt. 65, 42-54

A CHEMILUMINESCENT ASSAY FOR MYCOPLASMA IN CELL CULTURES

G. Bertoni[1], R. Keist[2], P. Groscurth[3], R. Wyler[1], R. Keller[2] and E. Peterhans[1]

[1]*Institute of Virology;*
[2]*Immunobiology Research Group, Institute of Immunology and Virology, and*
[3]*Institute of Anatomy, University of Zurich, Switzerland*

INTRODUCTION

Cell culture experimentation has a pivotal role in many areas of biological research, in particular in cell biology, e.g. cell differentiation, proliferation and functional aspects, including interaction between different cells and interaction between cells and exogenous particular matter such as viruses. Bacterial contamination represents a particularly critical complication of cell culture experimentation. As an example, mycoplasma have been shown to be a rather common, not always readily identified contaminant which is able to induce a wide array of alterations in cultured cells (1). We have recently found that the presence of mycoplasma on the surface of cultured cells confers on these cells the ability to induce a burst of CL in various effector cell populations (2,3). In this paper, we have evaluated whether CL measurement may be useful in the screening of cell cultures for mycoplasma contamination.

ISBN 0-12-426290-2

MATERIALS AND METHODS

Cells growing as monolayers were removed from tissue culture flasks by incubation with 0.15% EDTA in Ca^{2+} and Mg^{2+}-free phosphate-buffered saline. Before use in the experiments, the cells were washed twice in Hanks balanced salt solution (HBSS). Human polymorphonuclear leucocytes (PMNL) were isolated from blood using Ficoll-Hypaque and Dextran sedimentation. PMNL were suspended at 1×10^5/ml in phenol red-free HBSS buffered with 25 mM HEPES and supplemented with 7.5 µl/ml of 4% bovine serum albumin saturated with luminol. Aliquots of 0.75 ml of this cell suspension were dispensed in scintillation vials and CL was measured in a liquid scintillation spectrometer as described (2). Conditions for culture of mycoplasma were as in (3).

RESULTS and DISCUSSION

The present study has shown that mycoplasma-contaminated, but not mycoplasma-free, tumor and non-tumor cells maintained in continuous culture consistently triggered a CL burst in various effector cell populations. These included rat and mouse spleen cells, rat peritoneal cells activated with

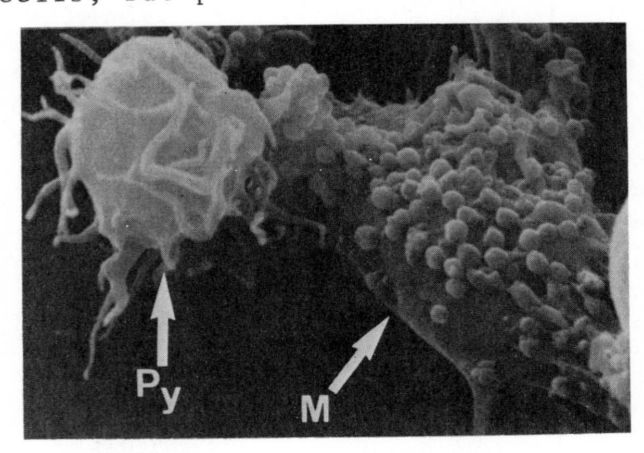

Fig. 1. Macrophage (M) with numerous round-shaped mycoplasma at the surface; note close contact to a Polyoma cell (Py). x 6600.

Corynebacterium parvum, bovine, horse and human PMNL. Electron microscopic studies revealed that mycoplasma organisms became associated with the surface of phagocytic cells and were internalized (Fig. 1 (3)). Of the cell populations that produced CL when exposed to mycoplasma-contaminated target cells, human PMNL consistently showed the highest light response. For this reason, human PMNL were used to further evaluate CL measurement as a mycoplasma screening assay. A typical example for the CL burst elicited by mycoplasma-contaminated target cells is shown in Fig. 2. BW cells evoked in PMNL a marked burst of CL beginning after a lag period of approximately 10 minutes and reaching peak activity between 30 and 40 minutes. Contamination of the target cells was verified by growth of colonies typical for mycoplasma on PPLO agar as described (3).

Subsequent screening of 52 cell lines for their capacity to elicit CL and for growth of mycoplasma

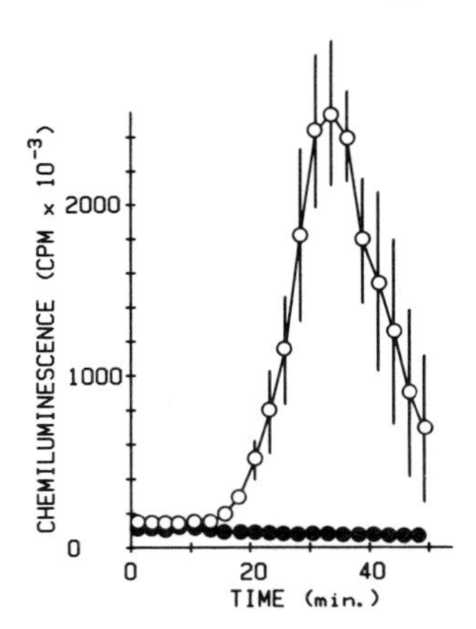

Fig. 2 CL induction in human PMNL by mycoplasma-contaminated BW cells (○). BW cells were added (final effector/target ratio 1:1) to 75,000 PMNL suspended in 0.75 ml HBSS and CL was measured in a LSC as described in (2); (●) PBS.

on PPLO agar has shown that 26 lines stimulated CL; in 22 of the 26 lines contamination was verified by growth of colonies typical for mycoplasma on PPLO agar. We are currently investigating whether a positive result in CL induction by cultures negative for mycoplasma on PPLO agar reflects a difference in the sensitivity of the two assays. Alternatively, other changes on the surface of the target cells, e.g. antigens of paramyxoviruses could lead to a positive result in the CL assay (Bertoni and Peterhans, unpublished observation). The fact that in no case mycoplasma were isolated from cell cultures that failed to evoke CL clearly demonstrates, however, that CL measurement is at least as sensitive as culture on PPLO agar in detecting mycoplasma contamination. Results are available within a few hours compared to 4 - 6 days with PPLO agar. The CL assay, as described in this paper, can also be performed in a conventional liquid scintillation spectrometer after switching off the coincidence circuit; it does not require additional equipment and can readily be performed in most research laboratories.

ACKNOWLEDGEMENTS

This work was supported by the Swiss National Science Foundation Grants No. 3.429.0.83 and 3.696.0.83. We thank Ms. T. Arnold and Mr. B. Scheier for expert technical and Ms. C. Gerber for secretarial assistance.

REFERENCES

1. McGarrity, G.J. *et al.* (1978). Mycoplasma Infection of Cell Cultures, Plenum Press, New York.
2. Peterhans, E. *et al.* (1984). *Eur.J.Immunol.* 14, 201-203.
3. Köppel, P. *et al.* (1984). *J.Immunol.*, in press.

MACROPHAGE CHEMILUMINESCENCE ACTIVATION BY BOVINE GAMMA-GLOBULINS

P. De Sole, S. Lippa, F. Meo, [+]G. De Sanctis, [+]P. Di Nardo, [+]A. Falera, [+]C. Giordano, [+]P. Massari

Department of Clinical Chemistry, [+]Department of Radiobiology-School of Medicine - Universita' Cattolica del Sacro Cuore - L.go A. Gemelli, 8 - 00168 ROMA (Italy)

INTRODUCTION

Immunological response involves macrophage activity. Mosier and Coppelson (1) reported evidence for a ternary cellular complex in the induction of antibody synthesis by certain antigens, and the role of macrophages in delayed-type hypersensitivity is well known (2).

Bovine gamma-globulins (BGG) induce a tolerance in several strains of mice through some effects on both humoral and cellular immunological factors (3). Lukic et al. (4) showed that different susceptibility to the induction of tolerance was related to macrophage function during the Xinduction phase of the immunological process.

Macrophages can be studied by means of their metabolic activation following the phagocytic stimulation and different indexes can be considered (phagocytosis, oxygen consumption, superoxide production, chemiluminescence).

The aim of the present paper is to study the effect of BGG peritoneal administration on the chemiluminescence (CL) response of mouse macrophages to phagocytic stimulus (opsonized zymosan).

MATERIALS AND METHODS

BGG (Fidia Pharmaceuticals Padua, Italy) and bovine serum albumin (BSA) (BDH Chemical Co.) were dissolved in phosphate buffer saline (PBS) at 0.5 mg/ml.

Thioglicollate medium (TH) (Difco Laboratories) was dissolved in H_2O (29.8 mg/ml) and autoclavated for 15

ISBN 0-12-426290-2

minutes at 120 `C.

Luminol (Sigma Chemical Co.) was dissolved in dimethylsulfoxide at 1.0 mM concentration.

Zymosan was prepared from baker's yeast according to Hadding et al. (5). Opsonization was achieved by adding 1 ml of heparinized human plasma from healthy donors to 0.05 ml (25 mg) of zymosan mother solution. After 45 minutes incubation time at 37`C, the suspension was washed with and resuspended in 1.0 ml of a modified Krebs-Ringer phosphate medium (KRP) (6).

Mice Swiss (mean weight 25 $\pm$ 3 g each) were divided in four groups of 10 animals each. Three groups were injected for a week with a daily 0.1 ml i.p. dose of PBS (group 1), BSA (group 2) and BGG (group 3). The fourth group was treated with a single 0.75 ml i.p. dose of TH 4 days before CL determination.

24 hours (groups 1-3) or 72 hours (group 4) after last i.p. injection the animals were killed and peritoneal cells were harvested washing the peritoneum of each mouse by means of 5 ml of PBS.

Macrophage chemiluminescence activity was evaluated by a modified method of De Sole et al. (6). Each scintillation vial contained about 0.5 x 10^6 cells, 10^{-5}M luminol, 0.5 mg of opsonized zymosan in 1.0 ml final volume with KRP. CL was analyzed both before and after the addition of opsonized zymosan with a beta counter (Packard Instrumets mod. 3385) in the out of concidence mode and in the tritium setting.

Specific activities (cpm/cell) were calculated as the ratio of cpm values at the plateau and the number of cells in each vial.

RESULTS

The number of cells recovered (ca. 10^7 cells/animal) was the same in all groups considered except in the case of TH group (30% higher). Macrophage percentage was the same in all series studied (ca. 70%).

Fig. 1 reports both resting and stimulated specific activities of different mice groups.

No significant difference was found in resting activity of the different groups of mice, while zymosan stimulated activity of the macrophages from BGG treated mice was notably increased.

Maximum values obtained after seven doses of BSA and BGG were 0.238 and 1.475 cpm/cell respectively. Macrophage activation obtained four days after a single i.p. dose of TH was not significantly different in respect to that obtained after seven doses of BSA.

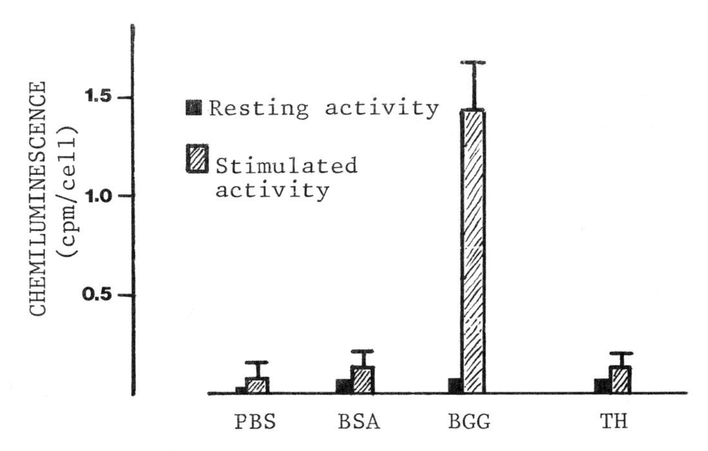

Fig.1 Resting and stimulated activities of peritoneal macrophages from different groups of mice. CL values are the mean of two different determinations for each mouse (10 mice/group).

DISCUSSION

Macrophage relevancy in the induction of the immunological response was extensively studied and a number of macrophage functions clarified (7). In particular Lukic et al. (4) described a lack of tolerance induction in mice pretreated with carrageenin, a macrophage toxic agent. Klaus (8) demonstrated that macrophage bound-antigens are much more immunogenic than the equivalent dose of the same soluble antigen.

On the other hand a relationship between gamma globulin and phagocytes has been reported. Lukic et al. in the already quoted paper (4) evidenced the importance of BGG in the tolerance induction of mice; Najjar (9) reported a stimulatory effect of gamma globulin on the phagocytic activity of polymorphonuclear leukocytes.

Our study was undertaken to focus the relationship between BGG administration and phagocytic activity of peritoneal mouse macrophages studied by their CL response to opsonized zymosan.

The results we obtained are indicative that all the treatments we used (BSA, BGG, TH) were unable to induce a macrophage stimulation "per se": in fact resting activities were not significantly different from control group.

On the contrary BGG treatment induces an enhancement of macrophage phagocytic and/or metabolic capability measured by the increased CL response to zymosan stimulation. BGG could thus be considered as activators of the metabolic capabilities of macrophages that can therefore respond in a more prompt and efficient way against external stimuli.

REFERENCES

1. Mosier D.E. and Coppelson, L.W. (1968) A three-cell interaction required for the induction of the primary immune response in vitro, Proc. Nat. Acad. Sci. U.S.A., 61, 542-547.
2. Eisen, H.N. (1973) Cell-mediated hypersensitivity and immunity. In "Microbiology" (Eds. B.D. Davis, R. Dulbecco, H.N. Eisen, H.S. Ginsberg, W. Barry Wood Jr.) pp. 557-595. Harper and Row, New York.
3. Chiller, J.M. Habicht G. S. and Weigle W.O. (1970) Interrelationships of tolerance and immunity. 1.Induction of immunologic tolerance in adult mice to bovine gamma-globulin purified in vivo, J. Immunol. 105, 370-388.
4. Luckic M.L., Cowing C. and Leskowitz S. (1975). Strain differences in case of tolerance induction to bovine gamma-globulin: dependence on macrophage function, J. Immunol. 114, 503-506.
5. Hadding U., Bitter-Suermann D. and Neichert F. (1970). A tool for the detection of C'6 deficencies. In "Protides of biological fluids".(Ed. H. Peeters) Vol.17 pp. 319-321. Pergamon Press, Oxford.
6. De Sole P., Lippa S. and Littarru G.P. (1982). Resting and stimulated chemiluminescence of polymorphonuclear leukocytes: a clinical approach. In "Biochemistry and function of fagocytes" (Eds. F. Rossi, P. Patriarca), pp.591-601. Plenum Press, New York.
7. Gordon S. S. (1978) Assays for macrophage activation and secrection products. In "Handbook of experimental Immunology" (Ed. D.M. Weir) Vol. 2, pp. 33.1-33.14. Blackwell Scientific Publications, Oxford.
8. Klaus G.G.B. (1974). Generation of thymus-derived helper cells by macrophages-associated antigen, Cell Immunol. 10, 483-488.
9. Najjar, W.A. (1974). The physiological role of gamma-globulin. In "Advances in enzymology" (Ed A. Meister). Vol. 41, pp. 129-178. An Interscience Publication, New York.

KINETICS OF LIPOXYGENASE REACTION MEASURED BY CHEMILUMINESCENCE

Esa-Matti Lilius, Pekka Turunen and Simo Laakso

Department of Biochemistry, University of Turku, SF-20500 Turku, Finland

INTRODUCTION

Lipoxygenase-1 from soybean is a dioxygenase or a peroxidase depending on the availability of O_2. Aerobically, fatty acids with the 1,4-cis,cis-pentadiene structure are converted to the corresponding conjugated hydroperoxides. In the absence of O_2, and in the presence of fatty acid substrate the product hydroperoxides are converted to a complex mixture of oxodienoic acids, n-pentane and fatty acid dimers (1). It has been proposed by several authors that oxidative radical species are formed during lipoxygenase catalysis. It is not therefore surprising that the lipoxygenase-catalyzed oxygenation of linoleic acid proved to be a source of chemiluminescence (CL) (2).

The present paper describes the kinetic characteristics of the linoleate-lipoxygenase-1 system measured luminometrically in the presence and absence of luminol.

MATERIALS AND METHODS

Soybean lipoxygenase-1 (EC 1.13.11.12) containing 170 U/mg protein was purchased from Sigma, purified further and tested for purity as described previously (3). 5-Amino-2,3-dihydrophtalazine-1,4-dione (luminol) was from Sigma. Linoleic acid (>99%) came from Fluka. Linoleate substrate (10 mM) was prepared in Tween 20 by the Surrey method (4). The content of hydroperoxide in the linoleic acid solution was tested spectrophotometrically at 234 nm. A molar extinction coefficient of 25000 was used in calculations. The measurements of CL were performed as described previously (2) using a LKB 1250 Luminometer. The assay mixture

ISBN 0-12-426290-2

contained in 1.0 ml of 0.2 M sodium borate buffer (pH 9.0) $1x10^{-8}$ - $2x10^{-6}$ mol of luminol and 0.1-100ug of lipoxygenase in a 3.5 ml-Ellerman plastic tube. The reaction was initiated by adding $1x10^{-7}$-$1x10^{-6}$ mol of linoleate.

RESULTS AND DISCUSSION

Since the emission of CL during the lipoxygenase-catalyzed oxygenation of linoleic acid is based on the formation of short-lived radicals also the temporal reaction curves measured luminometrically are far more complex than those obtained by spectrophotometry or by oxygraphy. Therefore, the amounts of the enzyme, the fatty acid and dissolved O_2 as well as the characteristics of the buffer influence not only on the intensity but also on the pattern of the light emission. Moreover, luminol enhances the emission by several hundreds of fold . Depending on the reaction conditions a multiphasic reaction curve with 1-3 maxima can be observed.

Phase 1.

When the concentration of linoleate is chosen below that of dissolved O_2 (about 240 uM at 25^oC) only one maximum is observed. If in addition the amount of the enzyme is low, (< 0.5 ug), the intensity of the light increases linearly with time, then levels to a constant emission or slowly decreases. Increments in enzyme concentration enhance both the formation and the decay of the light and the curve becomes flash-type at high enzyme concentrations. The rate of formation of light in phase 1 offers a basis for a sensitive lipoxygenase assay (2). The peak height is dependent on the fatty acid concentration providing a possibility for the assaying of picomolar quantities of 1,4-cis,cis-pentadiene fatty acids directly in aquous solutions (5). Luminol at its optimum concentration ($3x10^{-5}$M) enhances the intensity of the first phase emission about 400-folds. $1.5x10^{-4}$M luminol inhibits the emission by 50%.

Phase 2.

When the concentration of linoleate is higher than that of O_2 an emission curve with two maxima appears (Fig.1). The rate of increase of the emission in phase 2 is always lower than that in phase 1. Otherwise the rate of increase and decrease of the emission are in both phases dependent on the amount of entzyme and the peak heights are proportional to the concentration of linoleate. The maximum of phase 2 appears upon exhaustion of O_2. The emission is generally reduced very rapidly especially when compared to its rate of increase.

If the amount of the enzyme is high (> 20 ug), the aerobic reaction is fast enough to produce, instead of two maxima, only the peak of phase 2 with a shoulder or an inflection point remaining from the phase 1 reaction. On the other hand, if the amount of the enzyme is low (< 0.5 ug) the second maximum does not appear despite of the sufficient amount of linoleate.

Luminol enhances also the phase 2 emission by several hundreds of folds. However, the optimum concentration is around $2x10^{-4}$M, i.e. at a region where the phase 1 emission is partly inhibited.

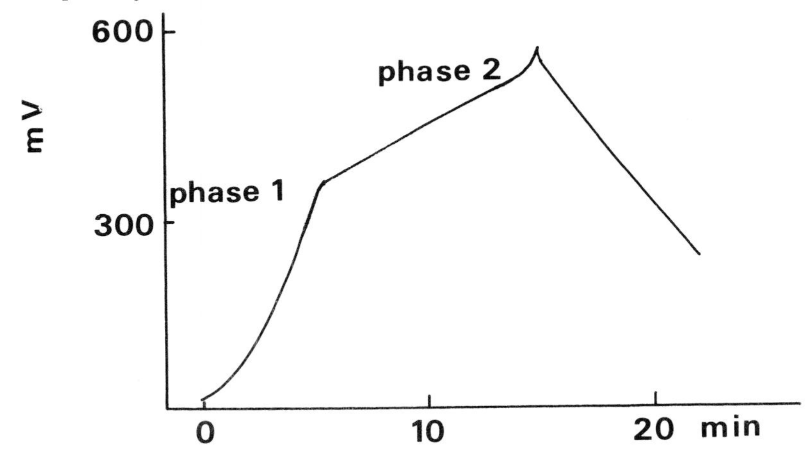

Fig. 1 CL emission from the linoleate-lipoxygenase-1 reaction. Phases 1 and 2. Reaction mixture: 2 ug of enzyme, 400 umol of linoleate and 40 umol of luminol in 1.0 ml of borate buffer, pH 9.0.

Phase 3.

The aerobic lipoxygenase reaction (phases 1and 2), is followed by a phase 3 emission at higher luminol concentrations (optimum $1.2x10^{-3}$M, inhibition of 50% at $2x10^{-3}$M), (Fig. 2). The height of this peak is also proportional to the initial concentration of linoleate and the rate of decrease of the emission is dependent on the amount of the enzyme The increase of the emission with time is often difficult to detect because the peak is more or less of a shoulder type. At very high enzyme or substrate concentrations the third peak is detectable even without luminol.

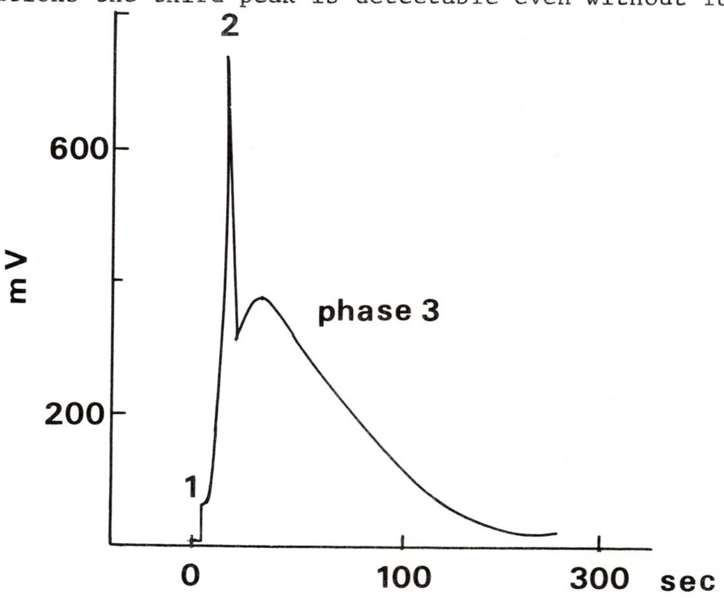

Fig. 2 CL emission from the linoleate-lipoxygenase-1 reaction. Phases 1,2 and 3. Reaction mixture: 40 ug of enzyme, 400 umol of linoleate and 1.4 mmol of luminol in 1.0 ml of borate buffer, pH 9.0.

REFERENCES

1. Veldink, G.A., Vliegenthart, J.F.G. and Bolding, J.(1977). Plant lipoxygenases, *Prog. Chem. Fats other Lipids* 15, 131-166.
2. Lilius, E-M. and Laakso, S. (1982). A sensitive lipoxygenase assay based on chemiluminescence, *Anal. Biochem.* 119, 135-141.
3. Axelrod, B., Cheesbrough, T.M. and Laakso, S. (1980). Lipoxygenase from soybeans, *Methods Enzymol.* 71, 441-451.
4. Surrey, K. (1964). Spectrophotometric method for determination of lipoxidase activity, *Plant Physiol.* 39, 65-70.
5. Laakso,S., Lilius, E-M. and Turunen, P. (1984). Determination of cis,cis-methylene interrupted polyunsaturated fatty acids in aqueous solution by lipo-oxygenase chemiluminescence, *J. Biochem. Biophys. Meth.* 9, in press.

CHEMILUMINESCENCE EMISSION FROM ENRICHED FRACTION OF HUMAN NATURAL KILLER CELLS

Esa-Matti Lilius[+], Kaija Laurila[+],
Jaakko Uksila[§] and Olli Lassila[§]

[+]*Department of Biochemistry, University of Turku, SF-20500 Turku, Finland*
[§]*Department of Medical Microbiology, University of Turku, SF-20500 Turku, Finland*

INTRODUCTION

Natural killer (NK) cells are identified as a morphological subpopulation of white blood cells called large granular lymphocytes. They are cytotoxic cells belonging to the cell class responsible for cellular cytotoxicity without prior sensitization. The characteristics of NK cells are reviewed in (1). Although nonadherent and nonphagocytizing they share a number of features with macrophages and polymorphonuclear leukocytes (PMNL), e.g. having some common cell surface markers and expressing surface receptors for the Fc portion of IgG. NK cells are thought to play a role in anti-tumor surveillance and they are supposed to have activity against microbial agents (virus-infected targets, parasites, fungi or bacteria). The question is posed (2), "Is the NK cell a true lymphocyte or is it nothing but a phagocyte in lymphocyte´s clothing?"

Triggering the activities of macrophages and PMNL by phagocytosable particulates, soluble immune complexes or non-specific soluble stimulants is known to generate chemiluminescence (CL) emission (3). In this work we wanted to find out if similar burst of CL could also be seen when NK cells bind to their target cells.

MATERIALS AND METHODS

Separation of mononuclear cells

ISBN 0-12-426290-2

Heparinized venous blood from healthy volunteers was diluted 1:2 with 0,9% NaCl and 30 ml of this suspension was carefully layered on 15 ml of Ficoll-Isopaque (Pharmacia) and centrifuged at 550xg for 20 min at 4°C. The interface containing mononuclear cells was collected.

Enrichment of NK cells

After washing mononuclear cells were suspended in RPMI 1650 supplemented with 10% fetal calf serum and incubated for one hour at 37°C on plastic culture dishes to remove adherent monocytes. Nonadherent cells were washed and suspended in HBSS buffer and fractionated using a 5-step discontinuous Percoll (Pharmacia) density gradient. The top layer consisted of 40% Percoll (v/v) diluted in RPMI 1640. Each 2.5-ml layer varied by 5% Percoll. The gradient was loaded with 10^{8} mononuclear cells and centrifuged at 550xg for 30 min at 20°C.

In latter experiments adherent cells were removed by using Sephadex G-10 column prior to fractionation with Percoll. $3x10^{7}$ cells were added in a 10-ml syringe containing 8 ml of Sephadex G-10 in RPMI. The suspension was incubated at 37°C for 30 min with mixing. The cells were eluted with 50 ml of warm (37°C) RPMI, washed twice with HBSS and fractionated further in Percoll gradient.

Target cells

K 562 tumor cells, a human erythroleukemic cell line, served as a target.

Cytotoxicity assay

$2x10^{6}$ K 562 cells were labeled with 300 uCi of ^{51}Cr-radioisotope (Radiochemical Center, Amersham) at 37°C for 60 min and washed twice in RPMI. The cytotoxicity assay was performed by using round-bottom Lindbro microplates. $5x10^{3}$ target cells/well and 4 different dilutions of effector cells producing effector:target cell ratios of 40:1, 20:1, 10:1 and 5:1 in a final volume of 0.2 ml in triplicate samples were incubated at 37°C for 4 hours. The supernatants were collected with a Titertec Supernatant Collection System and counted in a LKB Wallac gamma counter. Cytotoxicities were calculated from the following formula:

$$\% \text{ cytotoxicity} = \frac{\text{experimental release- spontaneous rel.}}{\text{maximal release - spontaneous release}} \times 100.$$

Maximal release was estimated by incubating labelled K 562 cells in 5% sodium dodecyl sulphate solution.

Chemiluminescence assay

NK cells in HBSS buffer (500 ul) and 20 ul of 10 mM luminol (Sigma, dissolved in 0.2M sodium borate buffer, pH 9.0) were incubated at 37°C for 15 min in 3.5 ml-Ellerman plastic tubes. A cold suspension of target cells (100ul) was added to the vial and CL was recorded using LKB Wallac 1250 Luminometer equipped with an LKB Bromma 2210 one-channel recorder held at 37°C. Each vial was measured several times at intervals over a period of 1 hour between keeping in the dark at 37°C.

RESULTS AND DISCUSSION

Fig. 1 shows typical results from the enrichment of NK cells by Percoll density gradient fractionation of nonadherent

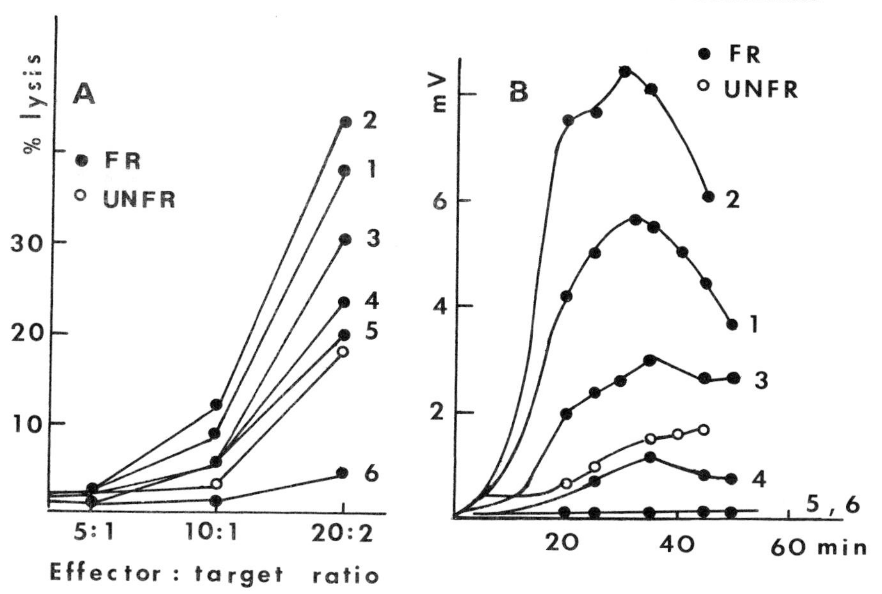

Fig. 1 The enrichment of NK cells by fractionation on Percoll density gradients. Cytotoxic activity (A) and CL emission (B) of each fraction together with the unfractionated cells. In CL experiments the effector:target ratio was 1:1 (10^6 cells).

mononuclear cells.There was a striking parallellism between the cytolytic activities and the CL levels of the fractions fraction number 2 yielding frequently the highest values in both assays. The maximum of the CL emission was usually reached after 30 min of incubation. It was observed that the maximum CL was linearly proportional to the effector:target ratio up to ratio of 20:1 then slowly smoothing when the number of target cells was held constant (10^6 cells). Effector:target ratio 1:10 shortened the time required to reach the maximum CL by about 40% as compared to the ratio of 1:1 (constant effector number of 10^6 cells). In the CL experiments shown in Fig. 1 the ratio was 1:1.

The NK cell fractions obtained in discontinuous Percoll density gradient centrifugations did not elicit CL at all during incubation with K 562 cells when the adherent cells were removed by eluting the mononuclear cell suspension obtained in the Ficoll-Isopaque enrichment through a Sephadex G-10 column. The cytolytic activities, however, were normal. Yet, these cells when incubated along with a cell fraction where the adherent cells were not removed this way caused a CL response twice as high as that of latter cell fraction alone. Similarly, the PHA stimulated cultured NK cells and NK cells enriched by Fluorescence-Activated Cell Sorter (gifts from Dr. T.Timonen,Helsinki) caused no CL emission during incubation with K 562 cells but increased the CL responses of other mononuclear cell preparations.

These results suggest that the CL emission can be used as a measure of the enrichment of NK cells. However, not all cell populations possessing the NK activity are able to generate a CL emission without a synergetic function with other mononuclear cell populations.

REFERENCES

1. Heberman, R.B. and Ortaldo, J.R. (1981). Natural killer cells: Their role in defences against disease, *Science* 214, 24-30.
2. Babior, B.M. and Parkinson, D.W. (1982). The NK cell: A phagocyte in lymphocytes clothing? *Nature* 298, 511
3. Baxter, M.A., Leslie, R.C.Q. and Reeves, W.G. (1983). The stimulation of superoxide anion production in guinea-pig peritoneal macrophages and neutrophils by phorbol myristate acetate, opsonized zymosan and Ig G2-containing soluble immune complexes, *Immunology* 48, 657-665.

A VERY SENSITIVE AND RAPID CHEMILUMINESCENCE METHOD FOR THE MEASUREMENT OF PHAGOCYTOSIS

Esa-Matti Lilius and Matti Waris

Department of Biochemistry, University of Turku, SF-20500 Turku, Finland

INTRODUCTION

Numerous advances have recently been made in the diagnosis of specific immunodeficiency disorders. Screening tests are available for each component of the immune system. Quantitative chemiluminescence (CL) is especially suited for evaluation of opsonization, leukocyte stimulation and phagocytosis where the use of CL measurements has several advantages over conventional methods (1-3). In order to facilitate the measurement of multiple samples in properly controlled conditions and reduce the operator time we have developed an automated luminometer setup for continuous and simultaneous measurements of up to 25 samples (4). By using this system we have investigated the CL emission response as a function of the number of cells from different leukocyte preparations and whole blood phagocytizing preopsonized zymosan.

MATERIALS AND METHODS

Reagents

Hanks balanced salt solution (HBSS buffer, pH 7.4) without phenol red was purchased from Flow Laboratories. Luminol (10 mM stock solution was prepared in 0.2 M sodium borate buffer, pH 9.0) and gelatin came from Sigma and dextran 500 from Pharmacia.

Zymosan (Sigma) was prepared by boiling 40 mg ($2x10^{10}$ particles) in 2 ml of HBSS for 20 min and washing twice with HBSS. Zymosan was opsonized by resuspending the cells to 2 ml of 65% serum of a healthy person in HBSS and incubating at 37°C for 45 min. The opsonized suspension was washed

ISBN 0-12-426290-2

twice with HBSS and resuspended to 20 mg/ml in HBSS. The suspension was kept at -70°C.

Phagocytic cells

The buffy coat of healthy blood donors obtained from the Finnish Red Cross Blood Transfusion Service, Turku, Finland was the source of human neutrophils (PMNL) and mononuclear cells (MNL) which were prepared as separate bands by using a modification of a single-step Percoll (Pharmacia) density gradient centrifugation method (5). Ten ml of buffy coat was added to 10 ml of 7.2% citrate in 0.9% NaCl. This suspension was carefully layered on the discontinuous Percoll gradient consisting of 10 ml of 76% stock isotonic Percoll (SIP) and of 10 ml of 60% SIP in 50-ml Falcon tubes (Becton Dickinson Co). SIP was prepared by mixing Percoll and 1.5 M NaCL 9:1. The desired dilutions of SIP were made in HBSS. The gradients were centrifuged in swing-out rotor at 400xg for 30 min at 8°C. After washing the cells were suspended in HBSS buffer containing 0.1% gelatin.

Unfractionated leukocytes were prepared from the buffy coat by sedimenting erythrocytes with 1.5% dextran at 37°C for 90 min. Leukocytes were separated from plasma by centrifugation and suspended in HBSS buffer after washings.

Heparinized whole blood from healthy donors were used in whole blood studies.

Measurement device

Automated luminometer setup (LKB Wallac 1251 Luminometer connected to Olivetti M 20 microcomputer) was used as described (4).

Assay procedure

Cold suspensions of reagents and cells were pipetted into the 4-ml polypropylene sample vials outside the luminometer. The reaction commenced when the contents of the vials warmed up in the temperature controlled sample carousel of the instrument. CL emission was measured at 37°C for 60 min in a volume of 500 ul of HBSS buffer including $4x10^{-4}$M luminol, 0.1% gelatin, 100 ug of preopsonized zymosan and different numbers of phagocytic cells.

RESULTS AND DISCUSSION

Our test system proved to be very sensitive since a sample

diluted to contain 50 PMNL (95% pure) gave rise to a reliably measurable CL signal (peak value 2x background). Maximum CL was linearly proportional to the cell number up to $5x10^5$ cells then slowly smoothing (Fig. 1). Mononuclear cells (98% pure) gave CL responses about 15% of those of PMNL corresponding to the portion of monocytes in the mononuclear cell population. Similarly, the responses obtained from unfractionated leukocytes corresponded to the proportion of

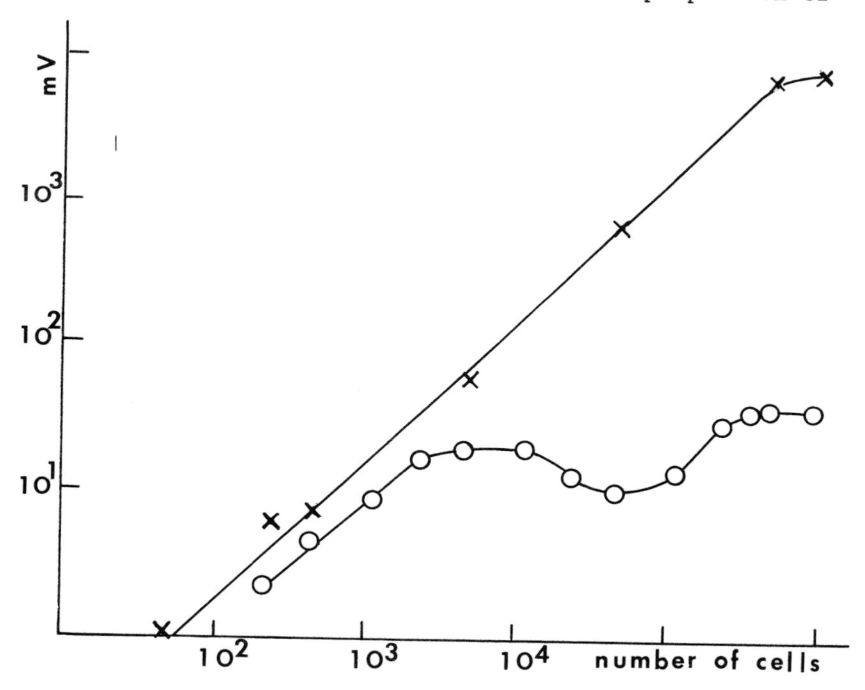

Fig. 1 Maximum chemiluminescence response of zymosan phagocytosis as a function of number of phagocytic cells.
(x - x - x) purified polymorphonuclear leukocytes
(o - o - o) whole blood samples

phagocytic cells in that cell population. The linear range was considerable large, at least 4 orders of magnitude, facilitating the measurement of samples with very differing numbers of responding cells.

On the contrary, the maximum CL response from whole blood samples corresponded to that of purified phagocytes only in higly diluted suspensionsincluding 2000 or less of phagocytic cells (Fig. 1). It was obvious that red blood cells quenched the CL emission causing the smoothing of the CL response vs. cell number curve. The behaviour of this curve at larger cell numbers could be at least partly due to the effects of plasma on the CL response. We have noticed that increments of serum to isolated leukocytes initially increased the CL responses of the cells until a serum concentration of 1% and about 5-fold activation had been reached, then the response declined. This observation parallels the results obtained with plasma (6). The effects of plasma and erythrocytes can be avoided by using highly diluted whole blood samples. Our test system allows even the use of nanoliter quantities of whole blood. Dynamic range is more than one order of magnitude rendering possible e.g. the distinction of carrier states from patients and controls in chronic granulomatous disease (7).

REFERENCES

1. Allred, C.D., Shigeoka, A.O. and Hill, H.R. (1979). Evaluation of group B streptococcal opsonins by radiolabeled bacterial uptake, *J. Immunol. Meth.* 26, 355-363.
2. Werkmeister,J., Helfland,S., Roder, J. and Pross, H. (1983). The chemiluminescence response of human natural killer cells. II. *Eur. J. Immunol.* 13, 514-520.
3. Allen, R.C., Stjernholm, R.L., Reed, M.A., Harper III, T.B., Gupta, S., Steele, R.H. and Waring, W.W. (1977). Correlation of metabolic and chemiluminescent responses of granulocytes from three female siblings with chronic granulomatous disease, *J. Infect. Dis.* 136, 510-518.
4. Lilius, E-M., Waris, M. and Lang, M. (1984). Automated luminometer setup for continuous measurement of 25 samples *In* "Proceedings of Third International Symposium on Analytical Applications of Bioluminescence and Chemiluminescence" this volume, Academic Press, New York.
5. Giudicelli,J., Philip, P.J.M., Delque,P. and Sudaka,P. (1982). A single-step centrifugation method for separation of granulocytes and mononuclear cells from blood using discontinuous density gradient of Percoll, *J. Immunol. Meth.* 54, 43-46.
6. Faden,H. and Maciejewski, N. (1981). Whole blood luminol-dependent chemiluminescence, *J. Reticuloendothel. Soc.* 30, 219-226.
7. Mills,E.L., Rholl, K.S. and Quie, P.G. (1980). Luminol-amplified chemiluminescence: a sensitive method for detecting the carrier state in chronic granulomatous disease, *J. Clin. Microbiol.* 12, 52-56.

DETERMINATION OF IMMUNE COMPLEXES BY CHEMILUMINESCENCE OF STIMULATED MURINE PHAGOCYTES: COMPARISON WITH OTHER QUANTIFICATION METHODS

J. Willems and M. Joniau

I.R.C. K.U.L. Campus Kortrijk ; B - 8500 Kortrijk (Belgium)

INTRODUCTION

Quantification and analysis of circulating immune complexes (CIC) has given valuable insight into the origin and evolution of many diseases. As yet several quantification methods for CIC have been described, many of them based on the specific interaction of their immunoglobulin or complement part with solid phase bound ligands in RIA or ELISA techniques.

Some years ago, the chemiluminescence (CL) produced by interaction of phagocytes with stimulator substances was observed to correlate with the quantity of ingested material.

Furthermore, the introduction of "chemilumigenic probes" (1) to enhance this naturally occuring phenomenon makes it very attractive to use this technique to quantitate small amounts of CIC, as they attach to phagocytes at their C_3bi and Fc receptors.

In the present study, we looked for optimal conditions of the phagocyte CL test to evaluate CIC levels in sera of tumor bearing mice and compared the results with those from another sensitive assay : the conglutinin RIA.

MATERIALS AND METHODS

1. Phagocytes

Peritoneal phagocytes were obtained at different times after prestimulation by i.p. injection of 1.5 ml glycogen solution (1 mg/ml). They were washed in buffered saline (pH 7.4) and,

ANALYTICAL APPLICATIONS OF
BIOLUMINESCENCE AND CHEMILUMINESCENCE

ISBN 0-12-426290-2

for some experiments, were separated into macrophage and granulocyte fractions after centrifugation on Percoll (2).

2. *Chemiluminescence test*

To 1 ml of veronal buffered saline (VBS) containing Ca^{2+} (2.10^{-4}M), Mg2+(5.10^{-4}M) BSA (0.2 mg/ml) and glucose (0.2 mg/ml), 100 µl of luminol (10^{-4}M) or lucigenin (10^{-4}M) and 100 µl of phagocytes (3 X 106 /ml) were added.
After 1 hr of dark adaption at 37°C, 25 µl of stimulus (CIC containing serum) was added and CL was measured in a liquid scintillation counter for 12 sec periods after suitable time intervals. All tests were run in duplicate.

3. *Sera*

A/J strain mice were inoculated with neuroblastoma (NB) tumor cells as described (3) and bled via retro-orbital plexus from 1 to 3 weeks post inoculum. Serum was allowed to clot at room temperature and frozen immediately after centrifugation. Blank serum of untreated A/J mice and sera of mice of other inbred strains were obtained and treated in the same way.

4. *Conglutinin RIA*

This test is described in detail in a previous paper (3).

RESULTS AND DISCUSSION

In a first set of experiments we examined different external parameters such as pH, temperature and enhancer, that could influence the test and found that incubations at pH 7.4 and at 37°C give optimal results. Furthermore the use of lucigenin, rather than the more popular luminol, as a chemilumigenic probe largely enhances the sensivity of this test. These results are summarized in table 1.

As to the source of phagocytes, we prefer peritoneal cells instead of peripheral blood cells since we noticed that prestimulation of phagocytes with glycogen gives excellent results, but can only be achieved in vivo. Indeed, i.p. injection of a glycogen solution prior to the isolation of the cells results in a biphasic increase of activity. After about 4 hrs, a first peak of activation, merely due to stimulated granulocytes - as evidenced after separation of the peritoneal cells on Percoll - is noted, whereas 3 to 5 days after injection a second peak, caused by stimulated macrophages arises.

TABLE 1

Effect of external parameters in relation to optimal conditions (100%)

1. pH (lucigenin enhanced)

pH	%
6.8	92 %
7.4	100 %
7.8	97 %
8.2	48 %

2. temperature

temperature	%
20°C	19 %
30°C	57 %
37°C	100 %
41°C	37 %

3. enhancer (final conc.)

enhancer	%
lucigenin	
10^{-4}M	100 %
5.10^{-5}M	64 %
luminol	
10^{-4}M	6 %
5.10^{-5}M	3 %

4. source of phagocytes

source	%
Balb/c 4 hrs prestimulated	100 %
C_3H "	108 %
A/J "	95 %
BXA "	103 %
Balb/c no prestimulation	13 %
A/J "	40 %

Since in this study we evaluated sera of tumor bearing A/J strain mice, we also checked if phagocytes obtained from different murine strains would respond in the same way to this CIC stimulus. We found that, following a 4 hrs prestimulation, A/J as well as Balb/c, C3H and BXA hybrid phagocytes all give a similar CL response. Without prestimulation however, A/J phagocytes were about 3 times as reactive as those from other strains (table 1).

In view of these observations we decided to run our tests at 37°C, using VBS pH 7.4, lucigenin as enhancer and murine phagocytes of either strain, prestimulated for 4 hrs with a glycogen solution. After dark adaption, 25 ul of serum from NB bearing mice is added. The rapid initial luminescence burst attains a maximum after 3 to 5 minutes, and the peak values of CL intensity are compared with Kg-RIA results, obtained with the same sera. These sera had previously been shown to have increasing quantities of CIC as a function of time after tumor challenge. Compared to the basal CIC levels of reference mice they yield values of up to 10 relative units, depending on the type of inoculum (3).

Using the present CL test, we found a good correlation with the previously obtained Kg-RIA results (fig.1)

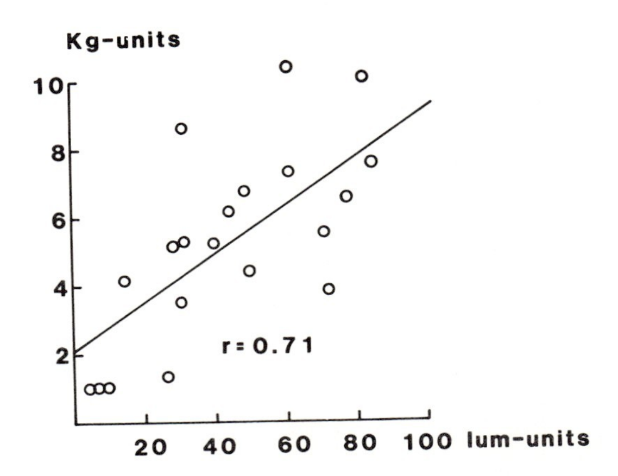

Fig.1: correlation between Kg-RIA on CL test results for sera of NB bearing mice.

Recently, we also applied this technique (using murine phagocytes) to evaluate CIC levels in pathological and normal human sera. Although only performed on a few sera, a satisfactory correlation with Kg- and Ciq-RIA is obtained.

In conclusion, these experiments show that, under appropriate conditions, the measurement of the chemiluminescence produced by interaction of prestimulated murine phagocytes with CIC, is a simple and easy means to evaluate and quantitate these complexes in sera.

AKNOWLEDGEMENTS

We wish to thank the Belgian N.F.W.O. for financial support.

REFFERENCES

1. Allen R.C. (1981) in : "Bioluminescence and Chemiluminescence" (De Luca M.A. and Mc Elroy W.D. Eds), p 63-73, A.P., New-York.
2. Coupland K. and Leslie R.G.Q. (1983) The expression of Fc receptors on guinea pig peritoneal macrophages and neutrophils. Immunology 48 , p 647-656.
3. Willems J. and Joniau M. (1983) Circulating immune complexes in murine neuroblastoma: quantification and shifts in immunoglobulin composition. Immunology letters 6 p 109-113.

INFLUENCE OF BACTERIAL SURFACE COMPONENTS ON THE STIMULATION OF PHAGOCYTE CHEMILUMINESCENCE

N. Topley, M.J. Harber and A.W. Asscher

Department of Renal Medicine, Welsh National School of Medicine, K.R.U.F. Institute, Royal Infirmary, Cardiff, U.K.

The interaction between bacteria and phagocytic cells is an important determinant of the outcome of infection and may play a role in the development of tissue damage and renal scarring (1,2). Measurement of respiratory burst-associated chemiluminescence (CL) provides a convenient tool for studying the response of phagocytic cells to bacterial stimulation. We have investigated the ability of two unopsonized urinary pathogens to stimulate luminol-dependent CL in human leucocytes and have related this to changes in bacterial surface structure induced by subculture in three different growth media. Fimbriation was assessed by standard haemagglutination reactions, glycocalyx production by electron microscopy with a ruthenium red stain, and surface hydrophobicity by a salt aggregation test (3). *Escherichia coli* 504 exhibited high surface hydrophobicity associated with expression of type 1 fimbriae when grown in static nutrient broth, and it stimulated a very strong CL with polymorphonuclear leucocytes (PMNL) and blood mononuclear cells (MNC). Fimbriation was suppressed, hydrophobicity reduced and cell-associated glycocalyx production promoted on growth in pooled human urine, and the CL responses were greatly diminished. Subculture in a gluconic acid medium (MVBM) was associated with high surface hydrophilicity, production of a secretory glycocalyx and a negligible CL response with both PMNL and MNC.

E.coli AB was non-fimbriate, hydrophilic and devoid of a cell-associated glycocalyx when grown in all culture media. However, this strain stimulated a moderate CL peak with PMNL when subcultured in MVBM.

These results indicate that hydrophobic interactions, and to a lesser extent unidentified factors, promote bacterial stimulation of phagocytes. Glycocalyx production by bacteria, which may occur *in vivo* (4), increases surface hydrophilicity and inhibits, but does not entirely prevent, bacteria-phagocyte interactions.

ISBN 0-12-426290-2

REFERENCES

1. Glauser, M.P., Lyons, J.M. and Braude, A.I. (1978). Prevention of chronic experimental pyelonephritis by suppression of acute suppuration. *J. Clin. Invest.* 61, 403-407.

2. Slotki, I.N. and Asscher, A.W. (1982). Prevention of scarring in experimental pyelonephritis in the rat by early antibiotic therapy. *Nephron* 30, 262-268.

3. Ljungh, A. and Wadstrom, T. (1982). Salt aggregation test for measuring cell surface hydrophobicity of urinary *Escherichia coli*. *Eur. J. Clin. Microbiol.* 1, 388-393.

4. Costerton, J.W., Irvin, R.T. and Cheng, K-J. (1981). The bacterial glycocalyx in nature and disease. *Ann. Rev. Microbiol.* 35, 299-324.

B LYMPHOCYTE SURFACE BOUND ICs STIMULATE PMNs TO PRODUCE OXYGEN DERIVATIVES WHICH IMPAIR T CELL RESPONSES

C. De Simone[+], M. Ferrari[*], V. Vullo[+], C. Mastropietro[+] and F. Sorice[+]

[+]*Clinica Malattie Infettive, Università di Roma "La Sapienza" I, 00185 Roma, Italia*

[*] *Laboratorio di Fisiopatologia, Istituto Superiore di Sanità, Viale Regina Elena 299, 00161 Roma, Italia*

INTRODUCTION

This paper reports "in vitro" experiments which suggest that PMNs (polymorphonuclear neutrophils) stimulated by the B lymphocyte surface-bound immune complexes present in the sera of patients with parasitic diseases produce and release highly reactive oxygen derivatives which impair T cell proliferative responses following phytohemagglutinin (PHA) stimulation.

MATERIALS AND METHODS

Serum samples were collected under sterile conditions from 58 adult subjects with amoebiasis (30), echinococcosis (20), schistosomiasis (3), malaria (3) or with filariasis (2).

The purity of the PMN suspensions employed in our experiments resulted always superior to 90%. Less than 3% of the T cell-enriched population and more than 76% of the T cell-depleted population reacted with fluorescinated anti-human immunoglobulin serum.

B cell (T cell-depleted cell suspension) were incubated at 37°C for 90 min with saline, normal human serum or parasitized patients' serum. After resuspension in PBS at a

ISBN 0-12-426290-2

final concentration of $1x10^6$ cells/ml, the B cell suspensions were mixed with 1 ml of PBS containing 10^{-7} M Luminol and 1 ml of PMNs ($1x10^6$ cells/ml). The specimens were counted for 0.1 min at 3-5 min intervals in a luminometer.

All the sera were examined for the presence of immuno complexes by the Raji cell enzyme immunoassay (1).

T cell-enriched lymphocyte suspensions were mixed to T cell-depleted lymphocyte suspensions treated with the test sera. The mixtures were then mixed with autologous PMNs and the tubes were incubated for 3 hours at 37°C. The lymphocytes were separated from the PMNs employing the Ficoll-Hypaque technique and cultured for 3,4,5 days in flat bottomed microtites plates in presence of PHA. 3H-thymidine was added to the culture wells 18 hours before harvesting.

RESULTS AND DISCUSSION

The data summarized in this paper show that in some of the sera from patients with parasitic diseases, "factors" are present which after interaction with T cell-depleted lymphocytes enable them to elicit a burst of oxygen consumption in PMNs (Table 1).

TABLE 1

Disease	n° patients	C.L. positive [a]	C.I.C. positive [b]	day [c]
Amoebiasis	30	16	19	5
Echinococcosis	20	5	7	4
Schistosomiasis	3	2	2	5
Malaria	3	0	0	3
Filariasis	2	0	0	3

[a] Serum was considered chemiluminescence (CL) positive when peak chemiluminescence counts exceeded 120.000 counts/0.1 min.

[b] Serum was considered circulating immunocomplexes (C.I.C.) positive when IgG $>5.1 \pm 2$ µg/ml.

[c] Day when occurred the peak proliferative responses. The control cultures have a peak on day 3 corresponding to 64.626 ∓ 19.028 c.p.m.

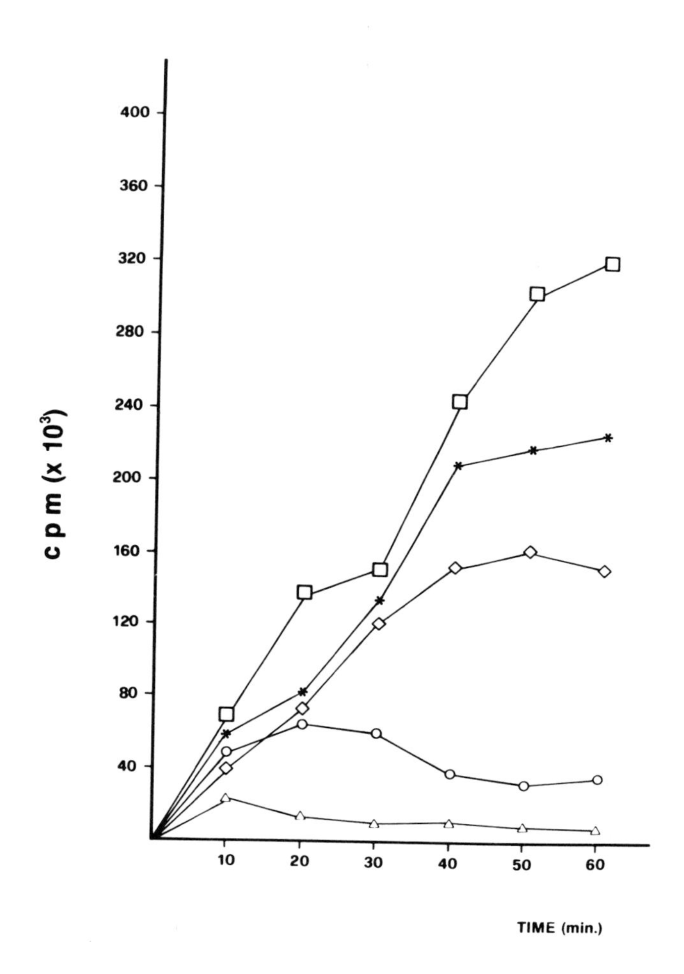

Fig. 1 PMN chemiluminescence following stimulation with B cells treated with: ▫ , *, ◇ C.I.C. positive sera; ∘ C.I.C. negative serum; ▵ saline.

The intensity of the reaction is related to the presence of $C3^+$, $FcIgG^+$ cells in the lymphoid cell suspensions employed. Raji cells incubated with the same sera were equally effective as T lymphocyte-depleted cell suspensions in eliciting PMN chemiluminescence. Aliquots of Raji cells incubated with the sera were also employed in the Raji cell enzyme immunoassay for the detection of circulating IgG containing immune

complexes (2).
All the chemiluminescence positive sera were positive for the presence of circulating immune complexes. On these grounds it was assumed that the antigen-antibody complexes were the "factors" present in some of the sera employed which, following incubation with $C3^+$ $FcIgG^+$ cells, induced PMN chemiluminescence (3). Mitogenic studies have shown that if the T cell-depleted lymphocyte suspensions had been incubated with the immune complex positive sera in presence of the PMNs, the T cells stimulated by PHA show delayed peak proliferative responses. This phenomenon is in part prevented by adding catalase (50 μg/ml) during the incubation step with the PMNs, an observation which focuses attention on the oxidative products released by the neutrophils, which may be responsible for the phenomenon.

REFERENCES

1. Rote, N.S. and Caudle, M.R. (1983). Detection of circulating immuno complexes with a Raji cell enzyme immunoassay, *J. Immunol. Methods* 56, 33-42.
2. Archibald, A.C., Cheung, K. and Robinson, M.F. (1983). The interaction of lymphocite surface-bound immune complexes and neutrofils, *J. Immunol.* 131, 207-211.
3. Starkebaum, G., Stevens, D.L., Henry, C. and Gavin, S.E. (1981). Stimulation of human neutrophil chemiluminescence by soluble immunocomplexes and antibodies to neutrophils, *J. Lab. Clin. Med.* 98, 280-291.

LIPOXYGENASE IS NOT A DIRECT SOURCE OF LIGHT IN SENDAI VIRUS-INDUCED CHEMILUMINESCENCE (CL)

Markus Grob and Ernst Peterhans

Institute of Virology
University of Zurich
CH-8057 Zurich, Switzerland

INTRODUCTION

Sendai virus, also referred to as Parainfluenza-I virus, induces in phagocytic cells an immediate, short-lived burst of luminol-dependent CL that is triggered by a interaction of the viral envelope glycoproteins with the cell membrane (1-3). Using metabolic inhibitors, we have shown that virus-induced luminol-dependent CL in mouse spleen cells is dependent on lipoxygenase but not cyclooxygenase activities (4). Both enzyme reactions in vitro have been shown to lead to CL (5,6) and results of other studies suggested that the decrease in light emission observed in intact cells treated with such inhibitors reflected a significant contribution of the respective pathways to luminol-dependent cellular CL (7,8). In the present work, we have investigated whether lipoxygenase could account for part of the CL evoked by *Sendai virus* in bovine polymorphonuclear leucocytes (PMNL). The results suggest that lipoxygenase activity is not a direct source of virus-stimulated CL. Our experiments do not rule out, however, that lipoxygenase may have a regulatory role in the activation of the respiratory burst from which reactive oxygen intermediates originate, leading to CL.

ISBN 0-12-426290-2

MATERIALS AND METHODS

Luminol, superoxide dismutase (SOD), lucigenin (bis-N-methylacridinium nitrate) and soybean lipoxygenase were purchased from Sigma. Eicosatetraynoic acid was a gift of F. Hoffmann-LaRoche, Basel, Switzerland. *Sendai virus* was grown, purified and titrated as described (3). Bovine PMNL were isolated from venous blood according to the method of Carlson and Kaneko (9) and suspended at 4 x 10^5/ml in phenol red-free Hanks' buffered salt solution (HBSS) containing 25 mM HEPES and 50 µM lucigenin or 7.5 µl/ml of 4% (w/v) bovine serum albumin saturated with luminol. CL was measured in a liquid scintillation counter (LSC) and recorded with a HP85 computer as described (3).

RESULTS AND DISCUSSION

We first verified that lipoxygenase in vitro was able to generate CL. Arachidonic acid was added to soybean lipoxygenase and CL was monitored in the presence of the chemiluminogenic probes, luminol or lucigenin.

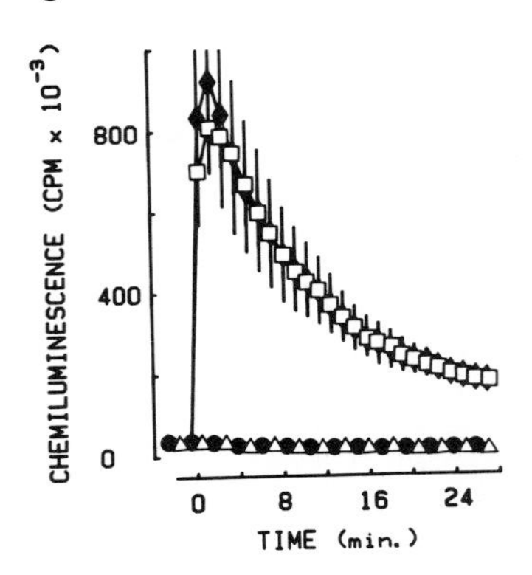

Fig. 1 CL generation by soybean lipoxygenase in the presence of luminol. To 1,000 units of enzyme suspended in HBSS was added arachidonic acid (final conc. 100 µM) and CL was measured in a LSC. □ arachidonic acid, ♦ arachidonic acid plus azide, △ azide alone (100 µM), ● PBS.

The addition of arachidonic acid to the enzyme in the presence of luminol, but not lucigenin resulted in a rapid and intense burst of CL lasting for several minutes. In the additional presence of azide in the reaction mixture, luminol-dependent CL was not decreased (Fig. 1).

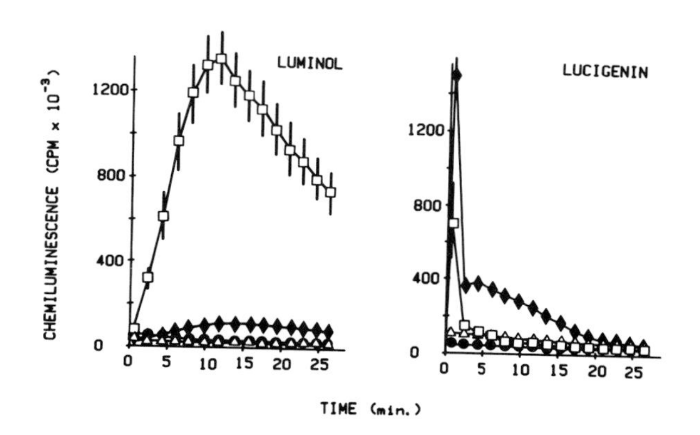

Fig. 2 The effect on PMNL CL of Sendai virus. PMNL were stimulated with 125 hemagglutinating units of Sendai virus. Zero time indicates addition of virus. The means ± standard deviations of three replicate samples are shown.
□ *virus alone,* ♦ *virus plus azide,* △ *azide alone (100 µM),* ● *PBS.*

On stimulation with *Sendai virus*, bovine PMNL emitted CL both in the presence of luminol and lucigenin (Fig. 2). Azide strongly inhibited luminol- but enhanced lucigenin-dependent CL. A similar effect on CL has been observed in human PMNL (10) and it was suggested that lucigenin-dependent CL reflects $O_2^{\cdot -}$-generation and luminol-dependent CL the production of a myeloperoxidase-dependent oxidant, possibly HOCl (11). In support for a similar mechanism in virus induced CL in bovine PMNL, SOD decreased lucigenin but had little effect on luminol-dependent CL (not shown). As noted earlier in mouse spleen cells (4), ETYA decreased CL also in bovine PMNL. Such inhibition of CL was observed with both luminol and lucigenin as the chemiluminogenic probes (not shown).

In summary, luminol-dependent cellular, but not lipoxygenase-generated, CL was inhibited by azide. Intact cells, but not lipoxygenase, produced lucigenin-dependent CL. This suggests that lipoxygenase is not a direct source of CL emitted by intact cells. Such interpretation is also supported by the recent observation that the CL-negative PMNL from patients with chronic granulomatous disease apparently synthesize lipoxygenase products similar to normal PMNL (12,13).

ACKNOWLEDGEMENTS

This work was supported by the Swiss National Science Foundation Grant No. 3.429.0.83. We thank Drs. Jörg, Wyler and Keller for discussion.

REFERENCES

1. Peterhans, E. (1979). *BBRC* 91, 383-392.
2. Peterhans, E. (1980). *Virology* 105, 445-455.
3. Peterhans, E. *et al.* (1983). *Virology* 128, 366-376.
4. Semadeni, B. *et al.* (1984). *In:* "Molecular Biology of Negative Strand Viruses" (Eds. D.H. L. Bishop and R.W. Compans) Academic Press New York, in press.
5. Marnett, L.J. *et al.*(1974). *BBRC* 60, 1286-1294.
6. Veldink, G.A. *et al.*(1977). *BBRC* 78, 424-428.
7. Smith, R.L., and Weidemann, M.J. (1980). *BBRC* 97, 973-980.
8. Cheung, K. *et al.*(1983). *J.Immunol.* 130, 2324-2329.
9. Carlson, G.P., and Kaneki, J.J. (1973). *Proc. Soc.Exp.Biol.Med.* 142, 853-856.
10. Williams, A.J., and Cole, P.J. (1981). *Immunology* 44, 847-858.
11. DeChatelet, L.R. *et al.* (1982). *J.Immunol.* 129, 1589-1593.
12. Feinmark, S.J. *et al.* (1983). *J.Clin.Invest.* 72, 1839-1843.
13. Henderson, W.R., and Klebanoff, S.J. (1983). *J.Biol.Chem.* 258, 13522-13527.

DIAGNOSIS OF CELIAC DISEASE IN CHILDREN USING LEUKOCYTE-ATTACHED IMMUNE COMPLEX LUMINESCENT ASSAY

Esa-Matti Lilius[+], Tuula Hämäläinen[+]
and Marja-Riitta Ståhlberg[§]

[+]*Department of Biochemistry, University of Turku, SF-20500 Turku, Finland*
[§]*Department of Paediatrics, University of Turku, SF-20500 Turku, Finland*

INTRODUCTION

Celiac disease is a disease of the small intestine characterized by functional disturbances causing malabsorption of food and nutrients and structural changes of partial or complete villous atrophy. Malabsorption is associated with stunting of growth in the child, loss of weight in the adult and various deficiency states. The disease expresses itself at any age, but most frequently in childhood. Ingestion of wheat gluten is shown to be the causative factor gliadin being the noxious component of gluten. The diagnosis is established on the basis of an abnormal jejunal biopsy. In celiac disease it becomes normal during one year's gluten exclusion and again abnormal in gluten provocation.

Our aim was to develop a test for circulating gliadin antibodies detected by leukocyte-associated chemiluminescence. The test is based on the following: Leukocytes bear membrane receptors for the Fc portion of immunoglobulins. The binding of aggregated or antigen complexed immunoglobulins to these recognition units on the leukocyte membrane is accompanied by a burst in oxidative metabolism of leukocytes. The cells produce reactive oxygen species that participate in light-emitting reactions generating CL. The natural CL of leukocytes can be enhanced by several hundreds of fold by adding luminol to the cell suspension. Gliadin is first incubated with sera from patients suspected to suffer from celiac disease or from controls. Leukocytes supplemented with luminol are then activated by these mix-

ISBN 0-12-426290-2

tures. CL emission from activated leukocytes is measured.

MATERIALS AND METHODS

Subjects

Children (ages between 1 and 16 yrs) suspected of suffering from celiac disease were send into the University Hospital for a small intestine jejunal biopsy done by one of the authors (M-R.S.). A blood sample for this study was taken at the same time.

Biochemistry students served as healthy controls for blood samples.

Serum collection

Blood was obtained by venipuncture and allowed to clot at room temperature for about one hour. Serum was collected by centrifugation and frozen in aliquots at $-70^{o}C$.

Leukocytes

Leukocytes were prepared from the buffy coat of healthy blood donors obtained from the Finnish Red Cross Blood Transfusion Service, Turku, by sedimenting erythrocytes with 1.5% dextran at $37^{o}C$ for 90 min. Leukocytes were separated from plasma by centrifugation and suspended in HBSS buffer supplemented with 0.1% gelatin.

Gliadin

Grude gliadin from wheat gluten (Sigma) was dissolved in 50% ethanol to a stock solution of 2 mg/ml which was kept at $-20^{o}C$ and diluted in HBSS buffer just before use.

Luminol

The stock solution of 10 mM was prepared by dissolving luminol (Sigma) in 0.2 M sodium borate buffer, pH 9.0. The solution was stored not more than 3 days.

Formation of antigen-antibody complexes and the CL assay

For the formation of gliadin-antigliadin complexes 50 ul of diluted sera (1:10 in HBSS buffer) and 50 ul of gliadin solutions containing 0,2,4,8 and 16 ug of protein were incubated at $37^{o}C$ for 30 min in 4 ml-polypropylene tubes and then cooled. Cold suspensions of separated leukocytes ($4x10^5$ cells in 400 ul of HBSS buffer including $4x10^{-4}$M luminol and 0.1% gelatin) were added to the vials. The CL emission commenced when the contents of the vials warmed up in the temperature controlled sample carousel of the luminometer. The CL emission was measured at 37°C for 60 min. Automated luminometer setup (LKB Wallac 1251 Luminometer connected to Olivetti M 20 microcomputer) for continuous and simultaneous measurements of up to 25 samples was used as described (1). In a luminometer run of celiac disease samples usually one vial containing 20 ug of preopsonized zymosan was included with the intention of controlling the competence of leukocytes.

RESULTS AND DISCUSSION

Mixtures of gliadin and sera preincubated at 37°C for 30 min caused the activation of leukocytes manifested as Cl emission (Fig. 1). If gliadin, sera and leukocytes were mixed

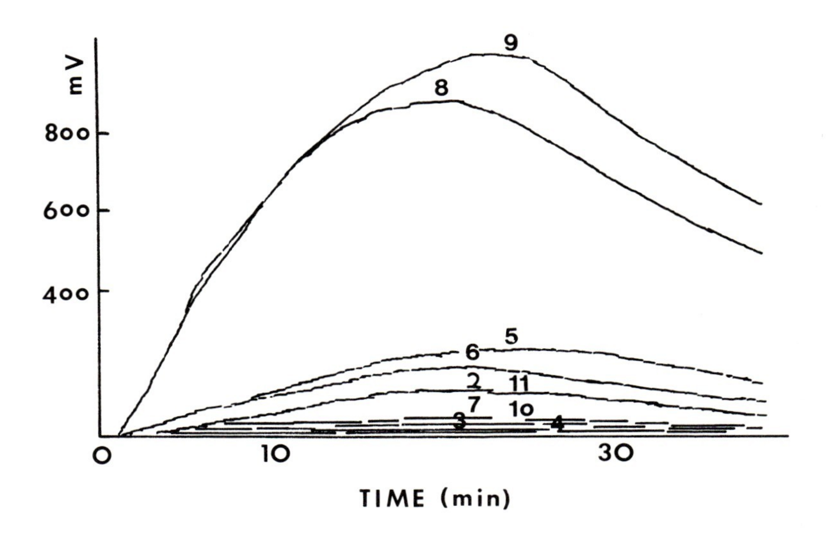

Fig. 1 CL emission traces from leukocytes activated by gliadin-antigliadin immune complexes. Serum samples from two patients with normal biopsy (1-3 and 4-6), from one patient with complete villous atrophy (7-9) and from a healthy control (10-11). The amounts of 0, 4 and 8 ug of gliadin were added. Vials 1, 4, 7 and 10 = 0 ug; Vials 2, 5, 8 and 11 = 4 ug; Vials 3, 6 and 9 = 8 ug.

together without preincubation of antigen and antiserum, the maximum CL emission from leukocytes was only 40% of that obtained with preincubation. Moreover, the time required to reach the peak CL value was lengthened by 30% suggesting that the formation of antigen-antibody complex is needed for the appearance of the CL emission and that the formation takes place also during the CL measurements. The maximum CL was linearly proportional to the amount of gliadin up to 4 ug larger amounts causing the reduction of the CL peak values. When different serum samples were assayed it was noticed that sera from patients with a complete villous atrophy in the small intestine gave very high responses in the CL assay the response being at least 5 times as high as that from patients with a normal jejunal biopsy or from healthy controls (Fig. 1). The serum dilution routinely used was 1:10. The dilution where the peak CL value was half of the maximum peak value, the "half titre", was found to be 1:30 in the sera of patients with complete villous atrophy. It was less than 1:10 in the sera of other patients.

These results suggest that our leukocyte-attached immune complex luminescent assay (LAILA) is sensitive enough to detect circulating gliadin antibodies in sera from celiac disease patients with total villous atrophy. The assay is simple and rapid to perform, a reply can be obtained in two hours after the venipuncture and multiple samples can be evaluated simultaneously. Thus, this assay may be helpful in the diagnosis of celiac disease in children.

REFERENCE

1. Lilius, E-M., Waris, M. and Lang M. (1984). Automated luminometer setup for continuous measurement of 25 samples *In* "Proceedings on Third International Symposium on Analytical Applications of Bioluminescence and Chemiluminescence" this volume, Academic Press, New York.

EVALUATION OF SERUM OPSONIC CAPACITY AGAINST *BORDETELLA PERTUSSIS* BY LUMINOL-DEPENDENT PHAGOCYTOSIS-ASSOCIATED CHEMILUMINESCENCE

Esa-Matti Lilius[+], Pekka Leivo[+], Matti Waris[+], Jussi Mertsola[§,&] and Olli Ruuskanen[§]

[+]*Department of Biochemistry, University of Turku, SF-20500 Turku, Finland*
[§]*Department of Paediatrics, University of Turku, SF-20500 Turku, Finland*
[&]*Department of Medical Microbiology, University of Turku, SF-20500 Turku, Finland*

INTRODUCTION

Whooping cough is an acute respiratory disease caused by *Bordetella pertussis*. Succesful immunization has reduced the incidence of the disease but there is controversy about the benefits and risks of the currently used killed whole-cell vaccines. The precise roles of cell-mediated and humoral immune mechanisms in pertussis are subjects of much current interest because of the need of better vaccines. In this work we have studied the serum opsonic capacity against *B. pertussis* by using quantitative chemiluminescence (CL) which is known to have several advantages over conventional methods in the evaluation of opsonization and phagocytosis (1).

MATERIALS AND METHODS

Subjects

Healthy controls: one unvaccinated infant of 2 months of age and 16 adults vaccinated in the childhood.
Patients: one culture-positive pertussis patient of 2 months of age and one adult whooping cough patient diagnosed by ELISA (IgM-IgA) serology. The serum of this patient was used to test the optimum conditions of the CL assay.

ISBN 0-12-426290-2

Blood was obtained by venipuncture and allowed to clot at room temperature for about one hour. Serum was collected by centrifugation and frozen in aliquots at -20°C.

Phagocytic cells

Human polymorphonuclear leukocytes(PMNL), mononuclear cells (MNL) as well as unfractionated lekocytes were prepared as described in another paper in this Symposium (2).

B. pertussis bacteria

Whole-cell suspension of the formalin-killed bacteria (a mixture of strains 1.2.3 and 1.2) obtained from the vaccine production of The National Public Health Institute (Helsinki, Finland) was washed twice in Hank's balanced salt solution (HBSS) and suspended to a concentration of $20x10^9$ cells/ml in HBSS supplemented with 0.1% gelatin.

Opsonization

For the opsonization 50 ul of diluted serum (different dilutions) and 50 ul of the cell suspension of *B.pertussis* ($1x10^9$ cells) were incubated in 4-ml polypropylene tubes at 37°C for 30 min and then cooled.

CL assay

Automated luminometer setup (LKB Wallac 1251 Luminometer connected to Olivetti M 20 microcomputer) for continuous and simulataneous measurements of up to 25 samples was used as described (2). Cold suspensions of separated leukocytes ($4x10^5$ cells in 400 ul of HBSS buffer including $4x10^{-4}$M luminol and 0.1 % gelatin) were added to the opsonization vials. The CL emission commenced when the contents of the vials warmed up in the temperature controlled sample carousel of the luminometer. The CL emission was measured at 37°C for 60 min.

Luminol

The stock solution of 10 mM was prepared by dissolving luminol (Sigma) in 0.2 M sodium borate buffer, pH 9.0. The solution was stored not more than 3 days.

RESULTS AND DISCUSSION

Assay conditions

CL emission was not observed when unopsonized *P. pertussis* ($1x10^9$ cells incubated in HBSS buffer instead of serum) was incubated with leukocytes. On the contrary, bacteria opsonized in diluted serum of a whooping cough patient caused the activation of leukocytes manifested as CL emission.

More than 10^8 bacteria was needed for a specific opsonization response in the CL assay. The maximum response was linearly proportional to the number of bacteria cells up to $5x10^9$ cells larger cell numbers causing a considerable background CL emission. The maximum CL response was also

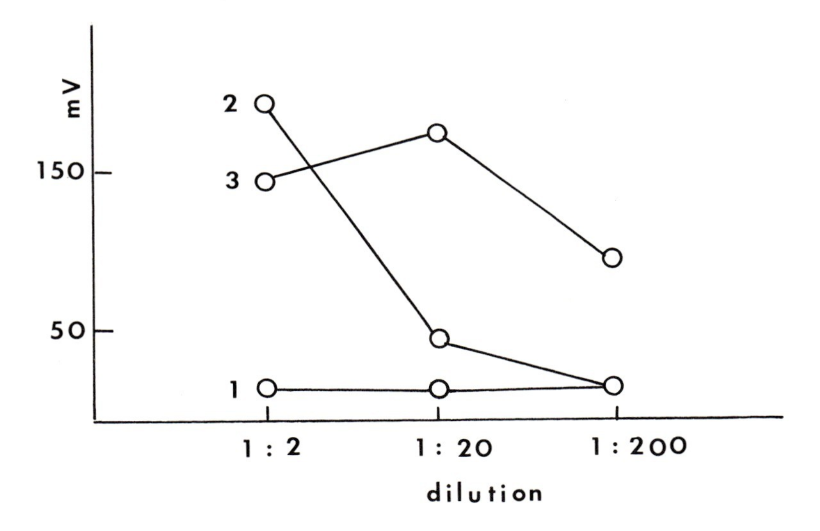

Fig 1 Opsonization of B. pertussis in different dilutions of sera from an unvaccinated child(1), a vaccinated adult (2) and a whooping cough patient(3).

dependent on the activity of leukocytes. The pattern of the CL response as a function of serum dilution (Fig.1) was, however, independent on the type (PMNL, MNL and unfractionated leukocytes) and on the number ($5x10^4$-$1x10^6$) of leukocytes suggesting that the evaluation of opsonins can be done by using any type of phagocytic cells.

Effects of vaccination and disease on the opsonization

Figures 1 and 2 show that sera from unvaccinated children did not contain opsonins against *B. pertussis*.On the contrary sera from adults vaccinated in the childhood gave

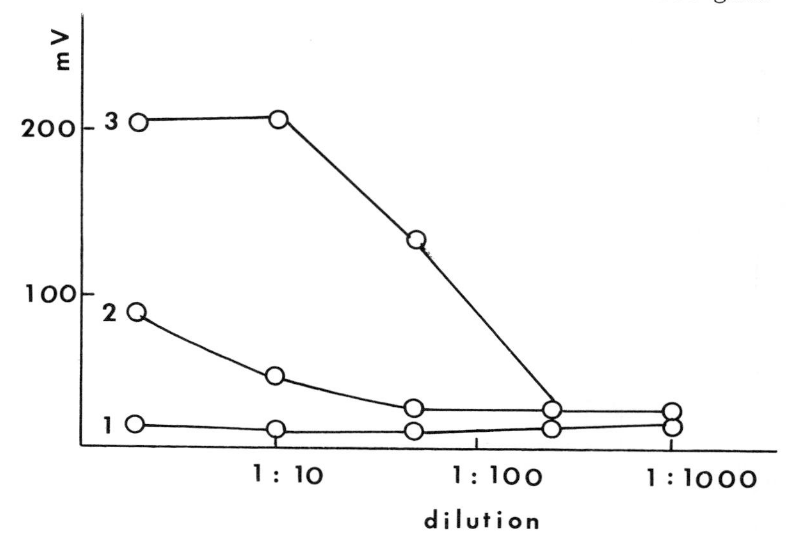

Fig. 2 Opsonization of B. pertussis in different dilutions of follow-up sera of a whooping cough patient. Sera were taken one week (1), two weeks (2) and six weeks (3) after the onset of the disease.

rise to a large emission of CL. Sera from whooping cough patients produced CL responses in more diluted solutions than those of vaccinated controls offering possibly a basis for the diagnosis of an acute inflammation. The CL assay is simple, rapid and reproducible offering new possibilities to evaluate immune mechanismis in whooping cough.

REFERENCES

1. Robinson, J.P. and Penny,R. (1982). Chemiluminescence response in normal human phagocytes. I. Automated measurements using a standard liquid scintillation counter, *J.Clin. Lab. Immunol.* 7, 215-217.
2. Lilius, E-M. and Waris, M. (1984). A very sensitive and rapid chemiluminescence method for the measurement of phagocytosis *In* "Proceedings of Third International Symposium on Analytical Applications of Bioluminescence and Chemiluminescence" this volume, Academic Press, N.Y.

LUMINOL DEPENDENT CHEMILUMINESCENCE OF GRANULOCYTES OF TUMOR PATIENTS (*)

M. Wurl, P. Schuff-Werner, A. Müller,
K. Gottsmann, S. Krull and G.A. Nagel

Dept. Internal Medicine, Div. Hematology/Oncology, University Clinic of Göttingen, Robert-Koch-Str. 40, D-3400 Göttingen (FRG)

(*) supported by BMFT-project no. 0384 100

INTRODUCTION

Recently granulocytes (PMN) of patients with malignant tumors became of interest because it could be demonstrated that PMN are involved in the cytotoxic defence of tumor cells (1;2). Although the mechanism of the tumor cell lysis is still unknown, the cytotoxic activity apparently depends on the activation of the respiratory burst of PMNs (1;2). Activated oxygen species seem to be responsible of as well bactericidal as tumoricidal activity and also generate chemiluminescence (CL) (1;2;3). This report provides additional evidence that peripheral blood PMNs of patients with different types of malignant tumors exhibite different enhancement of CL activity.

MATERIALS AND METHODS

Patient population. The patient population consisted of 89 tumor patients, 11 classified as having *Colorectal Ca.*, 6 *Sarcoma*, 11 *Mamma Ca.*, 7 *Melanoma*, 5 *Hodgkins Disease* and 9 *Non Hodgkin Lymphoma*. The control population consisted of 40 healthy donors.

Isolation of PMNs. Whole blood (heparinized at 50 U/ml) was obtained from patients or healthy donors by venipuncture and layered on Percoll den-

ISBN 0-12-426290-2

sity gradient, prior adjusted to physiological conditions. Centrifugation was performed at 400 x g for 30 min. The PMN fraction was harvested, washed three times with HBSS and purified from contaminating erythrocytes (RBC) by hypotonic lysis. After centrifugation at 200 x g for 10 min at 4^oC the PMN sediment was resuspended and adjusted to 5 x 10^5 cell/ml with HBSS.

Chemiluminescence measurement. (a) whole blood: to 100 µl LUMINOL (1.6 x 10^{-4} mol/l, final conc.), 300 µl diluted (1:1000 with HBSS) whole blood were added. *(b) isolated PMNs:* to 100 µl LUMINOL (10^{-5} mol/l, final conc.) 200 µl of PMN suspension were added (10^5 cells/tube, final conc.). After preincubation for 10 min at 37^oC in BIOLUMAT LB 9505 the CL reactions were started by adding 100 µl stand. opsonized ZYMOSAN (500 µg/tube).

RESULTS

Quench correction curve. An experimental quench curve (not shown here) was evaluated and linearized by calculating the linear regression lines of each segment (log. RBC 5-6,6-7,7-8). The equations of the regression lines were used for evaluating a quench correction table. Amounts of RBC smaller than 10^4 cells/tube do not influence the peak CL of PMNs.

CL activity of isolated PMNs. PMN fractions of all tumor patients were found to exhibite enhanced ZYMOSAN induced specific CL activity (peak CL/10^2 PMN) as compared to healthy donors. This effect is obviously demonstrated in *Fig. 1*.

CL activity of whole blood. PMNs of tumor patients as measured with whole blood samples also show increased ZYMOSAN induced CL activities. Subdivision in different types of cancer indicates preferentially augmented CL activities in the cases of *Colorectal Ca.* and Non Hodgkin Lymphoma (Fig. 2).

DISCUSSION

CL measurement with diluted whole blood samples, that is not affected by RBC, is a convenient

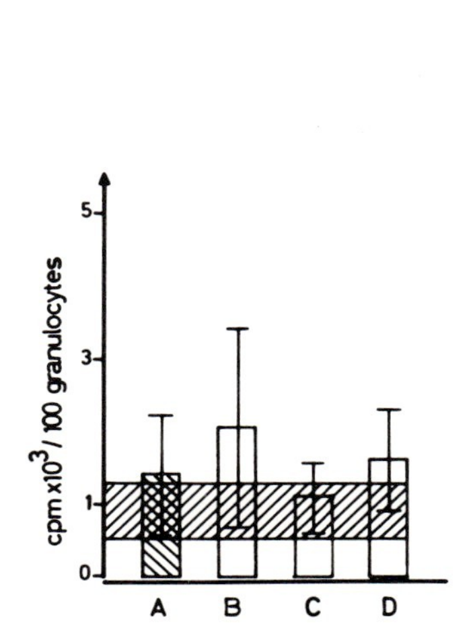

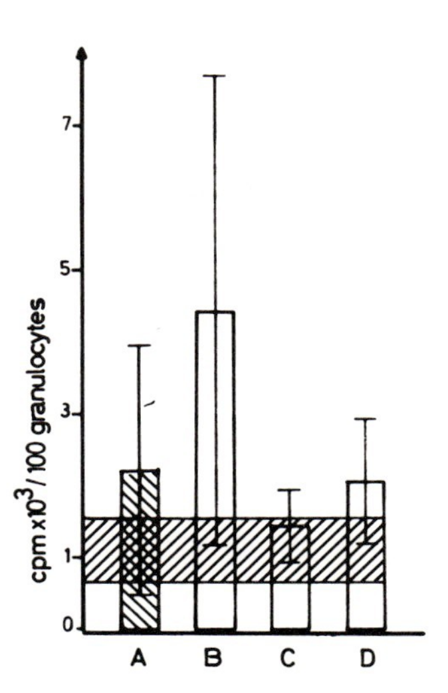

Fig. 1 Specific CL activities of isolated PMN not corrected (left) and corrected (right) for quench. Hatched horizontal section: healthy donors, A: all tumor patients, B: Sarcoma, C: Mamma Ca., C: Melanoma.

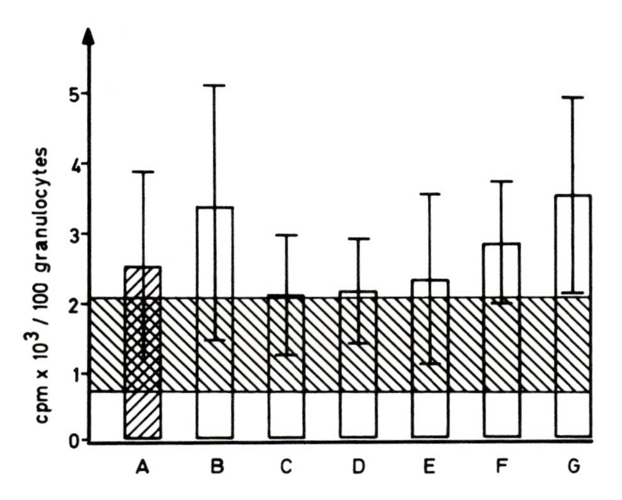

Fig. 2 Specific activities of PMN of tumor patients as measured with whole blood. Hatched horizontal section: healthy donors, A: all tumor patients, B: Colorectal Ca., C: Sarcoma, D: Mamma Ca., E: Melanoma, F: Hodgkins Disease, G: Non Hodgkin Lymphoma.

method, but does not distinguish cellular or humoral alterations. Comparison of specific CL activities evaluated by the whole blood method and the time consuming CL measurement of isolated PMNs respectively points out that increased CL activity may mainly depend on altered cellular events. Both methods are of low diagnostic value because of the high standard deviations that prevent discrimination of tumor and inflammation (3). So, increased CL activity of PMNs of tumor patients in general is possibly believed to depend on inflammatory processes produced by tumor margins. In the case of *Colorectal Ca.*, this augmentation effect might be provoked by the superinfection normally accompanied with this tumor, whereas patients with *Non Hodgkin Lymphoma* emphazise interest in further experiments, because of the possible relationship of enhanced CL activity to T-cell-derived factors.

REFERENCES

1. Hafeman, D.E. and Lucas, Z.J. (1979). Polymorphonuclear leukocyte-mediated, antibody-dependent, cellular cytotoxicity against tumor cells: dependence of oxygen and the respiratory burst, *J. Immunol.* 123, 55-62

2. Clark, R.A. and Klebanoff, S.J. (1977). Studies on the mechanism of antibody-dependent polymorphonuclear leukocyte-mediated cytotoxicity, *J. Immunol.* 119, 1413-1418

3. Heberer, M., Ernst, M., Allgöwer, M. and Fischer, H. (1982). Measurement of chemiluminescence in freshly drawn human blood, *Klin. Wochenschr.* 60, 1443-1448

WHOLE BLOOD CHEMILUMINESCENCE: INFLUENCE OF CELLULAR CONSTITUENTS, DILUTION, AND STIMULUS ON LUMINOL- AND LUCIGENIN-DEPENDENT CHEMILUMINESCENCE

J.Ennen, M.Ernst, and H.-D. Flad

Borstel Research Institute, D-2061 Borstel, FRG

INTRODUCTION

Measurement of chemiluminescence (CL) in whole blood was developed in order to investigate the CL response of phagocytic cells in an in vitro situation closely reflecting their in vivo environment (1). The present study was undertaken to characterize more precisely the CL phenomenon and thus the generation of activated oxygen species by polymorphonuclear granulocytes (PMN), eosinophils, and monocytes in dependence on other blood constituents (lymphocytes, erythrocytes, plasma components), phagocytosis stimulus (latex, zymosan), and CL amplifiers like luminol (5-amino-2,3-dihydro-1,4-phthalazinedione) and lucigenin (bis-N-methylacridinium nitrate). To evaluate the role of different oxygen species in generation of the CL response we used sodium azide as an inhibitor of myeloperoxidase (MPO), further potassium cyanide as an inhibitor of superoxide dismutase (SOD) and catalase in CL experiments. The influence of plasma components as a source of opsonic activity and as a scavenger of radicals was investigated by diluting whole blood.

MATERIAL AND METHODS

CL of whole blood

Heparinized blood (20 U/ml) of normal healthy adult volunteers was diluted with CL medium (Dulbecco's modified Eagle's medium containing 50 mM of HEPES, pH 7.4, lacking $NaHCO_3$ and phenolred, Boehringer, Mannheim, FRG) to ratios of 1:5, 1:50, and 1:500. Luminol (Boehringer, Mannheim, FRG) was dissolved (2 mg/ml) in PBS, supplemented with 8 µl

ISBN 0-12-426290-2

TABLE 1

WHOLE BLOOD CHEMILUMINESCENCE

Blood dilution	Amplifier	Stimulus	CL response (counts $x10^{-3}$/20min) total activity mean	total activity %SD	specific activity (total CL/10^3PMN) mean	specific activity %SD
1/5	Luminol	$Zym_{0.5}$	3 048	64	11.9	67
1/50	Luminol	$Zym_{0.5}$	423	57	16.2	46
1/500	Luminol	$Zym_{0.5}$	299	40	116.1	34
1/5	Luminol	$Zym_{0.17}$	2 977	72	12.0	38
1/50	Luminol	$Zym_{0.17}$	147	82	5.6	72
1/500	Luminol	$Zym_{0.17}$	123	56	47.7	51
1/5	Luminol	Latex	4 265	78	16.3	67
1/50	Luminol	Latex	3 429	70	129.5	59
1/500	Luminol	Latex	8 350	44	3 069.7	41
1/5	Lucigenin	$Zym_{0.5}$	7 565	81	28.6	69
1/50	Lucigenin	$Zym_{0.5}$	852	60	32.5	48
1/500	Lucigenin	$Zym_{0.5}$	264	41	102.2	34
1/5	Lucigenin	$Zym_{0.17}$	4 986	80	18.7	68
1/50	Lucigenin	$Zym_{0.17}$	432	85	16.3	72
1/500	Lucigenin	$Zym_{0.17}$	154	46	59.8	42
1/5	Lucigenin	Latex	3 815	84	14.2	71
1/50	Lucigenin	Latex	1 704	62	64.6	52
1/500	Lucigenin	Latex	1 743	34	674.7	29

Influence of blood dilution on Luminol- and Lucigenin-dependent chemiluminescence (CL) production measured in whole blood samples from 28 individuals in the presence of $Zymosan_{0.5}$ (=0.5 mg/400µl), $Zymosan_{0.17}$ (=0.17mg/400µl), and Latex (6 $x10^7$ particles/400µl) triethylamin per ml. Preparation of lucigenin was 5.1 mg in PBS (1 ml).Zymosan particles were suspended at stock concentrations of 17 mg/ml and 50 mg/ml, latex beads (1.1 µm diameter) were adjusted to 6 x 10^9 particles/ml CL medium. Lucigenin, zymosan, and latex were obtained from Sigma, FRG.

CL assay

For CL measurements a six-channel Biolumat (model LB 9505, Berthold, FRG) was used. Prior to CL assays diluted blood samples of each individual were kept at 37°C for 45 min to 200 min according to a standardized protocol. Six vials containing blood sample (0.4 ml) and 10 µl of CL amplifier were placed into counting chambers (37°C). CL response was initiated by 10 µl of phagocytosis stimulus and CL was continuously monitored. The enzyme inhibitors KCN and NaN_3 were added in a volume of 4 µl (final concentrations 1 mM, 10 µM respectively) one minute before phagocytosis was initiated.

RESULTS AND CONCLUSION

Whole blood dilution and CL activity

To evaluate the influence of whole blood dilution on CL we calculated the total CL activity during 20 min after initiation, furthermore the specific CL activity as the total CL related to 10^3 PMN. All stimuli increased the luminol- and lucigenin-dependent specific CL activity from dilution 1:5 to 1:500 (table 1). Considering the zymosan-initiated CL the two amplifying systems gave comparable amounts, but in the case of latex beads there was a much higher increase and yield of specific activity in the luminol- than in the lucigenin-amplified system. In contrast, zymosan induced total CL activities declined with blood dilution, whereas the latex-induced luminol-dependent CL activities showed a twofold increase at dilution 1:500 compared with dilutions 1:5 and 1:50. Latex as a non-activator of the alternative pathway of complement appeared to be the most sensitive stimulus for detecting cellular phagocytic activity by CL in an opsonin- and plasma-poor environment.

Cellular involvement in CL

In an attempt to discriminate between cellular and plasma effects on the CL response a multiple regression analysis was performed. The best correlation of the Cl signal to cellular constituents was obtained when total CL activity was related to the number of leukocytes, red blood cells, and the ratio of PMN to lymphocytes. There were better correlations at dilutions 1:50 and 1:500 than at dilution 1:5. The lucigenin-amplified system yielded better over-all correlations than the luminol-amplified system . The best correlation was obtained when the lucigenin-dependent latex-induced CL was calculated. In short, the dependence of CL on blood cells is best fitted when the ratio PMN/lymphocytes is taken into account. On the other hand, low correlation stands for high influence of plasma factors, which comprise both non-specific factors (e.g. complement, prostaglandins) and specific factors (immunoglobulins).

Influence of azide and cyanide on CL

The MPO inhibitor sodium azide inhibited the luminol-dependent zymosan-induced CL response to a residual value of 25% in a blood dilution independent manner, confirming the well known peroxidase dependence of luminol-amplified CL. Comparable amounts of azide had an enhancing effect on the lucigenin-dependent response (up to 130%). Since azide is known to increase the detectable amounts of superoxide anion and H_2O_2 generated following stimulation, increased amounts of reactive oxygen species are expected to potentiate the lucigenin-dependent CL. In both cases the role of plasma seems to be low. For the SOD inhibitor potassium cyanide the results were different. Depending on the dilution the luminol-amplified samples decreased from initially 121% (1:5) to 65% (1:500), for the lucigenin-dependent CL there was an increase from 147% up to 259% in respect of the dilution 1:500, this finding is in line with the notion that lucigenin preferentially detects superoxide anions.

SUMMARY

CL measurements in diluted whole blood reflect mainly cellular properties at dilutions 1:50 and 1:500 whereas at dilution 1:5 CL measurements are determined by humoral factors as well as by cellular features and interactions.

REFERENCES

1. Kato,T., Eggert,H., Wokalek,H., Schöpf,E., Ernst,M., Rietschel,E.Th., Fischer,H. (1981). Measurement of chemiluminescence in freshly drawn human blood. I. Role of granulocytes, platelets and plasma factors in zymosan-induced chemiluminescence. Klin.Wochenschr. 59, 203-211.

CHRONOBIOLOGICAL INVESTIGATION OF ZYMOSAN INDUCED AND LUMINOL AMPLIFIED CHEMILUMINESCENCE OF PMN LEUKOCYTES IN HEALTHY ADULTS

G. Harald Geelhaar, Michael Betzler, Georg F. Zimmermann, Hans Konrad Schackert, Hans-Peter Geisen, Lutz Edler[+], Christian Herfarth

Department of Surgery, University of Heidelberg D-6900 Heidelberg, W.Germany

[+]*Dep. of Biostatistic, German Cancer Research Centre D-6900 Heidelberg, W.Germany*

INTRODUCTION

Circadian rhythms have been found for several metabolic parameters. Heberer *et al.* (1) described a daytime dependence of zymosan induced and luminol amplified chemiluminescence of PMN leukocytes. The aim of this investigation was to review this data.

MATERIALS AND METHODS

In December 1983 twelve healthy volunteers with an average age from 23 years, ranging from 17-30, were examined for a period of 24 hours in our hospital. During the ensuing 24 hour-period, 15 ml of blood was drawn from the cubital vein, from each subject every two hours. Samples were processed immediately and CL was measured less than 30 min later. 1 ml of blood was diluted with 4 ml of DMEM (Dulbecco's modification of Eagle's medium). 10 U heparin/ml whole blood was used as an anticoagulant. White cells were counted and differentiated. Polymorphonuclear leukocytes and monocytes were considered as CL producing cells. 10 µl luminol were added to 500 µl diluted blood and incubated for 10 minutes at $37^{\circ}C$. Background activity was measured for 1 minute and the CL reaction was initiated adding .5 mg of non opsonized zymosan (suspended in 10 ul DMEM). Dynamic CL-activity (CLA) was measured in intervals of 30 seconds for 40 minutes using

ISBN 0-12-426290-2

a BIOLUMAT LB 9505 (Fa. Berthold, Wildbad, W.Germany). Data out of simultaneous six channel measurement were transferred to an apple II computer, stored on floppy discs and processed. The cortisol level was determined by radio-immuno-assay and the haemogram by a bloodcellcounter (ELT 800, Ortho Diagnostic System, USA).

The progress curve of each volunteer was standardized on the individual medians. This produced the possibility of comparing the individual curves, without being prejudiced by the initial values. The circadian curves were mathematically compiled for each volunteer in the following manner: for each measuring point used in these experiments a new median was found by the compiling of all the deviations of the individual medians of all test subjects and was then brought forth as the progress curve.

RESULTS

Some of the individual CL-activity curves showed a great variability. Others had a smooth profile with only slight fluctuations around the individual median. However some curves seemed to resemble to a circadian rhythm (Fig. 1).

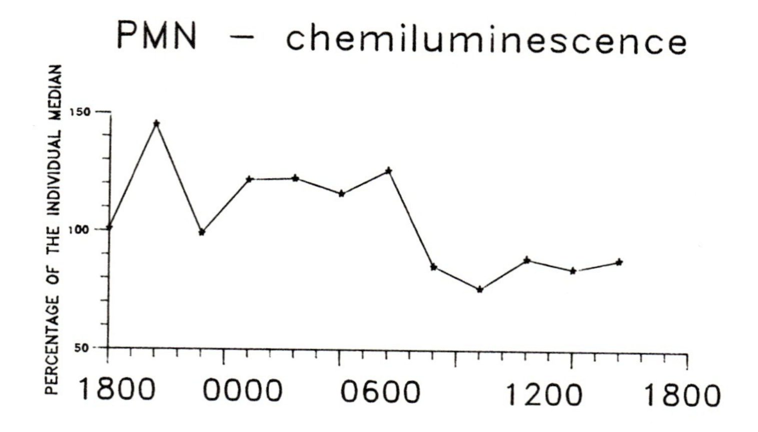

Fig. 1 Daytime dependence of CLA in one volunteer

Combining all individual curves (standardized on the median out of the twelve individual medians) there was no significant daytime dependency of CLA (Fig. 2).

The cortisol level showed, as has been described in many former tests a circadian rhythm with a significant increase ($p < .05$; Wilcoxon signed rank test) from the minimum between 0000 and 0200 to the peak between 0800 and 1000 (Fig. 2).

Total number of peripheral leukocytes showed a sinusoid characterization. High values in the time between 2000 and 0000 (median of all countings = 8250/µl) decreased during the early morning hours. The minimum (median = 6450/µl) was at 1000 ($p < .05$) (Fig. 2).

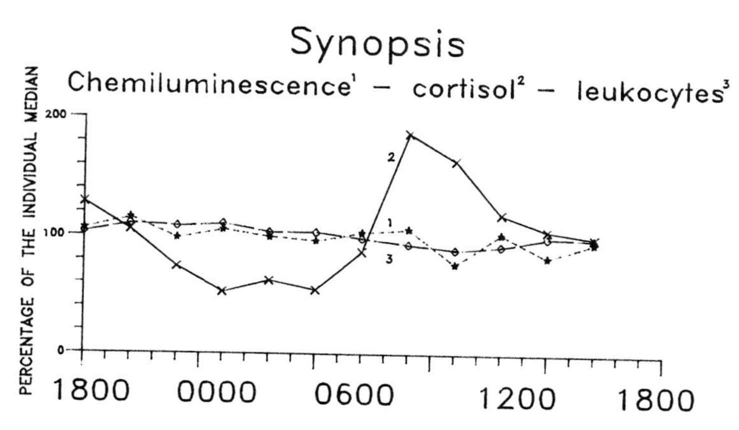

Fig. 2 Daytime dependence of chemiluminescence-activity, cortisol and leukocytes in 12 volunteers.

DISCUSSION

Using our method of determining CL-activity in humans, it seems not to be necessary to obtain blood samples on a predetermined schedule. Nevertheless a circadian rhythm of CLA may exist. Its investigation seems to require methodologies which are more sophisticated. Furthermore a future test design should avoid stress effects on the probands. Factors of disturbance may have been the short period of hospitalization and the repeated sleep interruption due to blood sampling.

ACKNOWLEDGEMENT

This study was supported by the TUMORZENTRUM HEIDELBERG/MANNHEIM "Colon-Rektum-Projekt"

REFERENCES

1. Heberer, M., Ernst, M., Dürig, M., Allgöwer, M., Fischer, H. (1982). Measurement of chemiluminescence in freshly drawn human blood, *Klin.Wochenschr.* 60, 1443-1448.

STUDIES OF GROUP B STREPTOCOCCAL (GBS) OPSONISATION USING LUMINOL-DEPENDENT CHEMILUMINESCENCE

M.J.G. Hastings,[1] J. Deeley,[2] K. Oxley[1] and C.S.F. Easmon[2]

[1]*Department of Medical Microbiology, University of Sheffield Medical School, Sheffield S10 2RX*
[2]*Department of Medical Microbiology, Wright-Fleming Institute, St. Mary's Hospital Medical School, London W2 1PG*

INTRODUCTION

GBS are now established as a major cause of sepsis in the neonatal period. The syndrome is characterised by a high colonisation-infection ratio suggesting either immunological deficiency in a minority of infants or a differential virulence on the part of the organism. Host defence against GBS is mediated by serum opsonisation which is dependent on type-specific antibody and complement(1). We have previously demonstrated variation in the opsonisation of different GBS isolates belonging to serotype III and proposed that resistance to serum opsonisation may act as a virulence factor(2). The main aims of the present study were to assess levels of opsonins to GBS type III in the sera of pregnant women and to analyse the opsonic requirement of a range of clinical isolates of GBS type III.

METHODS

Opsonisation

Pooled serum (PS) was collected from 6 healthy donors and patient serum samples were taken at antenatal clinic visits. All sera were kept at -70°C. Heated serum (HS; 56°C for 30 min.) reconstituted with complement (serum absorbed with GBS type III cells at 2°C) was used in some experiments. The following GBS type III isolates were used: NCTC 11080 (stock laboratory strain); various colonising strains from above patients; infecting strains from Dr. G. Coleman (Strep. Ref. Lab., Colindale). Organisms grown overnight

ISBN 0-12-426290-2

at 37°C in Todd Hewitt broth (BBL) were washed and resuspended in HEPES-buffered Hank's balanced salt solution (HBSS) without phenol red (Gibco Biocult) to a concentration of 10^9 colony-forming units per ml. Opsonisation was performed with 20% serum at 37°C for 20 minutes.

Luminol-dependent chemiluminescence (LDCL) assay(2)

Polymorphonuclear leucocytes (PMN) were prepared by dextran sedimentation followed by ammonium chloride lysis of remaining red cells. PMN were washed twice with HBSS and resuspended in Medium 199 without phenol red (Flow) supplemented with 1% foetal calf serum at 10^6 PMN/ml. Luminol (900μL, 10^{-5} Molar, Sigma), PMN (500μL) and opsonised bacteria (200μL) were added to the reaction vial in the luminometer chamber (Luminometer 1250, LKB Wallac). Peak LDCL was recorded.

RESULTS

Opsonins to GBS type III in sera from parturient women

Sera from 26 parturient women colonised with GBS type III and 50 non-colonised parturients were compared in their ability to opsonise the stock laboratory type III strain as measured by stimulation of PMN LDCL. There was no significant difference in mean opsonic capacity between the 2 groups (Figure).

Opsonisation of clinical isolates of GBS type III

Using pooled serum 25 infecting and 29 colonising strains of GBS type III were opsonised and compared for stimulation of PMN LDCL. The spectrum of response was similar; the majority of isolates being well opsonised. Of 6 isolates showing a marked resistance to opsonisation (<50% of PMN LDCL produced using control organism), 4 came from infecting sources and 2 were colonising strains.

Dependence on heat-labile and heat stable opsonins

When GBS type III isolates were opsonised using HS with varying amounts of complement PMN LDCL stimulation by some strains was markedly reduced. Other isolates were little affected even when only small quantities of complement were used. All strains were poorly opsonised in the complete absence of complement. The influence of type-specific antibody on the opsonisation of GBS type III strains was assessed by diluting HS prior to reconstitution with complement (50 L). Opsonisation of some strains was markedly reduced, whilst others were little unaffected.

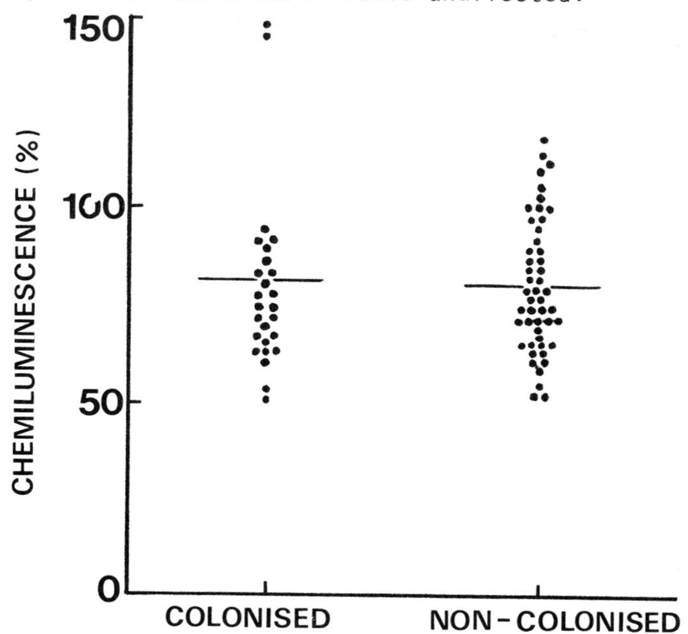

Fig. Opsonisation of GBS type III strain 11080 by sera from colonised (with type III) and non-colonised patients as shown by PMN CL. Each result expressed as a percentage of CL produced using strain 11080 opsonised with PS.

DISCUSSION

The study of the opsonisation of pathogenic organisms is often limited to the analysis of a single bacterial strain as established techniques are time-consuming. LDCL is a simple and rapid method for assessing opsonisation. Using PMN LDCL we have demonstrated that the majority of sera from a population of parturient women had adequate opsonic function against GBS type III. Colonised and non-colonised parturients had similar levels. Hemming *et al.*[1] using CL, found 85% of sera from parturients in their study to efficiently opsonise a GBS type III strain. Baker *et al.*[4] using a radioactive binding assay (RABA) found higher levels of GBS-type III type-specific antibody amongst pregnant women colonised with GBS type III than amongst non-colonised patients. This was not confirmed by Vogel *et al.*[5] using indirect immunofluorescence.

Whether a functional measure of serum activity towards the organism using a technique such as CL or a more specific assay of antibody concentration is a closer reflection of immunity to GBS is not clear. One recent study of GBS type III employing both approaches demonstrated a good correlation between a RABA and CL[6]. A limitation of our own work was the inability to quantitiate serum GBS type-specific antibody. Some valuation of the importance of heat stable opsonins was made by diluting HS prior to reconstitution with complement. The marked strain to strain variation observed may be related to complement activation.

Of the 54 GBS type III isolates tested 6 were resistant to serum opsonisation but correlation between the source of the strain and efficiency of opsonisation was poor. Two of the resistant strains were isolated from babies with surface colonisation alone. All isolates of GBS type III analysed showed an absolute requirement for complement. Variation in response was again apparent between type III strains when HS reconstituted with decreasing amounts of complement was used for opsonisation. The sensitive dependence on an intact complement system for efficient opsonisation shown by some type III strains may be relevant amongst neonates in whom decreased complement activity has been demonstrated.

REFERENCES

1. Anderson, D.C., Edwards, M,S. and Baker, C.J. (1980). Luminol-enhanced chemiluminescence for evaluation of type III group B streptococcal opsonins in human sera, *J. Infect. Dis.* 41, 370- .
2. Baker, C.J., Webb, B.J., Kasper, D.L., Yow, M.D. and Beachler, C.W. (1980). The natural history of group B streptococcal colonization in the pregnant woman and her offspring. II. Determination of serum antibody to capsular polysaccharide from type III group B Streptococcus, *Am. J. Obstet. Gynecol.*, 137, 39- .
3. Hastings, M.J.G. and Easmon, C.S.F. (1981). Variations in the opsonic requirement of group B streptococcus type III, *Br. J. Exp. Path.*, 62, 519- .
4. Hemming, C.F., Hall, R.T., Rhodes, P.G., Shigeoka, A.O. and Hill, H.R. (1976). Assessment of group B streptococcal opsonins in human and rabbit serum by neutrophil chemiluminescence, *J. Clin. Invest.*, 58, 1379- .
5. Vogel, L.C., Boyer, K.M., Gadzala, C.A. and Gotoff, S.P. (1980). Prevalence of type-specific group B streptococcal antibody in pregnant women, *J.Paediatr.*, 96, 1047-

This work was funded by a grant from the M.R.C.

EVALUATION OF ALTERNATIVE PATHWAY OF COMPLEMENT FOR OPSONIC ACTIVITY BY NEUTROPHIL CHEMILUMINESCENCE. ITS CORRELATION WITH THE PARTICLE COUNTER METHOD

P. Hindocha, Ruth Hill, J.D.M. Gould & C.B.S. Wood.

The Medical College of St Bartholomew's Hospital & the London Hospital Medical College, Academic Department of Child Health, Queen Elizabeth Hospital for Children, Hackney Road, London, E2 8PS.

INTRODUCTION

Opsonization by alternative pathway of complement is an important host defence mechanism. We are presenting a rapid, objective, semiquantitative test for yeast opsonization based on the principle of emission of light by phagocytosing neutrophils.

MATERIALS AND METHODS

Blood specimens were obtained from children investigated for suspected immunodeficiencies. Sera were stored at −70°C until analysis.

Live Baker's yeast *(Saccharomyces Cerevisiae)* was heat inactivated at 100°C in saline and washed 3 times in saline and stored concentrated at +4°C.

Polymorphonuclear leucocytes (PMNs)

Dextran separated PMNs after treating with 0.83% tris buffered ammonium chloride to remove contaminating erythrocytes were washed twice in Hank's balanced salt solution adjusted to 5 x 10^6/ml.

Opsonization of Yeast. HBSS washed yeast (3 x 10^8/ml) were rotated at 37°C with 100μl of serum and 100μl of HBSS to achieve the desired dilution of serum for the required length of time. Opsonization was stopped by addition of ice cold HBSS.

ISBN 0-12-426290-2

Chemiluminescence assay. Luminol dependent phagocytic chemiluminescence was measured at 37°C in a picolite luminometer (Packard Instruments). Light generated due to phagocytosis of opsonized yeast was recorded every 6 mins and a mean value obtained. CL results are expressed as a percentage of the result obtained using serum pool which was kept at −70°C. Serum pool was obtained from 10 normal individuals (5 males and 5 females).

RESULTS

Kinetics of Opsonization of Yeast

Fig. 1 shows optimal time for opsonization was 15 mins. Light emission increased with increasing serum concentration. 14% serum concentration was selected (somewhat arbitrarily) as a convenient concentration that produced a significant but not a maximum CL.

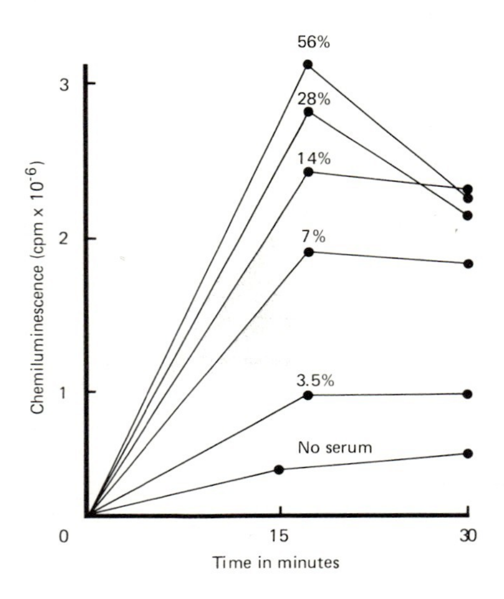

Fig. 1 Effect of serum concentration and time on opsonization of yeast.

Fig. 2 shows that peak CL was observed with granulocyte ratio of 1 and between 60–70 yeast particles. Ratio of 65 yeast particles to 1 PMN was used in subsequent experiment.

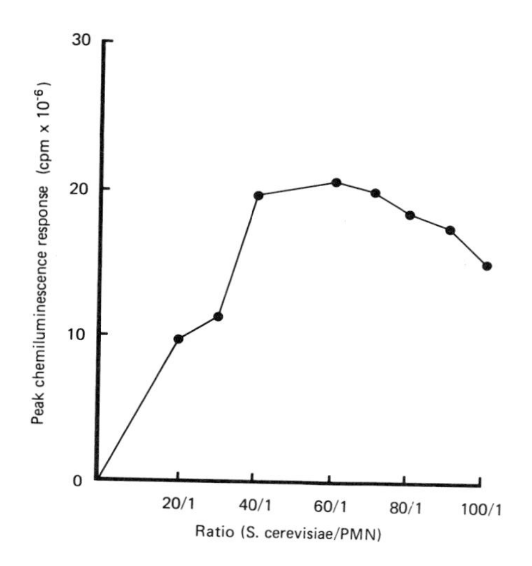

Fig. 2 Effect of yeast PMN ratio on kinetics of CL.

Fig. 3 shows that opsonization of yeast, which under normal circumstances is known to be effected almost entirely by the alternative pathway was little affected by the chelation of Ca^{++} using MgEGTA. In contrast, inactivation of factor B by heating the serum to 50°C for 20 minutes reduced opsonization considerably.

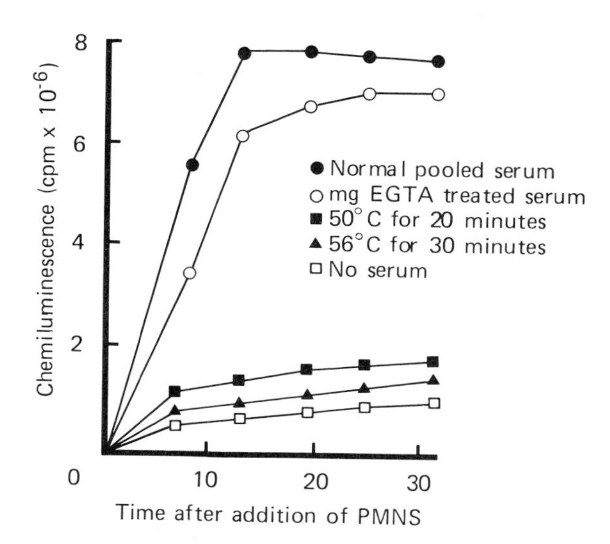

Fig. 3 Effect of MgEGTA chelation of serum, heating to 50°C for 20 mins and 56°C for 30 mins on kinetics of yeast opsonization.

Normal Values: The sera of 31 children aged between 5–12 years were studied. The mean percentage CL was 71 ± 32 (1 SD). Defective opsonization was therefore taken as below 39% (1 SD below the mean). 5 out of 31 children gave values below this.

Fig. 4 shows that when 31 sera were compared with the particle counter method, excellent correlation r=0.82 $p<0.001$ (Pearson's correlation test) is seen between two methods. One serum classified as normal by the particle counter method was found to be defective by neutrophil CL.

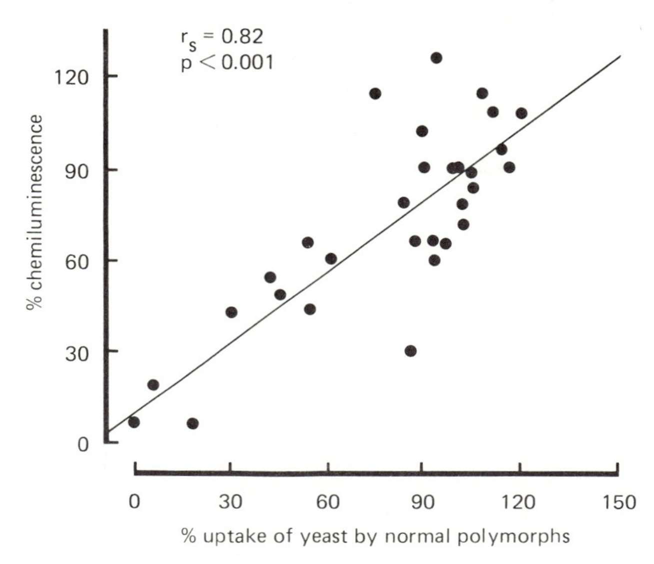

Fig. 4 Comparison with the particle counter method of yeast opsonization.

CONCLUSIONS AND RECOMMENDATIONS

The method of studying the alternative pathway of complement by neutrophil chemiluminescence is simple, objective and quantitative. The assay correlated with the particle counter method used for complement derived opsonins. Since opsonization is strongly dependent on serum concentration, in order to avoid getting false low and high results it is recommended that serum concentration is standardised.

REFERENCES

1. Levinsky, R.J. Harvey, B.A.M. and Paleja, S. (1978). Rapid objective method for measuring the yeast opsonization activity of serum. *J. Immunol. Methods* 24, 251–256.

AUGMENTATION OF THE LUMINOL-DEPENDENT CHEMILUMINESCENCE RESPONSE OF PURIFIED MONOCYTES TO STAPH.AUREUS BY AUTOLOGOUS LYMPHOCYTES (*)

A. Müller, P. Schuff-Werner, M. Wurl, and G.A. Nagel

Department of Internal Medicine, Division of Hematology/Oncology
University Clinic Göttingen
Robert-Koch Str. 40, D-3400 Göttingen (FRG)

INTRODUCTION

In another study presented at this symposium we could demonstrate, that chemiluminescence (CL) of purified monocytes upon stimulation with *Stap. aureus* decreases up to 25% as compared to a monocyte-lymphocyte suspension. In this study we systematically investigated this phenomenon in order to determine whether it was due to (a) a loss of CL-activity by different separation procedures or (b) a specific need of lymphocytes for an optimal CL-reactivity to *Staph. aureus*.

MATERIAL AND METHODS

Mononuclear cells (MNC) were obtained by Ficoll-Hypaque gradient centrifugation and further purified both by adherence to plasma coated gelatin and by Percoll TM density gradient centrifugation respectively. All MNC preparations were adjusted to the same number of monocytes by specific esterase stain (10^5 E^+ MNC).

For CL-measurement we used a six-channel luminescence analyzer (Biolumat, LB 9505, Berthold, Wildbad, FRG) equipped with six photomultiplier tubes allowing simultaneous recording of six samples. Experimental data were aquisited on-line by a microcomputer system (IMCA, Rosdorf, FRG) and recorded on 8" double density floppy disks.

ISBN 0-12-426290-2

The data were analyzed by different parameters concerning the time dependent kinetics of the CL reaction as peak-CL (maximal CL-activity), t_{max} and integrals. In addition the kinetics of the reaction was plotted as cpm/time diagram.

The CL-assay was performed as follows. The total volume of one test probe consisted of 100 µl purified monocytes or unfractionated mononuclear cells, always adjusted to the same number of 10^5 E^+ cells, 100 µl luminol (10^{-5}M), 100 µl stimulus and 100 µl lymphocytes at different concentrations.

RESULTS

In all our experiments it was evident that as well adherence purified as Percoll TM isolated MNC exhibited a decreased luminol-dependent CL-response towards opsonized *Staph.aureus* as compared to unseparated monocytes in a mononuclear cell suspension after Ficoll/Hypaque preparation.

Recombination of one part of purified monocytes with three parts of autologous lymphocytes was found to augment the CL-reactivitv , as compared to the signal of separated E^+ MNC (Fig. 1).

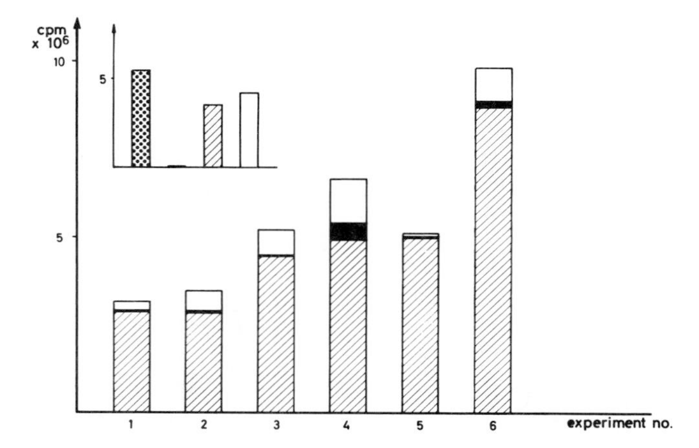

Fig. 1 Augmentation of the Staph.Aur. induced CL of 10^5 adherence purified monocytes by recombination with 3 x 10^5 autologuoes lymphocytes.
adherence purified monocytes; monocyte depleted lymphocytes; recombinated monocytes; unfractionated mononuclear cells

The recombination experiments of 10^5 monocytes with different concentrations of autologoues lymphocytes clearly demonstrated a CL-augmenting "helper effect", that depends on the number of lymphocytes. Additional preincubation experiments of lymphocytes stimulated by *Staph.aureus* over 20-30 min also showed an enhancement of the CL up to 50% as compared to identical experiments without preincubation.

DISCUSSION

From our data we suggest that enhancement of MNC-CL-activity by autologoues T-cells is dependent on a lymphocyte-derived factor. Immunological host-defense mechanisms are often based on the interaction of different immunological cell-types. For example there is a clear functional interdependence of monocytes, macrophages and lymphocytes.

It is well accepted, that sensitized T-lymphocytes release macrophage-activating factor(s) during contact with a specific antigen. In vivo this antigen initiates a differentiation of the macrophage into an activated state, associated by an increased intracellular bactericidy. This bactericidy is known to correlate with cellular-myeloperoxidase activity, that generates CL.

(*) supported by BMFT-project no. 0384 100

PHAGOCYTOTIC AND METABOLIC ACTIVITY OF GRANULOCYTES IN AIDS-, LYMPHADENOPATHY- AND HEMOPHILIA PATIENTS

L. Stöhr, *M.J. Sessler, P. Altmeyer, *I Scharrer, H. Holzmann, *W. Stille, *E.B. Helm, * M. Elbert, J. Gürenci

*Department of Dermatology and *Department of Internal Medicine J.W. Goethe University, Frankfurt/Main, F.R.G.*

INTRODUCTION

In patients with acquired immune deficiency syndrome(AIDS) the cellular immunity is suppressed resulting in opportunistic infections and malignancies. It is often asserted that frequent contamination of the immune system with foreign proteins is a possible cause of AIDS in homosexuals. On the other hand, hemophiliacs are treated with foreign proteins therapeutically and are, therefore, considered as possibly endangered for AIDS. Patients suffering from AIDS show a selective quantitative defect in their helper/suppressor subset T-cells resulting in a decreased helper/suppressor-ratio.

Our study aimed at revealing the function of granulocytes (polymorphonuclear leukocytes, PMNL) in AIDS-, lymphadenopathy- and hemophilia patients as the first step of defence against infections and not only the lymphocytes as participants of the cellular and humoral immune system.

MATERIALS AND METHODS

Granulocytes were isolated from heparinized human blood by sedimentation over Percoll^R (1) and adjusted to 2×10^6 cells/ml. Chemiluminescence (CL) was measured in a 6-channel "Biolumat" LB 9505 /Berthold,Wildbad,FRG) at 37°C. 100 µl of prepared cell suspension (2×10^5 cells) were mixed with 100 µl of diluted luminol (2) in a Lumac^R - cuvette. Phagocytosis was started by adding 100 µl Zymosan^R

ISBN 0-12-426290-2

suspension (3). The data continously obtained were recorded on line (Apple II) and CL-counts were plotted against time.

Stimulation of granolocytes was performed by using 100 µl cell suspension, 100 µl solution of luminol and, instead of Zymosan[R], 100 µl MEM-Dulbecco together with 10 µg concanavalin A (con A) were added.
Investigated subjects:
20 healthy individuals were compared to 20 homosexual patients suffering from hyperplastic lymphadenopathy and to 20 patients with hemophilia A of different severity, receiving antihemophilic globulines (factor VIII concentrate). In addition 4 patients with AIDS were examined before proceeding of any treatment. One of these patients together with another 3 AIDS-patients were examined three month after treatment with Bactrim[R] and Zovirax[R].

RESULTS

The integrals of the time activity curves of the five groups are listed in tab. 1.
Zymosan[R]-induced CL is significantly reduced to about 50 % in AIDS-patients without any treatment. After symptomatical treatment an overstimulation of phagocytosis occurs, exeeding normal values. Opsonization of Zymosan[R] with poolplasma can stimulate granulocyte activity to about 250 % in all the groups, by using autologous plasma,to about 200 %. In case of hemophiliacs, however, "autologous stimulation" is reduced, hinting at a serum factor in hemophiliacs impairing the phagocytotic activity of their granulocytes.
The ratio between con A-stimulated and resting granulocyte activity is shown in fig. 1: Control subjects and hemophiliacs are in the same range. According to the results obtained from the phagocytosis assay, untreated AIDS-patients show a significantly reduced cell activation with con A. After symptomatical therapy a slightly increased activity could be obtained, but is still within pathological limits. The metabolic response of granulocytes to con A was markedly suppressed in lymphadenopathy-patients but is still different from AIDS-patients.
No correlation could be found between CL and the T-helper/T-suppressor-ratio, the immunoglobulines (IgA,IgG,IgM), the serum level of components of the complement (C3c,C4) and the dosage of factor VIII concentrate given to hemophiliacs.

TABLE 1

Phagocytotic activity of polymorphonuclear neutrophil leukocytes, induced by Zymosan[R]

collectiva	not opsonized	opson.with poolplasma	opson.with autologous plasma
I	116 ± 55	406 ± 82	348 ± 90
IIa	65 ± 24	210 ± 35	164 ± 54
IIb	338 + 147	733 ± 295	637 ± 254
III	124 ± 70	433 ± 39	345 ± 132
IV	163 ± 67	499 ± 121	236 ± 137

I: control group, IIa:AIDS without any treatment, IIb:AIDS after treatment, II: homosexuals with hyperplastic lymphadenopathy, IV: hemophiliacs

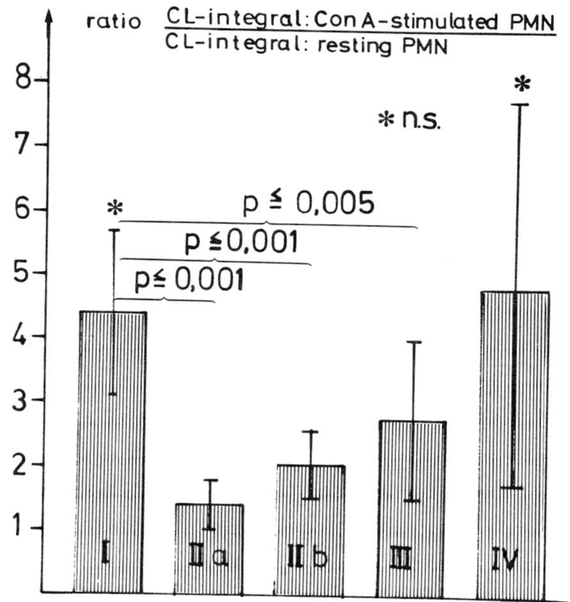

Fig.1: Stimulation of CL with concanavalin A (10 µg). Explanation of groups I to IV see tab. 1.

CONCLUSIONS

The significant decrease of PMNL-activity in AIDS-patients may indicate a depressive effect of a possible AIDS-causing agent. The impaired granulocyte activity itself could probably explain the reduced immunological defence in these patients.
The failure to release active forms of oxygen following stimulation with the mitogen con A (and F-meth-leu-phe , not shown here) could also explain the inability of neutrophils in AIDS to kill bacteria.
Summarizing one can say that there are at least two defects associated with AIDS and in part with AIDS-endangered homosexuals suffering from lymphadenopathy:

1. a defect of PMNL starting Zymosan[R]-induced phagocytosis with a subsequent release of oxygen radicals and
2. a defect of PMNL response to con A with metabolic events and changes in membrane potential and structure.

CL-measurement introduced a further parameter for studies on cellular immunity of AIDS- and AIDS-endangered patients.

REFERENCES

1. Hjorth,R.,Jonson,H.-K.,Vredblad,P. (1981): A rapid method for purification of human granulocytes using Percoll[R]. A comparison with dextran sedimentation. J.immunol. Methods 43: 95-101
2. Kato,T., Wokalek,H., Schöpf,E. et al. (1981):Measurement of chemiluminescence in freshly drawn human blood. Klin. Wochenschr. 59: 203-211
3. Cheson,B.D., Christensen,R.L., Sperling,R. et al. (1976): The origin of chemiluminescence of phagocytosing granulocytes. J.clin.Invest. 58: 789-796

VI INSTRUMENTATION

RECENT DEVELOPMENTS IN AUTOMATIC LUMINESCENCE INSTRUMENTATION

F. Berthold

Laboratorium Prof. Dr. Berthold
D-7547 Wildbad, FRG

ABBREVIATIONS

LIA:	Luminescence Immuno-Assay
Luminescence:	Bio-and Chemiluminescence
LSC:	Liquid Scintillation Counting
Luminometry:	The measurement of Bio-and Chemiluminescence
PM:	Photomultiplier

INTRODUCTION

LSC and luminometry have many things in common. Both types of instrumentation have to measure extremely low quantities of light.

In luminescence, single photon emission occurs at random.

LSC is the measurement of photon bursts occurring, in the case of liquid scintillation, within few nanoseconds.

To distinguish low-level scintillation events from single photons (chemiluminescence as a disturbing factor in LSC) and thermal electron emission (noise), two PMs operated in coincidence are used, discriminating photon bursts from randomly emitted single photons and noise electrons.

A LS-counter may be seen as two photon counters operated in coincidence. Summing the pulses from each PM, instead of rejecting all non-coincident events,transforms a LS-counter into a very sensitive luminometer.

ISBN 0-12-426290-2

LS-counters have, for a long time, been used for certain types of luminescence studies (1-4). However, LS-counters lack certain capabilities which are mandatory in some areas of luminometry:

1) reagent injection while the samples are in the measuring position in front of the PM-tube. This is required for reactions with fast kinetics (most chemiluminescence reactions as used in LIA).
2) Temperature stabilization. LS-counters frequently supply cooling, but never heating, as may be required for incubation or cell stimulation studies.
3) Standard LSC-vials are bigger than required for luminometry, giving rise to unnecessary color quench problems.

Combined LSC/Luminometer

It was the aim of this work to combine the features required for luminometry and LSC in a single instrument (Fig. 1).

Up to 500 samples are carried in a free-standing flexible chain, parts of which can be unhooked for centrifugation or other types of sample preparation.

An elevator lifts the sample from the carrier element upwards into the measuring station, providing a light-tight seal.

For luminometry, an activator reagent and/or other reagents may be injected into the sample vial placed in the measuring position. Instead of the injection ports, it is also possible to insert flow-through cells for either luminometry or LSC.

Two low-noise 2"-bialkali PM-tubes are normally used, but may be exchanged against tubes with increased red-sensitivity. To keep the background low (about 100 cps is a good value for a 2"-PM-tube),Peltier-cooling is applied to the cathodes.

The electronic circuitry has essentially all features common in modern LS-counters: fast coincidence (20 ns), pulse summation, two measuring channels for dual-label assays, quench-correction using an improved sample-channels ratio method. The instrument accepts 12 dia. x 75 mm tubes, the most common vials for RIA. Vials are open in luminometry, and capped for LSC.

Two-channel luminometer

Instead of adding the pulse counts from both PM-tubes, they may also be counted separately in two channels. If two different optical filters are placed between sample and photocathodes (Fig. 2), the luminometer is able to measure light at two different wavelengths simultaneously.

For applications in homogeneous immuno-assays with overlapping wavelength-ranges, a spill-over correction has to be applied. The ratio of the corrected intensities is then used as raw-data to be processed by essentially standard immuno-assay data reduction programs.

Software Aspects

An important aspect of our luminometer design is the future use for LIA in a clinical environment. The work of Woodhead and Weeks (5) has proved that LIA can achieve considerably higher sensitivities than RIA, as can be expected theoretically for immuno-luminometric assays.

We feel that great care has to be taken to obtain the same level of precision and accuracy as for RIA, and to allow the users to check the day-to-day performance of their assays.

To this end, extensive quality control programs are presently being developed, providing the user with at least 10 quality-control relevant data displayed as LEVI-JENNINGS plots, for at least 30 different assay-protocols.

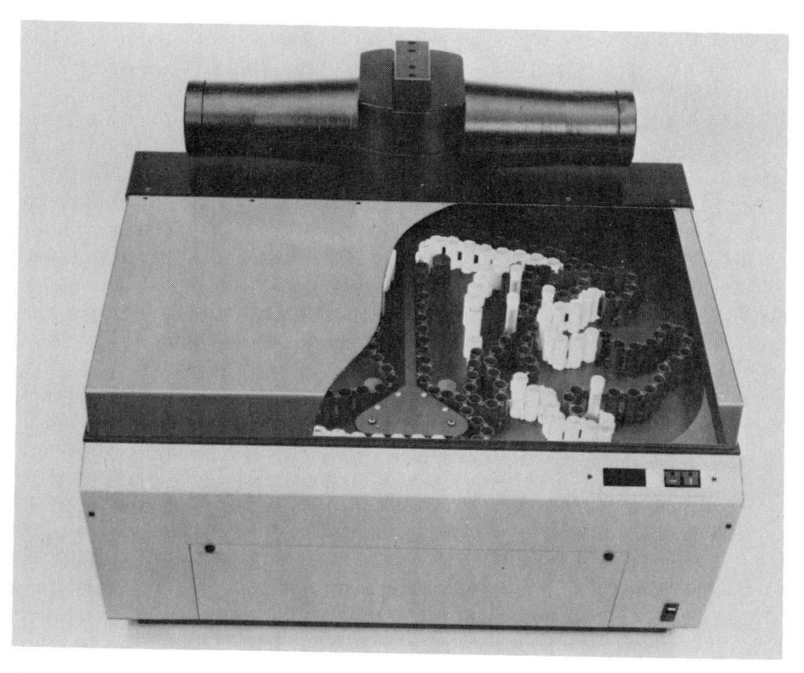

Fig. 1 Combined luminometer/LSC for 500 samples.

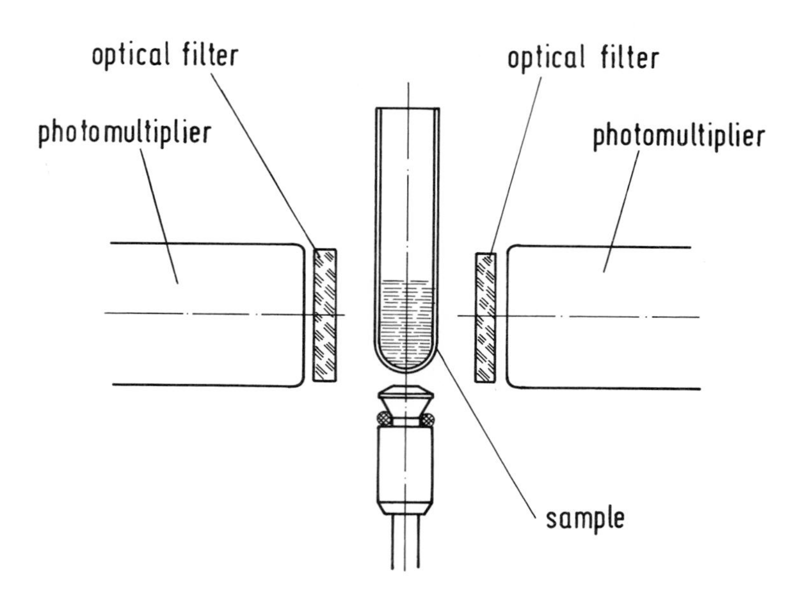

Fig. 2 Measuring station of dual-wavelength luminometer with 2 PM-tubes and 2 optical filters.

References

1. Horan, T.D., *et al.* (1980). Application of Liquid Scintillation Spectrometry in the Evaluation of Neutrophil Function. *In* "Liquid Scintillation Counting" (Eds A. Brown, *et al.*). Vol 2, pp. 321-340. Academic Press, New York.
2. Horan, T.D.,*et al.* (1980). Characterization of Neutrophil Chemiluminescence using a Liquid Scintillation Counter. *In* "Liquid Scintillation Counting" (Eds. C.T. Peng *et al.*) Vol. 2, pp. 341-353. Academic Press, New York.
3. Jederberg, W.W., and Krueger, G.G. (1980). Monocyte Chemiluminescence: Physical and Biological Parameters. *In* "Liquid Scintillation Counting" (Eds C.T. Peng *et al.*) Vol. 2, pp. 355-375. Academic Press, New York.
4. Allen, R.C. (1980). Chemiluminescence: An Approach to the Study of Anmoral-Phagocyte Axis in Host Defense against Infection. *In* "Liquid Scintillation Counting" (Eds C.T. Peng *et al.*). Vol. 2, pp. 377-393. Academic Press, New York.
5. Woodhead, S. and Weeks, I. Private Communication.

AUTOMATED LUMINOMETER SETUP FOR CONTINUOUS MEASUREMENT OF 25 SAMPLES

Esa-Matti Lilius[+], Matti Waris[+] and Matti Lang[§]

[+]*Department of Biochemistry, University of Turku, SF-20500 Turku, Finland*

[§]*EFLAB Oy, SF-00811 Helsinki, Finland*

INTRODUCTION

Applications of bio- and chemiluminescent analytical techniques have attained considerable interest in biosciences. Liquid scintillation counters, fluorimeters and self-built instruments have been used as luminometers and recently numerous sophisticated purpose-built instruments have become available on the market. Most of them, however, provide only a single vial measurement being unsuitable for rapid monitoring of multiple samples. We are interested especially in the opsonization, leukocyte stimulation and phagocytosis studies. In these processes the time dependent CL rate is usually observed over a period of 20-120 min. Temperature stabilization is necessary. Numerous samples (often over 20) must be measured. In order to achieve a temporal trace of luminescence simultaneously from multiple samples without handling each sample several times we have developed an automated luminometer setup for continuous and simultaneous measurements of up to 25 samples by connecting LKB Wallac 1251 Luminometer to Olivetti M 20 microcomputer.

MATERIALS AND METHODS

Instruments

LKB Wallac 1251 Luminometer has a capacity for 25 samples housed in a temperature controlled sample carousel. Temperature control (20-45°C) is stated to be within 0.1°C. The signal from the photomultiplier tube is in the form of anode

ISBN 0-12-426290-2

current. A time constant for signal smoothing as well as integration and delay times can be selected. Up to three optional peristaltic dispensers can be used to add automatically reagents to the cuvette. Mixing of the contents of the cuvette in the detector chamber can be accomplished by cuvette rotation either continuously or by pulses. The instrument has a built in microprosessor allowing the user to give commands and write programs to start and control operations.

The computer we use is Olivetti M 20 16-bit microcomputer with a system memory capacity of 128 Kb. It is equipped with two 5 1/4" disk drives with a saving capacity of 320 Kb each, a video and PR 1450 dot matrix printer with the graphics capability.

Measuring procedure

The luminometer is operated through the Olivetti by the programs written in Microsoft-BASIC. After the start command of the measurement the sample cuvettes are loaded into the luminometer sample carousel at time intervals at which the cuvettes are measured. The minimum time interval required to measure successive samples is 5.2 sec allowing the vials to

Phagocytosis of B. pertussis and zymosan by different numbers of leucocytes.
DATE 31 01.1984 TIME 16.15

01 -	0.025	00:01:36	02 -	0.022	00:01:43	03 -	0.027	00:01:51	04 -	0.019	00:01:59	05 -	0.017	00:02:06
06	0.001	00:02:14	07	0.008	00:02:22	08	0.061	00:02:29	09	0.032	00:02:37	10	0.273	00:02:45
11	0.630	00:02:52	12	2.660	00:03:00	13	4.623	00:03:07	14	11.260	00:03:15	15	0.036	00:03:23
16	0.104	00:03:30	17	0.530	00:03:38	18	1.166	00:03:46	19	4.623	00:03:53	20	7.976	00:04:01
21	14.450	00:04:08												
01 -	0.019	00:04:24	02 -	0.023	00:04:32	03 -	0.004	00:04:39	04 -	0.007	00:04:47	05	0.077	00:04:55
06	0.288	00:05:02	07	0.552	00:05:10	08	0.161	00:05:19	09	0.251	00:05:25	10	1.786	00:05:33
11	3.601	00:05:41	12	18.500	00:05:48	13	35.640	00:05:56	14	68.920	00:06:03	15	0.213	00:06:11
16	0.390	00:06:19	17	[illegible]	00:06:26	18	4.233	00:06:34	19	21.580	00:06:42	20	40.700	00:06:49
21	85.150	00:06:57												
01	[illegible]	00:07:12	02	0.018	00:07:20	03 -	0.009	00:07:28	04	0.013	00:07:36	05	0.177	00:07:43
06	0.379	00:07:51	07	0.691	00:07:59	08	0.535	00:08:06	09	0.633	00:08:14	10	4.204	00:08:22
11	8.311	00:08:29	12	42.280	00:08:37	13	86.320	00:08:44	14	158.500	00:08:52	15	0.413	00:09:00
16	0.843	00:09:07	17	4.422	00:09:15	18	8.049	00:09:23	19	40.430	00:09:30	20	79.800	00:09:38
21	169.600	00:09:46												
01 -	0.022	00:10:01	02 -	0.013	00:10:09	03	0.003	00:10:17	04	0.032	00:10:24	05	0.245	00:10:32
06	0.459	00:10:39	07	0.887	00:10:47	08	0.751	00:10:55	09	1.099	00:11:02	10	7.202	00:11:10
11	13.530	00:11:17	12	69.280	00:11:25	13	143.300	00:11:33	14	260.500	00:11:40	15	0.646	00:11:48
16	1.157	00:11:56	17	6.389	00:12:03	18	11.550	00:12:11	19	59.750	00:12:18	20	120.100	[illegible]
21	252.400	00:12:34												
01 -	0.010	00:12:49	02 -	0.009	00:12:57	03	0.019	00:13:04	04	0.058	00:13:12	05	0.347	00:13:20
06	0.686	00:13:28	07	1.267	00:13:35	08	1.032	00:13:43	09	[illegible]	00:13:50	10	10.330	00:13:58
[illegible]	[illegible]	00:14:06	12	96.640	00:14:14	13	196.400	00:14:21	14	354.000	00:14:28	15	0.764	00:14:36
16	1.571	00:14:44	17	8.044	00:14:51	18	14.660	00:14:59	19	77.400	00:15:06	20	157.000	00:15:14
21	321.900	00:15:22												
01 -	0.013	00:15:37	02 -	0.019	00:15:46	03	0.044	00:15:53	04	0.051	00:16:00	05	0.427	00:16:08
06	0.905	00:16:16	07	1.735	00:16:23	08	1.222	00:16:31	09	2.368	00:16:38	10	12.590	00:16:46
11	21.970	00:16:54	12	116.700	00:17:01	13	233.400	00:17:10	14	431.300	00:17:16	15	0.870	00:17:24
16	1.833	00:17:32	17	9.493	00:17:39	18	16.920	00:17:47	19	92.270	[illegible]	20	188.700	00:18:02

Fig. 1 A section of the data print-out

be remeasured at 130 sec intervals (the measuring cycle) if 25 vials are in, i.e. every sample is measured 27 times per

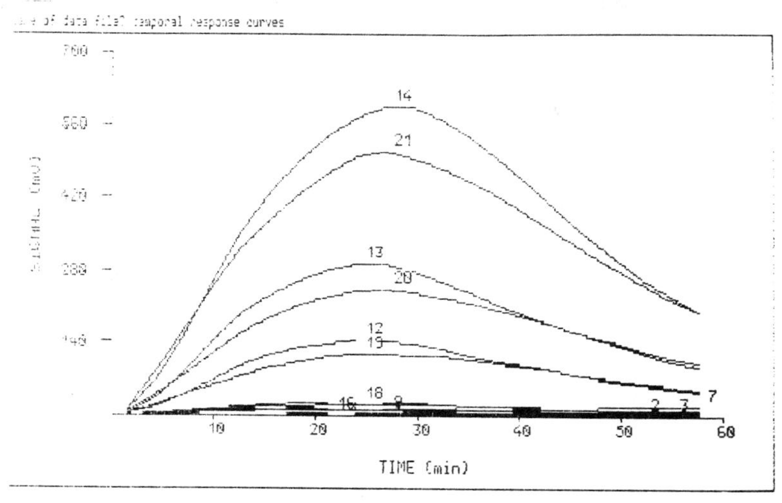

FIGURE FROM EXPERIMENT: **Phagocytosis of B. pertussis and zymosan by different numbers of leucocytes.**
DATE: **31.01.1984** **TIME: 16.15**

Vial	CL-peak mV	Time min	AVC mVmin	V-max mV/min	Time min
1	.03	57.65	.684067	6.78571E-03	5.8
2	.081	52.0333	1.32213	7.10059E-03	47.825
3	.48	54.8333	10.9019	2.89285E-02	45.0167
4	.969	57.65	21.4109	3.92857E-02	45.0167
5	4.8	57.65	115.37	.170299	39.425
6	9.616	57.65	221.332	.313928	45.0167
7	19.35	57.65	453.223	.640829	36.625
8	1.403	24.0167	49.9725	.209467	.8
9	2.703	24.8167	94.9271	.286071	14.2167
10	15.26	24.0167	535.37	1.11714	11.4167
11	26.05	24.0167	927.301	1.8529	8.60833
12	143.8	24.0167	5163.73	9.77143	11.4167
13	292.6	24.0167	10494.9	20.2296	8.60833
14	593.6	26.8167	21775.2	36.213	8.60833
15	1.253	21.2167	46.8988	8.27219E-02	8.60833
16	2.221	21.2167	83.0577	.161786	5.8
17	11.67	24.0167	459.316	.77	5.8
18	21.49	24.0167	841.364	1.36286	5.8
19	117.8	24.0167	4634.81	6.85917	8.60833
20	243.4	26.8167	9394.28	14.3077	8.60833
21	507	26.8167	19280.4	30.1607	5.8

Fig. 2 A print-out of processed data

hour. The measuring cycle can be lenghtened by increasing the delay time. The number of measuring cycles is unlimited. The data in the form of vial number, CL signal (mV) and time is saved on a disk in Olivetti as individual data files for each luminometer runs. After each complete run the data is automatically processed by the computer. The measurement data, a figure of temporal response curves as well as the peak, the integral and the maximum rate values of the CL responses are printed out in five minutes after the end of the luminometer program. Afterwards each saved data file can be processed separately for example to select individual samples for a figure print-out.

RESULTS AND DISCUSSION

Fig. 1 shows a section of a data print-out. It comprises the title of the experiment, the date and the real time, as well as vial numbers, the CL signals of each sample and the actual measuring times of the samples. If desired the program omits to print out this data (it is shown on the video, anyway) but prints out only the processed data as shown in Fig. 2. The temporal traces of luminescence of each cuvette are presented in one figure. The CL peak value and the time at which it is reached, the integral of the response (AUC = area under the curve) as well as the maximum rate of increase of the emission and its point of time are printed out. In order to facilitate the comparison of the peak values of different samples a block diagram of peaks is also automatically printed out (not shown). Moreover, individual samples can be selected to trace the emission in a more proper scale or a mean curve can be plotted (not shown).

Robinson and Penny (1) have presented a simple modification to any standard liquid scintillation counter that allows for the measurement of multiple samples automatically. Our system is superior to theirs because we can test simultaneously 25 samples instead of 10. Moreover, successive samples can be measured faster (5 sec vs. 16 sec) and the temperature as well as mixing are properly controlled.

REFERENCE

1. Robinson, J.P. and Penny, R. (1982). Chemiluminescence response in normal human phagocytes. I. Automated measurements using a standard liquid scintillation counter, *J. Clin. Lab. Immunol.* 7, 215-217.

A COMPARISON OF TWO SEMI-AUTOMATIC LUMINOMETERS SUITABLE FOR ROUTINE LUMINESCENCE IMMUNOASSAYS

W.G. Wood, A. Gadow and C.J. Strasburger
Klinik für Innere Medizin, Medizinische Hochschule Lübeck D-2400 Lübeck 1, FRG.

INTRODUCTION

Whether luminescence immunoassays can be performed routinely depends not only upon the availability of a robust assay system, but also upon an automatic and reliable luminometer. This article describes experience with two semi-automatic luminometers, the 25-sample LKB-1251 (LKB, Turku, SF) and the 250/300 sample LB 950 (Berthold, Wildbad, D). Both luminometers have been in routine use for over a year, during which time over 150,000 samples have been measured.

EXPERIENCES WITH THE LKB-1251 AND LB 950 LUMINOMETERS

The main features of both machines are given in Table 1. Both machines have been replaced under guarantee because of defects. The replacement luminometers functioned well and have stood up to the stress of daily routine use when up to 1100 samples and 6 or 7 operators are active in the laboratory. Several small alterations have been made to both instruments over the period under test, the majority of these to the LB 950. Both luminometers are described below as far as performance and alterations are concerned.

LKB-1251

The LKB 1251 was delivered with 3 peristaltic pumps and was connected to a Teletype printer as hardcopy output. All 3 peristaltic pumps dispensed directly into the measuring chamber. It was important to line up the nozzles of the delivery lines exactly over the cuvette position or the reagents splashed against the mirror assembly and ran down the cuvette-lifting mechanism, eventually leading to its blockage. The pumps themselves performed well, the silicone lines being replaced every 3-4 months.

The LKB-1251 used special cuvettes (51 x 12 mm) which

ISBN 0-12-426290-2

TABLE 1

A tabular comparison of the LKB-1251 and LB 950 Luminometers

Parameter	LKB 1251	LB 950
Capacity (no. of samples)	25^{+}	250/300
Display	LED	monitor
No of dispenser (max)	3	3
Max. volume/injection µl	35	350
Volume range with cv under 2% µl	15-35	120-350
Preincubation possibile	yes	yes
Temperature range °C	20-50	20-50
Cooling/Heating	water / thermostat	Peltier elements
Temp. Stability °C	0.1	0.1
Visual control of reaction kinetics	no	yes
Data printout	Teletype	Epson MX 82
Programmes	EEPROMS + do-it-yourself CK + CK-B and ATP measurement	Diskette with reaction kinetic ATP, phagocytosis programmes.
LIA data reduction	off-line	off-line (in preparation)
Precision obtainable* coefficient of variation (cv) %		
Tracer alone	2.0^{++}	1.5
Ferritin ILMA (n=412)**	5.9	4.2
Thyroglobulin ILMA (n=265)	6.6	5.1
Cortisol SPALT (n=270)	6.9	6.2
Transferrin SPALT (n=313)	7.2	6.8

Key:

$^{+}$24 samples can be loaded into the LKB-1251 if a non-cyclic operation is used as the luminometer needs an empty space as stop command. The variability of the LB 950 depends upon whether the incubation chamber is fitted or not.

*Data derived from compound precision profiles using means of duplicate values

$^{++}$Assays were carried out in parallel on both machines to compensate for reagent and operator variations.

**Full assay details are given in other articles appearing in this volume.

were available from LKB or from Sarstedt (Numbrecht, D). As can be expected, this increased the price per sample. The LKB cuvettes were available in polystyrene or polypropylene, the light transmission being almost identical in both cases.

To reduce light transmission along the delivery lines, a length of sticking plaster was placed over the rubber grommet where the delivery tubes entered the measuring chamber. The best effects were obtained when at least 15 cm of the delivery lines were covered.

Routine maintenance included dismantling the mirror assembly, checking the cuvette elevator and removing and cleaning the caroussel once a week. This took 15-20 min.

The instrument has been fitted with EEPROMS which allows the programmes written by the operator to be stored, even when the machine is disconnected from the mains. Even so, all immunoassay data reduction must be performed manually using a computer (CBM 8096).

LB 950

The alterations to the LB 950 were more extensive than to the LKB-1251. The luminometer was purchased with 3 dispensers with motor driven syringes and a Basis 108 desk-top computer, the latter being compatible with an Apple IIE computer. The sample transport was via a plastic chain and accommodated tubes between 37 and 55 mm in length and 11 and 12 mm diameter.

The original syringes with metal plungers were replaced by syringes with Teflon plungers (Hamilton, Bonaduz, CH), as the metal plungers seized up on the peroxidase line. The Hamilton syringes are still in use after more than 45,000 cycles. The Teflon valves in the dispensers proved not to be very resistent to NaOH and/or high pressure, the average lifespan being 3-4 weeks, a costly undertaking. The valves were replaced by magnetic steel ones which have been in use for over 22,000 cycles on the NaOH line (3 months) and over 12,000 cycles on the microperoxidase line. Fears that metal parts in the flow lines could cause interference in terms of worsened precision of measurement have not been confirmed at least for the oxidation system used in this laboratory, (NaOH, microperoxidase, H_2O_2). Problems arose with the transport chain when using 55 x 12 mm tubes. Each tube holder had a small plastic retaining spring which held the tube in place, but hindered it from falling back into place after measurment. The result was that the tube was not fully removed from the measuring chamber, thus blocking the luminometer. This problem was overcome by cutting out the retaining springs. The optimal tube size for the LB 950 would appear to be 51 x 11 mm, although with the alterations made,

1 blockage per 1000 tubes takes place using ordinary test-tubes of 55 x 12 mm (Sarstedt). After performing these alterations, the LB 950 was fully suitable for routine use, and has taken over this job in the authors' laboratory where all routine assays are run on this instrument.

Although the LB 950 is connected on-line to the Basis 108 at the time of writing, no immunoassay programme is available, so that data reduction must be performed manually as in the case of the LKB-1251. A LIA programme with quality control package should become available during 1984 according to the producer.

CONCLUSIONS

The problems of buying a product new on the market are that many small points have not been solved. This is often the case for new cars, and also for new luminometers! The experience of the authors is such that the manufacturer is only too pleased to co-operate with the user in order to optimise his product. Although fully automated systems are not yet available, the introduction of a dialogue-oriented immunoassay programme will bring the LB 950 up to the level of a RIA sample changer. Such a step is imperative when trying to convince laboratories that LIA is an alternative to RIA, as it cannot be expected that laboratories will take a retrograde step, especially where the comforts of automation are concerned! The market is now ready for the introduction of robust luminescence immunoassay kits, and only then can it be seen whether LIA can and will replace RIA in the routine laboratory.

ON-LINE COMPUTER ANALYSIS OF CHEMILUMINESCENT REACTIONS AND APPLICATION TO LUMINESCENT IMMUNOASSAYS.

M.Pazzagli, A.Tommasi, M.Damiani*, A.Magini & M. Serio.

Endocrinology Unit, University of Florence and
*Tecnocomputer, Siena, Italy.

INTRODUCTION

Materials of biological origin, such as plasma or urine, can significantly interfere the photon emission of chemiluminescent (CL) reactions reducing the reliability of the CL tracer measurement and thus of the LIA procedures. These interfering effects are usually associated with shape modifications of the kinetic light emission (1). In an attempt to assess a computerized analysis of CL reactions, a microcomputer has been interfaced on-line with a luminometer: i. to compute significant parameters of light emission and shape of CL reactions; ii. to study interferences from biological samples, before and after the pre-treatment of the biological sample with dextran coated charcoal; iii. to choose suitable light emission measurements; iiii. to investigate the possibility to reveal samples with a shape significantly different from those which are considered as standards.

LIGHT MEAUREMENTS AND INSTRUMENTATION

We have computed the following parameters of light emission: <u>TC</u> (total counts), which represents the integration value of the photon emission for the total time of counting; <u>PLI</u> (peak light intensity); <u>PI</u> (portion integration), an

ISBN 0-12-426290-2

integration value of counts for a fixed time interval, usually 0-10 sec or 2-10 sec.

Moreover we have investigated several possible "shape" indices for CL reactions: Tmax (value in sec), which represents the interval time from the oxidant injection and PLI; 2/10 ratio or 5/10 ratio which are the ratios of the PI values during the first 0-2 or 0-5 sec and the 0-10 sec time interval; Mean, skewness, kurtosis and the comparison Index (I). The latter parameters have been calculated following the method described by Wampler and coll. (2).

Measurements of light emission were made with a Packard Autopicolite Mod. 6500 Luminometer. The output light signal from the photomultiplier of the Luminometer has been interfaced to a microcomputer (a GENERAL PROCESSOR Mod. T/08, 48 kbytes read/write memory (RAM). The software includes a disk operating system CP/M* compatible, and a BASIC interpreter and compiler (Microsoft). Two different programs have been developed for this system:

i. DECODE. By using this program the light emission produced by the CL tracer is divided into sample intervals variable from 8 to 3,000 milliseconds and variable number of samples per reaction from 1 to 255. Data are stored on disk.

ii. GRAFST2. This program in Microsoft BAS80 language provides data processing and printing. A table is created, one row per sample, in which measurement parameters and "shape" indices are calculated togheter (Tab. 1.).

More details on instruments and programs are in Ref. 1. Reagents for Cl reactions have been described elsewhere (3).

RESULTS

Effects of plasma and urine on the CL light emission.

The effects of different amounts of a pool of plasma or urine on the CL light emission of the CL tracer and on the blank are reported in Fig. 1 under the experimental conditions described in the legend.

We have observed a reduction of the light emission of the CL tracer as well as an increment of the blank values for volumes exceeding 0.1 ul of plasma or 1 ul of urine. These

TABLE 1

Data processing using the program GRAFST6. Light emission parameters and shape indices are reported together, one row per CL reaction. In this table are reported data from a LIA method for urinary free cortisol (standards plus unknowns). The I(s,max) value obtained by the standard kinetics (standard curve points) was used for the shape test: * = rejected),

N.	PLI C/s	TOTAL COUNTS	PI 0–10s	PI 2–10s	T(max) s	RATIO 2/10	RATIO 5/10	MEAN s	SKEWNESS	KURTOSIS	I (RATIO) 2/10+5/10
blank	120	423	423	316	0.600	36.26	68.81	3.77	0.605	-0.806	4.211
ANS	180	645	645	451	0.500	36.30	69.05	3.69	0.646	-0.713	3.901
Total	7300	44041	44041	32217	0.600	37.72	70.27	3.64	0.667	-0.747	1.200
zero	5230	32764	32764	24297	0.500	37.63	70.79	3.61	0.665	-0.687	0.692
7.8 pg	5120	31592	31592	23175	0.500	38.55	71.98	3.55	0.708	-0.616	1.511
15.6 "	4880	29735	29735	21806	0.400	38.03	71.12	3.57	0.696	-0.631	0.052
......	....	.....	.....	.....	.....	.....	.....	....	.....	......	.....
500 pg	410	2490	2490	1836	0.500	37.51	70.73	3.66	0.689	-0.693	0.872
			MEAN		0.510	38.01	71.09	3.62	0.676	-0.670	I(s,max)
			± S.D.		0.050	0.54	0.44	0.03	0.016	0.028	2.141
0.62 ul	3730	21929	21929	16138	0.500	38.89	71.91	3.57	0.701	-0.680	1.746
1.25 ul	2800	17026	17026	12523	0.500	38.84	71.86	3.58	0.708	-0.627	1.643
2.5 ul	1840	10904	10904	7984	0.500	38.92	71.65	3.56	0.705	-0.628	1.478
5.0 ul	1140	6535	6535	4727	0.600	38.67	71.20	3.58	0.678	-0.617	0.736
10.0 ul	620	3638	3638	2655	0.600°	39.42*	72.35*	3.53*	0.725*	-0.584 *	2.737*

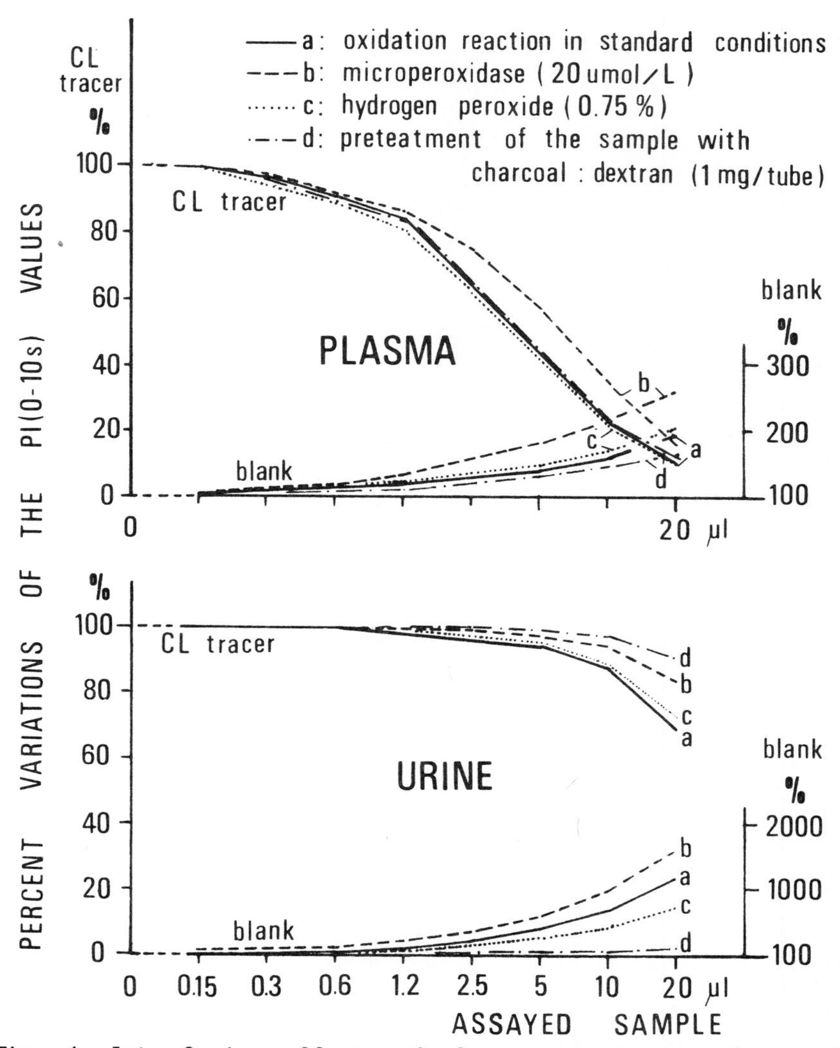

Fig. 1. Interfering effects of plasma or urine samples on the PI(0-10) values. CL reactions are performed in absence (blank) or in presence of 100 fmol/tube of cortisol-3-cmo-ABEI. Experimental conditions are: a. 0.1 ml of sample containing from 0.16 to 20 ul of the biological fluid, NaOH 2 mol/L, 0.1ml, microperoxidase 10 umol/L, 0.1ml, H202 0.2%, 0.1ml; b. as a but microperoxidase was 20 umol/L; c. as a but H202 was 0.75%; d as a but with the pretreatment of the sample with dextran coated charcoal.

oxidation reaction conditions (exp. b,c) resulted in some variations of the shape parameters (data not shown).

We have also pre-treated the biological samples with dextran coated charcoal (exp. d) and the results show that there was a significant reduction of the interfering effects in the urine whereas in plasma this effect was negligible. Also in this case shape parameters followed the interfering effects on the CL reaction.

Selection of suitable light emission parameters.

The reproducibility of various light emission parameters was studied at different levels of the CL tracer concentration. Results are reported in Fig. 2. The PI values showed better reproducibility than the PLI measurement.

Selection of kinetics with a shape significantly different from those considered as standards (the shape test).

Computer analysis of CL reactions was also employed to

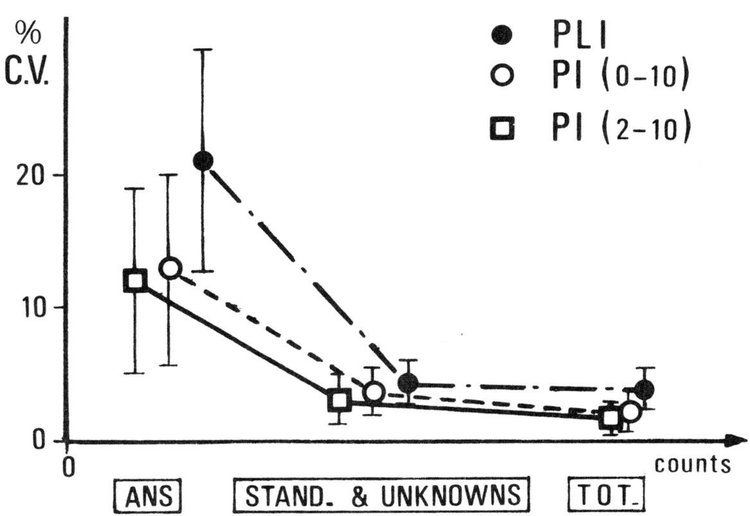

Fig. 2. Coefficient of variation values (mean ± S.D.; n=20) of light emission parameters at different levels of the CL tracer concentration.

perform a shape test. The comparison of the I values between standard and unknown kinetics can allow to reject those kinetics with shape indices significantly different from those of the standard kinetics. In Tab 1 an example of unknown samples (a dilution test of a LIA for urinary free cortisol) accepted or rejected by the shape test is reported. We have observed linearity of the dilution test reported in Tab. 1 only for the samples which were accepted by the shape test (data not shown).

DISCUSSION

In an attempt to improve the reliability of the CL tracer measurement, we suggest the use of computer analysis of CL reactions. This system can be applied: i. to define the maximum amount of biological sample to be assayed without interferences in the CL tracer measurement; ii. to evaluate if procedural steps which precede the CL tracer measurement (i.e. charcoal dextran) can remove the interferences of the sample; iii. to choice the most reliable parameter of light emission; iiii. to reveal kinetics with a shape significantly different from those which are assumed as standards.

ACKNOWLEDGEMENTS

This work is supported by a grant from CNR "Progetto Finalizzato Tecnologie Biomediche.

REFERENCES

1. Tommasi, A. et al (1984). On-line computer analysis of CL reactions and application to a LIA for urinary free cortisol. Clin. Chem., in press.
2. Wampler, J.E. et al (1979). Instrumentation and techniques for analysis of hydrogen peroxide and peroxide-producing reactions involving earthworm. Clin. Chem., 25, 1628-1634.
3. Pazzagli, M. et al (1983). Preparation and evaluation of steroid chemiluminescent tracers. J.steroid Biochem. 19, 407-412.

STABLE SINGLE PHOTON CALIBRATION STANDARDS AT DIFFERENT WAVELENGHTS FOR LUMINOMETRY

Oikari T.E.T, Hemmilä I.A., Soini E.J.

Wallac Oy, Box 10, Turku, Finland

INTRODUCTION

Calibration and standardization of luminometers has been a problem because of the lack of stable liquid single photon sources in the green-to-red region of spectrum where many luminescence reactions emit their light. A green or red light emitting diode (LED) fed by a small dc-battery can be constructed for the purpose but problems are faced in the non-uniform angular distribution of LED light. Another drawback is that the active light emitting geometry differs from that of actual luminescent samples. Conventional heavily quenched radioisotopic liquid scintillation systems (1) circumvent the difficulties associated with LEDs, but their emission is mainly at blue wavelengths.

We present four ^{14}C excited stable light sources for the green-to-red region with good single photon character. Two of them (Ia and Ib) are based on 3-aminofluoranthene and have broad emission spectra with maxima at $\sim$500 nm and $\sim$550 nm respectively. The other two (IIa and IIb) are based on lanthanide chelates of terbium (Tb) and europium (Eu) with narrow line spectra at $\sim$544 nm and $\sim$613 nm respectively.

Sources based on 3-aminofluoranthene (3AF)

A dye 3-aminofluoranthene (3AF) exhibits green fluorescence peaked at $\sim$ 500 nm when excited with ultraviolet radiation. We have now found that in organic solvents the excitation

ANALYTICAL APPLICATIONS OF
BIOLUMINESCENCE AND CHEMILUMINESCENCE

ISBN 0-12-426290-2

can be performed also with ionizing radiation (e.g. beta particles from ^{14}C) presumably analogously with liquid scintillation processes so that a green emitting photon source is obtained. Liquid paraffin is a suitable solvent since a high quantum yield is not necessary when a source with good single photon character is required. In fact conventional scintillation solvents such as toluene and zylene were tried, but photoluminescence problems were encountered with 3AF dissolved in them.

The emission spectrum of 3AF can be cut by adding a colour quencher, Sudan 1, that absorbs strongly below 540 nm. Thus a photon source with a peak at $\sim$560 nm can be obtained and, as a result of quenching, the single photon character of sources is improved.

Sources based on lanthanide chelates

A new way to prepare photon sources is to use lanthanide chelates in aromatic solvent as fluors excited with beta particles of ^{14}C. The used lanthanide ions were Tb^{3+} and Eu^{3+}. Their chelates or chelate solutions have narrowbanded emission lines with 544 nm peak dominating with Tb and 613 nm peak with Eu. Absorption is in the range 250-350 nm. Because of long lifetimes of their fluorescence (> 100 μs), lanthanide chelates inherently have good single photon characters.

Tb-chelate solution Tb-diketonate solution was prepared by adding $TbCl_3$, β-diketone (hexafluoroacetylacetone) and a synergistic compound tri-n-octyl-phosphine oxide to buffered solution, pH 5, and extracting the chelate in toluene.

Eu-chelate solution Eu-diketonato solution was prepared by adding $EuCl_3$ to a toluene solution containing β-diketone (2-naphtoyltrifluoroacetone) and a base, ortho-phenanthroline.

Properties of the sources

Sources of different intensities can be prepared by altering the amounts of ^{14}C activity (e.g. in the form ^{14}C-palmitic acid in toluene). Stabilities are good with drifts smaller than 2 % per month with properly prepared sealed samples.

The true emission spectra of the sources were difficult to determine because of the insensitive emission spectrometers available. However, the fluorescence spectra shown qualitatively in Fig. 1 could be easily measured and these can be assumed to be similar with the emission spectra. The instrument was a Perkin-Elmer 3000 Spectrofluorometer with excitation at 300 nm.

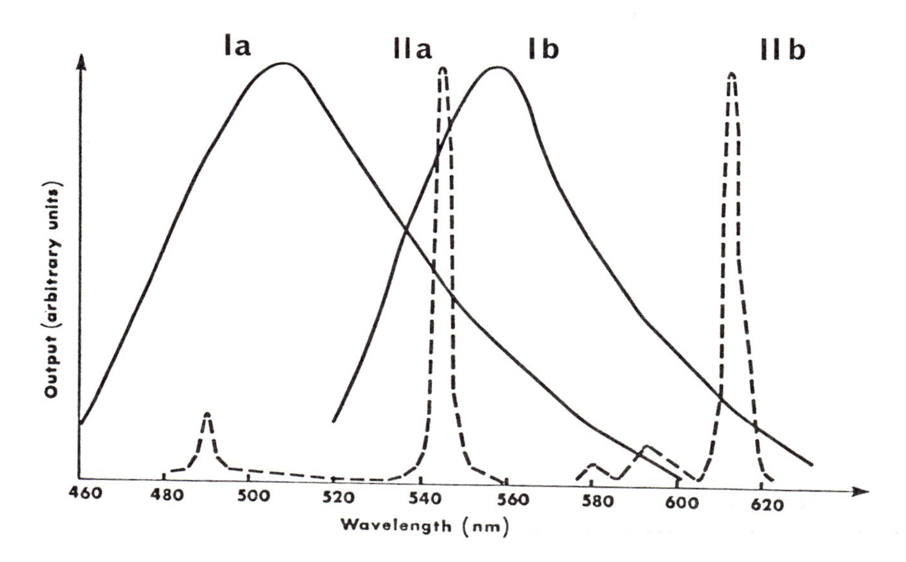

Fig. 1. Normalized fluorescence emission spectra of the photon sources. Solid lines (I): sources based on 3-aminofluoranthene; Ia without Sudan 1, Ib with Sudan 1. Dotted lines (II): sources based on lanthanide chelates; IIa Tb-chelate, IIb Eu-chelate.

The absolute photon emission of the sources can be determined e.g. with methods described by Seliger (2); this increases their usefulness even further. Procedures of this kind are under work.

REFERENCES

1. Schram E., Demuylder F, De Rycker J., Roosens H, *in* Liquid Scintillation Science and Technology, Ed. Noujaim A.A., Ediss C., Weibe L.I., Academic Press 1976, pp. 243.
2. Seliger H.H., *in* Methods in Enzymology, vol. LVII, Ed. De Luca M.A., Academic Press 1978, p. 560.

APPLICATION OF LIGHT GUIDES FOR ENHANCEMENT OF SIGNAL TO NOISE RATIOS AT LOW LEVELS OF LUMINESCENCE DETECTABILITY

S.E. Brolin,* P.-O. Berggren* and P. Naeser**

**Department of Medical Cell Biology, Biomedical Center, and **Eye Clinic, University of Uppsala, Uppsala, Sweden*

INTRODUCTION

Optical systems with high aperture have been used in bioluminescence analysis, but nevertheless only a minor part of the photons emitted reach the photocathode (1). A detailed evaluation of optical arrangements in luminescence has been presented by Malcolm and Stanley (2). Concerning the difficulties in attaining an effective light collection they draw attention to losses caused by reflexion and particularly to those which occur in the walls of the reaction vessel. Fiber optics have been employed for long distance conduction of optical signals which propagated with only minor losses of intensity. Light guides of coarser dimensions have proven to be of great value in various instruments in which preservation of intensity is essential but image formation of less significance. We describe tests with light guides designed for improved collection of photons, minimized reflection losses and allowance of a temperature gradient between the photocathode and the site of light formation.

EXPERIMENTAL DESIGN

The measuring equipment is shown in a block diagram (Fig. 1). The photomultiplier was operated in a photoncounting mode. The counts were stored at preset times in selected channel rows of the scalor and then exhibited in the memory display for evaluation of the time course and intensity of the light flux. Sample mixed with slowly fading luciferase were moved into the light guide through transparent vinyl tubing by the suction of a peristaltic pump with variable speed. The maximum value was used as measure of the emission intensity; integrals change with the rate of the flux. Several samples could be run in a row through the measuring

ISBN 0-12-426290-2

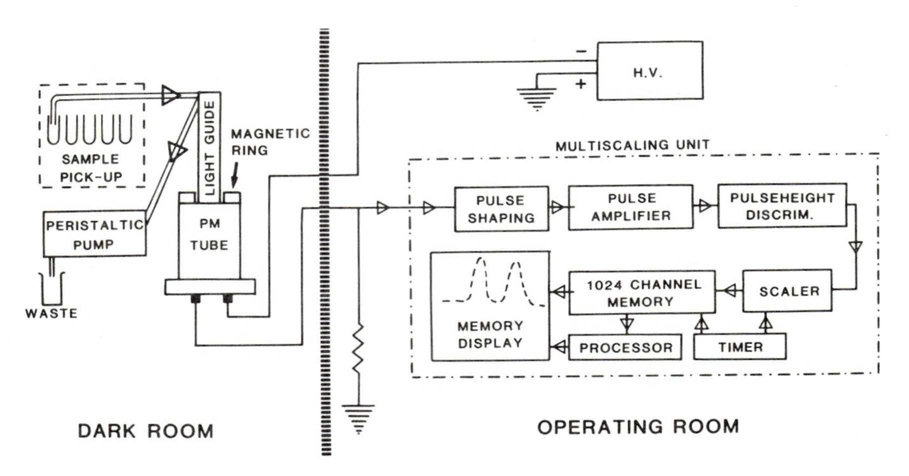

Fig. 1 Block diagram of the experimental assembley.

station as in automatic analysers used in clinical chemistry. Alternatively a more rapidly pumped rinse could be employed each time a sample had been measured.

Two varieties of light guides were produced. One of those was composed of a solid glass cylinder coated with a more refractive layer and a thin plain sivermirror on its top. The narrow space between the mirror and the top of the rod was filled with a transparent adhesive and traversed by transparent vinyl tubing, to each end of which was coupled black polyethylene tubing of a finer gauge to prevent passage of undesired light. If the small light emitting sample is centerred at the focus of a large parabolic or ellipsoid mirror, the light can be directed into a wide light pipe under favourable angular conditions. When the light comes from tubing almost as long as the diameter of the light guide only little would be gained by a curved mirror, particularly as the coating increases the real aperture. Hence the angle of the entering light is less important. The other light guide was a polished glass cylinder. Its top was ground to the segment of a cone with a hole in its centre for insertion of a small tube or a U-shaped tubing into a refractile liquid (Fig. 2). The surface of the light guide was coated with a silver layer which was protected by varnish. A long lasting emission was obtained with a firefly luciferase preparation (LKB Monitoring kit) after adding various amounts of ATP. All solutions were prepared sterile and kept covered but not completely protected from air pollution.

LIGHT MEASUREMENT

The lower limit of detection was about 7 fmole ATP (Fig. 3a).

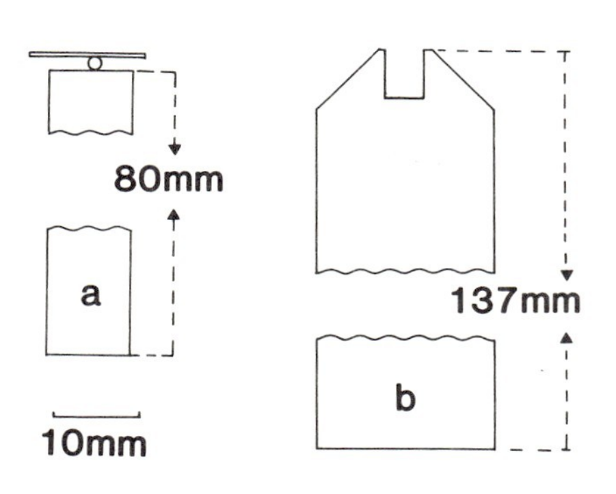

Fig. 2 Outlines of the light guides in longitudinal sections. a) Coated glass cylinder with a mirror and a small plastic tube attached to its top. b) Cylinder with cone shaped top for downward reflection of photons and a small chamber for insertion of the reaction vessel (U-shaped plastic tube or microtube)

Sterility maintained throughout the experiment would possibly reduce the present blank readings. Fig. 3b is a display of the signals from increasing amounts of ATP versus a blank. Samples transported by suction in a row were measured with acceptable reproducibility (Fig. 3c).Although we recommend a complete filling of the volume available for the collection of photons only a part of it can be used (Fig. 3d).

The silver coating prevented the loss of light which occurred at the misty surface of a chilled light guide. Another advantage is that less than 5 per cent of light is lost at each reflection at a surface coated by vaporized silver. A light guide can be brought into good optical contact with the light yielder and the photocathode which is essential for effective photon collection. A limitation as regards sample transport rate is the duration of fading even of a long lasting emission. Hence the combination of immobilized luciferase and a light guide should be of particular analytical advantage (3).

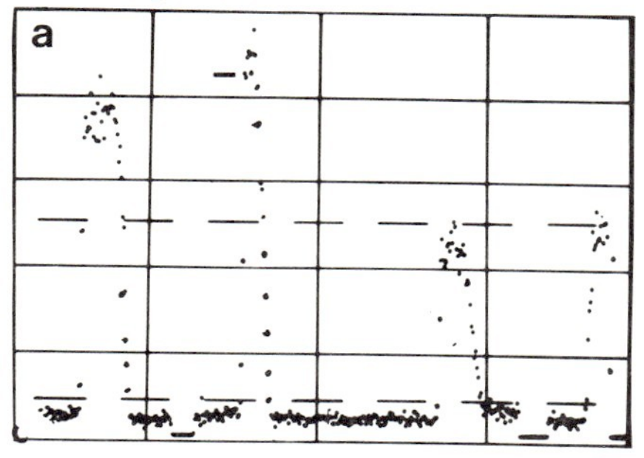

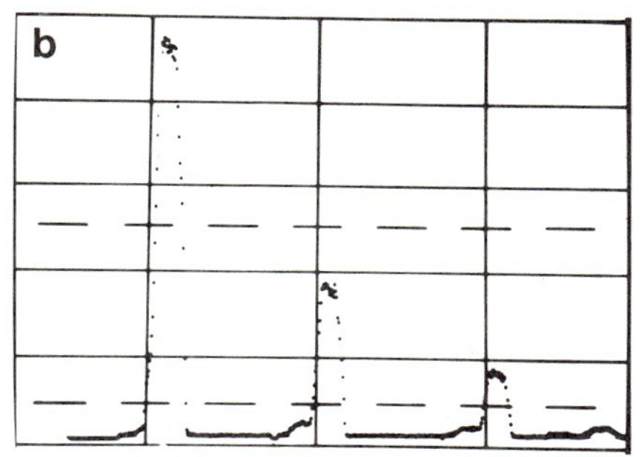

Fig. 3 Photon emission from ATP samples and blanks. Each dot represents the counts in one channel (0.5 s in a and 0.8 s in b,c and d). The light guide with the mirror on its top was used in a and the other set up in b, c and d. a) Two standards of 7.5 fmol and two blanks(10 ul) are displayed. b) The display shows three ATP samples (25 ul) of 2, 1 and 0.5 nmol followed by a blank.

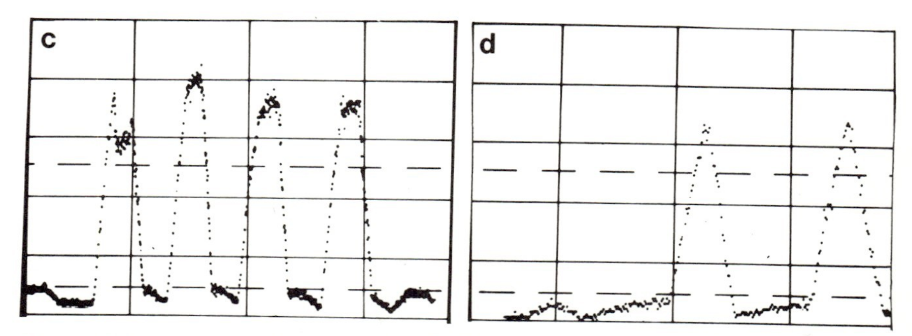

Fig. 3c) Test of reproducibility. Four samples (25 ul) of 8 fmol ATP are seen in the middle. The blanks are at each margin. The samples and blanks were separated by airbubbles and buffer solution and moved continuously in a row. An air bubble observed at the loading of the first sample reveales its presence in the display. c) The display shows three ATP samples (25 ul) of 2, 1 and 0.5 nmol followed by a blank. d) Display from two ATP samples (3 fmol) which are smaller (10 ul) than the volume of the tube (25 ul) at the measuring site for photon collection. The blank is to the left.

SUMMARY

Light guides permit convenient measurements of bioluminescence in the femtomole range. Our experiments points to the usefulness of light guides for collection and conduction of light to a photocathode surface which implies considerable potentialities in microanalysis.

REFERENCES

1. Wettermark, G., Stymne, H., Brolin, S.E. and Petersson, B. (1975). Substrate analyses in single cells. *Anal. Biochem.* *63,* 293-307.
2. Malcolm, P.J. and Stanley, P.E. (1976). A unified approach to the liquid scintillation counting process. *Int. J. Appl. Radiat. Isot.* *27,* 397-430.
3. Jablonski, E. and DeLuca, M. (1976). Immobilization of bacterial luciferase and FMN reductase on glass rods. *Proc.Natl. Acad. Sci. USA* *73,* 3848-3851.

Supported by the Swedish Medical Research Council (12x-00525), the Swedish Diabetes Association and Clas Groschinsky's Memorial Foundation.

VII BIOLUMINESCENCE AND BIOLUMINESCENCE ASSAYS

BINDING MECHANISM OF NUCLEOTIDE MIMICKING DYES TO FIREFLY LUCIFERASE

Sunanda Rajgopal and M.A. Vijayalakshmi

Laboratoire de Technologie de Separation
Universite de Technologie de Compiegne B.P. 233
60206 Compiegne Cedex, FRANCE

INTRODUCTION

Following the purification of firefly luciferase on immobilised blue dextran (a dextran conjugate of the monochlorotriazinyl polyaromatic dye, Cibacron blue F3G-A) (1) it was deemed necessary to clarify the nature of interaction between the enzyme and various nucleotide mimicking dyes of the series triazine and fluorescein which possess an anthraquinone and xanthene ring respectively. Earlier observations on triazine luciferase interaction (2) revealed a case of stereospecific ionic interaction. Any change in the polycyclic nature of the dye eg... cleavage of the azo linkage resulting in the removal of the sulphonated parts was found to result in weakened dye - enzyme interactions.

As early as 1969, De Luca (3) reported that certain naphthalene dyes, namely 1,5 Anilinonaphthalene sulfonate or the corresponding isomer, 2,6 Anilinonaphthlene sulfonate and 2,6 toluidinonaphthalene sulfonate bind to luciferase in a competitive manner with respect to luciferin. She attributed this binding to the hydrophobic nature of the active site of luciferase. It was also proposed that though the position of substituents on the naphthalene ring did not seem significant in the binding in the case of various isomers of ANS, 2,6 TNS was found to bind better than 2,6 ANS. (The only difference in the structure of these

ISBN 0-12-426290-2

dyes is a methyl group). This study was undertaken to determine if the entire dye molecule is required for the biospecific binding of luciferase, since it was earlier suggested that the dye molecule, 1 - amino - 4 (4' amino phenylamino) anthraquinone 2 - 3' - disulfonic acid) (ASSO) was a structural analogue of ATP and was the portion of CB3G-A responsible for its interactions with the enzyme (4). It was also intended to see if the steric isomer of CB3G-A, Cibacron Brilliant Blue BR-P (CBB-II) displayed similar binding propertiers. So, CB3G-A CBB-II, ASSO, Bromaminic acid and 2' 4' 5' 7' -tetraiodofluorescein (TIF) which has a xanthene ring) were compared kinetically as inhibitors of luciferase relative to its substrate ATP.

Materials and Methods

Luciferase was from Boerhinger, ASSO from Sandoz Ltd France. Ciba - Geigy provided CB3G-A and CBB-II. Dyes were purified by preparative thin layer chromatography. Dye concentrations were determined Spectrophotometrically at their max. using the absorption coefficients quoted elsewhere (4).

Luciferase activity was assayed by a Nucletimeter (2). The unit of activity is the maximum intensity of light recorded in millivolts per picogram of ATP per milligram of protein at 562 nm. Inactivation of firefly luciferase was carried out as described elsewhere (5).

Results and Discussion

The structure of the dyes used in our studies is given in Fig. 1.

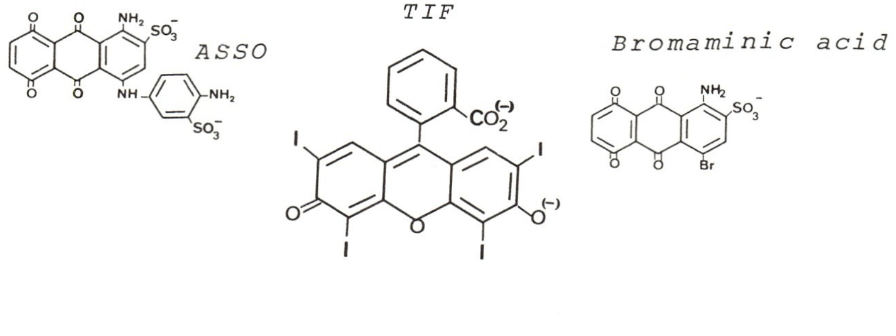

ASSO

TIF

Bromaminic acid

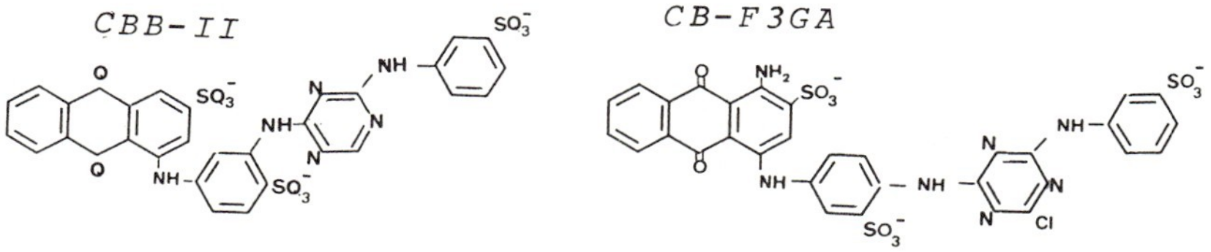

CBB-II CB-F3GA

Fig. 1 Structure of the dyes used in our studies.

K_D (the dissociation constant) K_I (the inhibition constant) and K_3 (the maximum rate of inactivation (min^{-1}) are calculated as per Clonis & Lowe (5) and listed in table 1. Fig. 2 represents the double reciprocal plot of $1/K_{obs}$ (observed rate of reaction constant) versus $1/_{(D)}$ (the concentration of the dye). This yields a straight line with a positive ordinate intercept which indicates saturation kinetics for the inactivation process.

Table 1 K_3, K_D and K_I of dyes dor firefly luciferase.

DYE	K_3 (MIN^{-1})	K_D (UM)	K_I (UM)
BROMAMINIC ACID	0.074	6.66	6.09
ASSO	0.083	2.56	16.06
CBF3G-A	0.006	2.13	111
CBB-II	0.004	4.16	108
TIF	0.014	16.66	400

Table 1 implies that the steric isomer CBB-II behaves very differently from the other dyes with a K_3 value of 0.004 min^{-1} (the least among the five), with a similar K_I value as that of CB3G-A but with a higher K_D value. The best inhibitor, however seems to be ASSO and CB3G-A.

Inactivation studies with TIF yields a K_D value of 16.66 uM, a K_I of 400 uM and a K_3 value of 0.1 (min^{-1}). The xanthene ring, thus seems to be less specific for luciferase.

From these studies and those reported by De Luca (3), we can conclude that dyes bind to luciferase in a hydrophobic and/or ionic manner,

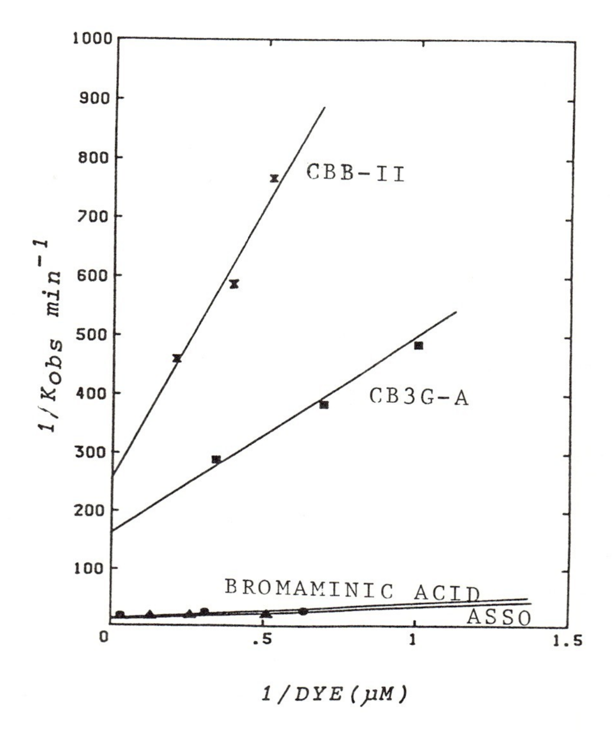

Fig. 3 K_D of dyes with respect to luciferase.

with stereospecificity playing a significant role when the dye structure is large enough to spread across the active site.

Conclusions

ASSO seems to be the essential part of the entire dye molecule which participates in the binding. For the rest of the dye binding to occur, it can be said that the better the arrangement of ionic groups of dye fits the arrangement of the corresponding functional groups of luciferase, the stronger is the complex binding. The hyperbolic dependence of inhibition on dye concentration is characteristic of an active-site directed process. The protection against inhibitor afforded by ATP argues that the substrate and dye interact with the same enzyme domain, namely the active site. This dye when immobilised could be fruitfully exploited in designing a rapid one - step method of purification for luciferase, which when correctly oriented can replace the comparitively expensive methods currently employed in the industrial purification of the enzyme.

References

1. Rajgopal, S and Vijayalakshmi, M.A (1982) Purification of firefly luciferase on Blue dextran columns Comparison of Sepharose and Silica as matrices, J. Chromatogr. 243, 164
2. Rajgopal, S and Vijayalakshmi, M.A (1983) Interaction of firefly luciferase with Triazine dyes, J. Chromatogr. 280, 77 - 84
3. De Luca, M (1969) Hydrophobic nature of the Active site of firefly luciferase, Biochemistry, 8, 160 - 166
4. Beissner, R.S. and Rudolph, F.B. Interaction of Cibacron Blue 3G-A and related dyes with nucleotide requiring enzymes (1978) Arch. Biochem. Biophys, 189, 76 - 80
5. Clonis Y.D, Goldfinch, M.J and Lowe, C.R. The interaction of yeast hexokinase with Procion Green H - 4G. (1981) Biochem. J., 197, 203 - 211.

EXTRACTION AND AUTOMATIC LUMINOMETRIC ASSAY OF ATP, ADP AND AMP

A.Lundin[1]

Research Centre and Department of Medicine, Karolinska Institute, Huddinge Hospital, S-141 86 Huddinge, Sweden and LKB-Wallac, SF-201 01 Turku 10, Finland

INTRODUCTION

Before being accessible to enzymatic analysis intracellular metabolites have to be extracted from the cell. This is usually achieved by addition of a chemical substance, i.e. an extractant. A reliable extractant fulfills three requirements: 1) It releases the entire intracellular pool of the metabolite to be assayed. 2) The extractant gives a complete, rapid and irreversible inactivation of all enzyme systems in the extracted cells that may affect the metabolite levels during extraction, storage or analysis of extracts. 3) The extractant may not inactivate the metabolite or the enzyme system involved in the analysis. The last two requirements impose contradictory properties on the extractant on one hand it should inactivate enzymes on the other hand it should not inactivate enzymes. In spectrophotometry this problem generally makes it necessary to remove the extractant before the assay, while in luminometry analytical interference from the extractant often can be avoided by dilution of samples. However, this is not always possible when working with low numbers of cells.

In a previous study (1) five bacterial strains were extracted by 10 different methods and the yields of ATP, ADP and AMP compared using luminescence analysis. It was

[1]Correspondence to A.Lundin: Reasearch Centre, Karolinska Institute, Huddinge Hospital, S-141 86 Huddinge, Sweden

ISBN 0-12-426290-2

concluded that only TCA (trichloroacetic acid) could be used for reliable extraction of all five strains and all three nucleotides. The present study was performed to see if this conclusion could be generalised also to other types of cells or if alternative extractants could be found. In particular the effect to extractant concentration and the presence of substances that may interfere with the extraction was studied. Furthermore an automatic luminometric assay of ATP, ADP and AMP based on Luminometer 1251 (LKB-Wallac, Turku, Finland) was developed.

MATERIALS AND METHODS

Preparation of Samples

Bacteria were grown overnight at 37°C in Nutrient Broth (Difco Laboratories, Detroit). Yeast cells *(Candida albicans)* were grown overnight at 37°C in a medium containing NaCl (5 g/l), Oxoid Code L29 (3g/l), Oxoid Code L34 (10 g/l) and glukos (20 g/l), pH adjusted to 5.6 with HCl. Algal cells *(Scenedesmus obtusiusculus)* were grown as described by Larsson and Olsson (2). Human blood cells were prepared at the Karolinska Hospital (Stockholm, Sweden) by standard blood bank techniques. Isolated human fat cells were prepared by the method of Rodbell (3) as modified by Smith et al. (4). Rabbit alveolar macrophages were isolated as described by Romert and Jensen (5). Thin slices (0.1 mm) of frozen human placenta were prepared by an Ultramicrotome (LKB-Produkter, Bromma, Sweden). Human lung fibroblasts were grown at 37°C in covered plates with 6 holes each.

Extraction Procedures

Most used extractants were TCA (trichloroacetic acid), PCA (perchloric acid), DTAB (dodecyl trimethyl ammonium bromide, Sigma Co), triton X-100 (Scintillation grade, Eastman Kodak Co) and saponin (Sigma Co). Extractants were dissolved in distilled water except for TCA and PCA also supplied with 4 mM EDTA, pH 7.0. All extractions were performed at room temperature rapidly pipetting an aliquot of sample into an equal volume of extractant solution or alternatively (blood cells) into nine volumes of extractant solution. Frozen slices of placenta were immersed into extractant solutions without previous thawing. Duplicate extracts were always prepared and assayed for ATP immediately after preparation.

ATP Assay

Luminescence was measured with a Luminometer 1250 (LKB-Wallac, Turku, Finland). One vial of ATP Monitoring Reagent (LKB-Wallac) was reconstituted with 50 ml of 0.1 M tris-acetate buffer containing 2 mM EDTA, pH 7.75. One vial of ATP Standard (LKB-Wallac) was reconstituted with 10 ml of distilled water to make a final ATP concentration of 10 uM. ATP assays were done by adding 10-20 ul of extract or with high concentrations of extractants 10-fold diluted extract to 1 ml of ATP Monitoring Reagent. After measuring the luminescence an equal volume (10 or 20 ul) of ATP Standard was added and the luminescence measured once again. From these measurements and from corresponding measurements on reagent blanks the ATP concentrations in the extracts were calculated. ATP concentrations were expressed as percent of the ATP concentration in the TCA extract with the highest yield of ATP.

Automatic Assay of ATP, ADP and AMP

Luminescence was measured with a Luminometer 1251 (LKB-Wallac) supplied with three dispensors and a personal computer. ATP Monitoring Reagent and ATP Standard were reconstituted as above. Equal volumes of 0.2 M phosphoenolpyruvate (pH adjusted to 6.0) and pyruvate kinase in 50 % glycerol (Boehringer-Mannheim) were mixed (PK-PEP reagent). Adenylate kinase (Boehringer-Mannheim) was dialysed overnight against tris-acetate buffer to remove ammonium sulphate. An AK-CTP reagent was prepared by addition of CTP (Boehringer-Mannheim) in a final concentration of 25 mM to the dialysed adenylate kinase.

Cuvettes containing ATP Monitoring Reagent (1 ml) and sample (10-20 ul) were loaded into the luminometer. Although the luminometer accepts 25 cuvettes not more than 10 cuvettes were assayed in each run in this particular assay to avoid the slight change of adenine nucleotide levels otherwise observed.

The luminometer automatically performes the following steps (steps 2-8 repeated with all cuvettes):

1) Preincubation (25°C, min) to achieve temperature equilibration.
2) Measurement of luminescence corresponding to ATP level.
3) Addition of PK-PEP reagent (10 ul) to convert ADP to ATP.
4) Measurement of luminescence corresponding to ATP+ADP level.
5) Addition of AK-CTP reagent (10 ul).

6) Measurement of luminescence corresponding to ATP+ADP+AMP level.
7) Addition of ATP Standard (10 ul).
8) Measurement of increase of luminescence due to addition of ATP Standard (standard addition technique).

In each series of assays duplicate reagent blanks were included. Each day duplicate assays of reagent blanks were performed including addition of ATP Standard also after measuring luminescence corresponding to ATP and ATP+ADP levels. In this way factors for correction of the slight inhibition of luciferase activity obtained at the additions of PK-PEP and AK-CTP reagents could be calculated. Details on preparation of reagents, computer programs for automatic control of luminometer and for calculation of results expressed as concentrations of ATP, ADP and AMP are available from the author.

RESULTS AND DISCUSSION

Extraction of ATP

In preliminary experiments the yield of ATP from *E.coli* was determined using a large number of extractants (strong acids, organic solvents, phenols, desinfectants, toxins, detergents, quaternary ammonium compounds). All extractants gave a similar or lower yield of ATP as compared to 5 % TCA, which was used as reference. The analytical interference from the extractants could be of two types: 1) At high concentrations of extractants the luminescence intensity was reduced to a lower level due to inhibition of the luciferase reaction. 2) With some extractants, particularly the quarternary ammonium compounds, an increased rate of the decay of luminescence was obtained presumably reflecting a continuous inactivation of luciferase. This effect can not easily be compensated for by the standard addition technique and extracts had to be diluted before the assay.

From the preliminary experiments on yield of ATP from *E.coli* and on analytical interference TCA, PCA and DTAB (a quarternary ammonium compound) were selected for further studies. The studies also included triton X-100 and saponin. These two detergents were expected to extract somatic cells but not microbial cells. Two fold dilutions of the five extractants were prepared. Extractant solutions containing DTAB, triton X-100 and saponin also contained 4 mM EDTA for inactivation of ATP converting enzymes in the extracted cells (1).

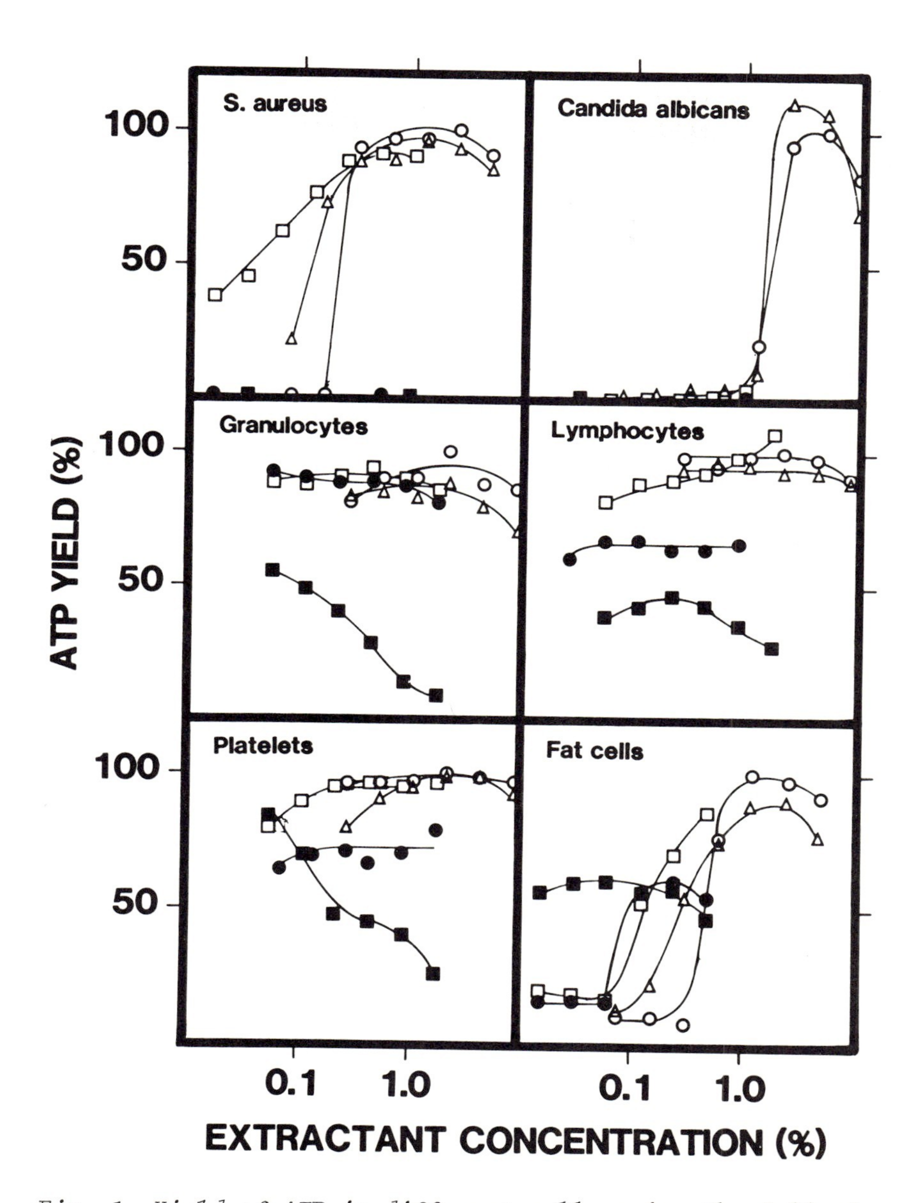

Fig. 1 Yield of ATP in different cells using the following extractants: ○ *, TCA;* △ *, PCA;* □ *, DTAB;* ● *, triton X-100;* ■ *, saponin.*

The extractant solutions described above were used to determine the yield of ATP as a function of extractant concentration. Results obtained with some of the cell types are shown in Fig. 1 (ATP yield expressed as percent of the maximum yield obtained with TCA). At extractant concentrations resulting in maximum ATP yield a similar yield was obtained by TCA and PCA in all cell types. In some cell types a similar yield was also obtained with DTAB and/or triton X-100. Furthermore with most of the cell types a high ATP yield was obtained in a fairly broad extractant concentration interval. An identical maximum ATP yield obtained in a broad extractant concentration interval and with several types of extractants strongly indicates that this maximum represents the total intracellular ATP concentration. In lymphocytes and platelets triton X-100 in a broad concentration interval resulted in a constant ATP yield lower than the maximum yield obtained with TCA, PCA and DTAB. This lower yield may represent a specific pool of ATP released by triton X-100. However, extracts particularly from lymphocytes were not stable and the lower yield as compared to extraction with TCA may be due to ATP degradation during and/or after extraction.

Table 1 summarises the maximum yields of ATP with the five extractants for all cell types tested so far. The extractant concentration resulting in this yield and the lowest extractant concentration resulting in >75 % yield are also shown. Except for yeast and algal cells a yield >75 % was always obtained with 5 % TCA and 5 % PCA and with few exceptions also with 1 % DTAB. Thus from Table 1 it is possible to see if a yield similar to the maximum was obtained within a narrow or a broad extractant concentration interval. Yields with TCA and PCA are on the average very similar. The higher yields obtained with PCA in *Klebsiella pneumoniae*, *Pseudomonas aeruginosa* and *Escherichia coli* were obtained at very low PCA concentrations and may be due to some artifact, e.g. increased ATP production before the cells lyse. Extraction with 2.5 % TCA or PCA resulted in >95 % (TCA) and >80 % (PCA) yields of ATP in all cell types except the algae. Extraction with 0.5 % DTAB resulted in >90 % yields in all cells except yeast and algal cells, erythrocytes, fat cells, alveolar macrophages, lung fibroblasts and human placenta.

It may be concluded from Table 1 that triton X-100 can only be used for extraction of whole blood, erythrocytes and granulocytes. However, 0.1 % triton X-100 can be used for lysis of somatic cells, since at this concentration a >50 % yield of ATP was obtained with all somatic cells except alveolar macrophages and human placenta. The only

Table 1 Yield of ATP with different cell types using TCA, PCA, DTAB, triton X-100 and saponin.

Cell type	Maximum ATP yield, extractant concentration for maximum and for >75% yield[1]														
	TCA			PCA			DTAB			Triton X-100			Saponin		
Microorganisms:															
Bacillus cereus	100	5,	0.31	89	1.25	0.31	101	1	0.25	2			2		
Staphylococcus aureus	100	2.5	0.31	97	1.25	0.31	91	0.5	0.13	0			0		
Klebsiella pneumoniae	100	2.5	1.25	125	0.16	0.16	99	1	0.03	1			0		
Pseudomonas aeruginosa	100	0.63	0.16	105	0.16	0.16	101	1	0.03	46			7		
Escherichia coli	100	1.25	0.16	118	0.31	0.16	108	1	0.06	3			3		
Candida albicans	100	5	2.5	111	2.5	2.5	4			0			0		
Scenedesmus obtusiusculus	100	10	5	98	5	5	2			0			0		
Somatic cells:															
Whole blood	100	2.5	<0.16	101	0.63	<0.16	107	2	<0.03	99	0.03	<0.02	95	0.03	<0.03
Erythrocytes	100	2.5	<0.31	91	0.31	<0.31	105	0.25	<0.06	94	2	<0.06	101	2	0.25
Granulocytes	100	2.5	<0.31	87	2.5	<0.31	94	0.5	<0.06	92	0.06	<0.06	55		
Lymphocytes	100	2.5	<0.31	96	0.63	<0.31	108	2	<0.06	67			47		
Platelets	100	2.5	<0.31	99	2.5	<0.31	96	0.5	<0.06	79	2	2	83	0.06	<0.06
Fat cells	100	1.25	0.63	90	2.5	0.63	86	0.5	0.5	61			60		
Alveolar macrophages	100	5	1.25	91	1.25	0.63	33			0			1		
Lung fibroblasts	100	2.5	0.16	99	2.5	0.16	100	1	0.13	97	0.5	0.13	60		
Human placenta	100	2.5	1.25	101	1.25	1.25	54			11			47		

[1] In each series of three figures the first gives the maximum yield obtained with the extractant in percent of the corresponding yield obtained with TCA, the second figure gives the extractant concentration by which the maximum yield was obtained and the third figure gives the least concentration by which >75 % yield was obtained. Figures have been rounded to two decimals (e.g. 0.015625 is given as 0.02). The symbol < means that the lowest extractant concentration used in the experiment gave >75 % ATP yield.

microorganism with a yield of ATP >3 % was Pseudomonas aeruginosa for which the yield was 46 % at 0.1 % triton X-100. Thus triton X-100 can be used for selective lysis of somatic cells in the presence of bacterial cells as previously described (6). The present work indicates that saponin can be used for the same purpose since 0.25 % saponin resulted in a >45 % yield of ATP with all somatic cells except alveolar macrophages and <5 % with all bacterial cells.

The time course of extraction of E.coli with TCA and DTAB was studied at room temperature. With 5 % TCA or with 0.25 or 1 % DTAB the extraction was completed within 5 min but with 0.5 % TCA only after 1 h. The stability of these extracts was also studied. Extracts prepared with 0.25 or 1 % DTAB or 0.5 % TCA could be kept in ice for 3 days without effects on the ATP level. Extracts prepared with 5 % TCA were stable for 1 day only when kept on ice and should be assayed within a few hours if kept at room temperature.

Interference with the extraction of *E.coli* and *Candida albicans* from NaCl (0.9 and 4.5 %), phosphate buffer (pH 7, 50 and 250 mM) and bovine serum albumin (1 and 5 %) present during the extraction was studied. Extractions were performed as in Fig. 1 and Table 1 using several dilutions of TCA and DTAB. The interference from NaCl was generally of little importance except in the extraction of *Candida albicans* with DTAB. A high degree of interference was observed with 5 % albumin in extractions of both organisms with both extractants. Phosphate buffer interfered with TCA extraction of both organisms and DTAB extraction of *E.coli* but improved extraction of *Candida albicans* with DTAB. The ATP yield with DTAB extraciton of E.coli was studied and shown to be independent of pH in the interval 5.5-8.5.

The choice of extractant is influenced not only by the yield of ATP but aslo of the degree of interference with the assay. Thus the interference of the extractants with the luminescence was studied as a function of extractant concentration. If extracts containing 2.5 % TCA or PCA were diluted 20 fold in ATP Monitoring Reagent, the inhibition of luminescence was approx. 25 % with TCA and approx. 75 % with PCA. With 0.5 % DTAB there was no inhibition but a slow continuous decay of the luminescence. At slightly higher concentration of DTAB the decay was rapid. The effect could be counteracted by addition of high concentrations of bovine serum albumin. However, the concentrations of albumin needed (2.5 - 10 %) inhibited the luminescence strongly. With triton X-100 and saponin inhibition of luminescence was low. Thus blood cells suspended in ATP Monitoring Reagent could be extracted by addition of 0.1 % triton X-100 compensating for the 10 % inhibition of luminescence by the standard addition technique.

Automatic Assay of ATP and AMP

The assay of ATP, ADP and AMP by monitoring the firefly luminescence during the conversions of ADP and AMP to ATP was published already in 1976 (7). A major problem with this assay has been the low rate of the adenylate kinase reaction (AMP+ATP⟶2 ADP) with low concentrations of both AMP and ATP. However, other nucleoside triphosphates can replace ATP in this reaction and should not interfere with the luciferase reaction (firefly luciferase is specific for ATP). Among the nucleoside triphosphates we decided to use CTP, since CTP could be commercially obtained in a purity resulting in the lowest reagent blank. The final concentration of CTP in the assay (0.25 mM) was optimized so that it should take less than 5 min to complete the entire assay. However, at this level the reagent blank from CTP (presumably due to ATP contamination) was as high as 30 nM. Studies will be performed to reduce the blank by purification of CTP. The reactions used in the new automatic assay are the following:

$$\text{AMP} + \text{CTP} \longrightarrow \text{ADP} + \text{CDP} \qquad (1)$$

$$\text{ADP} + \text{PEP} \longrightarrow \text{ATP} + \text{pyruvate} \qquad (2)$$

$$\text{ATP} + \text{luciferin} + O_2 \longrightarrow \text{AMP} + PP_i + \text{oxyluciferin} + CO_2 + \text{light} \qquad (3)$$

The new automatic assay was used for determination of yields of ATP, ADP and AMP in extracts of several types of cells using five concentrations of each of the following extractants: TCA, NaOH, ethanol and dimethyl sulfoxide. Results (not shown) confirm previous findings (1) that TCA can be used for reliable extraction of all three adenine nucleotides.

CONCLUSIONS AND RECOMMENDATIONS

In all cell types (except algae) included in the present study 2.5 % TCA resulted in a >95 % yield of ATP. With some cell types PCA, DTAB or triton X-100 could be used. Considering the degree of interference with the luminometric assay PCA and DTAB do not offer any advantages as compared to TCA in most cell types. However, in some cell types PCA, DTAB or triton X-100 can be of considerable interest. Interference with the extraction from salt, buffer and protein was observed requiring higher extractant concentrations to be used. Thus for each new type of cell or medium in which the cells are suspended it is recommended to compare ATP yields with several concentrations of TCA (e.g. 10, 5 2.5 and 1.25 %). If a too high inhibition of the luminescence is obtained with the optimum TCA concentration in a particular application it is necessary to compare yields also with other extractants (e.g. PCA, DTAB and triton X-100). A low ATP yield also indicates that the extraction will be sensitive to interferences from components in the sample. Thus it is strongly recommended not to adopt a new extraction method without comparison to TCA. If a suitable extractant can not be found TCA can be removed from the extracts by extraction with water satured ether (1). TCA can be used for extraction of all three adenine nucleotides as previously shown (1, 2) and also confirmed in the present study. Together with the new automatic assay of ATP, ADP and AMP this constitutes a strong analytical tool in the study of energy metabolism.

ACKNOWLEDGEMENTS

The present study was the combined effort of a large number of people. Special thanks are due to Drs Bo Lantz and Lennart Romert not only for providing cells but also for participating in performing extractions and assays. Thanks are also due to Drs Peter Arner, Anders Thore and Carl-Magnus Larsson for providing cells and to Rolf Eriksson and Ewa Forsberg for performing a majority of the extraction experiments and to Jörgen Persson for excellent technical assistance concerning the automatic assay of ATP, ADP and AMP and for developing the computer programs used in the calculations on this assay.

REFERENCES

1. Lundin,A. and Thore, A. (1975). Comparison of methods for extraction of bacterial adenine nucleotides determined by firefly assay. *Appl.Microbiol.* 30, 713-721.
2. Larsson, C.-M. and Olsson, T. (1979). Firefly assay of adenine nucleotides from algae: Comparison of extraction methods. *Plant & Cell Physiol.* 20, 145-155.
3. Rodbell, M. (1964). Metabolism in isolated fat cells. 1. Effects of hormones on glucose metabolism and lipolysis. *J.Biol.Chem.* 239, 375-380.

4. Smith, U. Sjöström, L. and Björntorp, P. (1972). Comparison of two methods for determining human adipose cell sizes. *J.Lipid Res.* 13, 822-824.
5. Romert, L. and Jensen, D. (1983). Rabbit alveolar macrophage mediated mutagenesis of polycyclic aromatic hydrocarbons in V79 Chinese hamster. *Mutation Res.* 111, 245-252.
6. Thore, A., Ånséhn, S., Lundin, A. and Bergman, S. (1975). Detection of bacteriuria by luciferase assay of adenosine triphosphate. *J.Clin.Microbiol.* 1, 1-8.
7. Lundin, A., Rickardsson, A. and Thore, A. (1976). Continuous monitoring of ATP-converting reactions by purified firefly luciferase. *Anal.Biochem.* 75, 611-620.

BIOLUMINESCENT DETERMINATION OF FREE FATTY ACIDS: NEW TOOL IN OBESITY RESEARCH

E. Wieland and H. Kather

Clinical Institute of Infarction Research of
The Medical Clinic of The University of Heidelberg
Bergheimerstrasse 58, D-6900 Heidelberg West-Germany

INTRODUCTION

Bioluminescent assays are sensitive and specific. We developed a bioluminescent assay for determination of free fatty acids using bacterial NADH-linked luciferase, which is applicable to unextracted serum or plasma and determination of lipolysis in microsamples of isolated human fat cells.

PRINCIPLE

As originally described by Shimizu et al. fatty acids are activated in the presence of ATP and CoA by acyl-CoA-synthetase (1). The pyrophosphate formed is used to phosphorylate fructose-6-phosphate in a reaction catalyzed by the enzyme pyrophosphate-fructose-6-phosphate-1-phosphotransferase (2). The triosephosphates produced from fructose-1-6-bisphosphate by aldolase are oxidized by NAD in the presence of arsenate to 3-phosphoglycerate as previously published (3).

MATERIALS AND METHODS

10 ul of native serum or plasma were stopped by adding 10 ul of 1N HCl. After one minute the sample was neutralized by adding 10 ul of 1N NaOH and diluted 100 times with deionized water containing 0.25% Triton-X-100.

ISBN 0-12-426290-2

Lipolysis was followed in microsamples of human fat cells (200 - 1000 cells/ml). Before estimation of free fatty acids the deproteinized media were diluted 2 - 10 fold with deionized water containing 0.25% Triton-X-100.

50 ul of diluted sample were added to an equal volume of a medium composed of 23 mmol/l Hepes, pH 8.0, 1.1 mmol/l KCl 20 mmol/l Na_3AsO_4, 1.1 mmol/l dithiotreitol, 2.9 mmol/l MgCl 8 mmol/l NAD, 0.065 mmol/l CoA, 0.53 umol/l EDTA, 33 U/ml triosephosphateisomerase, 14 U/ml glyceraldehyde-3-phosphate-dehydrogenase, 0.28 U/ml aldolase, 0.10 U/ml PP_i-phosphofructokinase, 0.04 mmol/l fructose-6-phosphate and 6.7 mU/ml acyl-CoA-synthetase.

After incubation for 180 minutes at 25°C the samples were further diluted (10 times) and 0.03 - 0.05 ml aliquots were assayed for NADH content.

Luciferase (1.5 mU/ml) and NAD(P)H:FMN oxidoreductase (1.7 U/ml) were dissolved in a potassium-phosphate buffer (0.2 mol/l) containing 0.4 mmol/l dithiotreitol and 67 mmol/l raffinose. Tetradecanal (4.7 mmol/l) was dissolved in a solution containing 50 g/l of bovine serum albumin, essentially fatty acid free, and 10 g/l Triton-X-100, pH 7.0 at 50°C.

Solutions of tetradecanal and of luciferase were stored frozen in small portions. FMN (0.11 mmol/l) was dissolved in a potassium-phosphate buffer (2 mmol/l, pH 7.0) and kept in a dark bottle on ice. The solution was made daily.

The assay cocktail (100 ml) contained 0.5 mmol/l tetradecanal, 1.1 umol/l FMN, 5.3 mU/l luciferase and 83 U/ml NAD(P)H:FMN oxidoreductase. Portions (0.5 ml) of the assay cocktail were brought to 25°C by preincubation for 20 minutes. Bioluminescent assays were started by addition of 0.03 - 0.05 ml of sample. Production of light was measured by Berthold Biolumat 9500T. Tetradecanal and raffinose were from EGA Chemie, Steinheim, FRG and Merck AG, Darmstadt, FRG respectively. CoA fatty acids and buffer reagents were from Serva GmbH, Heidelberg, FRG; ATP, NAD, triosephosphateisomerase and enzymes for luminescence from Boehringer Mannheim, FRG. Acyl-CoA-synthetase, enzymes for free fatty acid conversion and bovine serum albumin, essentially fatty acid free were from Sigma Chemical Co., St. Louis, USA.

CONCLUSIONS

Light production is linear to fatty acid concentrations up to 30 umol/l. Based on a signal to noise ratio of 1 the detection limit is about 5 umol/l (Fig. 1).

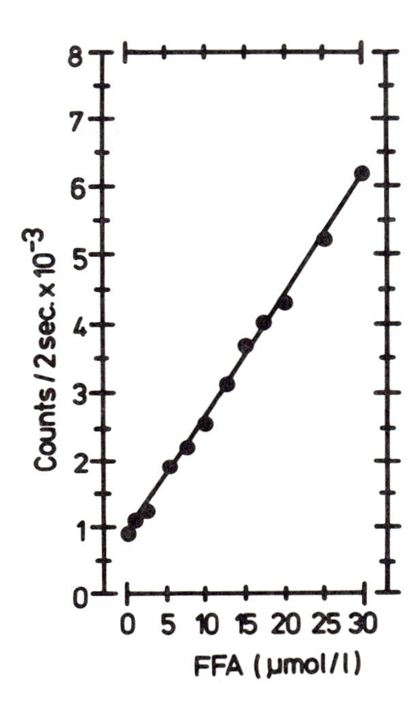

Fig. 1 Standard curve for palmitic acid

The excellent sensitivity of the method permits systematic studies on human fat cell lipolysis with mg amounts of adipose tissue. At present 200 cells are minimally needed.

Precision of the method was estimated by comparison with two commercially available kit methods. The linear regression data show excellent correlations for the chemical colorimetric method (Boehringer Mannheim GmbH), r = 0.96, Y intercept = -4.103, slope = 0.92, n = 12 and the enzymatic colorimetric procedure (Wako NEFA C) r = 0.99, Y intercept = -4.75, slope = 1.082, n = 20.

Fig. 2 shows inhibition of free fatty acid and of glycerol release of human adipocytes.

The method was also successfully applied for determination of serum free fatty acids. Even if small samples (5 - 10 ul) are employed serum fatty acid concentration exceeded the upper detection limit of the method. In order to be in the dynamic range sera have to be diluted 50 - 100 fold prior to fatty acid determination.

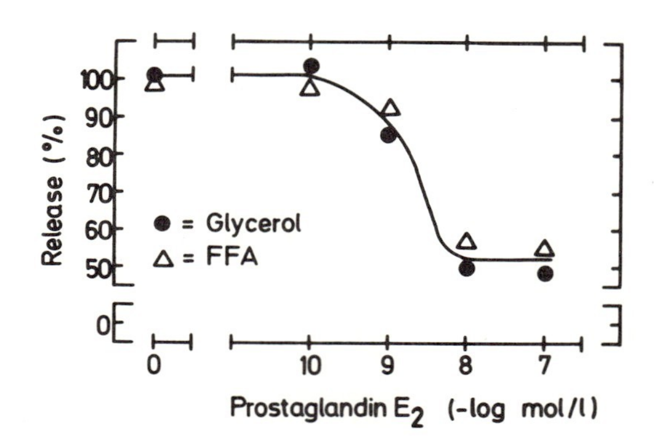

Fig. 2 *Effect of ascending PG E_2 concentrations on the FFA and Glycerol release of isolated human fat cells.*

REFERENCES

1. Shimizu, S., Inoue, K., Tani., Yamada, H. Tabata, M. and Murachi, T. (1980). Enzymatic Determination of Serum Free Fatty Acids: A Colorimetric Method, *Anal. Biochem.* 107, 193-198.
2. O´Brien, W.E. (1976). A Continous Spectrophotometric Assay for Argininosuccinate Synthetase Based on Pyrophosphate Formation, *Anal. Biochem.* 76 , 423-430
3. Kather, H., Schröder, F. and Simon, B. (1982). Bioluminescent Method for Determinating Microquantities of Glycerol: Application for Measurements of Lipolysis in Isolated Human Fat Cells. *In* "Luminescent Assays: Perspectives in Endocrinology and Clinical Chemistry" (Eds. M. Serio and M. Pazzagali), pp. 53 - 56, Raven Press, New York

KINETIC ANALYSIS OF STEROIDS WITH BACTERIAL BIOLUMINESCENCE

T. Lövgren*, R. Raunio+, J. Lavi*, K. Kurkijärvi*

*Wallac Biochemical Laboratory, Box 10, Turku, Finland
+Dept. of Biochemistry, Turku University, Turku, Finland

INTRODUCTION

Steroid hormones are determined in clinical laboratories for several reasons. The commonly used methods in steroid determinations are based on immunoassay techniques. Already thirty years ago it was demonstrated (1) that steroids can be determined with certain NADH linked dehydrogenase enzymes, but the method has not been used because the sensitivity of conventional photometric assays of NADH is not high enough to detect steroids in biological fluids. The bacterial bioluminescence reaction is well documented in analysis of NAD(P)H dependent enzymes and their substrates (2,3). The sensitivity of the bioluminescent method is usually 1000-fold as compared to photometric assays. A bioluminescent assay of estrone and estradiol based on transhydrogenase cycling system (4) has recently been published. In the present work a direct enzyme kinetic assay for estradiol and progesterone was developed, exploiting a stable light emitting bacterial bioluminescence reagent and appropriate hydroxysteroid dehydrogenases.

MATERIALS AND METHODS

Reagent

Bioluminescent NADH Monitoring Kit (LKB-Wallac, Turku, Finland); 3α-20β-hydroxysteroid dehydrogenase, progesterone and estradiol standards (Sigma Chem. Co., St. Louis, Missouri, U.S.A.). 17_{β}-estradiol dehydrogenase was purified

ISBN 0-12-426290-2

from human placenta by affinity chromagraphy (5). All other reagents were analytical grade.

Assay principle

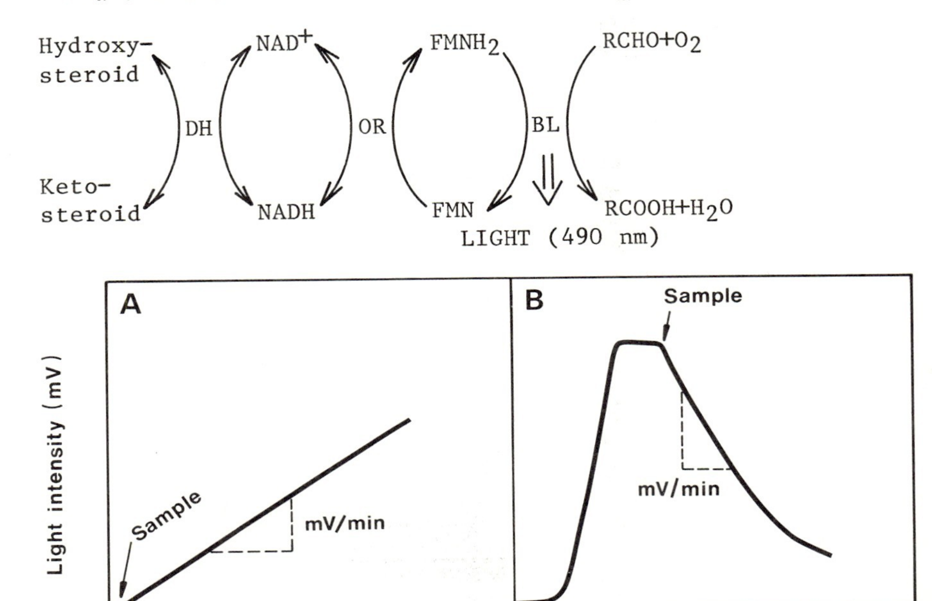

Fig. 1. Time course of estradiol (A) and progesterone (B) assays. A: NADH is produced in the estradiol assay system, B: NADH is consumed in the progesterone assay system.

Assay procedure

Estradiol 1 ml of serum was extracted twice with 2 ml of diethyl ether. The ether phase was washed twice with 0.1 M $NaHCO_3$ followed by evaporation, and the residue was dissolved in 500 μl of 0.1 M sodium phosphate buffer pH 7.0, containing 10 μl of absolute ethanol. 100 μl of reconstituted NADH Monitoring reagent, 20 μl 17β-estradiol dehydrogenase (20 mU), and 20 μl NAD (4 mg/ml) were added into 400 μl of the previous solution. The standards were measured similarly omitting the extraction procedure.

Progesterone 1 ml of serum was extracted twice with 2 ml of hexane. The organic phase was evaporated and the residue was dissolved in 500 μl of 0.1 M sodium phosphate buffer pH 7.0 containing 10 μl of absolute ethanol. 100 μl of NADH Monitoring reagent, 20 μl $5x10^{-6}$M NADH, and 10 μl 3α-20β-hydroxysteroid dehydrogenase (30 mU) were added into 400 μl of the previous solution. Standards were measured similarly without any extraction. Light emission was measured with an automated 1251 luminometer (LKB-Wallac, Turku, Finland).

RESULTS

A linear response was obtained for estradiol and progesterone from 0.15 to 50 pmoles/assay and from 2 to 1000 pmoles/assay, respectively (Fig. 2). The sensitivites of the assays are high enough for serum samples when extracting the steroids from 1 ml of serum.

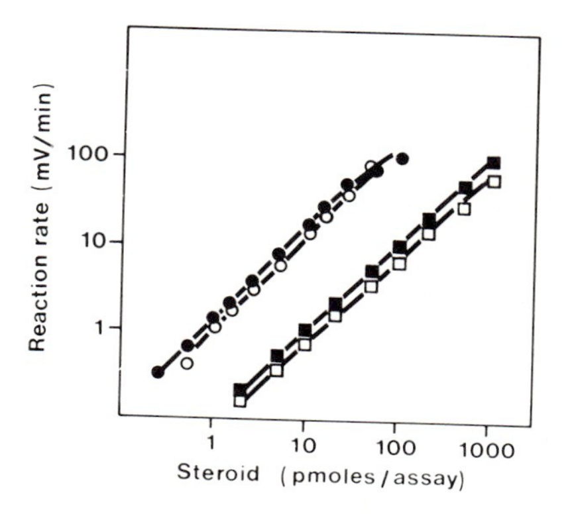

Fig. 2 Standard curves (estradiol●, progesterone■) and recovery curves for the steroids (estradiol○, progesterone□).

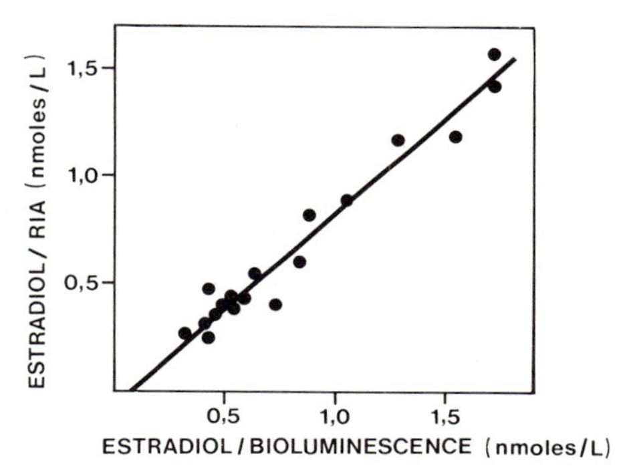

Fig. 3 Correlation between RIA and bioluminescent determination of estradiol.

Table 1 shows the specificity of the steroid dehydrogenases used in the measurements. 17β-estradiol dehydrogenase is highly specific to estradiol. In addition to estradiol only some estradiol derivatives act as substrates with a significant reactivity. The interference from OH-progesterone, deoxycorticosterone, and cortisone is avoided because they are not extracted into hexane in a significant amount (below 5 %).

Table 1. Interference index for various steroids in luminescent progesterone (A) and estradiol (B) determinations

Steroid	%	Steroid	%
Progesterone	100	17β-estradiol	100
deoxycorticosterone	73	estriol	4
OH-progesterone	91	3-methoxy-estradiol	71
cortisone	36	2-OH-estradiol	16
cortisol	10	6α-OH-estradiol	79
corticosterone	1	15α-OH-estradiol	76
		5-androstene-3β,17β diole	1

The bioluminescent steroid assays correlated well with the corresponding RIA-methods (r=0.88-0.96) although present enzyme kinetic methods showed little higher results than RIA (slope=0.80-0.90). Fig. 3 presents a correlation curve between RIA and bioluminescent assay for serum estradiol.

CONCLUSIONS

This work presents direct enzyme kinetic assays for estradiol and progesterone. The assays are rapid; neither incubation time (immunoassays) nor accumulative amplifying reaction (4) is needed. The most critical and time consuming point of the assays is the extraction procedure which should be performed very carefully. Single step chromatography by commercial columns might be simpler and more straightforward.

REFERENCES

1. Härkönen, H., Adlercreutz, H., Groman, E.V. (1974). Enzymatic techniques in steroid assay, *J. Steroid. Biochem.* 5, 717-725
2. Lövgren, T., Thore, A., Lavi, J., Styrelius, I., Raunio, R. (1982). Continuous monitoring of NADH-converting Reactions by bacterial bioluminescence, *J. Appl. Biochem.* 4, 103-111
3. Kurkijärvi, K., Raunio, R., Lavi, J., Lövgren, T. (1984). Stable light emitting bacterial bioluminescence reagents in bioluminescence and chemiluminescence: instrumentation and applications (Ed. K. Van Dyke) (in press) CRC Press, Boca Raton, U.S.A.
4. Nicolas, J.C., Boussinoux, A.M., Boularan, A.M., Descomps, B., De Paulet, A.C. (1983). Bioluminescent assay of ferritomole levels of extrone and estradiol, *Anal. Biochem.* 135, 141-145
5. Kurkijärvi, K., publication pending

A RAPID BIOLUMINESCENCE ASSAY FOR NON-CYCLIC PHOTOPHOSPHORYLATION OF ISOLATED CHLOROPLASTS

F. Van Assche and H. Clijsters

Dept. SBM, Limburgs Universitair Centrum
B - 3610 Diepenbeek, Belgium
I.W.O.N.L.

INTRODUCTION

Firefly bioluminescence has been used to measure ATP production of photosynthetic bacteria (1) and of cyclic photophosphorylation of chloroplasts (2). However, it is not commonly applied for non-cyclic photophosphorylation measurements. This could be due to the following reasons : a) a stable bioluminescence output is necessary to permit stepwise (light/dark) reaction procedures; b) several components of the current reaction cocktails severely quench bioluminescence; c) the ATP concentration range of non-cyclic photophosphorylation reactions is relatively high for bioluminescence.

We describe a rapid (within 2 min) and convenient assay for the direct measurement of non-cyclic ATP production in type C chloroplasts, using commercial firefly bioluminescence reagents and equipment. The technique is designed to be combined with polarographical electron transport measurements.

MATERIAL AND METHODS

1) electron transport

Photosynthetic electron transport activity of a suspension of type C chloroplasts from *Phaseolus vulgaris* was measured at 22°C in a Clark-type oxygen electrode (Rank Brothers, Cambridge, UK) in the following cocktails :
- Photosystem (PS) 2 : Tricine 0,01 M pH 7,8; K_2HPO 2mM;

ISBN 0-12-426290-2

Mg Cl_2 5mM; diaminodurene (DAD) 0,4 mM; and $K_3Fe(CN)_6$ 1,2mM.
- PS 1 : Tricine 0,01 M pH 7,8; K_2HPO_4 2mM; NaCl 2mM; $MgCl_2$ 5mM; methylviologen (MV) 0,25mM; dichlorophenyl-dimethylurea 0,25 µM; NaN_3 0,25mM; DAD 0,1mM and Na-ascorbate 1mM;
- PS 1+2 : Tricine 0,01 M pH 7,8; K_2HPO_4 2mM; NaCl 2mM;$MgCl_2$ 5mM; MV 0,2mM; ADP 1mM.

Finally, 100 µl of a chloroplast suspension were added (±50 µg of chlorophyll). The total reaction volume was 2 ml. After equilibration in darkness,white light with a photon flux density of 2000 $\mu mol.m^{-2}s^{-1}$ was provided.

2) ATP measurement

Bioluminescence was measured in an LKB 1250-001 luminometer thermostated at 22°C. Freeze-dried ATP monitoring reagent LKB 1250-121 was diluted in 10 ml bidistilled water, from which separate samples were frozen, stored at -20°C, and thawed just before use.

Non-cyclic photophosphorylation and electron transport measurements were combined in the following way : during dark equilibration of the reaction mixture in the oxygen electrode, the bioluminescence cocktail consisting of 290 µl of the corresponding electron transport cocktail, 100 µl ATP monitoring reagent solution and 50 µl ADP 5mM, was prepared in a LKB 2174-086 plastic vial. About 15s after illumination of the oxygen electrode, a 50 µl sample (±1 µg of chlorophyll) was taken from the reaction mixture through the capillary in the stopper, while electron transport proceeded. This volume was injected in the luminometer vial. From that moment, all manipulations were carried out in dim green safety light. After a short equilibration (±10s) the vial was illuminated for exactly 10s with a light source as described above. Finally, an internal standard of 10 µl 5 x 10^{-5} M ATP was added.

RESULTS AND DISCUSSION

The phosphorylation assay was carried out within the range of 0,1 - 2,0 µM ATP, corresponding to a linear bioluminescence response of 0,5-10 V. A signal decay of 0,5% min^{-1} was neglected. A typical recorder trace of the assay, performed stepwise within 1 min, is shown in Figure 1.

Step A. ATP contamination of the cocktail. An excess ADP is required to assure non-limited phosphorylation. The purest ADP preparation available, resulting in a limited ATP-signal, should be used (e.g. Boehringer 236675).

Step B. ATP production by the chloroplasts in darkness.

At time 0 chloroplasts were injected in the reaction vial. This type of chloroplast preparation produces ATP in darkness, probably due to the presence of some bound adenylate kinase. This dark reaction proceeds at a constant rate throughout the assay. It is registered for 10s and extrapolated graphically during the following steps of the assay.

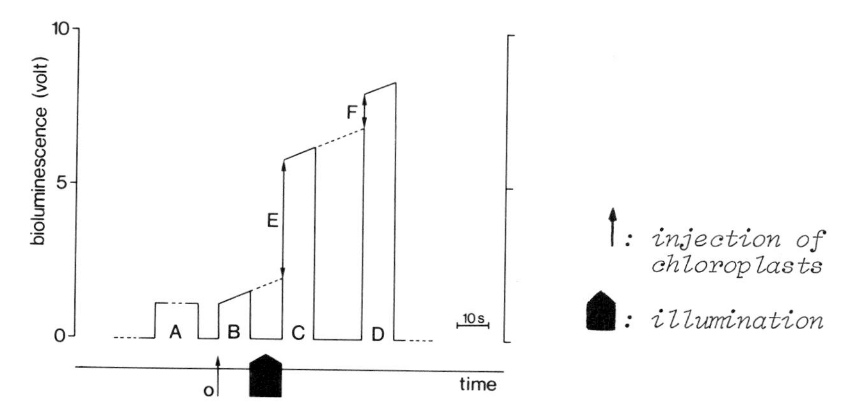

Fig. 1 Time chart of phosphorylation assay

Step C. Phosphorylation. While registration continued, the reaction vial was removed from the measuring chamber, illuminated for exactly 10s, and immediately returned to the apparatus. The increase E of the bioluminescence signal corresponds to the ATP produced during illumination. As can be seen on the figure, this ATP production is easily corrected graphically for the ATP produced in darkness.

Step D. Internal ATP-standard. Before calculating the ATP production from the standard curve, we added an internal standard at the end of the assay. This allows correction for two phenomena : 1) decrease of the bioluminescence capacity of the ATP monitoring reagent during storage at the assay temperature; 2)quenching of bioluminescence by several components of the reaction cocktails, including the chloroplast suspension. This quenching was reduced by diluting these components as much as possible without changing electron transport supporting capacity of the cocktail (Table1).

Two mean differences exist between the assay mixture for electron transport and photophosphorylation : (a) the bioluminescence cocktail contains the ATP-monitoring reagent; (b) its chlorophyll concentration is much lower. However, the reagent proved to have no effect on electron transport, and the response of the electron transport activity to the amount of chlorophyll was linear far beyond the concentra-

TABLE 1

Quenching (%) due to components of electron transport cocktails. (no quenching of $MgCl_2$, DAD, DCMU, NaN_3, was observed at the concentrations described in Mat & Meth.

component	current composition	% quenching	modified composition	% quenching
Tricine	0,05 M	40	0,01 M	10
K_2HPO_4	0,1 M	55	0,002 M	0
NaCl	0,5 M	75	0,005 M	0
Na-ascorbate	0,1 M	95	0,001 M	15
K_3FeCN_6	1,2 mM	75	0,125mM	30

tions used. Underestimation of electron transport to photophosphorylation, due to selfshading of chloroplasts in the oxygen electrode, was therefore excluded.

The ATP/2 electron ratios, obtained with this combined assay (PS 1+2:1,5±0,3; PS 1:0,7±0,1; PS 2:0,7±0,1) are in good agreement with values, found with other methods (3). A tendency of our values to be slightly higher could be a result of the short duration of the assay, minimizing denaturation of PS 2. Attention is drawn to the fact that in this coupled assay, electron transport and related ATP production are measured on the same chloroplast preparation, by taking a subsample for bioluminescence from the oxygen electrode. In this way any discrepancy resulting from differences in storage time of the chloroplasts or from the intrinsic variability of the preparation itself, is avoided.

REFERENCES

1. Lundin, A., Baltscheffsky, M. and Höijer, B. (1979) Continuous monitoring of ATP in phosphorylating systems by firefly luciferase. *In* "Proc. Int. Symp. on Analytical Application of Bioluminescence and Chemiluminescence" (Eds E. Schram and P. Stanley). pp. 367-391. State Printing & Publ. Inc. California.
2. Sigalat, C. and de Kouchkovsky, Y. (1979). Short-time ATP synthesis in isolated chloroplasts measured by the luminescence of firefly extracts. *ibid.* pp. 367-391.
3. Reeves, S.G. and Hall, D.O. (1978). Phosphorylation in chloroplasts. *Biochim. Biophys. Acta* 305, 45-126.

THE SORBITOL DEHYDROGENASE REACTION IN BIOLUMINESCENCE ASSAY

S.E. Brolin*, P. Naeser** and P.-O. Berggren*

*Department of Medical Cell Biology, Biomedical Center and **Eye Clinic, University of Uppsala, Uppsala, Sweden

INTRODUCTION

Among the sugar alcohols sorbitol predominates in the metabolism of mammalian cells but occurs normally only at low concentrations. In this analytical study of sorbitol dehydrogenase (SDH; EC 1.1.1.14). We describe its use in nucleotide assay, determination of its activity and measurements of its substrate. SDH catalyzes the reaction Sorbitol + $NAD^+ \rightleftharpoons$ Fructose + NADH. The analyses are designed for applications in studies of altered sugar metabolism in experimental diabetes. A possible simultaneous enzymatic oxidation of tracer amounts of other alcohols is of minor importance for these applications.

ANALYTICAL PERFORMANCEE

The analyses are carried out either in a single step (1-4), using continuous formation of NADH within the light yielding solution, or for enhanced detectability by accumulation of NADH in a preceding reaction. The light emission is elicited both with fast reacting and alternatively with slowly fading luciferase solutions, see Table 1. Accumulated NADH elicits a flash of light with a fast light yielder, but gives a more durable and weaker intensity with the slow variety. In the single step performance the emission intensity is nearly constant using the fast luciferase (1), but increasing with the slowly reacting preparation (5). The rate of this increase serves as measure of the concentrations of either NAD^+ or sorbitol, or the activity of SDH. The light emission was measured mainly as described previously, as mV (6) or photon counts (7). The supernatants of centrifuged homogenates of lyophilized organ pieces are used in the analytical applications.

ISBN 0-12-426290-2

TABLE 1

Composition of Solutions at the Moment of Reaction

Incubation for NADH formation	Fast reacting light yielder	Slowly fading light yielder
Tris 0.1 M, pH 8.6	Sodium phosphate	Potassium phosphate
NAD^+ 1.4 mM	buffer 75 mM, pH 6.8	buffer 27 mM, pH 7.0
SDH .8 U/ml	Glycerol 92 mM	Dithiothreitol .11 mM
Sorbitol 6 mM	Dithiothreitol .4 mM	FMN reductase 43 U/l
When assayed,	Luciferase (vibr. fish.	Luciferase (Ph. fish.
sorbitol and	Sigma) 1.4 mg/ml	Boehringer) 5.4 mg/ml
SDH respectively	FMN 0.42 mM	FMN 2.7 μM
were excluded	Tetradecanal 20 μM	Tetradecanal 20 μl

After incubation for 30 min at 38°C 10 ul samples were added to 100 ul light yielding solution. In single step assay the light yielder contained NAD^+, 1 mM; sorbitol, 50 mM and SDH, 1 U/ml with exclusion of the component to be assayed. The LKB luciferase is also useful.

ASSAY OF NAD^+

Single step analysis of NAD^+ can be accomplished by continuous conversion to NADH in the light yielding solution. For this purpose addition of sorbitol and SDH is suitable. They cause only an insignificant change of the background emission and the equilibrium is more in favour of NADH formation than in the malate dehydrogenase reaction (1). The performance of the assay is described in Table 1 and calibration curve is shown in Fig. 1.

DETERMINATION OF SDH ACTIVITY

The SDH activity is determinable in a single step with a bacterial luciferase solution supplemented with sorbitol and NAD^+ (Fig. 1b). If the luciferase preparation is designed to react with slow fading, the coupled SDH reaction generates an accumulation of NADH which results in an increasing emission. The rate of this increase serves as measure of the SDH activity. However, a prolonged accumulation of NADH may be desirable. Particularly if tissue samples are diluted in order to restrain disturbing side reactions. This demand is met by a preceding incubation step in which the accumulation is terminated. Thereafter the amount of NADH obtained is measured in a second step using the maximum value of the emission intensity either of slow or fast reaction luciferase (Fig. 1c). An example of such a tissue assay is given in Fig 1d.

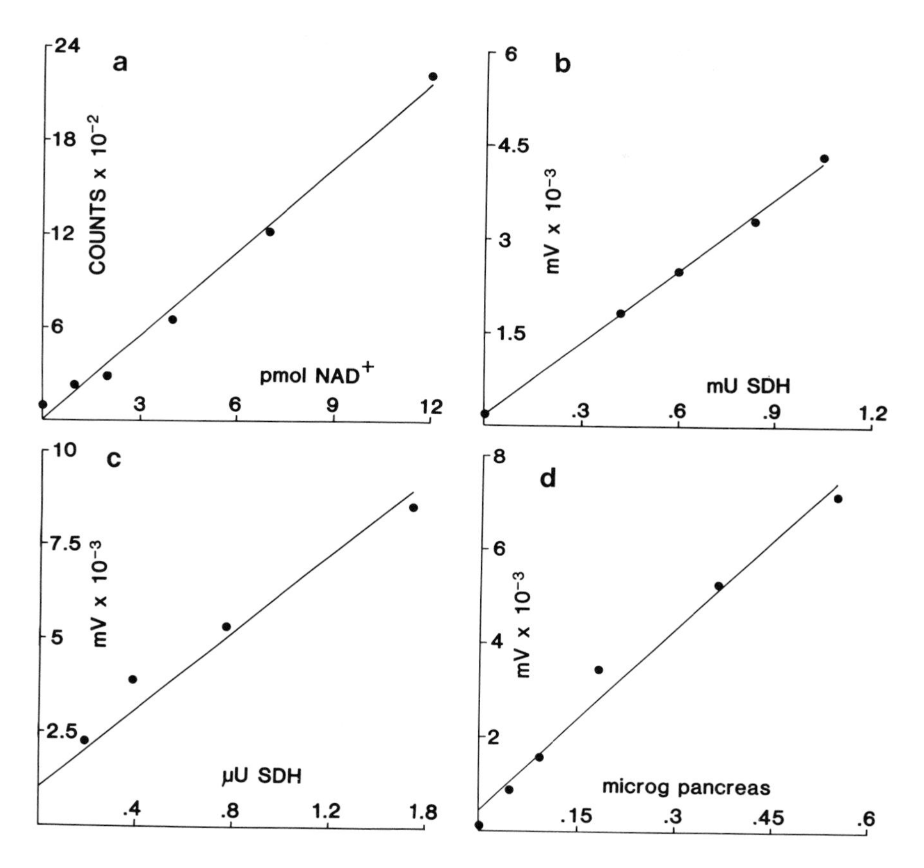

Fig. 1 Examples of analytical applications, presented as means of duplicate or triplicate determinations.
a) A calibration curve for assy of NAD^+ in a single step, using a slowly fading luciferase solution supplemented with sorbitol and SDH. Abscissa: Amount NAD^+. Ordinate: Increase in emission rate given as rise in photon counts after 2 min.
b) Single step assay of SDH with a fast reacting luciferase solution containing NAD^+ and sorbitol. Abscissa: Enzyme amount. Ordinate: Near maximum intensity (2 min) in mV.
c) Assay of SDH, using a slowly fading light yielder after a preceding accumulation of NADH. Abscissa: Enzyme amount. Ordinate: Maximum intensity.
d) Assay of SDH in diluted homogenates of a lyophilized mouse

pancreas accomplished as in c which may serve as a calibration curve. The curves show a satisfactory linearity which is sustained by the correlation coefficient (.99 in a,b and d and .97 in c).

ASSAY OF SORBITOL

Fig. 2a shows a calibration curve for analyses of sorbitol in the single step performance, but assay of small tissue samples requires a preceding accumulation of NADH. In attempts to estimate low concentrations of sorbitol in various organs, tissue blanks revealed considerable disturbances. The analytical need of a reasonable sorbitol concentration precludes dilution to the same extent as in the assay of SDH. Liver samples inhibit the light formation as does diluted fresh bile from guinea pigs which receive a regular addition of vitamin C to their diet. Bilirubin and bile acids lacked this effect but the ascorbic acid might be responsible. Storage of frozen homogenized liver samples abolishes the inhibition. Tissue blanks which are processed together with the samples but lacking addition of SDH provoked a very disturbing emission of light.

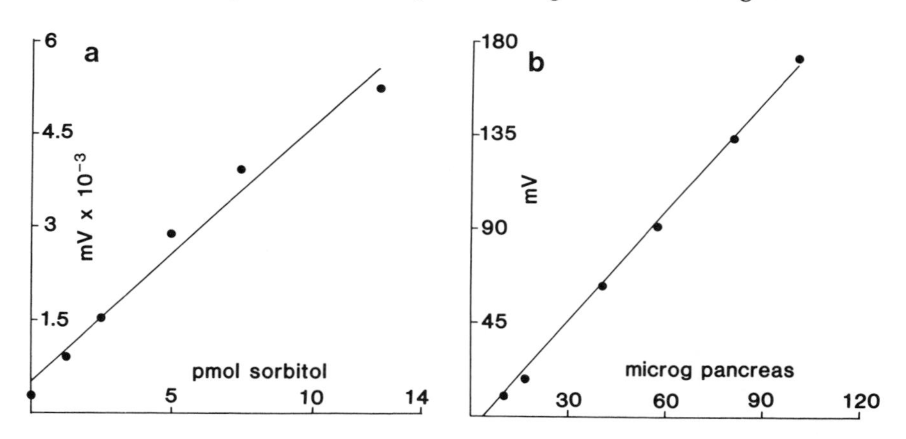

Fig. 2 Analyses of sorbitol.
a) Single step assay with a fast light yielder. Abscissa: Amount of sorbitol. Ordinate: Maximum intensity (2 min).
b) Sorbitol amounts in diluted homogenates of a lyophilized mouse pancreas. Assay with a slowly fading light yielder after a preceding accumulation of NADH. The correlation coefficient is .99 in a and b.

Such emission was caused by small particles and among these probably mitochondria, which remained suspended in the supernatant at ordinary centrifugal force. After complete removal of the particles by ultrafiltration or high speed centrifugation with an airfuge, the additional light production was entirely abolished. In this way the analysis of sorbitol is applicable to tissue samples as exemplified in Fig. 2b.

If a few particles remain in the supernatant or pass the filter, high blank readings disturb the measurements. Since it cannot be taken for granted that the extraneous light formation is caused by SDH in the particles it must be eliminated. This is achieved by immersing the tubes for 5 min in boiling water. If the concentration of sorbitol permits a moderate dilution of a homogenate, boiling alone may suffice, but the effect on tissue blanks has to be checked. No light-inducing sugar alcohols were split off from larger molecules as evidence by experiments with the high molecular weight fraction from a G-25 Sephadex column.

SUMMARY

The sorbitoldehydrogenase reaction is suitable for converting NAD^+ to NADH in assay of the nucleotide and in determinations of sorbitol concentrations or the activity of the enzyme itself.

REFERENCES

1. Ågren, A., Brolin, S.E. and Hjertén, S. (1977). Simplified luciferase assay of NAD^+. *Biochem. Biophys. Acta* 500, 103-108.
2. Brolin, S.E. (1977). Attempts to simplify the analytical performance in microassays of metabolites with bacterial luciferase. *Bioelectrochem. Bioenerg.* 4, 257-262.
3. Stanley. P.E. (1978). Quantitation of malate, oxaloacetate and malate dehydrogenase. *In* "Methods of Enzymology"(Ed M.A. DeLuca). Vol. LVII, pp. 181-188. Academic Press, New York.
4. Brolin, S.E. (1983). Single-step bioluminescence analyses of enzymes, using cibacrone blue chromatography for removal of interferring dehydrogenases. *Mol. Cell. Biochem.* 55, 177-182.
5. Lövgren, T., Thore, A., Lavi, J., Styrelius, I. and Raunio, R. (1982). Continuous monitoring of NADH-conventing reactions by bacterial bioluminescence. *J. Appl. Biochem.* 4, 103-111.
6. Brolin, S.E., Borglund, E., Tegnér, L. and Wettermark, G. (1971). Photokinetic microassay based on dehydrogenase reactions and bacterial luciferase. *Anal. Biochem.* 42, 124-135.
7. Wettermark, G., Stymne, H., Brolin, S.E. and Peterson, B. (1975). Substrate analyses in single cells. *Anal. Biochem.* 63, 293-307.

Supported by the Swedish Medical Research Council, the Swedish Diabetes Association and Clas Groschinsky's Memorial Foundation.

A NOVEL ASSAY FOR AGENTS CAUSING MEMBRANE PERTURBATION

D.E. Jenner, M.J. Harber and A.W. Asscher

Department of Renal Medicine
Welsh National School of Medicine
K.R.U.F. Institute, Royal Infirmary, Cardiff
CF2 1SZ, Wales, U.K.

INTRODUCTION

Bacterial agents which damage membranes are likely candidates as virulence factors in urinary tract infections (UTI). It has been shown in this laboratory (1) and elsewhere (2) that bacteria are phagocytosed by bladder epithelial cells, and production of minute quantities of a membrane damaging agent in the microenvironment of the phagosome could enable them to escape into the cytoplasm. In the upper tract, haemolysin production appears to be an important attribute for pyelonephritic strains of *Escherichia coli* (3) when it may serve to increase the availability of iron for bacterial growth by releasing haemoglobin from erythrocytes (4). A sensitive assay for agents causing membrane perturbation could, therefore, be of value in assessing the virulence of urinary isolates. However, existing methods (5,6) are cumbersome, insufficiently sensitive or prone to non-specific interference.

In this paper we describe a novel assay for measuring the biological activity of a variety of agents which interfere with membrane structure or function. This assay is based on measurement of light emission from a calcium sensitive photoprotein (obelin), enclosed within artificial lipid vesicles, stimulated by a calcium influx induced by the membrane perturbating agents.

MATERIALS AND METHODS

Obelin was the generous gift of Dr. A.K. Campbell, Department of Medical Biochemistry, W.N.S.M., Cardiff. Phosphatidyl choline (PC) was obtained from Lipid Products, U.K.; cholesterol, phospholipase C (PL) and haemoglobin were obtained from Sigma Chemical Co., U.K.; ionophore A23187 was a kind gift from Ely Lilly and Co. The buffer used for preparing liposomes was composed of 154 mM NaCl, 10 uM EDTA and 10 mM Tris, pH 7.4. These and all other reagents were A.R. grade from BDH Chemicals, U.K.

ISBN 0-12-426290-2

Preparation of Liposomes
Multilamellar vesicles were prepared according to the method of Dormer *et al.* (7) with 50 mg PC and 18 mg cholesterol in a final volume of 2.0 ml buffer. These liposomes contained 1 mg bovine serum albumen/ml and gave 4.8×10^8 relative light units (RLU) from encapsulated obelin measured over 10 sec following lysis of liposomes. Liposomes were diluted 1 in 10 in preparation buffer before use.

Bacteria
Five clinical isolates of *Escherichia coli* were used. ER2, NK1 and T1 were pyelonephritic strains obtained from the National Bacteriological Laboratory, Stockholm, Sweden. Strains 504 and SC were cystitis isolates, 504 being from the Department of Medicine, Charing Cross Hospital, London, and SC from our own laboratory. In addition a haemolytic strain of *Staph. albus* isolated in Cardiff from a patient with pyelonephritis was used.

Bioluminescence was measured on a Lumac Biocounter 2010 multijet and counts were expressed as RLU over 10 sec following injection.

Two Step Obelin Assay
Diluted liposomes (10 ul) were suspended in 790 ul assay buffer (0.1 mM EDTA, 200 mM Tris, pH 7.0) in a polystyrene luminometer cuvette and pre-incubated at 37°C for 10 min. Injection of 100 ul of 250 mM $CaCl_2$ stimulated light emission from non-entrapped obelin and this was used as a measure of the accuracy in aliquoting liposomes. The membrane perturbating agent was then added and incubated at 37°C for 1 hr after which the liposomes were lysed by injection of 100 ul of 1% (v/v) Nonidet P40 to measure unreacted entrapped obelin. The percentage of obelin that had reacted due to increased influx of calcium ions could be calculated using results from blanks containing no perturbating agent.

Haemolysin Assays
Bacterial haemolysin secretion was promoted by growth in Davis and Mingioli minimal medium supplemented with 1 mg bovine haemoglobin/ml. Cell free extract (100 ul) was then incubated with human erythrocytes for 10 min and the E420 was used as a measure of haemoglobin release (5). Haemolytic activity was also assessed on agar containing 7% human blood.

RESULTS

Maximum sensitivity was obtained with a high level of obelin entrapment during preparation of the liposomes and with a small volume (10 ul) of liposomes per cuvette. An incubation time of 1 hr provided the optimum balance between inactivation of obelin due to natural influx of calcium ions and the time necessary for the membrane damaging agents to have an effect. Obelin laden vesicles remained active in excess of three months after preparation but sensitivity to PL was decreased in older liposomes. Unilamellar vesicles were more stable on storage but were also less susceptible to perturbation. As little as 0.001 units of PL could be detected with this assay (Fig. 1a) whilst with ionophore A23187 the limit of sensitivity was 0.01 ng (Fig. 1b).

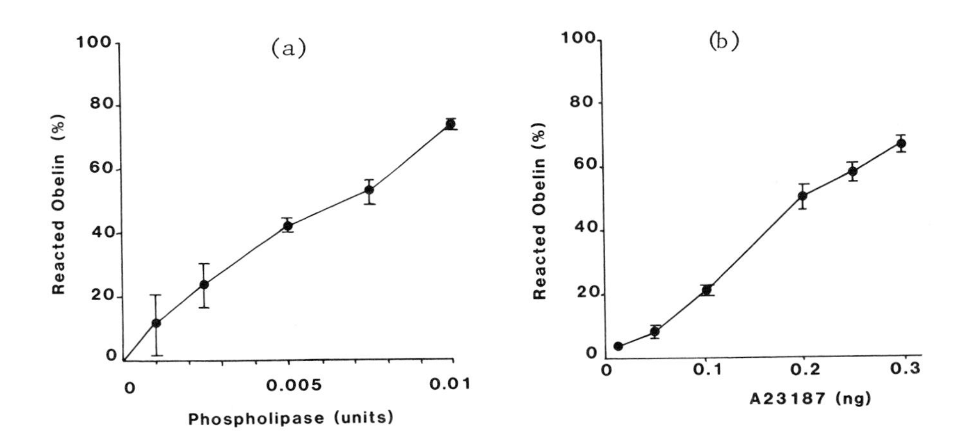

Fig. 1 Assay of (a) phospholipase C, and (b) A23187.

Results from the obelin assay correlated with those from the haemolysin assay for 5 of the 6 bacterial strains tested (Table 1), while the haemolysin assay showed agreement with the agar haemolysis test for only 3 of the strains. Staphylococcal toxin was expressed only on agar plates and not in assays using the cell free extract.

DISCUSSION

The obelin assay is at least as sensitive as other phospholipase assays, and compares favourably with alternative methods for detecting haemolysin production. Liposomes provide a well defined system whilst still being a non-specific substrate for membrane damage. The method is applicable to a variety of membrane perturbating agents and may provide a useful means of screening bacteria for potential pathogenicity.

TABLE 1

Comparison of obelin assay with reference haemolysin assays

Bacterial strain	Human blood agar	Haemolysin assay	Obelin assay
ER2	++	+	+
NK1	++	+	+
T1	–	+	–
504	+	+	+
SC	–	+	+
S. albus	+	–	–

REFERENCES

1. Harber, M.J., MacKenzie, R. and Asscher, A.W. (1984). Virulence factors of *Escherichia coli. In* "Pyelonephritis" (Eds. H. Losse, A.W. Asscher, A. Lison and V.T. Andriole), Vol. V, pp 43-50. Thieme, Stuttgart.
2. Fukushi, Y., Orikasa, S. and Kagayama, M. (1979). An electron microscopic study of the interaction between vesical epithelium and *E. coli*. *Invest. Urol.* 17, 61-68.
3. Brooks, H.J.L., O'Grady, F., McSherry, M.A. and Cattell, W.R. (1980). Uropathogenic properties of *Escherichia coli* in recurrent urinary tract infection. *J. Med. Microbiol.* 13, 57-68.
4. Waalurijk, C., MacLaren, D.M. and de Graaff, J. (1983). In vivo function of haemolysin in the nephropathogenicity of *Escherichia coli*. *Infec. Immun.* 42, 245-249.
5. Springer, W. and Goebel, W. (1980). Synthesis and secretion of haemolysin by *Escherichia coli*. *J. Bacteriol.* 144, 53-59.
6. Wightman, P.D., Dahgren, M.E., Hall, J.C., Davies, P. and Bonney, R.J. (1981). Identification and characterisation of a phospholipase C activity in resident mouse peritoneal macrophages. *Biochem. J.* 197, 523-526.
7. Dormer, R.L., Newman, G.R. and Campbell, A.K. (1978). Preparation and characterisation of liposomes containing the Ca^{2+} activated photoprotein, obelin. *Biochem. Biophys. Acta.* 538, 87-105.

MEASUREMENT OF GLUTAMATE, AMMONIA, AND UREA WITH BACTERIAL BIOLUMINESCENCE

T. Heinonen*, R. Raunio*, J. Lavi+,
T. Lövgren+, K. Kurkijärvi+

**Dept. of Biochemistry, Turku University, Turku, Finland*
+Wallac Biochemical Laboratory, Box 10, Turku, Finland

INTRODUCTION

Glutamate dehydrogenase (GlDH) catalyzes the following reaction:

$$\alpha\text{-ketoglutarate} + NH_3 + NADH \rightleftharpoons \text{L-glutamate} + NAD^+$$

Any of the substrates can be measured by following the consumption or formation of NADH. Stable light emitting bacterial bioluminescence reagents consisting of an NADH:FMN oxidoreductase (OR) and a luciferase (BL) have been found to be excellent for measuring sensitively NADH converting reactions (1, 2).

In this work a direct enzyme kinetic assay for L-glutamate, ammonia, and urea was developed and optimized by coupling the GlDH reaction to the bacterial bioluminescence system. Urea was hydrolyzed with urease to ammonia before determination.

MATERIALS AND METHODS

Reagents

Bioluminescent NADH Monitoring Kit (LKB-Wallac, Turku, Finland); GlDH, urease, spectrophotometric L-glutamate and ammonia/urea kits (Boehringer-Mannheim GmbH, Mannheim, W. Germany). All other reagents were analytical grade.

ISBN 0-12-426290-2

Assay principle

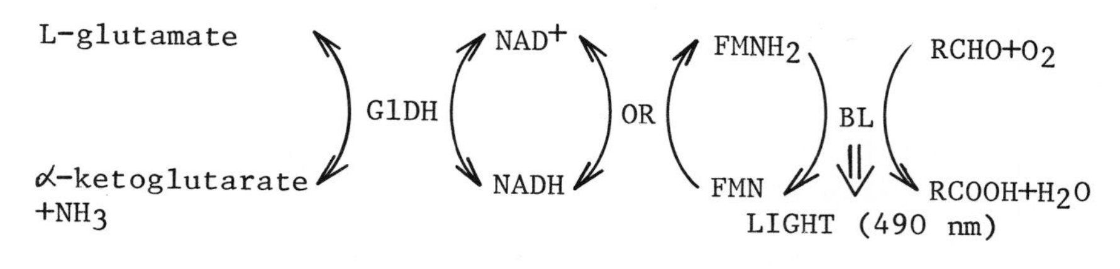

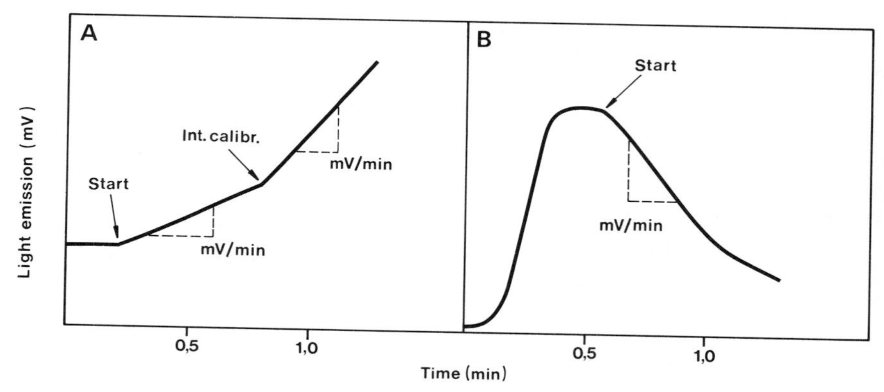

Fig. 1. Time course of L-glutamate (A) and ammonia (urea) (B) assays

Assay procedures

Ammonia/urea

350 μl 0.1 M sodium phosphate buffer, pH 7.6
5-20 μl sample or standard
(10 μl urease, 2 IU, and incubation for 3 minutes if ureasample)
20 μl GlDH (10 mU)
50 μl NADH Monitoring Reagent
70 μl $5x10^{-6}$M NADH
5 μl 0.2 M α-ketoglutarate starts the reaction

L-glutamate

340 μl 0.05 M Tris-HCl buffer pH 8.2
20 μl sample or standard
20 μl GlDH (400 mU, ammonia free)
100 μl NADH Monitoring Reagent
20 μl NAD^+ (500 μg/assay), starts the reaction

The light emission was measured with an automated 1251 luminometer (LKB-Wallac, Turku, Finland)

RESULTS

Reaction conditions were optimized for maximal reaction rate both in the GlDH and the light reaction. The GlDH reaction is difficult to drive in the direction of L-glutamate to α-ketoglutarate and ammonia; the pH should be above 8.0 and the concentration of NAD^+ quite high. However, the optimum pH of the bioluminescent NADH Monitoring reagent is 7.0 and high NAD^+ concentrations inhibit light production. Fig. 2 presents the effect of pH and Fig. 3 the effect of NAD^+ concentration on the response in the glutamate assay.

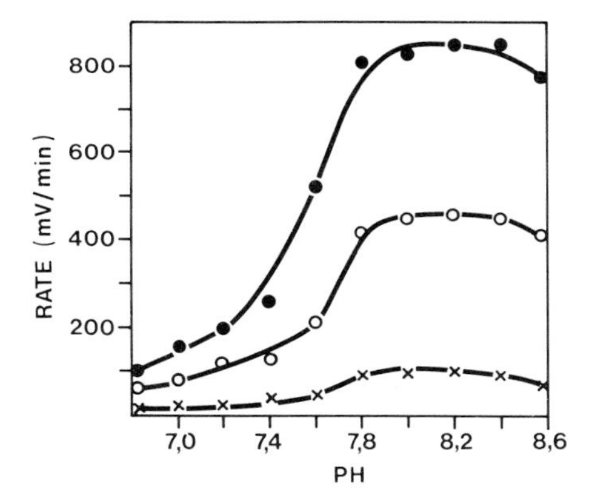

Fig. 2. The effect of pH on the response in the glutamate assay: -●- 6 nmol L-glutamate, -o- 3 nmol L-glutamate, -x- 0.6 nmol L-glutamate.

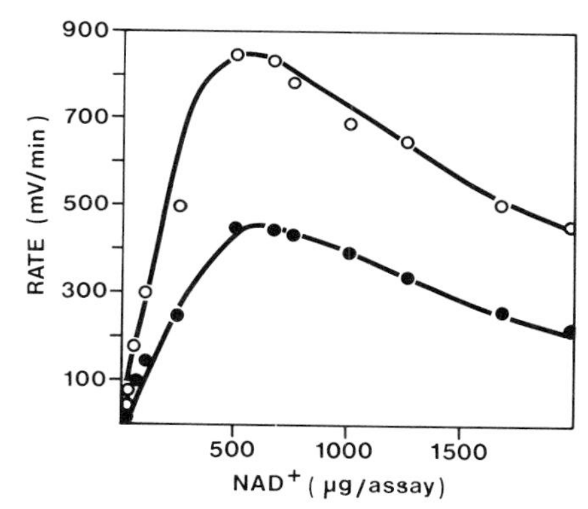

Fig. 3 The effect of NAD^+ concentration on the response in the glutamate assay pH 8.2: -o- 6 nmol L-glutamate, -●- 3 nmol L-glutamate

In the ammonia determination the pH 7.6 used is not as critical as in the L-glutamate assay. Saturating amounts of α-ketoglutarate cannot be used because a concentration above 2 mM inhibits the light reaction. The initial NADH concentration is 10^{-6} M as the NADH Monitoring reagent gives a linear response up to this concentration.

The linear ranges for L-glutamate and ammonia are 10^{-12} - $5x10^{-9}$ moles/assay and 10^{-10}-10^{-7} moles/assay, respectively (Fig. 4). The precision of both assays was excellent (CV $<$ 3 %).

The urease catalyzed hydrolysis of urea in serum specimens was complete in three minutes at room temperature. The

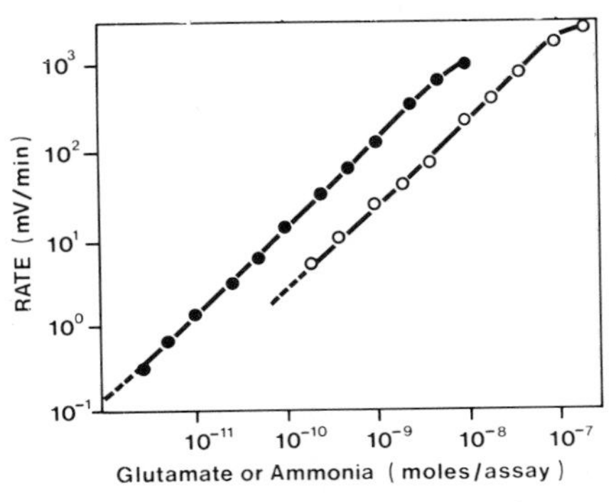

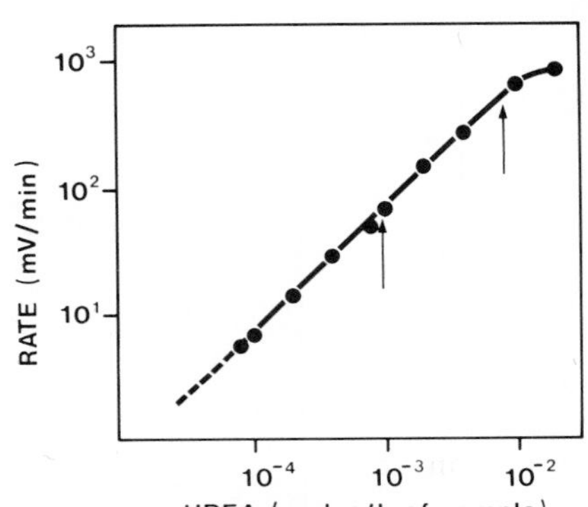

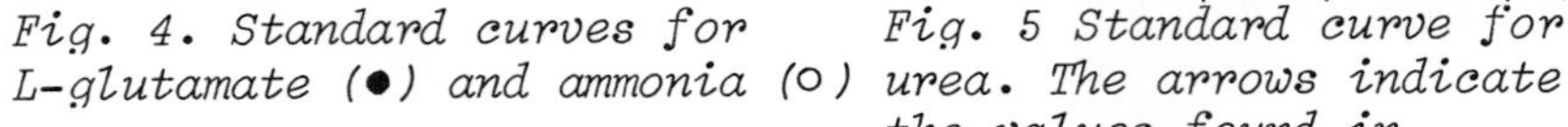

Fig. 4. Standard curves for L-glutamate (●) and ammonia (○)

Fig. 5 Standard curve for urea. The arrows indicate the values found in standard serums

assay was tested with two stadard serums containing urea 1 and 8.6 mmol/l (Fig.5, Table 2). Boehringer-Mannheim Ammonia/Urea kit (Cat.No. 542 946) gave similar results.

Table 2. Urea content in serum

Sample	No. of replicates	known amount (mmol/l)	found amount (mmol/l)	CV%	recovery
1	5	1.0	0.95	3.11	95
2	5	8.6	8.0	2.30	93

CONCLUSIONS

This work presents direct enzyme kinetic assays for ammonia L-glutamate and urea. The assays are rapid, sensitive and accurate. The stable light emitting reagent allows the use of internal calibration when following NADH formation and to correct for sample interference in case of NADH consumption. The assays are suitable for any biological samples.

REFERENCES

1. Lövgren, T., Thore, A., Lavi, J., Styrelius, I., Raunio, R. (1982). Continuous monitoring of NADH-converting reactions by bacterial bioluminescence, *J. Appl. Biochem.* 4, 103-111
2. Kurkijärvi, K., Raunio, R., Lavi, J., Lövgren, T. (1984). Stable light emitting bacterial bioluminescence reagents in bioluminescence and chemiluminescence: instrumentation and applications (Ed. K. Van Dyke) (in press) CRC press, Boca Raton, U.S.A.

BIOLUMINESCENT ASSAYS FOR ENZYMES METABOLIZING LONG CHAIN FATTY ACID DERIVATIVES

D. Morse, R. Szittner and E. Meighen

Department of Biochemistry, McGill University
Montreal, Canada H3G 1Y6

INTRODUCTION

The bacterial luciferase assay is ideal for the direct analysis of long chain aldehydes in aqueous solution. The light producing reaction can be used to measure very low levels (∿ 1 pmol) of fatty aldehydes (1) produced in different enzyme reactions (Fig. 1, I-III). Using suitable coupling systems, reactions producing precursors of fatty aldehydes can also be assayed with the luminescent assay. Thus esterases (IV) producing fatty acids or fatty alcohols can be coupled to luciferase by using a fatty acid reductase

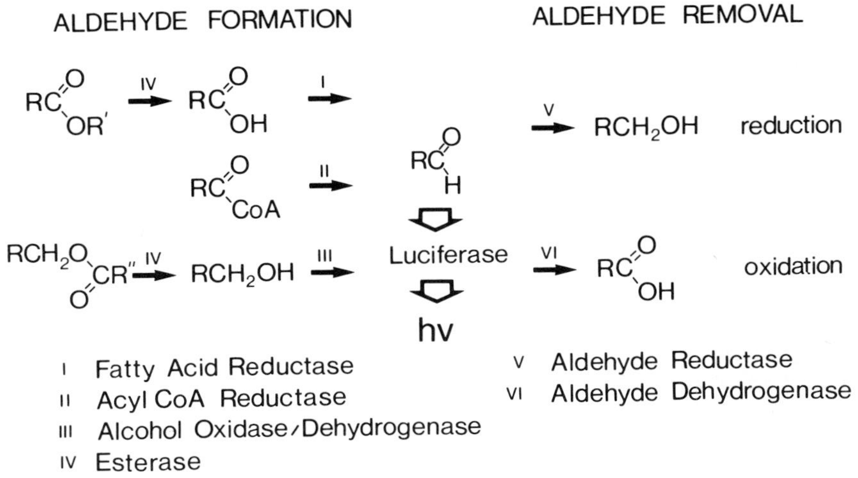

Fig. 1 Enzyme reactions producing and using fatty aldehyde.

ISBN 0-12-426290-2

(I) or alcohol dehydrogenase (III), respectively, to convert the product to aldehyde. The assay is also adaptable for measurement of enzymes catalyzing the removal of aldehyde (V, VI) from solution (2).

RESULTS

Although the maximum luminescence response to nonsaturating concentrations of aldehyde occurs with tetradecanal, the luciferase reaction can be used to detect aldehydes of different chain lengths with the response depending on the specific bacterial luciferase (Fig. 2). Luminescence is

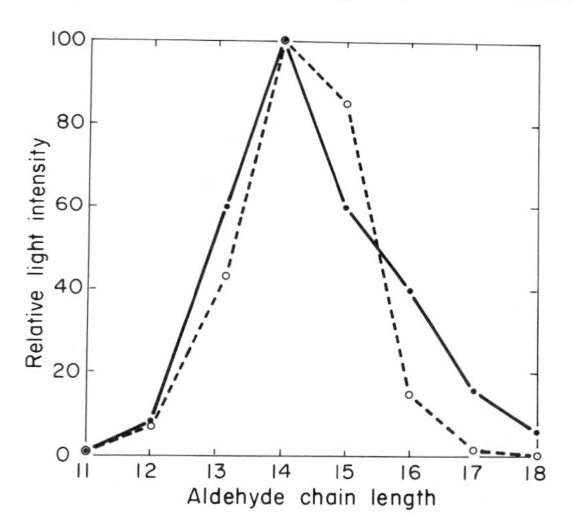

Fig. 2 Relative luminescence responses of V. harveyi (●) and P. phosphoreum (○) luciferase to 100 pmols aldehyde.

proportional to the amount of aldehyde over a wide range ($0.1\text{-}5\times10^3$ pmol) with the sensitivity limited only by the background response (1). The assay can also be used to measure monounsaturated aldehydes which are often found in nature. In some cases (e.g. cis-9-hexadecenal), the luminescent response is higher than with the corresponding saturated aldehyde.

The kinetics of product formation for two enzymes (III, IV) that have been measured by the luminescent assay are illustrated in Fig. 3. From these plots, the initial velocity can be determined leading to characterization of their kinetic properties. The presence of the lipid substrate at high concentrations ($\sim$ 1 μM) does not interfere with analysis of much lower concentrations of the aldehyde

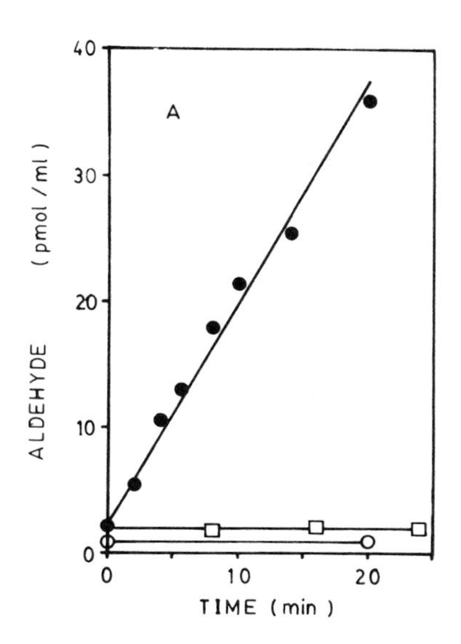

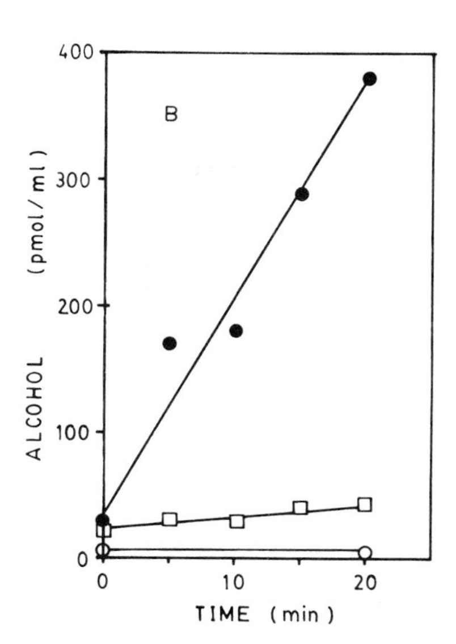

Fig. 3 Kinetics of product formation by an alcohol oxidase (A) from tetradecenol and by an acetate esterase (B) from tetradecenyl acetate (●). Controls lack either substrate (○) or enzyme (□).

product (0.01 µM). The esterase activity (IV) was measured by converting the fatty alcohol product into aldehyde by a 5 sec reaction with 10 µg/ml horse liver alcohol dehydrogenase and 0.5 mM NAD. Other enzymes assayable with luciferase (I, II) involve measurement of aldehyde produced from fatty acid (+ATP, +NADPH) or acyl-CoA (+NADPH) (3).

The analysis of the disappearance of aldehyde with time is more difficult since a sufficient percentage of substrate must be removed for accurate determination of the change in luminescence response. Thus, the initial velocity of the reaction cannot be directly measured in most cases. However, for an irreversible reaction, an integrated form of the Michaelis-Menten equation can be used for analysis of enzyme activities:

$$\ln S_o/S_t + (S_o - S_t)/K_m = (V_m/K_m)t$$

This equation simplifies to a first order exponential at low substrate concentrations (i.e., $S_o < K_m$):

$$\ln S_o/S_t = (V_m/K_m)t$$

The enzyme activity is obtained from the slope ($-V_m/K_m$) of a plot of ln S_t vs time. Shown in Fig. 4 is the measurement of aldehyde dehydrogenase activity at a substrate concentration (14 nM) well below that used in most assays.

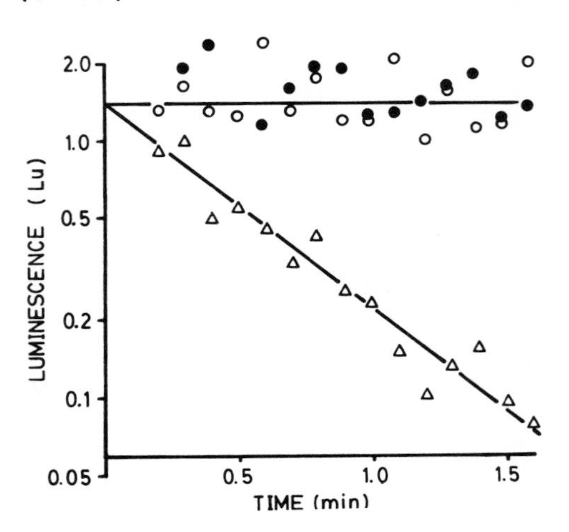

Fig. 4 Kinetics of an aldehyde dehydrogenase reaction with 0.5 mM NAD (△). Controls lack NAD (●) or enzyme (○).

CONCLUSION AND RECOMMENDATIONS

Bacterial luciferases can be routinely used for the analysis of enzymes metabolizing fatty acid derivatives that can be converted into long chain aldehydes. This assay is particularly advantageous for measuring low levels of enzyme activities as the luminescent assay is about 100 to 1000 times more sensitive than most spectrophotometric or chemical assays.

REFERENCES AND ACKNOWLEDGEMENTS

Supported by the Medical Research Council of Canada.

1. Meighen, E.A., Slessor, K.N. and Grant, G.G. (1982). Development of a bioluminescent assay for aldehyde pheromones of insects, *J. Chem. Ecol.* 8, 911-933.
2. Morse, D. and Meighen, E.A. (1984). Detection of pheromone biosynthetic and degradative enzymes *in vitro*, *J. Biol. Chem.* 259, 475-480.
3. Riendeau, D., Rodriguez, A. and Meighen, E.A. (1982). Resolution of the fatty acid reductase from *P. phosphoreum* into acyl protein synthetase and acyl-CoA reductase activities, *J. Biol. Chem.* 257, 6908-6915.

DETERMINATION OF 100 CHEMICALS BY THE IMPROVED BIOLUMINESCENCE TEST FOR MUTAGENIC AGENTS

S. Ulitzur, I. Weiser, B.Z. Levi & M. Barak

Department of Food Engineering & Biotechnology
Technion, Haifa, Israel.

INTRODUCTION

Although over 100 different short term *in vitro* tests have been proposed for testing the carcinogenic potential of chemical compounds, only a small number of these methods are currently in routine use. Several factors contribute to the selective process. The most important are the predictive value of the suggested test and the nature of the end-point detected. The simplicity and the cost of each test have also to be considered.

The Bioluminescence-Test for mutagens (BLT) uses dark mutants of luminous bacteria and determines the capacity of the mutagen to restore the ability of the dark mutant to emit light (1-3). The BLT was described as a forward mutation test in which any mutation that affects either the formation or the stability of the luminescence system repressor will restore the luminescence. The high sensitivity of the BLT stems also from the ability to score mutations that are accompanied by lethal events. Technically the test is very simple: Sterility of the assay mixture is not essential, volatile agents could be tested and the toxicity of the tested chemical is determined concomitant with its mutagenicity. The BLT can be run automatically and up to 50 chemicals can be tested in one day.

The present study presents a list of chemicals that were tested by the old version of the BLT together with new data on the potential of a newly selected dark mutant (Pf-13) to detect genotoxic agents. The improved BLT is suggested as a first prescreening test for carcinogenic agents.

ISBN 0-12-426290-2

MATERIALS AND METHODS

Photobacterium leiognathi dark variant SD18 is described by Ulitzur et al. (1). *Photobacterium fischeri* (Pf-13) dark mutant was selected after mutagenesis with MMS (Methyl-methane sulfonate). The dark strains were lyophilized and stored in 4°C for up to two years.

The BLT. The bioluminescence test for mutagens was run as described by Ulitzur et al. (1). The full description of the improved BLT will be described elsewhere (Ulitzur, in preparation). An agent giving a positive response is a chemical that results in an increase in luminescence to a level at least four times higher than the light level of the controls.

RESULTS

Table 1 sums up the results with about 100 chemicals that were tested with the dark strains SD18 and Pf-13. The chemicals were grouped according to their structure or functional groups. The lowest detected concentration in μg/ml of each chemical is given (in brackets) together with the requirement for microsomal activation (asterisks). The comparative mutagenic activity determined in the Ames Test is given as + or - for each compound for which such information was available. All the chemicals that are known to be active in the Ames Test gave a positive response in the BLT. The minimal detected concentrations were always much lower in the BLT than in the Ames Test. The improved BLT used a newly selected dark mutant of *P. fischeri* that was shown to be much more sensitive and discriminatry than the spontaneous dark variant *P. leiognathi* SD18 which was used in the old version of the BLT (1,2,3). The improved BLT allowed the detection of many carcinogens that showed weak or no activity with SD18. Table 1B lists the chemicals that were preferentially or solely detected with Pf-13 strain. Using this strain we were able to determine different carcinogenic agents that were not detected in the Ames Test, e.g. beryllium, nickel, lindane, safrole, O-tolidine and paratolylhydrazine. Not all the chemicals that increased the luminescence of the dark mutants are carcinogens or mutagens, but all of them are known as genotoxic agents. Most of the non-proven carcinogens that respond positively in the BLT were the DNA-intercalating agents whose genotoxic affects are controversial.

TABLE 1

The Activity of Different Chemical Compounds in the BLT.

(A) *Chemical agents that were determined with SD18*

Quinacrine(0.5)+, Ethidium bromide(2.5)+, Acriflavin(0.1)+, 2-aminobiphenyl(25)+, Proflavin(0.2)+, 9-aminoacridine(0.2)+, Acridine orange(0.2)+, N-methyl -N-nitro-N-nitrosoguanidine (0.002)+, 2-nitrofluorene*(0.7)+, Mitomycin-C(0.002)+, Hydrazine(0.07)+, Isoniazide(650)+, Phenyl-hydrazine(0.32)+, 2-nitrophenylhydrazine(25)+, 2-4 nitrophenylhydrazine(6)+, p-tolylhydrazine(1)+, Benzylhydrazine(12)+, m-hydroxybenzyl-hydrazine(100)+, 4-nitroaniline(6)+, Caffeine(20)−, Theophylline(20)−, Methylmethane sulfonate(6)+, Ethylmethane sulfonate(6)+, Ethylamino sulfonate(0.005)+, Ethylisopropyl amino sulfonate(150)+, Skatol(60)−, Norharman(3)−, Trp-p-2 (3)+, Glu-p-1(12)−, Glu-p-2(3)−, Desulfiran(0.15)+, Hydroxylamine(0.1)+, Ascorbic acid* (300)+, Emodin(3)+, Coumermycin(2)+, Fistein* (30)+, Kampferol*(1.5)+, Quercetin* (1.5)+.

(B) *Chemical agents preferentially or solely detected with Pf-13 strain.*

4-nitroquinoline-N-oxide(0.007)+, 1,3-propansultone*(6)+, 1,3-propan sulfonate*(6)+, Epichlorohydrin(50), Nitrosobenzene*(6), Niridazole(0.1)+, O-tolidine(1)−, Thio-Tepa(50)+ 9-10-dimethylbenzo(a) anthracene*(25)+, Benzo(a)pyrene*(0.1)+ Adriamycin(50)+, 2-anthramine(0.5)+, Indol(2), Nitrogen mustard(1.5)+, Methothroxate(2)−, Vincristine(1)−, Bleomycin (0.1)−, Vinyl chloride(gas)+, 6-mercaptopurine(1)+, Cisplatinum-II-diamino dichloride(20)+, Phenol(20)−, Safrole(7)−, Shikimic acid(25)−, Lindane(50)−, Aflatoxin B_1*(0.01)+, Novobiocin(0.02)−, Malonyl aldehyde(0.5), 4-aminophenol(6)−, Na_2WO_4(4000)−, $ZnCl_2$(30)−, KH_2AsO_4(100)+, $BeCl_2$(1.5)−, $K_2Cr_2O_7$(2)+, $MnCl_2$(20.000)−, Ni-acetate(10)−, $HgCl_2$(1)−, $CoCl_2$(40)−, $CdCl_2$(50)−.

(C) *Chemical agents that were not active with either SD18 or Pf-13 strains*

Chloramphenicol(−), Actinomycin D(−), Patulin(−), Penicillic acid(−), Acetamide(−), Flavone(−), Chrysen(−), Chloroform(−), $FeSO_4$(−), $Pb(NO_3)_2$(−), $NaAsO_2$(−), $RbCl_2$(−), $CuCl_2$(−), $SnCl_2$(−), $NaSeO_2$(−).

REFERENCES

1. Ulitzur, S., I. Weiser, and S. Yannai (1980). A new sensitive and simple bioluminescence test for mutagenic compounds. *Mutation Res.* 74, 113-124.

2. Ulitzur, S. and I. Weiser (1981). Acridine dyes and other DNA-intercalating agents induce the luminescence system of luminous bacteria and their dark variants. *Proc. Nat. Acad. Sci (USA)* 78,3338-3345.

3. Weiser, I., S. Ulitzur and S. Yannai (1981). DNA-damaging agents and DNA-synthesis inhibitors induce luminescence in dark variants of luminous bacteria. *Mutation Res.* 91, 443-450.

ACKNOWLEDGEMENT

This work was supported in part, by the Israeli National Council for Research and Development.

BIOLUMINESCENT ASSAYS OF ESTROGENS AND ANDROGENS

J.C. Nicolas, A.M. Boussioux, A.M. Boularan, B. Colomb
and A. Crastes de Paulet

*I.N.S.E.R.M. Unité 58, 60 rue de Navacelles
34100 Montpellier, France*

INTRODUCTION

The enzymatic assay of estrogens using the transhydrogenase activity of the estradiol dehydrogenase of human placenta has been previously described (1). However, this method requires several hours of incubation, to accumulate enough NADH to be determined spectrophotometrically at 340 nm. Androgen can be also determined after their specific conversion to estrogens.

The bioluminescent assay of NADH is sufficiently specific and it is a thousand more sensitive than spectrophotometric determination and it can be used to determine the amount of NADH accumulated during the estradiol transhydrogenase reaction. We propose a sensitive two step assay of estrogens, the first reaction is the transhydrogenase reaction (reaction 1) in which the hydrogen of a trace amount of NADPH (2.10^{-7}M continously regenerated by a NADPH regenerating system) is transferred to an excess of NAD (2.10^{-4}M) via repeated cyclic interconversion of estrone and estradiol. In the presence of a large excess of the enzyme (estradiol 17 β dehydrogenase) with respect to the steroid, the rate of NADH formation is a linear function of the concentration of estrone and estradiol. As the substrate is continuously regenerated the formation rate is constant for several hours. The second step is the bioluminescent assay of the NADH accumulated after 15 min, 30 min or 1 h of the transhydrogenase reaction (reactions 2 and 3). Transformation of androgens to estrogens is obtained after one hour incubation at 37°C in the presence of microsomal aromatase, the specificity of this conversion can be increased by using anti androgens antibodies to block unwanted conversion.

ISBN 0-12-426290-2

G 6 P		NADP		E_2		NAD
3.10^{-3}M		2.10^{-7}M				2.10^{-4}M
	*		0		0	
G-Lactone 6 P		NADPH		E_1		NADH

* Glucose 6 phosphate dehydrogenase
0 Estradiol transhydrogenase *(Reaction 1)*

$$NADH + FMN \xrightarrow{\text{FMN-oxidoreductase}} NAD + FMNH \quad (Reaction\ 2)$$

$$FMNH_2 + R\text{-}CHO + O_2 \xrightarrow{\text{luciferase}} FMN + RCOOH + h\nu \quad (Reaction\ 3)$$

Fig. 1 Bioluminescent assay of estrone and estradiol.

MATERIAL AND METHODS

The transhydrogenase reaction was performed with the Biomerieux kit or with purified estradiol dehydrogenase as described previously (1). The aromatase preparation was obtained from human placenta microsomes which were sonicated several minutes, then endogenous steroids were removed completely from the 10,000 g supernatant by chromatography on a XAD2 column. The NADH assays were done in a fully automated luminometer (LKB 1251), the bioluminescent reagent was luciferase 20 mg per liter, NADH oxidoreductase 3 I.U. per liter, decanal 50 μM, FMN 10 μM, in phosphate buffer 0.1 M, pH 7.2. This reagent was kept in the dark and in ice box at 4°C. In this condition the bioluminescent reagent can be used for a week. 200 μl of bioluminescent reagent was added to 250 μl of transhydrogenase incubation, the bioluminescence was read 15 s after addition of the reagent.

Estrogens were diluted after hydrolysis (urine) or extraction (serum) in Tris-HCl buffer pH 7.4. Transhydrogenase buffer was 0.2 mM NAD, 0.2 μM NADP, 5 mM glucose 6 phosphate, 0.02 UI G 6 PDH and 0.01 UI estradiol dehydrogenase/ml buffer. This buffer was prepared immediately before addition to the various samples. In general three different tubes were necessary, one blank, one assay and one standard tube containing the sample plus a known amount of estradiol. In the assay and standard tubes 25 μl of sample, or sample plus 20 or 50 pg estradiol were added to 200 μl transhydrogenase buffer and incubated 15 to 90 min depending on the sensitivity. The NADH accumulated during the incubation was determined by bioluminescence, 200 µl of bioluminescent reagent were added to the tubes and 30 s integration value of luminescence measured. The blank was prepared by adding the transhydrogenase buffer to the sample just before the bioluminescence measurement.

The aromatisation reactions were done in 100 µl Tris-HCl pH 7.5 containing 1 µM NADPH, 1 mM glucose 6 phosphate, 0.1 UI/ml glucose 6 phosphate dehydrogenase and between 5 to 10 µg microsomal aromatase. After one hour incubation, the reaction was stopped by heating at 70°C for 5 minutes and 25 µl of the medium were added to 200 µl of transhydrogenase buffer. All unsaturated androgens (testosterone, androstenedione, dehydroepiandrosterone and androstenediol) can be quantitatively convert to estrogens. To increase the specificity of the conversion anti-testosterone antibodies were added to block the aromatisation of testosterone. The amount of antibodies required is 10 times that necessary for standard radioimmunoassay, the antibodies were incubated with the sample 10 min before addition of the microsomal aromatase.

RESULTS

The NADH bioluminescent assay allows the detection of less than one picomole of reduced nucleotide however in our assay such sensitivity is not required since the transhydrogenase buffer alone produces small amount of reduced nucleotide. NADPH is the main reduced nucleotide but in our system the reactivity of this nucleotide is less than 5% of the reactivity of NADH. The initial background bioluminescence of the buffer is to due NADPH but also NADH produced by the yeast glucose 6 phosphate dehydrogenase, this background level increases slowly with the incubation time (0.01 V/min). In presence of small amount of estradiol or estrone the NADH production increases rapidly (0.160 V/min) for a 10^{-10} M estradiol concentration (less than 6 picogrammes in the assay). Table 1 shows integration bioluminescence values obtained for varying amounts of plasma. After a one hour incubation with microsomal aromatase androgens are assayed using the same method. Table 2 shows values of androgens and testosterone obtained with different serums.

CONCLUSION

Bioluminescence is a useful tool to increase the sensitivity of enzymatic determination of estrogens and androgens values of less than one picogramme can be determined in one

TABLE 1

Bioluminescent Assay of Estrone and Estradiol in Varying Amounts of Plasma

	Plasma (µl)				
	2.5	5	10	20	40
Blank	1.772	1.778	1.782	1.790	1.800
Assay	2.180	2.640	3.512	5.290	8.905
Standard	10.310	10.785	11.440	13.420	17.012
Estrone + estradiol (pg)	0.50	1.05	2.20	4.30	8.76

TABLE 2

Enzymatic Assay of Androgens and Testosterone

Serum	Blank (L1)	Arom. + Ab T (L2)	Arom. (L3)	Arom. + 20 pg T (L4)	$\frac{L3-L1}{L4-L3}$ Andro.	$\frac{L3-L2}{L4-L3}$ Testo.
O 2.5 µl	6.331	15.520	23.580	37.290	25 pg	11.7pg
♀ 5 µl	7.376	19.170	20.590	33.490	20.5pg	2.2pg

T, testosterone; Ab T, anti-testosterone, Arom, aromatase

hour. Interferences of biological sample are insignificant when the sample is obtained after extraction or heat denaturation. The method allows determination of estrone, estradiol and after reaction with 1 M hydrozine of estradiol alone. All unsaturated androgens which are transformed to estrogens can be determined by this enzymatic method. Anti-testosterone antibodies avoid specifically the aromatisation of this steroid and thus increase the specificity of the assay.

1. Nicolas, J.C., Boussioux, A.M., Descomps, B., and Crastes de Paulet, A. (1979) *Clin. Chim. Acta* 92, 1-9.

THE METABOLIC ACTIVITY OF BACTERIA UNDER CONDITIONS OF LOW WATER ACTIVITY - LUMINOUS BACTERIA AS A MODEL.

A. Reinhertz, I.J. Kopelman and S. Ulitzur*

Department of Food Engineering & Biotechnology, Technion - Haifa, Israel.

INTRODUCTION

The metabolic activity of microorganisms in intermediate moisture foods (IMF) is extremely low. This and different technical obstacles (high density, lack of continuous water phase) are the main reasons for the limited available data concerning the metabolic activity of micro-organisms under conditions of low water activity (a_w)

The luminous bacteria offer a unique model for studying the effect of low a_w, on the metabolism of gram-negative bacteria. The expression of the luminescence system depends on different metabolic activities including protein and lipid synthesis, and formation of ATP, $NADH_2$, and $NADPH_2$. The light, as a physical parameter, is readily determined in a wide range of intensities in a continuous and non-destructive manner.

The effect of low a_w on the metabolism of luminous bacteria has been studied in three systems: (1) In solutions containing high concentrations of glycerol, as a model for a permeable solute; (2) In solutions containing high concentrations of sucrose, or NaCl, as a model for solutes that are not permeable to bacteria and (3) under lyophilized conditions in the absence of a continuous water phase.

* To whom all correspondence should be addressed.

ISBN 0-12-426290-2

RESULTS

Fig. 1 shows that the addition of glycerol to luminous bacteria suspensions resulted in a biphasic decrease in luminescence and in viability, as well as in the capability of treated cells to recover in a glycerol-free medium. The magnitude of the changes taking place in each of these phases was dependent on the glycerol concentration and on the temperature. Low temperature largely protected the bacteria at low a_w. At 30°C, but not at 0°C, irreversible damage to the cells' membrane was observed, as could be judged by the drop in the cellular ATP pool, leakage of labeled compounds from the cells and increased permeability toward impermeable substrates of the luminous system. The effect of high concentrations of nonpermeable solutes (NaCl or sucrose) on the luminescence of *Vibrio cholerae* var *albensis* is shown in Fig. 2. The transfer of the cells to 0.75 M sucrose solution resulted in plasmolysis that was accompanied by a drastic decrease in luminescence. This stage may be followed by a stage of deplasmolysis that required the presence of K^+ and the amino acids glutamic acid or proline. The uptake of the amino acids into the palsmolysed cells was dependent on the uptake of K^+ ions by the cells, a process that was accompanied by an equivalent extrusion of H^+ from the cells. This process required energy and occured only under aerobic conditions. Luminous *Vibrio cholerae* var *albensis* as well as cloned luminous *Escherichia coli* responded similarly to *Photobacterium leiognathi* cells to low water activity obtained by glycerol, sucrose or NaCl.

The luminescence of lyophilized preparations of *P. leiognathi* cells was studied at different water activities (a_w 0.4 - 1) in skim milk powder. The maximal stability of the preparation was achieved at a_w of 0.5 The *in vivo* luminescence at a given a_w is much higher in powders than in solutions.

REFERENCES

1. Ulitzur, S., Weiser, I., Yannai, S. (1981). DNA damaging agents and DNA-synthesis inhibitors induce the luminescence in dark variants of luminous bacteria. *Mus. Res.* 91: 443-450.

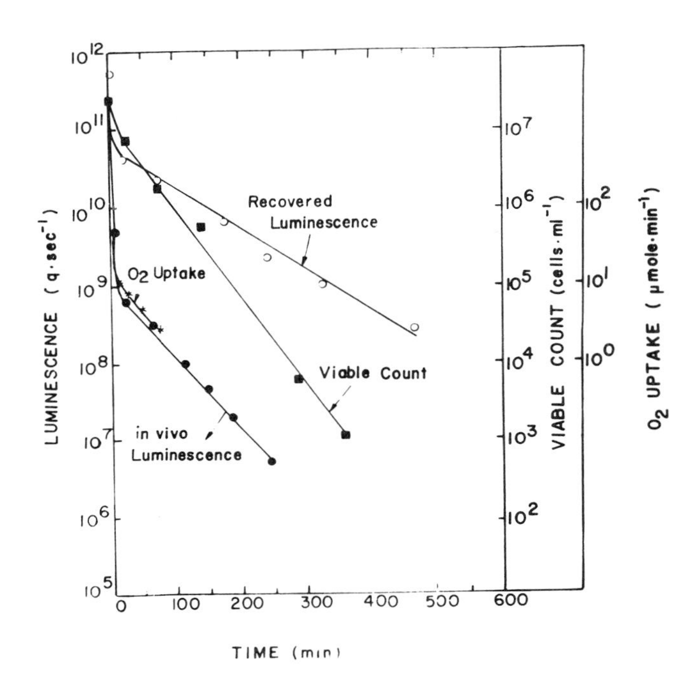

Fig. 1. *Effect of water activity (a_w) of 0.85 on viability, respiration and luminescence of P. leiognathi (BE8) cells.*

Photobacterium leiognathi (BE8) cells were grown in complex medium ASWRP (1) at 30°C to final density of $2.5x10^8$ cells ml^{-1}. The culture was diluted 1:30 into 1 ml of ASWRP liquid medium containing 44.7% glycerol (that corresponds to a_w of 0.85). The culture was incubated with shaking at 30°C and the *in vivo* luminescence was determined. To determine the maximal recovery of the *in vivo* luminescence 20 volumes of artificial sea water (ASW) containing MOPS buffer (pH 7.0) were added to a vial containing 0.5 ml culture and the luminescence of the whole vial was continuously determined. The oxygen uptake of the cells in the glycerol solution was determined using a Clark-type oxygen electrode. To determine the viable count aliquots were withdrawn with time, diluted in ASW and spread over ASWRP solid medium containing 0.3% yeast extract. The counting was made after 24 hours of incubation at 30°C.

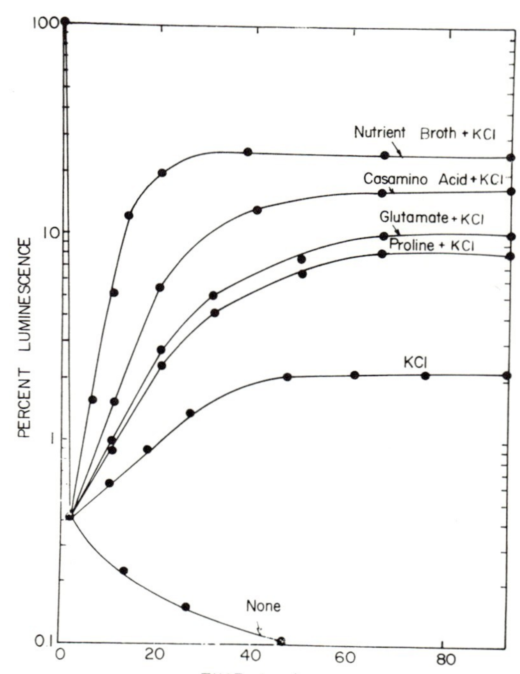

Fig. 2. *The effect of potassium ions, amino acids and nutrient medium on the luminescence of V. cholerae cells in sucrose containing medium.*

V. cholarae cells were grown in nutrient broth containing 1% NaCl. The cells were washed three times in cold solution of 1% NaCl containing sodium phosphate buffer and resuspended to give a final density of 5000 KU_{66}. Samples of 0.3 ml were diluted into 20 ml of the buffer containing 0.75 M Sucrose in the presence of KCl 50 (mM) and (1) L-proline (0.05 mg/ml) (2) Sodium glutamate (0.05 mg/ml) (3) Casamino acids (Difco) (1mg/ml) (4) Nutrient broth (Biolife) (8 mg/ml) or without KCl. The cultures were shaken at 35°C and the changes in the luminescence were determined with time.

ACKNOWLEDGEMENT

This study was supported by the Israeli National Council for Research and Development.

AUTOMATIC LUMINOMETRIC ASSAY OF GLYCEROL FOR STUDIES ON LIPOLYSIS

A.Lundin[1], J.Hellmér and P.Arner

Research Centre and Department of Medicine,
Karolinska Institute, Huddinge Hospital
S-141 86 Huddinge, Sweden

INTRODUCTION

Lipid mobilization from fat cells (lipolysis) plays a key role in the regulation of the energy balance in man. During lipolysis triacylglycerols are degraded to free fatty acids and glycerol. Glycerol release is generally used as an index of the rate of lipolysis, since glycerol in contrast to free fatty acids are not re-utilized by fat cells (1). The rate of lipolysis is 10-100 times lower in man than in rat (1). The amount of adipose tissue that can be taken from patients is often rather low. This amount should generally be used for assays of the rate of lipolysis under a variety of conditions using several hormons or drugs. Thus for investigations of lipolysis in human fat cells highly sensitive assays of glycerol are needed.

Luminescent assays of glycerol using bacterial as well as firefly luciferase have been published (2-4). All these assays have a high and similar sensitivity. In our group several hundreds of assays of glycerol are performed each week. Thus convenient and reliable assays are of utmost importance. It was decided to try to further develop the firefly assay based on ATP degradation in the glycerol kinase reaction (4) so that it could be performed automatically on Luminometer 1251 (an automatic luminometer from LKB-Wallac with temperature controlled sample

1 Correspondence to A.Lundin: Research Centre, Karolinska Institute, Huddinge Hospital, S-141 86 Huddinge, Sweden.

ISBN 0-12-426290-2

carrousel, addition of reagents by up to three dispensors and mixing of sample cuvettes in measuring position). The purpose of the present paper is to describe the new method also including a simplified pretreatment of samples avoiding a previously used inactivation step.

MATERIALS AND METHODS

Reagents and Instrumentation

Glycerol kinase from *Candida Mycoderma* (Boehringer-Mannheim) was purified by gel filtration as described previously (4). In manual assay of glycerol ATP Monitoring Reagent (LKB-Wallac) was reconstituted in 10 ml of distilled water (concentrated reagent) and luminescence measured with a Luminometer 1250 (LKB-Wallac). In automatic assay of glycerol ATP Monitoring Reagent was reconstituted in 45 ml of 0.1 M Tris-acetate buffer with 2 mM EDTA, pH 7.75 (ready-made reagent) and luminescence measured with a Luminometer 1251 (LKB-Wallac). ATP Standard (LKB-Wallac) was reconstituted in 10 ml distilled water to obtain 10 uM ATP.

Preparation of Samples

Subcutaneous adipose tissue was taken from patients by open biopsy during routine surgery. Isolated fat cells were prepared according to the method of Rodbell (5) as modified by Smith et al. (6). Incubation of fat cells (2 % v/v, $37^{o}C$, 2 h with mixing) was done in 0.2 ml of Krebs-Ringer bicarbonate buffer (pH 7.4) containing bovine serum albumin (40 mg/ml) and glucose (1 mg/ml) with a gas phase containing 95 % O_2 and 5 % CO_2. After the incubation fat cells were allowed to float to surface and an aliquot was taken from the bottom of the tube. This aliquot (0.1 ml) was added directly to the cuvette used in the luminometer, sealed with a stopper and stored in the freezer. Alternatively the aliquot (0.15 ml) was inactivated by mixing with an equal volume of hot distilled water (4) and stored in the refrigerator.

Manual Assay of Glycerol

The assay was performed essentially as previously described (4). Each cuvette was supplied with 0.2 ml ATP Monitoring Reagent (concentrated reagent) 0.6 ml of 0.1 M Tris-Cl buffer (pH 8.0), 20 ul of purified glycerol kinase and 10 ul of ATP Standard. After measuring the blank slope due to contamination of reagents 0.2 ml of inactivated sample (or glycerol standard solution) was added and the initial slope measured. After compensating for the inhibition of luciferase activity by sample addition (4) the increase of slope was calculated. From these increases obtained with glycerol standard solutions a standard curve was plotted and the glycerol concentration in unknown samples estimated.

Automatic Assay of Glycerol

Cuvettes containing 0.1 ml of frozen sample were allowed to attain room temperature. After adding 0.9 ml of ATP Monitoring Reagent (ready-made reagent) and loaded in Luminometer 1251. Alternatively 0.2 ml of inactivated samples were added to cuvettes containing 0.8 ml of ATP Monitoring Reagent (ready-made reagent) and loaded in Luminometer 1251. After loading up to 25 cuvettes the luminometer automatically performs the following steps (program can be obtained from the authors):

1) Preincubation (25°C, 10 min) to achieve temperature equilibration.
2) Measurement of luminescence for check-up that excessive lysis of fat cells has not occured.
3) Addition of ATP Standard (10 ul) with mixing of cuvette.
4) Measurement of luminescence for check-up that luminescence decay is negligible.
5) Addition of purified glycerol kinase with mixing.
6) Measurement of luminescence every 0.1 min during 1.1 min.

Each series of assays included the assay of blanks and glycerol standard solutions. From results obtained in step 6 the first-order rate constant of ATP degradation were calculated by taking the difference in natural logarithm of luminescence intensity divided by the time, e.g. $(\ln I_{0.1\ min} - \ln I_{1.1\ min})/(1.1-0.1)$. From the first-order rate constants obtained with glycerol standard solution a standard curve was plotted and the glycerol concentration in unknown samples estimated. Calculations can be done manually or by connecting a computer to the luminometer.

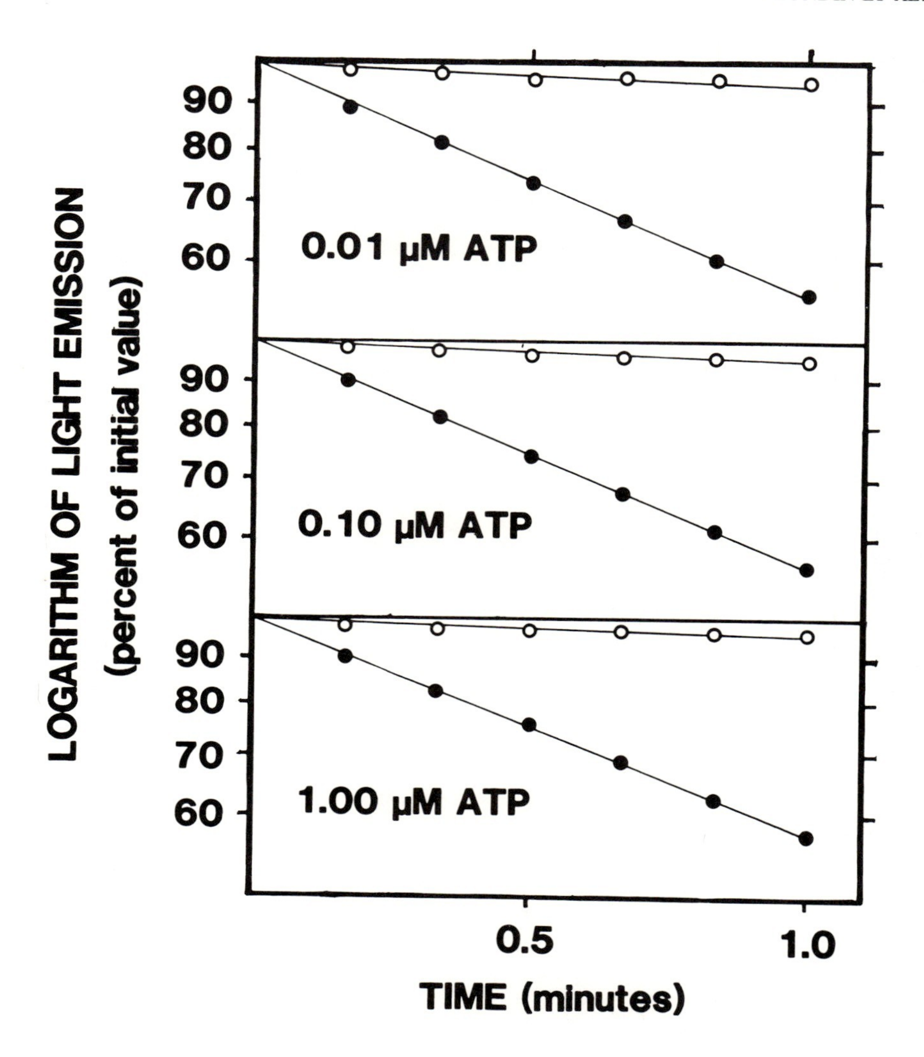

Fig. 1 Kinetics of ATP degradation in glycerol kinase reaction as measured by the automatic method. At three initial ATP concentrations (1.00, 0.10 and 0.01 uM) the luminescence was measured from reagent without added glycerol (○) and with 10 uM glycerol (●) after starting the reaction by addition of glycerol kinase.

RESULTS AND DISCUSSION

Bioluminescence ATP monitoring of the glycerol kinase reaction (ATP+glycerol——>ADP+glycerol-1-phosphate) can be used for end-point or kinetic assay of glycerol. An end-point assay has actually been developed (results not shown) but requires too high concentrations of glycerol kinase to be economically feasible. The manual and automatic assay used in the present paper are kinetic methods measuring the rate of ATP degradation. During the assay only a very low proportion of the glycerol is consumed (the ATP concentration and the rate of the reaction are low). Thus the rate of ATP degradation (-dS/dt) would be expected to depend on ATP concentration (S) according to the Michaelis-Menten equation: $-dS/dt=V_{max}*S/(K_m+S)$. Since $S<<K_m$ this can be approximated to $-dS/dt=k*S$, where $k=V_{max}/K_m$, i.e. the first-order rate constant. Rearranging to $-dS/S=k*dt$ and integrating from t_1 to t_2 gives the formula: $\ln S_1-\ln S_2=k*(t_2-t_1)$. Thus a plot of the natural logarithm of ATP concentration versus time will give a straight line with the slope equal to the first-order rate constant.

In bioluminescence ATP monitoring the intensity of luminescence is proportional to ATP concentration. Thus, as experimentally confirmed in Fig. 1 using three initial ATP concentrations (1.00, 0.10 and 0.01 uM ATP) and the automatic assay, the logarithm of luminescence intensity versus time will give a straight line. In the manual assay as previously described (4) no use is made of this fact and the calculation of rate of ATP degradation was actually an effort to extrapolate to time zero, i.e. the addition of sample. However, the errors introduced by this procedure are rather small provided that the assay is strongly standardized.

With the new method used in the automatic assay first-order rate constants are calculated. Such constants are independent of ATP concentration as shown in Fig. 1 but linearly related to glycerol concentration and activity of glycerol kinase as shown in Fig. 2. The intercept with the x-axis in Fig. 2 corresponds to a reagent blank of approx. 0.5 uM glycerol. Without purification of reagents, assays of glycerol concentrations below this level can not be achieved by increasing the level of glycerol kinase.

The previous method (4) included inactivation of samples by heating with distilled water. This may introduce sources of error (pipettings, evaporation) and it was decided to see if this step could be omitted. Thus aliquots were transferred directly and without heating to the cuvettes

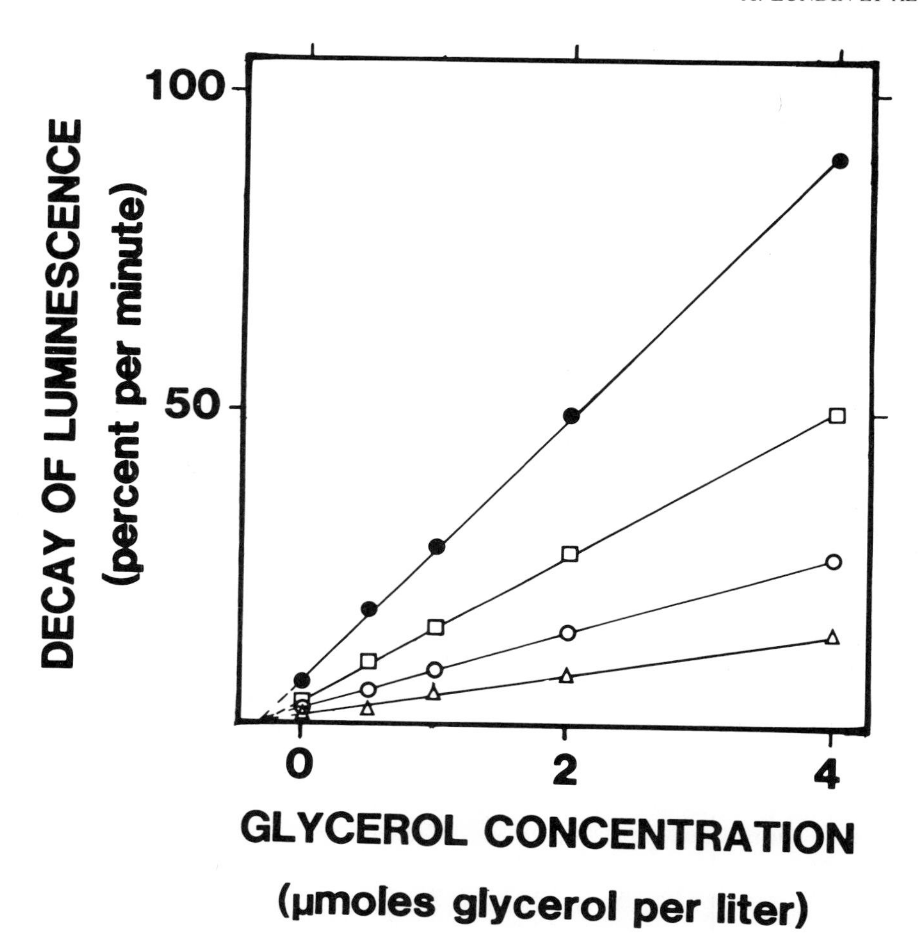

Fig. 2 Glycerol standard curves obtained with 5 (△), 10 (○), 20 (□) and 40 (●) ul purified glycerol kinase.

used in the luminometer and stored in the freezer until being assayed. As shown in Fig. 3a results on frozen and inactivated samples obtained with the new automatic method correlated well. A stability study on eight non-inactivated samples was also performed. In samples stored in the freezer during 6 months the change of the glycerol concentration was insignificant (0.2 $\pm$ 1.1 percent increase per month). In samples stored at room temperature during 20 hours a minor change was observed (1.9 $\pm$ 1.0 percent increase per hour). In samples stored at room temperature in the presence of ATP Monitoring Reagent during 4 hours the change was insignificant (0.04 $\pm$ 1.4 percent decrease

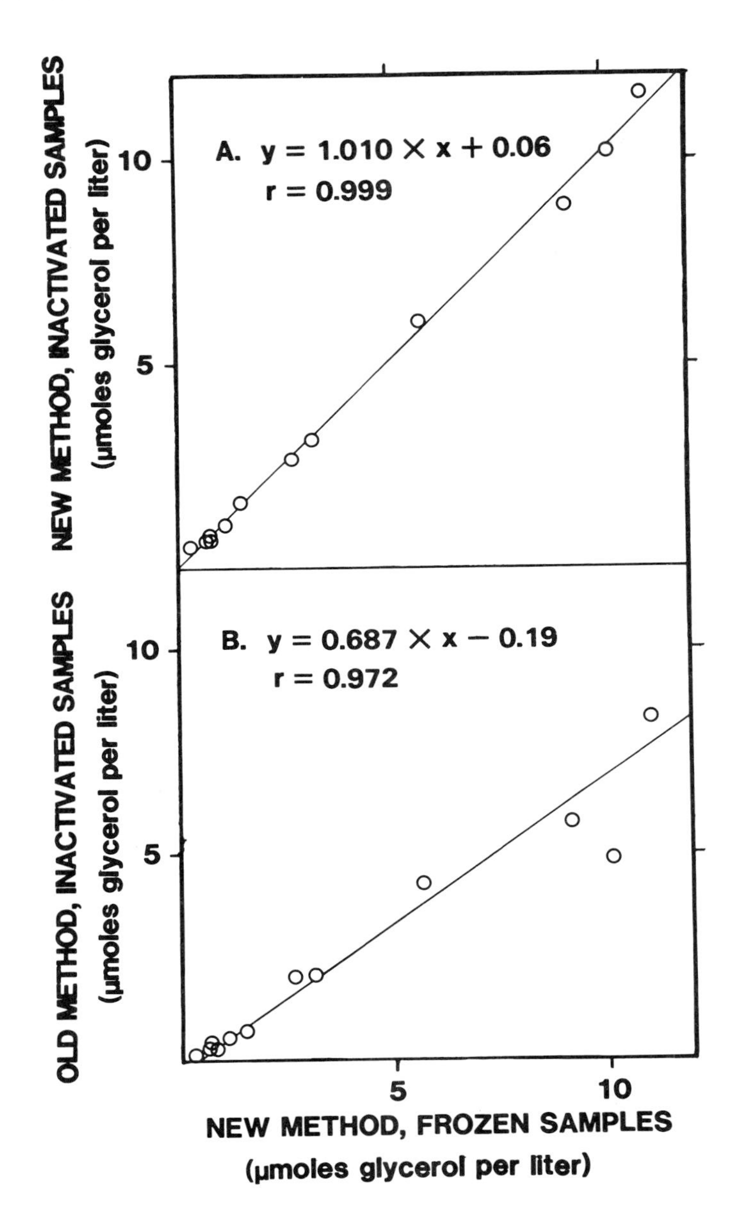

Fig. 3 Correlation between results on frozen and inactivated samples obtained with new automatic method (A) and between results on frozen samples obtained with new automatic method and results on inactivated samples obtained with old manual method (B).

per hour). Thus omitting the inactivation step should not introduce any new sources of error.

The correlation between the old manual method using inactivated samples and the new automatic method using frozen samples is shown in Fig. 3b. The correlation is somewhat lower than in Fig. 3a presumably due to the lower accuracy of the manual method. However, we can not explain the lower glycerol concentrations obtained with the manual method. The effect can not be due to the inactivation step, since the effect did not appear in Fig. 3a. Neither would a systematic error in any of the methods be expected. Both methods were calibrated with glycerol standard solutions of similar compositions as the samples and also pretreated in the same way (inactivated or frozen).

CONCLUSIONS

In studies on lipolysis aliquots from the incubation medium can be directly transferred to the cuvette used for luminometric assay of glycerol based on monitoring of the ATP degradation in the glycerol kinase reaction. The automatic assay described in the present paper takes advantage of the entire time-course of the luminescence and is considerably more reliable and convenient than the previously described manual method. With the automatic method 25 samples can be assayed within 60 min including calculations. The detection limit is determined by a reagent blank corresponding to 0.5 uM glycerol. Reducing the reagent blank requires further studies.

REFERENCES

1. Hales, C.N., Luzio, H.P. and Siddle, K. (1979). Hormonal control of adipose tissue lipolysis. *Biochem. Soc. Sym.* 43, 97-135.
2. Kather, H., Schröder, F. and Simon, B. (1982). Microdetermination of glycerol using bacterial NADH-linked luciferase. *Clin.Chim.Acta* 120, 295-300.
3. Lavi, J. (1984). A sensitive kinetic assay for glycerol using bacterial bioluminescence. Anal.Biochem (in press).
4. Björkhem,I., Arner, P.,Thore, A. and Östman, J. (1981). Sensitive kinetic bioluminescent assay of glycerol release from human fat cells. *J.Lipid Res.* 22, 1142-1147.
5. Rodbell, M. (1964). Metabolism in isolated fat cells. 1. Effects of hormones on glucose metabolism and lipolysis. *J.Biol.Chem.* 239, 375-380.
6. Smith, U., Sjöström, L. and Björntorp, P. (1972). Comparison of two methods for determining human adipose cell sizes. *J.Lipid.Res.*13, 822-824.

VIII CHEMILUMINESCENCE AND CHEMILUMINESCENCE ASSAYS

REVIEW OF PEROXYOXALATE CHEMILUMINESCENCE DETECTION FOR LIQUID CHROMATOGRAPHY

Robert Weinberger

Kratos Analytical Instruments
Ramsey, N.J. 07446 USA

INTRODUCTION

Originally developed by American Cyanamid as a chemical light source, peroxyoxalate chemiluminescence (CL) has recently been employed for fluorophor detection in liquid chromatography (LC) (1-8). Unlike most other forms of CL, where the CL reagent produces the emitting species, peroxyoxalate CL provides a chemical excitor intermediate that can chemically excite an interactive fluorophor.

The CL advantage in fluorophor detection arises from eliminating the relatively unstable and background producing light source. In this regard, the luminometer or conventional fluorescence detector (source off) can be operated at very high sensitivity. In fact, low attomole limits of detection have been reported for efficient CL energy acceptors (4,5).

This paper will review methods for the effective production of CL and discuss present and future applications for LC/CL.

POST-COLUMN CL SYSTEM

Reagents necessary for CL production are added post column using a system configured as in Figure 1. Since the CL kinetics may be rapid, (4.5s half-life) the length of tubing connecting the mixers to the detector should be short. Low pulse pumps such as the Kratos Spectroflow 400 should be employed for pumping all reagents to minimize noise. Because of its efficient light collection ability

ISBN 0-12-426290-2

and sensitive photomultiplier, the Kratos Spectroflow 970 fluorescence detector has been employed for virtually all CL measurements of this class.

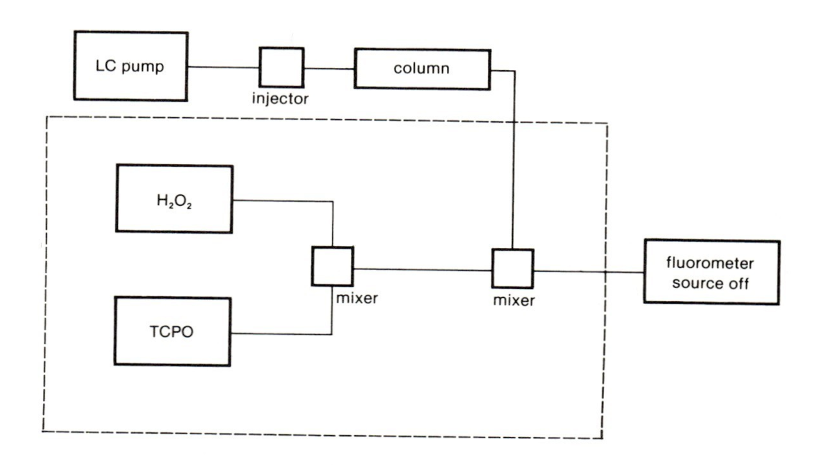

Figure 1 Chemiluminescence System

REAGENTS

Bis-(2, 4, 6- trichlorophenyl) oxalate (TCPO) synthesized conventionally (9) is dissolved in ethyl acetate, (2.4 mg/ml) and pumped at a flow rate of 0.6ml/min.

The hydrogen peroxide is prepared by mixing 10mL of 30% reagent grade peroxide in 40mL isopropanol and 50mL ethyl acetate. The flow rate for this reagent is 1.2mL/min.

Conventional reversed phase mobile phases with methanol or acetonitrile modifiers at flow rates of 1.0mL/min are generally employed. Tris buffer, 1.2 g/L at p^H 7.4 is often useful in optimizing the CL kinetics.

At mobile phase concentrations below 50% modifier, TCPO precipitation is possible. This can be accomodated by lowering the mobile phase flow rate or increasing the peroxide flow.

APPLICATIONS

Kobayashi et. al reported 10 and 25 femtomole limits of detection for dansyl amino acids (1) and fluorescamine (2) derivatives of catecholamines. Koziol, et. al., (8) have employed derivatization with dansyl hydrazine for the trace determination of keto-steroids in blood plasma.

Since most compounds, including some strong fluorescers are not good CL energy acceptors, pre-column tagging is often necessary.

Both Birks (5) and Grayeski (10) have shown that fluorophors with low oxidation potentials will have strong native CL properties. This further supports the electron transfer mechanism for the reaction.

Sigvardson, et. al. reported amino polycyclic aromatic hydrocarbons to be excellent CL energy acceptors and employed this to illustrate the selectivity of the technique in complex shale oil matrices (5). With pre- or post-column on-line solid phase reduction, the method was expanded to include nitro polycyclic aromatic hydrocarbons (7).

Further research with compounds whose molecular properties point to good CL should extend the scope of LC/CL methods to areas where ultratrace analysis is important.

REFERENCES

1. Kobayashi, S. and Imai, K. (1980). Determination of Fluorescent Compounds by High Performance Liquid Chromatography with Chemiluminescence Detection, Anal. Chem. 52, 424-427.
2. Kobayashi, S., Sekino, J., Honda, K. and Imai, K (1981). Application of High Performance Liquid Chromatography with a Chemiluminescence Detection System to Determine Catecholamines in Urine, Anal. Biochem. 112, 99-104.
3. Sigvardson, K.W. and Birks, J.W. (1983). Peroxyoxalate Chemiluminescence Detection of Polycyclic Aromatic Hydrocarbons in Liquid Chromatography, Anal. Chem. 55, 432-435.
4. Weinberger, R., Mannan, C.A., Cerchio, M. and Grayeski, M.L. (1984). Noise and Background in Peroxyoxalate Chemiluminescence Detection for Liquid Chromatography, J. Chromatogr. 288, 445-449.
5. Sigvardson, K.W., Kennish, J.M. and Birks, J.W. (1984). Peroxyoxalate Chemiluminescence Detection of Polycyclic Aromatic Amines in Liquid Chromatography, Anal. Chem. In press.

6. Weinberger, R., Mannan, C.A. and Cerchio, M. (1984), Solvent and pH Effects on Peroxyoxalate Chemiluminescence Detection for Liquid Chromatography, presented at the 35th Pittsburgh Conference on Analytical Chemistry and Applied Spectroscopy, March 1984, Atlantic City, N.J., Paper No. 058.
7. Sigvardson, K.W. and Birks, J.W. (1984), Detection of Nitro Polycyclic Aromatic Hydrocarbons in Liquid Chromatography by Zinc Reduction and Peroxyoxalate Chemiluminescence Detection, Presented at the Eight International Symposium on Column Liquid Chromatography, May 1984, New York.
8. Koziol, T., Grayeski, M.L. Weinberger, R., (1984), Approaches for the Determination of Trace Levels of Steroids in Blood Plasma by Liquid Chromatography with Chemiluminescence Detection, Presented at the Eight International Symposium on Column Liquid Chromatography, May 1984, New York.
9. Mohan, A.G. and Turro, N.J. (1974), A facile and Effective Chemiluminescence Experiment. J. Chem. Ed. 51, 528-529.
10. Personal Communication.

A LUMINOL-MEDIATED ASSAY FOR OXIDASE REACTIONS IN REVERSED MICELLAR SYSTEMS

A.J.W.G. Visser and J.S. Santema

Department of Biochemistry, Agricultural University, De Dreijen 11, 6703 BC Wageningen, The Netherlands

INTRODUCTION

Oxidase reactions involving flavoenzymes lead to the formation of hydrogen peroxide, which can be coupled to luminol chemiluminescence (CL) detection. Both the mechanism and the optimized reaction conditions have been thoroughly investigated in homogeneous solution (1,2). Catalysts and/or cooxidants like hemoglobin and peroxidase will stimulate CL, whereas enzymes like catalase and superoxide dismutase will quench CL. The only limitation of the sensitive assay consists of the difference in pH optima between enzyme and base catalyzed luminol reaction. The method has been successfully applied to immobilized oxidases (3). The assay can also be extended to another heterogeneous phase system, namely oxidases entrapped in inverted micelles (4). The stabilization of glucose oxidase in water droplets enclosed by CTAB reversed micelles and the activity measurements via luminol CL is described.

MATERIALS AND METHODS

Glucose oxidase (GOD), grade I from *Aspergillus niger* was obtained from Boehringer. The following stock solutions were prepared: 1 mg/ml GOD in 10 mM KPi, pH 7.5; 0.3 M cetyltrimethylammonium bromide (CTAB, Sigma) in $CHCl_3$/n-octane (6:5 v/v); 5.65 mM luminol (Sigma) in 10 mM NaOH; 0.5 M glucose in water. Inverted micelles were prepared by adding under stirring to a 2 ml CTAB solution, in a flat

ISBN 0-12-426290-2

bottom cylindrical cuvette, small aliquots of luminol and buffer until a clear solution was obtained, which required several minutes. The final volume was about 2.10 ml corresponding to a certain water/surfactant concentration ratio (ω_o). The vial with reaction mixture was put in a photometer, which design and calibration have been described elsewhere (5,6). To initiate the reaction a very small aliquot of GOD was squirted in followed by gently stirring. The amplified photomultiplier signal from the generated CL was measured on a strip chart recorder.

RESULTS AND DISCUSSION

The encapsulation of GOD in CTAB reversed micelles is feasible, when the enzyme is added to a clear micellar solution containing a predetermined volume of water. GOD has an isoelectric point of 4.2, which may enable positively charged surfactant molecules forming a protective coat for facile solubilization in a hydrocarbon solvent. We assume that the CL product 3-aminophthalate will reside in the aqueous phase of the micelle. In reversed micelles no additional catalyst is required for full development of CL, while addition of

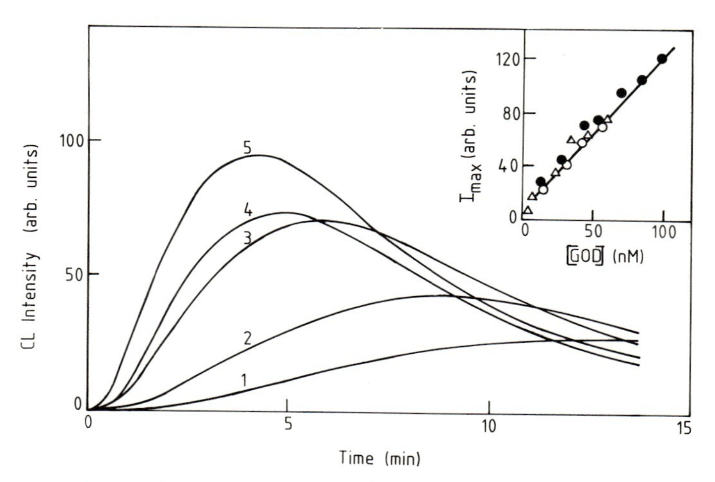

Fig. 1 The time course of chemiluminescence in reversed micelles. The assay mixture consisted of 0.3 M CTAB, 5.95 mM glucose and 67 μM luminol, all based on a total volume of 2.1 ml. The numbers 1-5 correspond with 5, 10, 15, 20 and 25 μl additions from a stock solution of 5.9 μM glucose oxidase, respectively. The |water|/|surfactant| ratio was kept constant at ω_o = 9.3. The different symbols in the inset indicate measurements carried out in triplicate.

superoxide dismutase does not result in quenching of CL. These observations suggest that the CL reaction depends on the partitioning of the reactants among the different phases, the available micelles and on the compartmentalization of the proteins, *i.e.* if a micelle contains an enzyme, it is occupied by a single protein molecule. The layer of surfactants around the enzyme may act as a putative catalyst for generation of CL.

We can conclude from the time course of CL formation and decay, that both the time for reaching maximum intensity (t_{max}) and the peak intensity (I_{max}) are functions of either enzyme or glucose concentrations. In figure 1 the experiments with varying enzyme concentrations are given. When the GOD concentration decreases we see an increase of t_{max}, while the integrated intensity is similar for all GOD concentrations. The inset of Figure 1 shows a linear dependence of I_{max} on GOD concentration. Extra addition, after some hours, of glucose to the reaction mixture resulted in repeatedly induced CL, indicating that the enzyme remains stable.

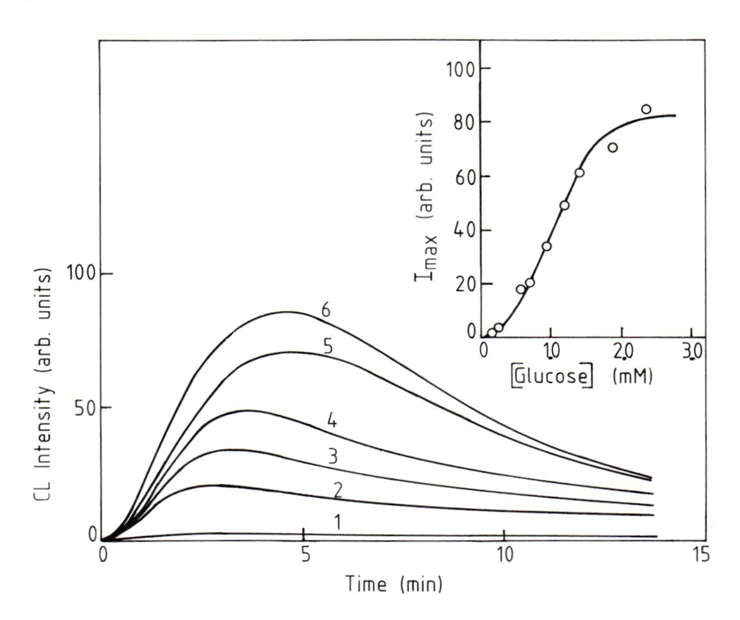

Fig. 2 The time course of chemiluminescence in reversed micelles. The glucose oxidase concentration was always 70 nM, numbers 1-6 correspond with 0.24, 0.71, 0.95, 1.19, 1.90 and 2.38 mM glucose respectively. See legend of Fig. 1 for other conditions. The inset shows the dependence of I_{max} on |glucose|.

Figure 2 accounts for the effect of varying substrate concentration. Both t_{max} and I_{max} decrease with decreasing glucose concentration. The inset of Figure 2 shows that I_{max} saturates at high substrate concentration.

The influence of the size of the water droplet (*i.e.* ω_o) on the CL reaction was also investigated. The maximal intensity remains constant for $5<\omega_o<11$ and decreases progressively up to $\omega_o = 25$. These experiments suggest that the size of the enzyme containing water pool is determined by the volume taken up by the protein. The conditions were chosen such that micelles having no enzyme enclosed are in large excess over enzyme-occupied micelles. Addition of water would increase the size of micelles occupied with substrate molecules. A decrease in the amount of light can then be visualized as a dilution of glucose in the water pool, thus lowering the effective concentration near the catalytic center of the enzyme.

CONCLUSIONS

CL is a very sensitive method to detect oxidase activity in reversed micellar systems. The boundary layer between surfactants and enzyme may function as catalyst for generation of CL. The mechanism awaits elucidation. At low enzyme concentration some indication of the size of the enzyme-occupied micelle can be obtained. The integrated intensity of CL is a good measure of total activity.

ACKNOWLEDGEMENTS

We thank Dr. C. Laane and Prof. F. Müller for helpful comments, Prof. J. Lee for making available the portable photometer and Mrs. J.C. Toppenberg-Fang and Mr. M. Bouwmans for assistance in the preparation of the manuscript. This work was supported in part by The Netherlands Foundation for Chemical Research (S.O.N.) with financial aid from the Netherlands Organization for the Advancement of Pure Research (Z.W.O.).

REFERENCES

1. Roswell, D.F. (1978). The chemiluminescence of luminol and related hydrazides, *Methods Enzymol.* 57, 409-423.
2. Seitz, W.R. (1978). Chemiluminescence detection of enzymically generated peroxide, *Methods Enzymol.* 57, 445-462.

3. Hinkkanen, A., Maly, F.-E. and Decker, K. (1983). Quantitation of immunoadsorbed flavoprotein oxidases by luminol-mediated chemiluminescence, *Hoppe-Seyler's Z.Physiol. Chem.* 364, 407-412.
4. Fendler, J.H. (1982) "Membrane Mimetic Chemistry", Wiley-Interscience, New York, NY.
5. Anderson, J.M., Faini, G.J. and Wampler, J.E. (1978). Construction of instrumentation for bioluminescence and chemiluminescence assays, *Methods Enzymol.* 57, 529-540.
6. Lee, J. and Seliger, H.H. (1972). Quantum yields of the luminol chemiluminescence reaction in aqueous and aprotic solvents, *Photochem.Photobiol.* 15, 227-237.

ENHANCEMENT OF LUCIGENIN CHEMILUMINESCENCE WITH CYCLODEXTRIN

Mary Lynn Grayeski and Eric Woolf

Department of Chemistry
Seton Hall University
S. Orange, New Jersey

INTRODUCTION

The chemiluminescence reaction of lucigenin with hydrogen peroxide has been studied recently in the presence of organized media. Micellar solutions have been shown to increase the quantum yield by a factor of 4 (1). Solutions of didodecyl dimethyl ammonium bromide lamellar aggregates increase the efficiency of the lucigenin reaction by a factor of 12 (2). It has been postulated that the gain in quantum yield is due to an increase in the efficiency of the excitation step of the reaction, caused by solubilization of an intermediate in the less polar environment of the organized media.

We are investigating aqueous solutions of cyclodextrins as another form of organized media. Cyclodextrins are macrocyclic carbohydrate molecules produced by a bacterial degradation of starch. They consist of 6, 7, or 8 glucose monomers arranged in a torus. The cyclodextrins are designated alpha, beta, or gamma depending on the number of monomers present. The coupling of the glucose monomers through alpha 1,4 glycosidic bonds gives cyclodextrins a rigid conical molecular structure with a hollow interior. The size of the interior is dependent upon the number of glucose monomers making up the cyclodextrin. The interior cavity, which is hydrophobic in nature, contains one or more waters of crystalization. This situation is thermodynamically unfavored, in part, because of polarity repulsion effects. Thus, the cyclodextrin will expel the water molecules in order to include less polar species

ISBN 0-12-426290-2

satisfying the size criterion of fitting at least partially into the interior. The inclusion complexes formed as a result of this have been shown to affect spectroscopic and chemical phenomena by enhancing fluorescence quantum yields and catalyzing reactions (3).

Because of these properties we are investigating the effect of cyclodextrin solutions on lucigenin chemiluminescence.

MATERIALS AND METHODS

Chemiluminescence intensity time curves were obtained as follows: 800 µL of lucigenin solution in either water, cyclodextrin, or 1-0-methyl alpha, D glucopyranoside (MGP) solutions were pipeted into a 1.6 mL polypropylene cuvette. Added to this was 50 µL of .1% aqueous hydrogen peroxide and finally 50 µL of 5 M sodium hydroxide solution. After a 1 second delay, the intensity was integrated for a 2 minute period with a Turner Designs model 20 photometer.

Relative intensities for comparison of cyclodextrin or sugar solutions with water were obtained by dividing the integrated intensity time curve of the reaction in cyclodextrin or MGP by that in water.

RESULTS

Enhancement is observed with increasing concentrations of all three forms of cyclodextrin with the greatest effect being observed for beta cyclodextrin. (Table 1) Presumably alpha and gamma cyclodextrins show minimal chemiluminescence enhancement because of size limitations for adequately accommodating the necessary species.

A change in the shape of the Intensity time curve is observed for beta cyclodextrin in that the maximum intensity increases but there is a decrease in the half life (i.e. time for maximum intensity to decay to half its original value). This indicates that the chemiluminescence reaction is catalyzed.

Table 1

Relative Intensities of Lucigenin Chemiluminescence In Cyclodextrin Solutions

Concentration of Cyclodextrin (M)	alpha	beta	gamma
10^{-2}	2.24	20.2	4.79
10^{-3}	1.58	2.35	1.65
10^{-4}	1.48	1.49	1.65
10^{-5}	1.33	1.33	1.34
10^{-6}	1.21	1.23	1.20

To determine whether the increase in relative intensities and kinetics is due to an inclusion effect or a chemical reaction between the lucigenin and the sugar units of the cyclodextrin, the chemiluminescence intensity of the lucigenin reaction in .07 M MGP, a cyclodextrin monomer analog was determined. MGP was chosen instead of glucose because lucigenin is known to react with reducing sugars (4). In MGP solution, the relative intensity of the reaction is 2.29, considerable less than the relative intensity of 20.2 in 0.01 M beta cyclodextrin solution indicating that the enhancement is largely due to an inclusion effect.

Since inclusion in cyclodextrin is known to enhance fluorescence efficiencies, this effect was investigated for reactant and product. No change in fluorescence efficiency was detected for either the lucigenin or N methyl acridone, the primary emitter, indicating that the observed enhancement is primarily a kinetic effect.

These results suggest that the increased chemiluminescence is due to inclusion of a reaction intermediate. According to the kinetically derived mechanism of the lucigenin reaction (5), a hydroperoxide intermediate is produced which forms a dioxetanedione resulting in chemiluminescence. The hydroperoxide may undergo competing non-chemiluminescent decomposition reactions.

Beta cyclodextrin is known to form inclusion complexes

with hydroperoxides (6) and to stabilize their decomposition. Possibly the hydroperoxide group of the lucigenin intermediates included in the cyclodextrin cavity stabilized by hydrogen bonding and decomposition through the dioxetane intermediate is favored. Inclusion of the dioxetane intermediate in the hydrophobic environment of the cyclodextrin cavity would also result in a higher efficiency as postulated for the micellar enhancement. (1)

CONCLUSION

Beta cyclodextrin has been shown to enhance chemiluminescence intensity of the lucigenin-hydrogen peroxide reaction. This increase is attributed to an increase in excitation efficiency and rate of the reaction through the inclusion of a reaction intermediate in the cyclodextrin cavity.

REFERENCES

1. Paleos, C.M., Vassilopoulos, G. and Nikokavouras, J. (1982). Chemiluminescence in Oriented Systems: Chemiluminescence of 10,10'-dimethyl-9,9'-biacridinum nitrate in micellar media, *J. Photochem.* 18, 227-34.
2. Nikokavouras, J., Vassilopoulas, G. and Paleos, C.M. (1981). Chemiluminescence in oriented systems. Chemiluminescence of lucigenin in model-membrane structures, *J. C. S. Chem. Comm.* 1082-3.
3. Bender, M.L. and Komiyama, M. (1978). "Cyclodextrin Chemistry". Springer-Verlag, New York.
4. Trotter, J.R. (1975) Light production in alkaline mixtures of reducing agents and dimethylbiacridinum nitrate, *Photochem. Photobiol.* 22, 203-11.
5. Maskiewicz, R., Sogah, D. and Bruce T.C. (1979). Chemiluminescent reactions of lucigenin. 1. Reactions of lucigenin with hydrogen peroxide, *J. Amer. Chem. Soc.* 101, 5347-54.
6. Matsui, Y. et al. (1970) Formation of inclusion compounds of β-cyclodextrin with hydroperoxides, *Bull. Chem. Soc. Japan* 43, 1909.

IN SITU GENERATION OF CHEMILUMINESCENT SUBSTANCES THROUGH HYDROXYLATION

G. Merényi and J. Lind

Department of Physical Chemistry and Department of Nuclear Chemistry, Royal Institute of Technology, S-100 44 Stockholm, Sweden

The detection in biological systems of hydroxylating agents, especially OH· radicals, is a challenging problem not yet solved satisfactorily (1). Usually, the occurence of the above species is implied from measurements of typical end-products. Apart from specificity problems these methods are rather insensitive and therefore require a relatively large degree of conversion. In contrast to this, chemiluminescent measurements are inherently sensitive. Many chemiluminescent substances, especially luminol, are known to generate light in the presence of free radicals. However, precisely because *e.g.* luminol responds quite indiscriminately to many free radicals it cannot be used to determine the nature of the radicals. The situation can be improved if a poorly chemiluminescent substance can be transformed into a highly effective chemiluminescer by the direct action of the free radical of interest. In a previous study (2) we have observed that phthalic hydrazide is hydroxylated by OH· radicals to yield 3- and 4-hydroxy phthalic hydrazides. Contrary to the parent (3), the hydroxy derivatives have chemiluminescence efficiencies comparable to that of luminol (4). The present work aims at quantifying the above hydroxylation reactions and the increase in chemiluminescence thus obtained.

RESULTS

OH· radicals were generated through γ-radiolysis of aqueous solutions saturated with a 1/1 mixture of O_2/N_2O. In such a system the following reactions operate:

ISBN 0-12-426290-2

TABLE 1

Irradiated at pH	[a]Converted PHTH $mol \cdot dm^{-3}$	The sum of hydroxylated products $mol \cdot dm^{-3}$	[b]Light yield a.u.
3.3	1.1×10^{-4}	1.0×10^{-4}	138
5.0	1.1×10^{-4}	1.0×10^{-4}	135
7.0	1.0×10^{-4}	8.5×10^{-5}	119
9.3	9.0×10^{-5}	6.0×10^{-5}	81
11.6	8.9×10^{-5}	6.1×10^{-5}	83
unirradiated	5×10^{-4} PHTH (99% purity)	-	1
	10^{-4} luminol	-	900

a) The concentration of PHTH prior to γ-irradiation equalled 5×10^{-4} $mol \cdot dm^{-3}$.

b) All light measurements were performed with the $S_2O_8^{2-} + H_2O_2$ reagent[3] after adjusting the pH to 11.6.

1 $H_2O \rightarrow OH\cdot, e^-_{aq}, H\cdot$ yield of radicals 100%

2 $e^-_{aq} + N_2O \rightarrow OH\cdot + N_2$ ∿85%

3 $e^-_{aq}(H\cdot) + O_2 \rightarrow O_2^-(HO_2)$ ∿15%

The experiments were performed by imparting to solutions containing 5×10^{-4} mol·dm^{-3} phthalic hydrazide an irradiation dose of 230 Gy. After irradiation the samples were analyzed by HPLC. Subsequently, the samples were mixed with a chemiluminescent reagent and the light yield was measured in a luminometer (LKB).

Several chemiluminescent reagents were tested such as hemin + H_2O_2, $Fe(OH)_6^{3-}$ with and without H_2O_2, ClO^- with and without H_2O_2 and $S_2O_8^{2-}$ + H_2O_2 all in a 0.1 mol·dm^{-3} carbonate buffer at pH 11.6. For the purpose of comparison the last reagent proved the most suitable one due to the relatively stable (minutes) light level it gave. The results are presented in Table 1.

If, instead of the O_2/N_2O mixture the solutions are irradiated in the presence of pure N_2O, the light yields obtained are much lower (by a factor of 4). This is a paralleled by a similar decrease in the yield of the hydroxylated phthalic hydrazides, even though the consumption of the parent PHTH slightly increases.

These findings are consistent with the following reaction scheme:

4 $PHTH + OH\cdot \rightarrow PHTH(OH)\cdot$

5 $PHT^- + OH\cdot \rightarrow PHT(OH)\bar{\cdot}$

6 $PHT^- + OH\cdot \rightarrow PHT\cdot + OH^-$

7 $2PHTH(OH)\cdot \rightarrow OH - PHTH$ + products

8 $PHTH(OH)\cdot + PHT\cdot \rightarrow$ products

9 $PHTH(OH)\cdot + O_2 \rightarrow PHTH(OH)(OO\cdot)$

10 $PHTH(OH)(OO\cdot) \rightarrow OH - PHTH + O_2^-$

PHTH and PHT^- denote phthalic hydrazide and its monoanion.

The crucial steps are the addition of the OH· radical followed by oxidation of the adduct by O_2, thus producing the potent chemiluminescent compounds 3- and 4-hydroxyphthalic hydrazide. The reactions 5, 6, 9 and 10 have been described in Refs. 2 and 5. Irradiation of 3-methyl phthalic hydrazide instead of PHTH results in a similar increase in light yields (about 50% of that obtained with PHTH). This finding indicates that attachment of an alkyl side chain essentially retains the chemiluminescent potentiality of hydroxylation.

The observations reported in this work immediately point to the possibility of detecting and quantifying hydroxyl radicals in biological matter.

Since we have in no way optimized the chemiluminescent conditions we feel that further work could greatly improve the sensitivity. In particular, purification of the PHTH is expected to result in a dramatic decrease of the background (3).

REFERENCES

1. See for instance "Oxygen and Oxy-Radicals in Chemistry and Biology". Academic Press, New York, 1981.

2. G. Merényi and J. Lind, The Fifth Tihany Symposium on Radiation Chemistry, 1983, ed. Dobó, Akadémiai Kiadó 103.

3. M. M. Rauhaut, A. H. Semsel and B. G. Roberts. *J. Org. Chem.* 31, 1966, 2041.

4. H. D. K. Drew and F. H. Pearman. *J. Chem. Soc.* (London), 1937, 586.

5. J. Lind, G. Merényi and T. E. Eriksen. *J.Am.Chem.Soc.* 105,1983,7655

THE IDENTIFICATION OF GAMMA IRRADIATED FOODSTUFFS BY CHEMILUMINESCENCE MEASUREMENTS

W. Bögl and L. Heide

Institute for Radiation Hygiene of the Federal Health Office, Neuherberg, FRG

INTRODUCTION

The exposure of foodstuffs to ionizing radiation has again become the centre of lively discussion when a joint committee of experts of the FAO/IAEA/WHO at a meeting on 27 October - 3 November 80 had passed a recommendation stating that the irradiation of all foodstuffs with an average total dose up to 10 kGy would not result in any toxicologic effects (1). In the Federal Republic of Germany, contrary to many other countries, it is at present not permitted to expose foodstuffs to ionizing radiation for the purpose of germ-reducing (sterilization).
In order to ensure the observance of the food law analytical procedures have become necessary for identification of irradiated foodstuffs.

EXPERIMENTAL

Samples of cinnamon-, curry-, red pepper- and milk powder were exposed to a Co-60 source with radiation doses up to 10^4 Gy (5.5 kGy/h). The subsequent reaction of the irradiated foodstuffs with a luminol solution (125 mg $\hat{=}$ 0.7 mM luminol; 2.5 mg $\hat{=}$ 3.8 μM hemin; 1.25 g $\hat{=}$ 11.8 mM Na_2CO_3; ad 1000 ml H_2O; pH 10 adjusted with 1M HCl; 3 - 4 mg food sample per analysis) resulted in light emission (chemiluminescence). This effect may be used as an indicator of radiation exposure (2,3).

RESULTS

The most important results can be summarized by means of the example of cinnamon powder (a typical glow curve is illustrated in fig. 1) as follows:
follows:

a) The chemiluminescence intensities of the irradiated food samples are obviously dependent on the absorbed radiation dose (fig. 2)
b) In some cases radiation treatment can be identified even 2 months after radiation exposure (fig. 3)

ISBN 0-12-426290-2

c) The radiation treatment can still be identified following a 16-hour heat treatment at various temperatures in a drying oven (fig. 4)
d) Following a water-vapour treatment (2 h at 25°C in a saturated water vapour atmosphere; after the treatment the samples were sticky and could no longer be dispersed) it is impossible to identify prior irradiation (fig. 5)
e) An exposure of untreated samples to UV-radiation results in an increase of the chemiluminescence intensity. As the form of the light-emission-curve varies, it is sometimes possible to make a distinction between γ- and UV-irradiation (fig. 6)
f) Curry-, red pepper- and milk powder, compared to cinnamon powder, react in a similar way, however the following exceptions:
   - the chemiluminescence of red pepper powder is difficult to measure as, due to the extraction of dyestuffs from the pepper in the luminol solution, an absorption of the emitted light takes place.
   - milk powder which had not been γ-irradiated may show a great chemiluminescence; the reason for this may be seen in the existence of fat oxidation products.

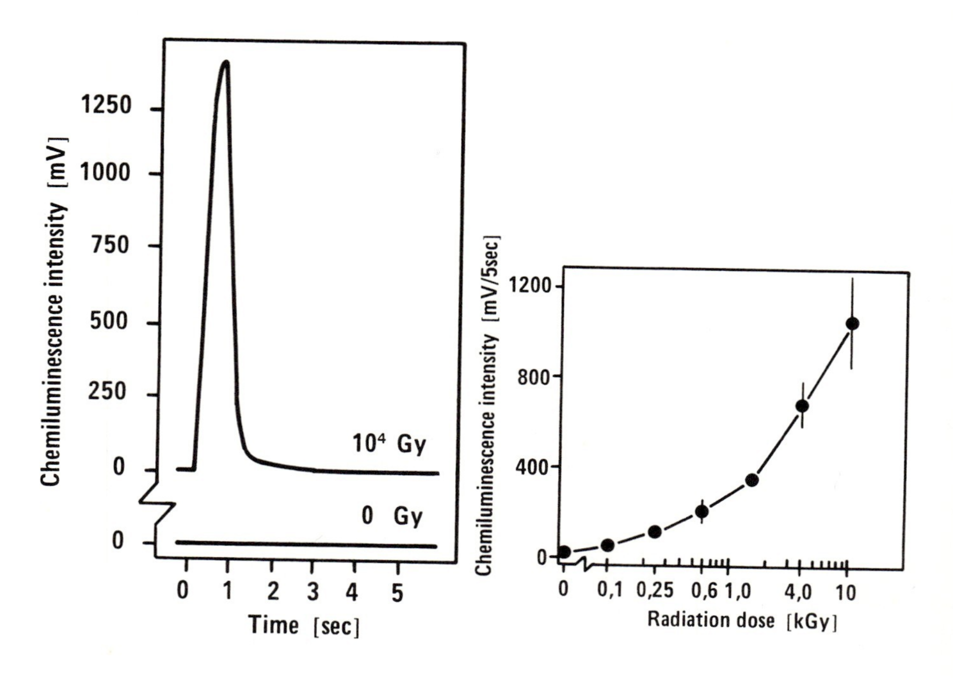

Fig. 1 Chemiluminscence measurements of unirradiated and irradiated (10^4 Gy) cinnamon powder.

Fig. 2 Chemiluminescence intensity of irradiated cinnamon powder as a function of radiation dose.

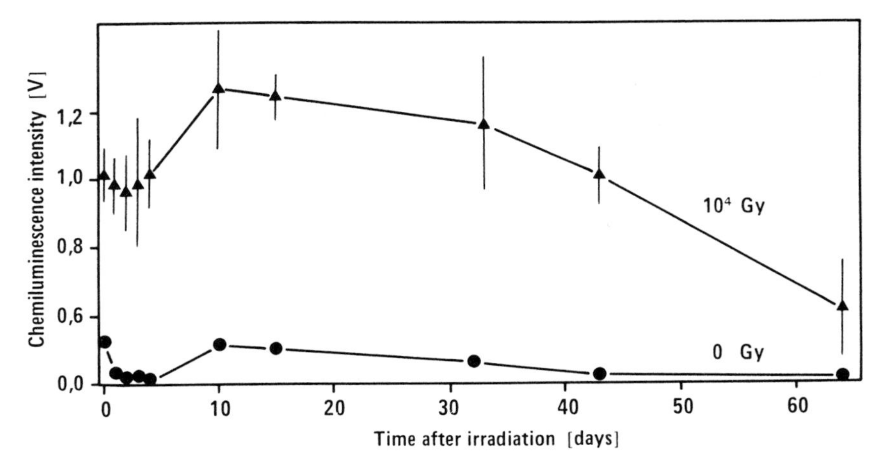

Fig. 3 Chemiluminescence intensity of irradiated cinnamon powder as a function of storage time after irradiation.

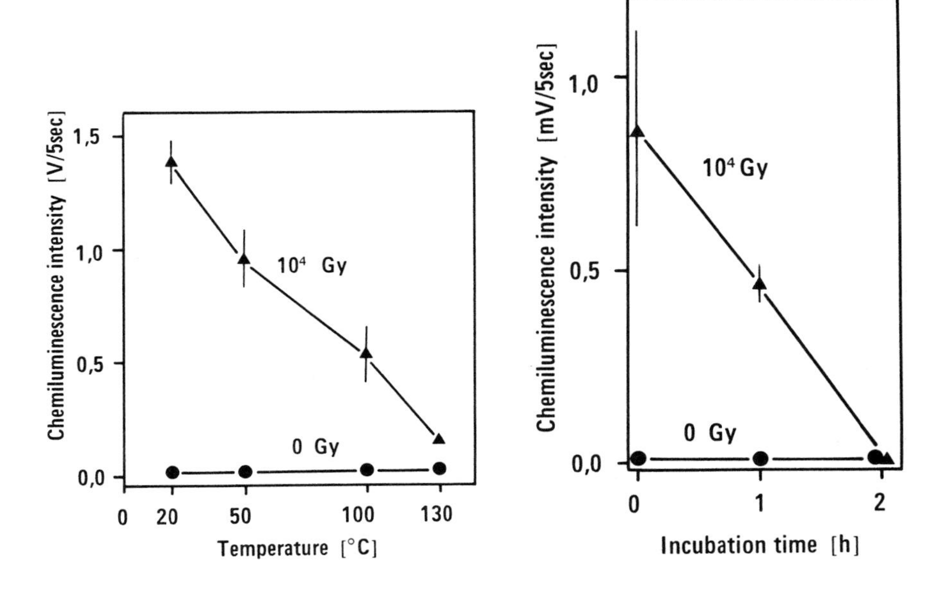

Fig. 4 Chemiluminescence intensity of irradiated cinnamon powder as a function of the temperature of the heat treatment after irradiation.

Fig. 5 Chemiluminescence intensity of irradiated cinnamon powder (10^4 Gy) as a function of the duration of a watervapor treatment.

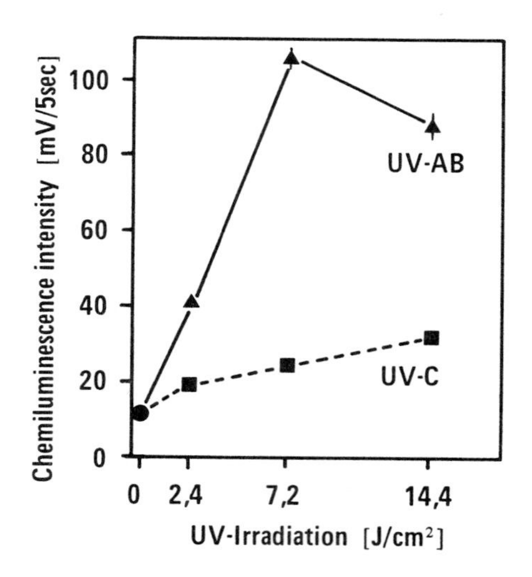

Fig. 6 Chemiluminescence intensity of untreated cinnamon powder after UV-irradiation.

In sum, the chemiluminescence measurement appears to be an appropriate method to quickly and reliably identify some foodstuffs treated with ionizing radiation.

REFERENCES

1. World Health Organization:
   Wholesomeness of irradiated food.
   Report of a Joint FAO/IAEA/WHO Expert Committee.
   Technical Report Series 659, World Health Organisation, Geneva 1981
2. Ettinger, K.V. and Puite, K. J. (1982). Lyolumiscence dosimetry part I. principles, *Int. J. Appl. Radiat. Isot.* 33, 1115 - 1138
3. Bögl, W. and Heide, L. (1983). Die Messung der Chemilumineszenz von Zimt-, Curry-, Paprika- und Milchpulver als Nachweis einer Behandlung mit ionisierenden Strahlen.
   Report ISH-Bericht 32, Nov. 1983, Institute for Radiation Hygiene of the Federal Health Office.

CHEMILUMINESCENCE METHOD FOR ESTIMATION OF AUTOXIDATION IN FOODS: INTERFERING REACTIONS

Bjørn E. Børdalen

Central Institute for Industrial Research
P.O. Box 350, Blindern, 0314 Oslo 3, Norway

INTRODUCTION

Several authors have measured "ultraweak" chemiluminescence (CL) in order to study autoxidation (1-3). In some foods, such as oils and fats, the intensity of CL increases with storage time, and this increase can be correlated to the increasing rate of autoxidation.

In many other foods, however, the relationship between intensity of CL and autoxidation is less clear, and in some cases, this intensity is observed to fall off during storage. This pertains especially to fried or baked products. Frying and baking in some way give rise to a high level of CL, which interferes seriously with any attempt at estimation of autoxidation by CL measurements (3). In the present work, this interfering CL is investigated by model experiments involving heat treatment of such food components as sugars and amino acids, and of some natural foods for comparison.

MATERIALS AND METHODS

The instrument used, and the experimental methods, were as described earlier (3). The results for CL intensity have been corrected for dark current and are given in relative units (1 unit corresponding to 12.5 pulses per second). Briefly, the experiments were performed as follows: an aqueous solution of the test substance(s) was added to filter paper in a glass dish (or flask), which was then put in an oven for heat treatment. After one hour for cooling to 20°C, the CL intensity was measured. Besides ordinary filter paper, a cellulose-free quality, ChromAR Sheet 1000 (Mallinckrodt Chemical Works, St.Louis, USA) has been used.

ISBN 0-12-426290-2

RESULTS AND DISCUSSION

The CL observed after heat treatment of various compounds is given in Table 1. It should be noted that ordinary filter paper in this test gives a quite high blank value, exceeding the value obtained with some of the test compounds added. Nevertheless, these experiments may be of some interest, since cellulose is an important constituent of vegetable foods. Under the conditions of this test, cellulose is not inert. Proline is noteworthy for giving rise to such a high level of CL in this test.

Results of experiments with heat treatment of sugars and amino acids together are also given in Table 1. The results show that in mixtures where Maillard reactions occur, the levels of CL are quite considerable. During storage at room temperature, the level will decline gradually.

TABLE 1

Chemiluminescence after heat treatment (130°C, 10 min) of various compounds and mixtures on filter paper[1])

	n mmol	CL		n mmol	CL
				0.25	73
Water (blank)		7 (63)	L – Proline	0.5	(597)
D – Fructose	1	13 (55)	L – Hydroxy-		
D – Galactose	2	15	proline	0.25	45
D – Glucitol	2	13	DL – Threonine	0.5	22
D – Glucose	2	15 (22)	L – Tryptophane	0.13	13
Lactose	1	17	L – Tyrosine	0.13	12
Sucrose	1	27	DL – Alanine-	0.25	107 (140)
D – Xylose	2	12	D – Glucose	0.25	
DL – Alanine	1	12 (115)	DL – Alanine-	0.25	22 (63)
L – Cysteine	0.5	11	Sucrose	0.13	
L – Cystine	0.25	(79)	L – Tryptophane-	0.13	190
L – Glutamine	0.5	18	D – Glucose	0.25	
Glycine	1	(73)	L – Proline-	0.25	149
DL – Lysine. HCl	0.5	(61)	D – Glucose	0.25	

1)
Column marked n gives the quantity used.
Column marked CL gives the intensity of chemiluminescence (relative units); numbers in parenthesis are for ordinary filter paper and those not in parenthesis for ChromAR.

In model systems where a sugar and two amino acids were present together, some interesting interactions were observed (Table 2). Addition of cysteine to a solution of glucose and alanine, resulted in a lower level of CL. Cysteine also showed a CL-lowering effect when added to milk and to potato juice. Sodium thioglycollate had a similar effect, whereas cystine was less effective. This may indicate that the CL-lowering effect of cysteine is related to the presence of the thiol group. Tyrosine showed a CL-lowering effect in some systems, but increased CL in others. Ascorbic acid showed a remarkable CL-lowering effect in all cases tested.

TABLE 2

Effect of adding certain compounds to foods or a Maillard model system, as observed by chemiluminescence obtained after heat treatment (130°C, 10 min)

Compound added [1]	n mmol	Chemiluminescence (relative units)			
		Milk	Potato juice	Wort	Glucose-alanine [2]
None		38	52	26	140
Ascorbic acid	0.25	10	7 (x2)	10	9
L – Cysteine	0.25	9 (x4)	10	31	64
L – Cystine	0.5		42		
Sodium thioglycollate	0.5	17	9		41
L – Tyrosine	0.125	50	76 (x2)		36

1) The quantity added (to 2 ml liquid) is given in the column marked n, except that symbols such as (x2) mean that the double quantity has been added.

2) 0.25 mmol of each; the results in this column are from experiments on ordinary filter paper, but the other results are from experiments on ChromAR.

The experiments described so far were made with free access of air. An experiment was made with a mixture of glucose and alanine under nitrogen (both during heat treatment for 10 min at 130°C, subsequent cooling, and CL measurement). The CL intensity found was 8, against 106 for a control experiment in air. When air was admitted to replace the nitrogen, the intensity rapidly rose to 71; new purging with nitrogen for ten minutes lowered it to 10. This is in general agreement with theory, since only oxidation reactions, or reactions involving peroxy compounds, would be expected to have sufficient free energy to give rise to CL.

It is known that free radicals are formed in Maillard reactions. It would seem a possibility that some of these may give peroxy radicals which, in reacting together, give CL in a similar way as peroxy radicals which form during autoxidation of lipids.

When using CL as a method for estimation of autoxidation, the CL which is related to Maillard reactions, is only one of several possible interferences. For instance, it has earlier been pointed out that vegetables in salads may introduce sufficient interfering CL that the autoxidation of the mayonnaise constituent is difficult to estimate in this way (3). The fluorescer and quencher effects of food components, especially coloring matter of various kinds, also introduce uncertainties.

CONCLUSION

Although intensity of CL may be used as a measure of autoxidation in some foods, this seems difficult or impossible in others, especially in fried or baked products, in which the CL is mainly due to other reactions. Experiments with model systems indicate that Maillard reactions play a considerable role in such cases.

REFERENCES

1. Usuki, R., Kaneda, T., Yamagishi, A., Takyu, C. and Inaba, H. (1979). Estimation of oxidative deterioration of oils and foods by measurement of ultraweak chemiluminescence, *J. Food Sci.* 44, 1573-1576.
2. Timms, R.E. and Roupas, P. (1982). The application of chemiluminescence to the study of the oxidation of oils and fats, *Lebensm.-Wiss. u. -Technol.* 15, 372-377.
3. Bøordalen, B.E. (1984). Chemiluminescence method for evaluation of oxidative changes in fats and foods, *In* "Proceedings of the 12th Scandinavian Symposium on Lipids, Mariehamn, Finland, June 1983" (Ed R. Marcuse). Lipidforum, in press.

DETERMINATION OF CHOLINE AND ACETYLCHOLINE BY A CHEMILUMINESCENT METHOD

L.F.J. Woods* and P. Neaves+

*Applied Microbiology Section, Leatherhead Food Research Association, Randalls Road, Leatherhead, Surrey KT22 7RY, England.
+Milk Marketing Board, Microbiology Laboratory, Technical Division, Thames Ditton, Surrey KT7 OEL, England

INTRODUCTION

The method of detection and identification of *Clostridium botulinum* in foods and pure culture is usually dependent upon detection of its potent protein neurotoxin (1). The only specific and sensitive assay for toxin is the mouse injection test, presumptive positive reactions being confirmed by immunological protection of further mice injected with mixtures of sample and specific antisera. However, there is a strong desire to reduce the number of animal experiments which are performed and as a result several alternative methods of detection have been investigated. For example, botulinal toxin can be detected immunologically and enzyme-linked immunoassays have been used routinely (2). However, the cross-reactions and variable sensitivity observed with immunological methods (3) are difficult problems to overcome.

Botulinal toxin is known to inhibit the release of acetylcholine in the neuromuscular junction (4) and specific binding sites for the toxin have been identified in motor nerve terminals (5). We have been attempting to develop a method of detection based on the possible physiological effect of botulinal toxin on nerve tissue culture cells. In order to investigate the effect of toxin on the choline and acetylcholine balance of the tissue culture cells, a chemiluminescent method was developed for the measurement of these compounds based on the method described by Israel & Lesbats (6).

MATERIALS AND METHODS

Choline/acetylcholine assay

The assay system consisted, in a 1-ml volume, of: 0.2M-Tris/HCl buffer, pH 8.6, 0.17mM-5-amino-2,3-dihydro-1,4-phthalazinedione (luminol), 2.9

 ISBN 0-12-426290-2

purpurogallin units horse-radish peroxidase and 0.4 units choline oxidase. Luminescence was measured in an LKB Luminometer 1250 (LKB Wallac, Finland) after initiation of the reaction by addition of the choline oxidase. Acetylcholine was measured by pre-incubation of samples with 0.1 unit/ml acetylcholinesterase for 30 min at 30°C. All chemicals were obtained from Sigma (London) Chemical Company Ltd, Poole, Dorset.

Culture conditions

Mouse neuroblastoma cells (NB41A3, Flow Laboratories, Irvine, Scotland) were grown in Hams F-10 medium containing 10mM-sodium bicarbonate, 2mM-glutamine, 20mM-HEPES, 2.5% foetal calf serum and 15% horse serum, pH 7.3 at 37°C. Cells were cultured on microcarrier beads (0.5 g dry wt, Cytodex 1, Pharmacia) suspended in growth medium (500 ml) and stirred using a microcarrier stirrer system (MCS-104S, Techne (Cambridge) Ltd, England); incubation was for 5–7 days at 37°C until beads were evenly covered in cells. The beads were then decanted with a small volume of medium and washed with balanced salts solution containing 10mM-sodium bicarbonate, pH 7.3 (Flow Labs). Crystalline botulinal neurotoxin was added to bead suspensions and incubated at 37°C for 18 h; after incubation, samples were detoxified by placing in a boiling water bath for 15 min. Cell numbers were determined by trypsinisation and counting in a haemacytometer chamber (Gallenkamp, London).

RESULTS

A graph showing the relationship between peak luminometer deflection and amount of added choline chloride standard in the assay system is given in Fig. 1.

The limit of sensitivity was in the region of 2–3 nmol choline.

No changes were observed in the amounts of extracellular acetylcholine during incubation of the tissue culture cells with botulinal toxin. However, an increase in extracellular choline was measured during incubation, as shown in Table I.

TABLE I

Effect of botulinal type A toxin on extracellular choline compared with a control

Toxin concentration (Mouse LD_{50}/ml)	1000	500
Increase in extracellular choline less control (fg choline/cell)	4.4	0.4

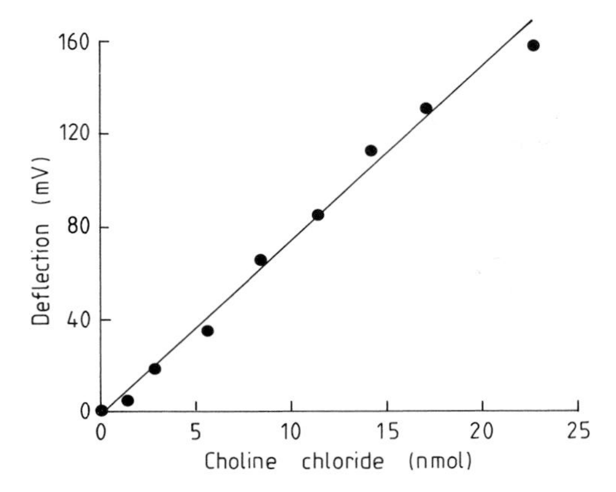

Fig. 1 Standard curve for choline assay.

DISCUSSION

This paper describes a chemiluminescent method for the measurement of choline and acetylcholine; the sensitivity of the assay is similar to that for a colorimetric assay using choline oxidase described by Capaldi & Taylor (7). Choline oxidase has also been used in immobilised form to produce an enzyme electrode for choline measurement (8).

Investigations of choline or acetylcholine release and uptake with various neural preparations generally use radiolabelled choline or acetylcholine (9, 10, 11, 12) but we have found the chemiluminescent method rapid and simple to use. With this method we have measured an increase in extracellular choline during incubation with botulinal type A toxin (Table I) which might reflect an inhibition of choline uptake as has been shown with rat brain synaptosomal fraction using (^{3}H) choline (9). We are extending this observation with a view to developing an assay system for botulinal toxin which does not utilise animals. Such work may also give further insight into the mode of action of this toxin.

ACKNOWLEDGEMENTS

This work was supported by the Ministry of Agriculture, Fisheries and Food.

Purified botulinal type A crystalline neurotoxin was kindly supplied by the Centre for Applied Microbiological Research, Porton Down.

REFERENCES

1. Hobbs, G., Crowther, J.S., Neaves, P., Gibbs, P.A. and Jarvis, B. (1982). Detection and isolation of *Clostridium botulinum*. *In* "Isolation and Identification Methods for Food Poisoning Organisms". (Eds J.E.L. Corry, D. Roberts and F.A. Skinner), pp. 151–164. SAB Technical Series No. 17, Academic Press, London.

2. Notermans, S. and Dufrenne, J. (1981). Effect of glycerylmonolaurate on toxin production by *Clostridium botulinum* in meat slurry. *J. Fd Saf.*, 3, 83–88.

3. Guilfoyle, D.E. and Mestrandrea, L.W. (1980). Problems encountered with the capillary tube immunodiffusion method for detection of botulinal toxin. *Appl. environ. Microbiol.*, 40, 847–848.

4. Bigalke, H., Ahnert-Hilger, G. and Habermann, E. (1981). Tetanus toxin and botulinum A toxin inhibit acetylcholine release from but not calcium uptake into brain tissue. *Naunyn-Schmiedeberg's Arch. Pharmacol.*, 316, 143–148.

5. Dolly, J.O., Black, J., Williams, R.S. and Melling, J. (1984). Acceptors for botulinum neurotoxin reside on motor nerve terminals and mediate its internalization. *Nature, Lond.*, 307, 457–460.

6. Israel, M. & Lesbats, B. (1980). Détection continue de la libération d'acetylcholine de l'organe électrique de la Torpille à l'aide d'une réaction de chimiluminescence. *C.R. Acad. Sc. Paris*, 291, 713–716.

7. Capaldi, D.J. and Taylor, K.E. (1983). A new peroxidase color reaction: oxidative coupling of 3-methyl-2-benzothiazolinone hydrazone (MBTH) with its formaldehyde azine. Application to glucose and choline oxidases. *Analyt. Biochem.*, 129, 329–336.

8. Matsumoto, K., Seijo, H., Karube, I. and Suzuki, S. (1980). Amperometric determination of choline with use of immobilized choline oxidase. *Biotechnol. Bioeng.*, 22, 1071–1086.

9. Habermann, E., Bigalke, H. and Heller, I. (1981). Inhibition of synaptosomal choline uptake by tetanus and botulinum A toxin. *Naunyn-Schmiedeberg's Arch. Pharmacol.*, 316, 135–142.

10. Massarelli, R., Wong, T.Y., Harth, S., Louis, J.C., Freysz, L. and Dreyfus, H. (1982). Possible role of sialocompounds in the uptake of choline into synaptosomes and nerve cell cultures. *Neurochem. Res.*, 7, 301–316.

11. Vyas, S. and Marchbanks, R.M. (1981). The effect of ouabain on the release of (^{14}C) acetylcholine and other substances from synaptosomes. *J. Neurochem.*, 37, 1467–1474.

12. Rowell, P.P. and Duncan, G.E. (1981). The subsynaptosomal distribution and release of (^{3}H) acetylcholine synthesized by rat cerebral cortical synaptosomes. *Neurochem. Res.*, 6, 1265–1282.

AUTOMATIC LUMINOMETRIC ASSAY OF CHOLINE AND ACETYLCHOLINE

[1]A.Lundin, [2]I.Andersson, [2]H.Nilsson and [2]A.Sundwall

[1]LKB-Wallac, SF-201 01 Turku 10, Finland and Research Centre and Department of Medicine, and [2]Department of Pharmacology, Karolinska Institute, Huddinge Hospital, S-141 86 Huddinge, Sweden

An automatic assay of choline and acetylcholine based on the horseradish peroxidase catalysed luminol reaction, with H_2O_2 formed in the choline oxidase reaction has been developed. Acetylcholine esterase is used for release of choline from acetylcholine. The assay has been optimised with respect to choice of buffer and concentrations of luminol, peroxidase, choline oxidase and acetylcholine esterase. Aliquots of samples (eg, extracts of cholinergic tissues or media used for incubation of such tissues) are added to cuvettes containing luminol, peroxidase and buffer. Up to 25 cuvettes can be loaded into the sample carousel of the luminometer and the entire assay is automatic. In the first cycle of the carousel, choline oxidase is added and the choline concentration determined. In the second cycle acetylcholine esterase is added and acetylcholine concentration determined. When the second cycle starts the light emission from the first step has decayed and does not interfere with the second light measurement. Results can be calculated either with respect to a standard curve or by using the standard addition technique. The detection limit (a few picomoles) is presently determined by both reagent blanks and the intensity of the light emitted in the luminol reaction. The sensitivity of the method makes it highly competitive with alternative methods for assay of choline and acetylcholine, particularly for large numbers of samples.

*Correspondence to A. Lundin: Research Centre, Huddinge Hospital, Karolinska Institute, S-141 86 Huddinge, Sweden

ISBN 0-12-426290-2

SPONTANEOUS CHEMILUMINESCENCE OF SMOKER'S BLOOD

Binkoh Yoda*, Ryuzo Abe*, Yoshio Goto*, Akio Saeki+
Choichi Takyu† and Humio Inaba†

**Department of Medicine, Tohoku University School of Medicine, Sendai 980, +Tohoku Electronic Industrial Co. Ltd., Sendai 982 and †Research Institute of Electrical Communication, Tohoku University, Sendai 980, Japan*

INTRODUCTION

An extremely sensitive single photon counting system has been developed in a commercially available form. Application of this equipment to the measurement of extra-low level photon emission from various types of faint light sources like medical and biological crude materials is now revealing several interesting findings. As a unique instance, a phenomenon of elevated levels of the chemiluminescence of smoker's blood, especially those of plasma, will be described in this paper.

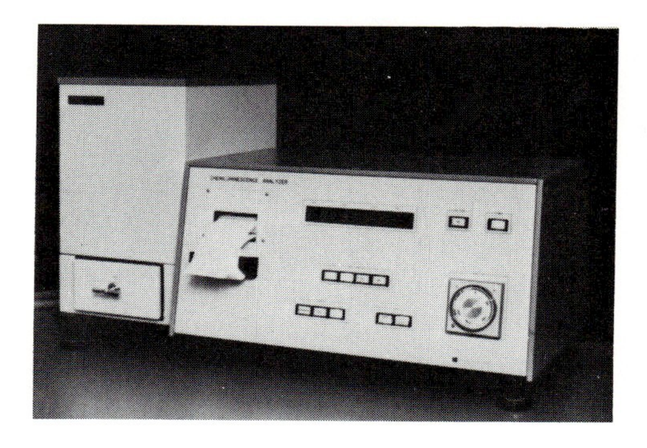

Fig 1 High performance single photon counter " Chemiluminescence Analyzer OX-7 "

ISBN 0-12-426290-2

METHODS

Quantitative detection of the ultra-weak chemiluminescence arising from blood samples was performed with a single photon counter of ultra-high sensitivity (Chemiluminescence Analyzer OX-7, Tohoku Electronic Industrial Co., Ltd., Sendai, Japan) (Fig. 1). This equipment uses carefully selected low noise photomultipliers such as HTV R 878, R 550, R 375, etc. with a 50 mm diameter photocathode (Hamamatsu Photonics Co., Japan). The photocathode is cooled by a thermoelectric cooler to minimize noise (1).

Smokers and nonsmokers tested in this study were all male and healthy on medical examinations. The ages of the smokers and nonsmokers were 47.2 ± 7.1 years (n=27)(mean ± SD) and 46.1 ± 6.9 (n=24), respectively, with no statistically significant difference.

Blood samples were taken by venous puncture from cubital veins with a small amount of heparin and plasma portions were separated from blood cells by centrifugation.

Two ml blood samples were placed in a stainless steel dish type cell of 50 mm diameter and single photoelectron pulses were counted at 37°C with a continuous air flow.

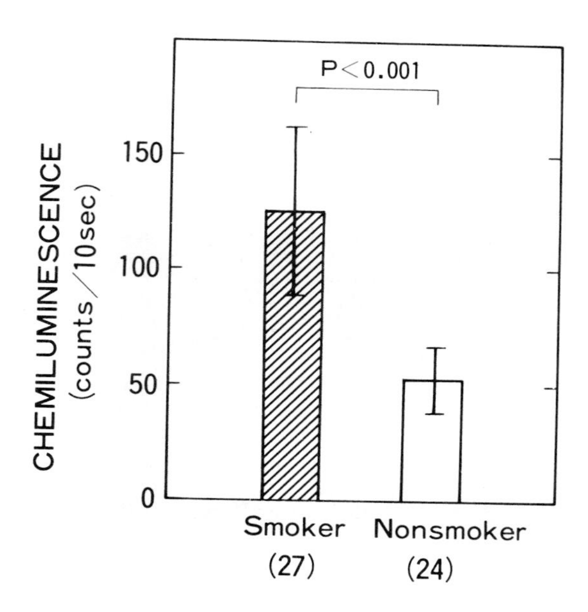

Fig. 2 Chemiluminescence levels of the plasma samples of smokers and nonsmokers. Numerals in parentheses represent the numbers of subjects and vertical bars stand for SDs.

Results obtained were arbitrarily expressed by the numbers of observed photoelectrons per 10 sec.

Statistical analyses were based on the Student's t-test.

RESULTS

Preliminaly experiments indicated that most of the spontaneously emitted, weak chemiluminescence of human blood comes up from the plasma portion of the blood and only minimal amount from the blood cells. As the result, all data shown in this report are those of plasma samples.

Fig. 2 shows the chemiluminescence intensities of the smoker's and nonsmoker's plasma samples. Smoker's plasma samples gave approximately twice higher chemiluminescence levels than nonsmoker's plasma samples, i.e., smoker's plasma showed the chemiluminescence intensities of 125.2 ± 36.9 counts/10 sec (mean ± SD), while nonsmokers 53.5 ± 14.0. The difference between these values was statistically highly significant ($p<0.001$).

Changes in plasma chemiluminescence after the withdrawal of cigarette smoking was tested in a 40 years old volunteer who has been smoking about 20 cigarettes a day for 20 years.

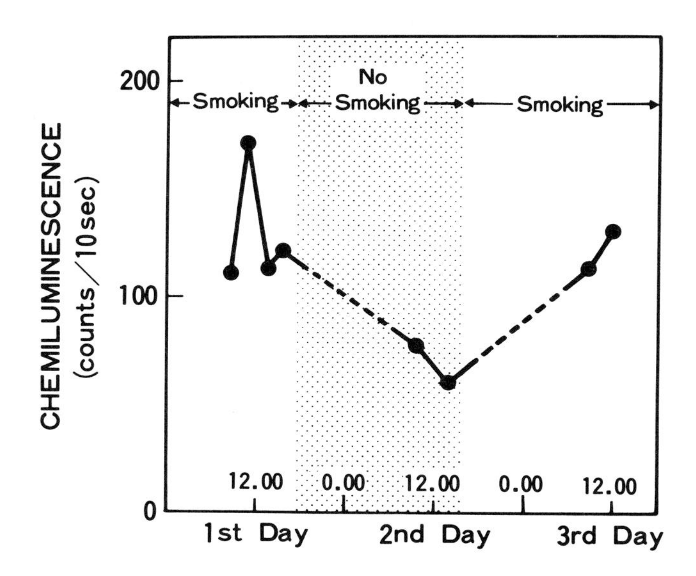

Fig. 3 Changes in the chemiluminescence levels of the plasma of a volunteer after the cessation and resumption of cigarette smoking.

Plasma chemiluminescence levels dropped relatively quickly to the nonsmoker's range by the cessation of smoking (Fig.3). Resumption of cigarette smoking again elevated the plasma chemiluminescence levels.

Separate experiments showed the chemiluminescence of cigarette smoke itself and cigarette smoke absorbed into organic solvents. Requirement of oxygen for the chemiluminescence was also noticed.

DISCUSSION

Though the biological significance of the present finding is not known, it is fascinating to assume that the higher chemiluminescent property of smoker's blood might be somehow related to cigarette smoking associated diseases.

Free radicals abundantly residing in cigarette smoke would have some contribution to the observed chemiluminescence.

The roles of free radicals and active oxygen species in aging process, carcinogenesis and other physical disorders are already widely mentioned.

So far several instances of extremely faint chemiluminescence from biological materials such as animal liver (2) and human breath (3) have been reported.

Easy avai ability of a ultra-sensitive single photon counter may prompt further investigation of this field.

ACKNOWLEDGEMENTS

This study was supported in part by a Grant-in-Aid for Co-operative Research (No.00539002) from the Ministry of Education, Science and Culture, Japan and a fund from the Watanabe Memorial Foundation for the Advancement of Science and Technology.

REFERENCES

1. Inaba, H., Yamagishi, A., Takyu, C., *et al.* (1982). Development of an ultra-high sensitive photon counting system and its application to biomedical measurements, *Optics and Lasers in Eng.* 3, 125-130.
2. Boveris, A., Cadenas, E., Reiter, R. *et al.* (1980). Organ chemiluminescence: Noninvasive assay for oxidative radical reactions, *Proc. Natl. Acad. Sci. USA* 77, 347-351.
3. Williams, M.D. and Chance, B. (1983). Spontaneous chemiluminescence of human breath, *J. Biol. Chem.* 258, 3628-3631.

THE USE OF LUMINOUS BACTERIA TO STUDY THE MODE OF ACTION AND THE ACTIVITY OF AMINOGLYCOSIDE ANTIBIOTICS

S. Ulitzur and A. Naveh

Department of Food Engineering & Biotechnology
Technion - I.I.T., Haifa, Israel.

INTRODUCTION

Luminous bacteria offer a unique system to determine the activity of aminoglycoside antibiotics on two levels: (1) The misreading effect of aminoglycosides is determined by the aid of a missense dark mutant of luminous bacteria; the misreading agents increase the luminescence of the dark mutant over 100 fold (2). The activity of the aminoglycosides as protein synthesis inhibitors was applied to determine the concentrations of antibiotics in serum and milk in a short term (45 min) test.

MATERIALS AND METHODS

Bacterial Strains and Conditions for Growth

Photobacterium leiognathi strains BE8 and its spontaneous dark variant SD18 are described by Ulitzur (1). The dark mutant *P. leiognathi* BE52D was selected from the BE8 cells after mutagenesis with methyl methane sulfonate (MMS). The cells were grown on solid medium ASW-3 or liquid medium ASWRP as described elsewhere (1). NaR buffer consisted of 3% (w/v) NaCl, 20 m mol/l MOPS buffer (Sigma). NARP medium consisted from NaR with 0.5% (Difco) peptone.

Bioluminescence Determination. *In vivo* and *In vitro* luminescence determinations were carried out as described by Ulitzur (1).

ISBN 0-12-426290-2

Chemicals. All the chemicals(reagent)grade) and antibiotics were purchased from Sigma.

RESULTS

1. *Determination of misreading effect of antibiotics with the aid of luminous bacteria.*

Streptomycin and other aminoglycosides are known to result in misreading of mRNA under both *in vivo* and *in vitro* conditions (2,3). The currently used techniques to demonstrate the misreading effect are quite complicated and laborious In order to develop a bioluminescence assay for misreading agents we looked for a dark mutant of luminous bacteria having either a nonsense or a missense mutation in a key amino acid of the luciferase gene. Using MMS we isolated the dark mutant BE52D that showed 1/5000 of the *in vivo* luminescence of the wild type. The BE52D cells did not respond to neither long-chain aldehyde nor to the natural inducer. SDS gel-chromatography revealed that BE52D cells have normal level of luciferase with its characteristic molecular weight distribution. Thus, the primary lesion of BE52D seems to be a missense mutation in a key amino acid of the luciferase, although a nonsense mutation at a distal end of the enzyme was not excluded.

Fig. 1 shows the effect of various protein synthesis inhibitors on the luminescence of BE52D cells. Streptomycin, neomycin, amikacin, gentamicin and kanamycin increased the luminescence by a factor of ten or more. Protein synthesis inhibitors that are not known as misreading agents such as spectinomycin, lincomycin, clindamycin as well as chloramphenicol, tetracycline and erythromycin resulted in a decrease of luminescence. Chloramphenicol (10 μg/ml) completely abolished the streptomycin promoting effect on the luminescence of BE52D cells. Besides certain antibiotics the misreading effect of ethanol (2-5%) was also demonstrated with this mutant. In all cases the increase in the *in vivo* luminescence was accompanied by a corresponding increase in the *in vitro* luciferase activity

2. *Determination of antibiotics that inhibit protein synthesis*

A new, rapid and sensitive bioluminescence test for antibiotics that inhibit protein synthesis has recently been described by us (4). The ability of the tested antibiotics to inhibit the de-novo synthesis of the enzymes participating in the luminescence system is determined by means of SD-18 cells that undergo prompt induction of the luminescence in the presence of proflavin. Upon induction, the *in vivo* luminescence is increased by more than 50 fold within 30 min. Antibiotics that block the *de-novo* synthesis of protein limit the development of luminescence at a level that was found to be a function of the antibiotic concentration. Fig. 2 shows the effect of different concentrations of several antibiotics in milk on the proflavin induced luminescence of SD18 cells. The same test was applied for determination of gentamicin in human serum.

Lyophilized cultures of luminous bacteria were applied successfully in the assays described.

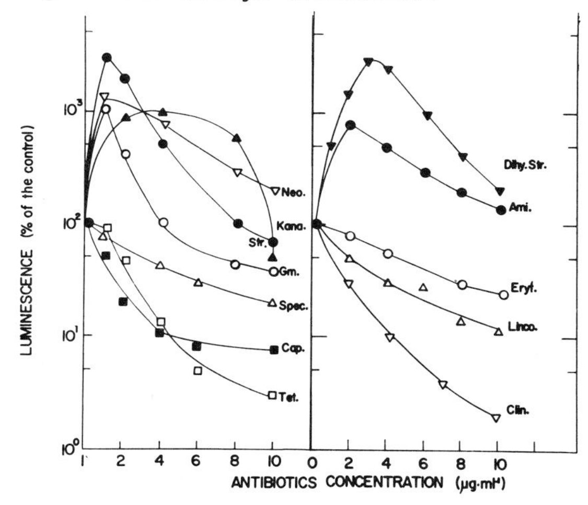

Fig. 1. The effect of different antibiotics on the in vivo luminescence of BE52D cells.

1 ml aliquotes of washed cells of BE52D (2.10^8 cells/ml) in NaR buffer containing proflavin (0.5μg/ml) were placed in a set of scintillation vials. To each vial different concentrations of various antibiotics were added. The luminescence was determined after 60 min of incubation at 30°C. The figure shows the luminescence as a percentage of the controls without antibiotics. Neo-neomycin, Kana-kanamycin, Str-streptomycin, Gen-gentamicin, Spec-spectinomycin, Cap-chloramphenicol, Tet-tetracycline, Dihy-Str-dihydrostreptomycin, Ami-amikacin, Eryth-erythromycin, Linco-lincomycin, Clin-clindamycin.

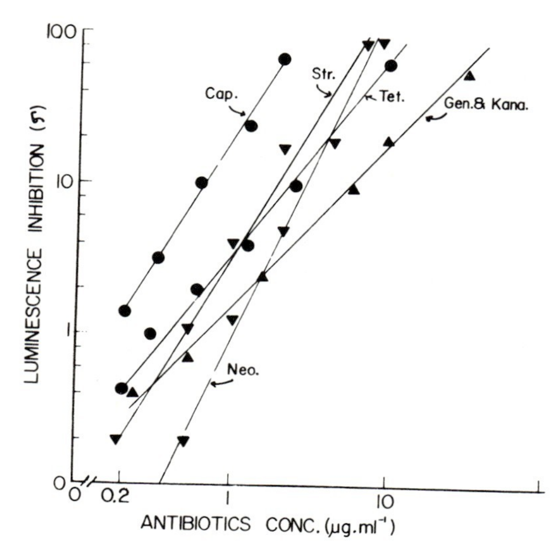

Fig. 2. Bioluminescence test for antibiotics in milk.

Different concentrations of the tested antibiotics were added to 1 ml milk containing 2% NaCl and 50mM MOPS buffer. The test for tetracycline was done a pH 6.0 in the presence of 25μg/ml proflavin. The tests for the other antibiotics were done at pH 8.0 in the presence of 1.25μg/ml proflavin. To each vial 10^7 SD18 cells/ml were added. The figure shows the percentage of inhibition of luminescence after 45 minutes of incubation at 30°C. Antibiotic symbols as in Fig1.

REFERENCES

1. Ulitzur, S. (1982) A bioluminescence test for genotoxic agents. *Trends in Anal. Chem.* Vol 1, 329-333.
2. Gorini, L and Kataja, E. (1964) Phenotypic repair of streptomycin of defective genotypes in *E. coli*. *Proc. Nat. Acad. Sci. U.S.A.* 51, 487-492.
3. Davis, J., Gilbert, W. and Gorini, L. (1964) Streptomycin, supression and the code. *Proc. Nat. Acad. Sci. USA* 51, 883-887.
4. Naveh, A., Potasman, I., Basan, H. and Ulitzur, S (1984) A new rapid and sensitive bioluminescence assay for antibiotics that inhibit protein synthesis. *J. Appl. Bacteriol.* 56.

INDEX

A

Acetyl choline, CL assay, 581, 585
Acridinium ester as label, 149, 159, 185
Activated oxygen, 315
Acute phase proteins, 219
Acyl CoA reductase, 105
Adenosine, 175
Adenosine diphosphate, assay, 116, 493
Adenosine monophosphate assay, 493
Adenosine monophosphate, cyclic, assay, 175
Adenosine triphosphate, *see* ATP
Adenylate energy charge, 36
Adenylate kinase, assay, 116
Adherence of bacteria
  plaque, 10
  polystyrene, 10
Adhesin, 327
Aequorea, 273
Aequorin, 273
AIDS, 451
Albumin, effect on luminol CL in PMN, 311
Alcohol dehydrogenase, assay by BL, 116
Aldehyde biosynthesis, 88, 105
Aldehyde, effect of chain length, 530
Alga, ATP extraction, 492
Alpha-fetoprotein, assay, 187, 215, 236
Alveolar proteinosis, 307
Aminofluoranthene, for standard, 475
Ammonia, assay by BL, 525
Ampicillin, 67
Androgens, assay by BL, 537
Antibiotic, *see also* MIC
  ampicillin, 66
  assay, 8, 73
  chloramphenicol, 73
  clindamycin, 17
  dapsone, 30
  erythyromycin, 17
  fusidic acid, 17
  gentamicin, 73, 239
  ketoconazole, 63
  membrane damaging, 63
  methicillin, 17
  protein synthesis inhibition, 592
  susceptibility testing, *see* Susceptibility testing
  tioconazole, 63
Apyrase, 5, 8, 25, 63
Aspartate, assay by flow BL assay, 126
ATP (adenosine triphosphate)
  assay in
    biomass, *see* Biomass
    food, 53, 57
    fruit juice, 53
    rapid microbiology, *see* Rapid microbiology
    soil, 33
    susceptibility testing, *see* Susceptibility testing
    using light guides, 479
    water, 33
  automatic assay of ATP, ADP, and AMP, 491
  Candida albicans, 63, 492
  chloroplasts, ATP production in, 511
  detection limit, 481
  Escherichia coli, 27, 35, 67, 75
  extraction, *see* Extraction
  in immunoassay, 189
  in photophosphorylation, 511
  Klebsiella edwardsii, 27, 73
  leakage from yeast, 63
  Mycobacterium bovis, 81
  Mycobacterium leprae, 29
  Sarcina lutea, 73

Staphylococcus aureus, 17, 73
Streptococcus faecalis, 27
yeast, 63, 75
ATPase, *see* Apyrase
Australia antigen, assay, 223
Automated assay of ATP, ADP, AMP, 491
estriol glucuronide immunoassay, 171
immunoassays, 465
phagocytosis, 289, 401, 423, 457, 461
Automation, 289, 457, 461, 465
Autoinducer, 89
Autoxidation of food, CL, 577
Avidin, 149

B

Bacteraemia, 7, 25
Bacterial
ATP, *see also* ATP; Biomass luciferase, *see* Luciferase, bacterial
Bacteriuria screening by BL
ATP threshold, 22, 71
commercial reagents and kits, 6, 75
interpretation of results, 4, 20, 71, 75
methods, 3, 71, 75
types of, 2
B cells, 411
Bile acids, BL assay, 118, 129, 133, 141
Biomass, by ATP assay
ATP/gram, 34
bacterial, 35
compost, 43
C/ATP ratio, 34
drinking water, 41
fungal, 35
pollutants, 39
relation to culture age, 35, 51
sand, 34
sediments, 38
sludge, 43
soils, 34, 36
textiles, 49
water, 41
Biotinylated antigen/antibody, 149, 154
Bis(2,4,6-trichlorophenyl)oxalate
CL in HPLC, 556
in immunoassay, 253
Blood
CL of smokers, 587
whole, CL of, 403, 431, 435
Bordetella pertussis, 423
Botulinal toxin, 581

C

Calcium sensitive photoproteins, 273, 521
Candida albicans, 63, 492
Celiac disease, 419
Cell sorter, 326
Chemiluminescence (CL), *see also* Immunoassay; Phagocytosis; and compound concerned
assay for oxidases, 559
autoxidation of food, 577
cigarette smokers, blood CL, 587
gamma irradiation of food, 573
generation of compounds by hydroxylation, 569
mitogen-stimulated mononuclear cells, 331
Chloramphenicol, assay, 73
Chloroplasts, 511
Choline, CL assay, 581, 585
Cigarette smoking, blood CL, 586
Circulating immune complex, 405
Clindamycin, 17
Cloned genes (bacterial luciferase), 90, 101, 108
Clostridium botulinum neurotoxin, 581
Coimmobilised enzymes
activities recovered, 115
flow assays, 120, 126
in assay of
bile acids, 118, 129, 141
NADH, 133
NADP, 133
TNT, 133
limit of detection of analytes with, 115
luciferase
bacterial, 115, 273
firefly, 116
mode of action, 113
nylon, 129
Sepharose-4B, 111, 134
Coliforms, assay for, 41, 42
Concanavalin A, 381
Cortisol
immunoassay, 227, 239
phagocytosis and chronobiology, 435
Cream, post-pasteurization contamination, 61
Creatine kinase assay, 116
Creatine phosphate, assay, 116
Cuvettes, growing cell lines in, 300
Cyclic-AMP, assay, 175
Cyclic photophosphorylation, 511

Cyclodextrin, enhanced CL with lucigenin, 565
Cytokines, 331

D

Data processing, 159, 461, 469
Detector, light, 289, 457
Dextran–coated charcoal separation, 157, 159
Dyes, firefly luciferase binding, 485

E

Ecdysteroids, 249
Energy transfer CL assay, 175
Enhanced luminescence, 243, 565
Erythromycin, 17
Escherichia coli, 41, 67, 409
  cloning of genes, 91, 101, 105
Estradiol
  assay by BL, 537
  immunoassay, 150
Estriol, immunoassay, 150
Estriol glucuronide immunoassay, 171
Estrogens
  assay by BL, 537
  immunoassay, 167
Ethanol, flow BL assay, 126
Estrone
  assay by BL, 537
  immunoassay, 168
Extraction
  AMP from bacteria, 477, 478
  ADP from bacteria, 477, 478
  ATP from
    alga, 492
    bacteria, 25, 33, 67
    blood, 25, 492
    E. coli, 67
    fat cells, 477
    fungi, 34, 35, 49
    lung fibroblasts, 486
    macrophages, 492
    Mycobacteria, 29, 81
    placenta, 492
    platelets, 492
    soils, 33
    somatic cells, 492
    yeast, 63
  comparison, 34, 37, 81, 477, 491
  reagents for, 33, 81, 477, 492

F

Fat cells
  ATP content, 492
  lipolysis in, 503, 545
Fatty acid reductase, 88, 105
Fatty acid, assay by BL, 503
Fatty acid derivatives, assay by BL, 529
Ferritin, immunoassay, 185, 189
Ficoll gradient separation, 331, 381, 447
Firefly luciferase, *see* Luciferase, firefly
Flow systems of analysis, *see also* Coimmobilised enzymes; Luciferase
  assays, 120, 125, 130
  bacteriuria, 23
  bile acids, 129
  cells, 135
  effect of flow rate, 127
  limit of sensitivity of assays, 137
  metabolites and enzymes, 115, 116, 125
  rapid microbiology, 23, 41
Fluorescein isothiocyanate, 265
Food
  autoxidation, CL, 577
  gamma irradiation, CL, 573
  rapid microbiology, 53, 57
Forskolin, 175
Free radical, in smoke, 587
Fungi, ATP content, 35, 49
Fusidic acid, 17

G

Genes
  bacterial luciferase, 87, 101
  Gonyaulax, 95
Gentamicin, 18, 73, 239
Gliadin, 419
Glucose
  assay, 116
  by flow analysis, 126
Glucose oxidase, assay, 559
Glucose-6-phosphate, assay, 121
Glucose-6-phosphate dehydrogenase, assay, 116
Glutamate, assay, 126, 525
Gluten, 419
Glycerol, assay, manual and automatic, 545
Gonadotropin
  assay, 154
  therapy, 153
Gonyaulax, molecular genetics, 95
Granulocytes, 331

H

HCG, assay, 155
Hexokinase, 115
Hybridomas, 297
Hydrophobicity and bacterial surfaces, 10, 409
HBsAg assay, 265
Herpes, 324
Hormones, *see* Immunoassay
HPLC, CL detection by peroxyoxalates, 555
Humoral immune mechanisms, 279
Hydrogen peroxide, 265
Hydroxylation, CL production, 569
Hydroxyprogesterone assay, 193, 253
Hydroxysteroid assay, 116, 118, 129, 133, 141, 193, 253
Hypochlorite, 265

I

Immobilized enzymes, *see* Coimmobilized enzymes
Immunoassay by bio- and chemiluminescence
- acridinium ester, 149, 154, 159, 185
- acute phase proteins, 219
- aequorin, 273
- alpha fetoprotein, 215, 236
- Australia antigen, 223
- bacteria, 11
- bioluminescence based, 150, 156, 206, 269
- bis(2,4,6-trichlorophenyl)oxalate fluorescent dyes, 253
- calcium-sensitive photoproteins, 273
- CELIA, 190, 209
- ceruloplasmin, 257, 261
- coating with antibody, *see* Immunobeads; Polyacrylamide; Polystyrene; Microcrystalline cellulose
- coefficient of variation of assay, 211, 258
- competitive binding, 154
- cortisol, 179, 227, 239
- cyclic AMP, 175
- data processing, on-line, 289, 457, 461, 469
- dextran-coated charcoal, 151
- ecdysteroids, 249
- EELIA, 190
- energy transfer assay for cyclic AMP, 175
- estradiol, 152, 159
- estriol, 167
- estriol glucuronide, 152, 171, 179
- estrogens, 167
- estrone glucuronide, 152
- ferritin, 186, 257
- fluorescein isothiocyanate, 265
- gentamicin, 240
- gonadotropin, 154, 185
- HBsAg, 265
- hydroxyprogesterone, 193, 253
- ICMA, 186
- ILMA, 211, 215, 219, 257
- ILSA, 190
- Immunobeads, 151, 179
- isoluminol, 149, 159, 209
- labelling,
  - acridinium ester, 154
  - aequorin, 273
  - biotin, 149, 151
  - fluorescein isothiocyanate, 265
  - isoluminol, 149, 159, 185, 193, 209, 249
  - microperoxidase, 233, 469
  - NAD analogue, 150, 269
  - peroxidase, 150, 243
  - procedures for, 150, 159, 185, 193
  - pyruvate kinase, 150, 189
- LH, 154
- luciferin, enhanced CL, 243
- microcrystalline cellulose, coating with antibody, 189
- microperoxidase, 168, 200, 233, 239, 469
- peroxidase, 11, 149, 189, 243
- polyacrylamide tubes, coating with antibody, 152, 236
- polystyrene, coating with antibody, 149, 189, 200, 205, 215, 227, 235, 239, 262, 265
- precision, 152, 160, 187, 199, 219, 241, 244, 465
- pregnanediol glucuronide, 152, 159, 179
- progesterone, 149, 160, 163, 269, 273
- prolactin, 186
- prostaglandin, 152
- review, 149
- Sac-Cel separation, 159
- solid phase, 154, 159, 180, 189, 205, 209
- SPALT, 190, 209, 219, 233, 239, 257, 261
- testosterone, 160, 199
- testosterone glucuronide, 179
- thyroglobulin, 191, 209
- thyroglobulin antibody, 209
- thyroid parameters, 209
- thyrotropin, 186, 209
- thyroxine, 209, 245, 253
- thyroxine binding globulin, 209
- transferrin, 189, 257, 261
- triiodothyronine, 209
- two site assays, 156, 186
- using bacterial luciferase, 156, 162, 269

virus, 11
without extraction (progesterone), 163
Immobilised enzymes, *see* Coimmobilised enzymes
Immunobeads, coating with antibody, 151, 179
Insect hormones, immunoassay, 249
Instrumentation
automation, 289, 457, 461
comparison of automated unit, 465
dual wavelength, 458
incorporating light guides, 479
liquid scintillation counter, 458
multichannel, 289, 348, 457, 587
Interleukin, 331
Internal standard, 514
Iso–luminol, label for immunoassay, 149, 159, 185, 193, 209, 249

K

Ketoconazole, 63
Klebsiella edwardsii, 73
Kinetics
immobilized enzymes, 117
in immunoassay, 470
in opsonization of yeast, 443
in phagocytosis, 293, 461

L

Labels for immunoassay, *see* Immunoassay
Lactate, assay, 116
Lactate dehydrogenase, 115
Lanthanide chelate, for standard, 475
Latex particles and phagocytosis, 431
Lectin, 381
Leprosy, 29
LH, asay, 155
Lidocaine, effect on CL of cell lines, 303
Light guides, 479
Lipolysis, assay, 503, 545
Liposomes, 522
Lipoxygenase
polymorphonuclear leukocytes, 415
soybean, 393, 415
Luciferase bacterial
aldehyde and chain length, 530
assay for
ammonia, 525
androgens, 537
aspartate, 125
bile acids, 118, 129, 133, 141
estradiol, 508
estrogens, 537
ethanol, 125
fatty acids, 503
fatty acid derivatives, 529
glucose, 125
glutamate, 125, 525
long-chain aldehydes, 529
NADH, 125, 129, 133, 507, 515, 525, 537
NADH conjugate, 269
NADP, 133
oxaloacetate, 125
progesterone, 269, 508
sorbitol dehydrogenase, 480, 515
steroids, 509
testosterone, 537
TNT, 133
urea, 525
autoinducer, 89
flow analysis, 121, 125
gene, *see* Lux genes
immobilised, 111, *see also* Coimmobilised enzymes
induction, 89
in immunoassays, 156, 162, 269
molecular genetics, 89, 101
subunits, 87, 101
Luciferase, firefly, *see* ATP; Biomass; Rapid microbiology
assay for
ATP formed in chloroplasts, 511
glycerol in lipolysis, 545
photophosphorylation, 511
binding to dyes, 486
immobilised, 111, *see also* Coimmobilised enzymes
importance of purification, 36
purification on blue dextran, 485
Luciferin, enhanced luminescence immunoassay, 243
Lucigenin
in CL of cells, 281, 293, 297, 331, 343, 406, 415, 419, 431
CL with cyclodextrin, 565
Luminol, *see also* Immunoassay, label, luminol
choline assay using, 581, 585
CL of cells, 281, 293, 297, 311, 315, 327, 335, 343, 369, 389, 406, 417, 419, 427, 431, 435, 439, 443, 447
enhanced luminescence with luciferin, 243
in lipoxygenase assay, 393
in oxidase assay, 559
Lux genes, 87, 103

Lymphocytes, 411
Lymphokines, 297

M

Macrophage
  activation of, 279, 300
  alveolar, 327
  peritoneal, 307, 389
Malate, assay, 116
Malate dehydrogenase, 114
Membrane perturbation, 521
Methicillin susceptibility test, 17
MIC, 17, 69
Micellar solution in enhanced CL, 565
Micelles, reversed, 559
Microbial ATP, *see also* ATP; Biomass; Extraction
  effect of culture age, 35
  extraction efficiency, 34, 477
Microbiology, *see* Rapid Microbiology
Microcrystalline cellulose, coating with antibody, 189
Microperoxidase, *see also* Immunoassay
  in immunoassay, 168, 200, 233, 239, 469
  preparation, 233
  use as label, 233
Microperoxidase deficient PMN, 336
Milk, antibiotics in, 594
Minimum inhibitory concentration, *see* MIC
Mitogen-stimulated cells, 331
Monocytes, 315, 343, 381, 431, 447
Mutagenic agents, assay with BL, 533
Mutants (Photobacterium and Vibrio), 88, 101, 105
Mycobacterium
  bovis, ATP, 81
  leprae, ATP, 29, 81
  lepraemurium, ATP, 81
Mycoplasma, CL assay, 385
Myeloperoxidase, 279, 335

N

NAD assay, 116, 134, *see also* Luciferase, bacterial, assay
NADP assay, 116, 121, 134, *see also* Luciferase, bacterial assay
Natural killer cells, 316
Nylon tube for immobilised enzyme, 127, 129

O

Obelin, enclosed in artificial vesicles, 521
Open heart surgery, CL of PMN, 369
Opsonization
  Bordetella, 424
  C3 mediated, 299
  luminous bacteria, 377
  Pseudomonas aeruginosa, 283
  Streptococcus, 439
  yeast, 443
Oxaloacetate, BL flow assay, 126
Oxidase, assay with luminol, 559
Oxidoreductase, NADH:FMN, 88, 114, 125
Oxygen, activated, 315
Oxygen, consumption PMN, 348
Oxygen radical production, 175
Oxygen radical scavengers, 311
Oxygen, singlet, 297

P

Percoll gradient separation, 26, 175, 316, 385, 447, 453
Peroxidase, *see* Immunoassay, labelling procedures
Peroxyoxalate CL, HPLC use, 555
Phagocytosis, *see also* Polymorphonuclear leukocytes; Macrophages; Monocytes
  AIDS, 453
  automatic data collection and processing, 159, 289, 457, 461, 469
  circadian rhythm, 435
  CL, 279ff
  CL probes, 280, 298
  effect of azide and cyanide, 431
  hemophilia, 453
  instrumentation to measure CL, 289, 348, 457, 461, 469
  lymphadenopathy, 463
  Mycoplasma, 385
  use of luminous bacteria, 373
  whole blood CL, 283, 431
Phorbol myristate acetate, 282, 293, 307, 332, 343
Phosphoenolpyruvate, assay, 116
Phosphogluconic acid, 134
Phospholipase, assay, 521
Photobacterium, 87, 105, 529
  in study of
    antibiotics, 591
    low water activity, 541
    mutagenic agents, 533
Photomultipliers, 289, 348, 458
Photon counting, 289, 457, 479, 587
Photon standard, 475

Photophosphorylation, 511
Phthalic hydrazides, hydroxy-, 569
Plasmids, 102
Platelets, 339
Polymorphonuclear leucocytes (PMN), 246ff
  cyclic AMP in, 175
  tumour monitoring, 427
Polyacrylamide tubes and beads, antibody coating, 152, 236
Polystyrene
  adhesion of bacteria, 10
  coating with antibody, *see* Immunoassay, polystyrene
  cuvettes, *see* Cuvettes
Post-pasteurization contamination, assay for, 61
Probes, 91, 300
Progesterone, assay, 149, 153, 159, 163, 269, 508
Protein acylation, 105
Protein biosynthesis, 591
Pyruvate kinase
  assay, 116
  label, 189

Q

Quenching
  by inhibitors, 36, 514
  by plasma or urine, 5, 168, 171, 470

R

Radiolysis and CL, 569
Rapid microbiology
  bacteraemia, 7, 25
  bacteriuria, 5, 21, 71, 75
  coliforms, 41
  cream, 61
  E. coli, 41, 67
  environmental, 33
  food, 53, 57
  leprosy, 29
  meat, 53, 57
  Mycobacterium, 29
  review
    clinical, 3
    environmental, 33
  sand, 34
  sludge, 43
  soil, 33, 36
  spore suspension, 49
  susceptibility testing, 7, 17, 29, 67
  textiles, 49
  use of phages, 41
  use of resin for separation, 53
  vitamin assays, 8
  water, 41
  yeast, 53, 63
Recombinant library, 91
Resin, ion exchange, in rapid microbiology, 53
Respiratory burst, 381

S

Sac-Cel separation, 159
Sarcina lutea, 73
Salivary
  bile acids by BL, 129
  cortisol assay, 227
  steroids, 160, 227
  testosterone, 199
Sample changers, 291, 457
Sand, microbial ATP in, 34
Seminal plasma, 261
Sendai virus, 415
Separation
  Ficoll gradient, 381, 385, 447
  microbial ATP in food, 53, 57
  Percoll gradient, 26, 175, 316, 453
  Resin, ion exchange, 53
  Sac-Cel, 159
  yeast by filtration, 75
Sepharose 4B, 111, 125, 134
Shape index, in CL immunoassay, 469
Signal to noise ratio enhancement, 479
Smokers, CL of blood, 587
Soils, microbial ATP in, 34
Sorbitol dehydrogenase, assay, 515
Soybean, lipoxygenase, 393
Standard light source, 294, 475
Staphylococcus aureus, 17, 73
Steroid, *see* Immunoassay and steroid concerned
  assay by BL, 507, 537
Streptococcus serotypes, 365
Superoxide, 297
Surface antigens, 301, 323, 409
Susceptibility testing
  of *Escherichia coli,* 67
  of *Mycobacterium leprae,* 29
  of *Staphylococcus aureus,* 7, 17
Synovial fluid, 319

T

Target cells, 315
Temperature control of instruments, 292, 458

Testosterone
  assay by BL, 537
  immunoassay, 160, 199
Textiles, rapid microbiology, 49
Theophylline, effect on cellular CL, 321
Thrombin-activated platelets, 339
Tioconazole, 63
TNT assay, 135
Toxicity testing, 43
Transcription, assay, 91
Transferrin, assay, 189
Triton, ATP extraction, 26
Tumour monitoring, 215, 427

U

Urate crystals, CL and PMN, 319
Urea, assay by BL, 525
Urinary pathogens, 409

V

Viability of yeast, 63
Vibrio cholerae in phagocytosis, 375, 377
Vibrio fischeri, 90, 101, 107, 530, *see also* Photobacterium
Virus, Sendai, 415
Vitamin assay by ATP, 8

W

Wash system for immunoassay, 207
Water
  activity, with luminous bacteria model, 541
  assay for microbes in, 33
  flow cell to assay coliforms, 41
Wavelength, standard source, 475
Wheat germ agglutination, 381

Y

Yeast
  ATP leakage, 63
  effect of antibiotics on ATP, 63
  in fruit juice, 53
  in urine, 75
  opsonization, 443

Z

Zymosan, 300, 339, 431, 435, 453